W0106649

The Application of Polymeric Reinforcement in Soil Retaining Structures

NATO ASI Series

Advanced Science Institutes Series

A Series presenting the results of activities sponsored by the NATO Science Committee, which aims at the dissemination of advanced scientific and technological knowledge, with a view to strengthening links between scientific communities.

The Series is published by an international board of publishers in conjunction with the NATO Scientific Affairs Division

A Life Sciences **B Physics**	Plenum Publishing Corporation London and New York
C Mathematical and Physical Sciences **D Behavioural and Social Sciences** **E Applied Sciences**	Kluwer Academic Publishers Dordrecht, Boston and London
F Computer and Systems Sciences **G Ecological Sciences** **H Cell Biology**	Springer-Verlag Berlin, Heidelberg, New York, London, Paris and Tokyo

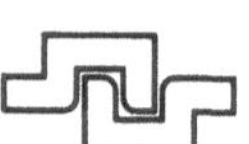

Series E: Applied Sciences - Vol. 147

The Application of Polymeric Reinforcement in Soil Retaining Structures

edited by

Peter M. Jarrett

Department of Civil Engineering,
Royal Military College of Canada,
Kingston, Ontario, Canada

and

Alan McGown

Department of Civil Engineering,
University of Strathclyde,
Glasgow, Scotland, U.K.

Kluwer Academic Publishers

Dordrecht / Boston / London

Published in cooperation with NATO Scientific Affairs Division

Proceedings of the NATO Advanced Research Workshop on
Application of Polymeric Reinforcement in Soil Retaining Structures
Kingston, Ontario, Canada
June 8–12, 1987

Library of Congress Cataloging in Publication Data

NATO Advanced Research Workshop on the "Application of Polymeric
Reinforcement in Soil Retaining Structures" (1987 : Kingston, Ont.)
The application of polymeric reinforcement in soil retaining
structures / editors, Peter M. Jarrett, Alan McGown.
p. cm. -- (NATO ASI series. Series E, Applied sciences ; no.
147)
"Proceedings of the NATO Advanced Research Workshop on the
'Application of Polymeric Reinforcement in Soil Retaining
Structures,' Kingston, Ontario, Canada, June 8-12, 1987"--T.p.
verso.
"Published in cooperation with NATO Scientific Affairs Division."
Includes index.
ISBN-13: 978-94-010-7128-4
1. Geosynthetics--Congresses. 2. Soil stabilization--Congresses.
I. Jarrett, P. M. II. McGown, Alan, 1941- . III. North Atlantic
Treaty Organization. Scientific Affairs Division. IV. Title.
V. Series.
TA455.G44N37 1987
624.1'64--dc19 88-14805
CIP

Published by Kluwer Academic Publishers,
P.O. Box 17, 3300 AA Dordrecht, The Netherlands.

Kluwer Academic Publishers incorporates the publishing programmes of
D. Reidel, Martinus Nijhoff, Dr W. Junk, and MTP Press.

Sold and distributed in the U.S.A. and Canada
by Kluwer Academic Publishers,
101 Philip Drive, Norwell, MA 02061, U.S.A.

In all other countries, sold and distributed
by Kluwer Academic Publishers Group,
P.O. Box 322, 3300 AH Dordrecht, The Netherlands.

ISBN-13: 978-94-010-7128-4 **e-ISBN-13: 978-94-009-1405-6**
DOI: 10.1007/978-94-009-1405-6

CONTENTS

EVALUATION OF MATERIAL PROPERTIES

ANALYTICAL TECHNIQUES AND DESIGN METHODS

CONSTRUCTION METHODS AND ECONOMICS

RESEARCH NEEDS

PARTICIPANTS

INDEX

PREFACE

Polymeric materials are being used in earthworks construction with ever increasing frequency. The term "Geosynthetics" was recently coined to encompass a diverse range of polymeric products designed for geotechnical purposes. One such purpose is the tensile reinforcement of soils. As tensile reinforcement, polymers have been used in the form of textiles, grids, linear strips and single filaments to reinforce earth structures such as road embankments, steep slopes and vertically faced soil retaining walls. A considerable number of retaining structures have been successfully constructed using the tensile reinforcing properties of "geosynthetics" as their primary means of stabilization. Despite such successes sufficient uncertainty exists concerning the performance of these new materials, their manner of interaction with the soil and the new design methods needed, that many authorities are still reticent concerning their use in permanent works.

This book represents the proceedings of a NATO Advanced Research Workshop on the "Application of Polymeric Reinforcement in Soil Retaining Structures" held at the Royal Military College of Canada in Kingston, Ontario from June 8 to June 12, 1987. The initial concept for the workshop occurred during the ISSMFE Conference in San Francisco in 1985 when a group of geotextile researchers mooted the idea of holding a "prediction exercise" to test analytical and design methods for such structures. It was decided that members of the group and others would attempt to predict in advance the behaviour of two, 3m high, polymerically reinforced soil walls to be constructed and tested in the laboratories at the Royal Military College. After the tests were completed the group would meet to discuss and compare predictions and behaviour in order to see where the results pointed concerning our analytical abilities. The "prediction exercise" began. In seeking the means to get this international group together it was recognized that a NATO Advanced Research Workshop might be appropriate. Early in our approach to the NATO Scientific Affairs Division it was realized that their sponsorship would allow us to broaden the scope of and attendance at the workshop. Thus the "prediction exercise" became the initial focus and springboard for a group of experts to study and thoroughly discuss the state of the art of this new form of construction. This volume contains the results of the prediction exercise, the papers developed by the leaders of five sessions in which all aspects of this method of construction were addressed and summaries of certain of the session discussions including the research needs session.

The workshop ultimately drew together 37 researchers, builders, designers and manufacturers from 12 countries. The workshop fell under the Double Jump program and was attended by six persons representing manufacturers of geosynthetic products. The harmonius manner in which this diverse group worked together was quite outstanding and went a long way to ensuring that the scientific objectives of the workshop were met. The extension of the working cooperation to the social aspects of life was equally outstanding leading for example to an impromptu international "soccer" match amongst participants that showed up some unexpected yet quite prodigious skills. With such spirit and congeniality it is felt that the additional objectives of the NATO Scientific Affairs Division for the establishment of international understanding and cooperation amongst scientists were also fulfilled at this workshop.

To reach the position where the scientific group could meet in such productive circumstances and have a book produced from the proceedings demands considerable effort from a number of people and financial and material support. The Scientific Director for the meeting and the editors of this volume wish to acknowledge with sincere thanks the following:

The NATO Scientific Affairs Division for their financial support of the Advanced Research Workshops. This far-sighted aspect of NATO is now far better appreciated by the group of scientists involved in this meeting.

The Department of National Defence of Canada who through the Royal Military College of Canada provided the meeting facilities and considerable material and administrative support to enable the workshop to be organized.

The Workshop Organizing Committee of Drs. Andrawes, Leflaive and Koerner who leant weight and enthusiasm to the original proposal and wise counsel throughout.

The session leaders and authors for their co-operation in the presentation of papers and submitting them by the deadlines or at least reasonably close to deadlines in some cases.

Mr. J.E. DiPietrantonio of RMC for his considerable work and skill in preparing, repairing and mounting of drawings in many of the final edited papers.

Miss Martina M. Lahaie of RMC for her considerable secretarial and administrative skills and her long term enthusiasm to see the workshop succeed. Her contributions prior to and during the workshop and in the editing of the manuscripts for this volume were the keystones of the administrative process.

In conclusion, the workshop occurred through the administrative efforts described above but it suceeded scientifically through the cooperative efforts of all the participants. They were a pleasure to do business with!

Peter M. Jarrett
Royal Military College
Canada

Alan McGown
University of Strathclyde
Scotland

Introduction

REINFORCED SOIL RETAINING STRUCTURES AND POLYMERIC MATERIALS

F. SCHLOSSER and P. DELAGE

Ecole Nationale des Ponts et Chaussées, CERMES
B.P. 105 - 93194 NOISY-LE-GRAND CEDEX

1. INTRODUCTION

Soil has been used since the most ancient civilizations as a construction material. However, due to insufficient tensile strength, designers had to improve its resistance by using mechanical processes (compaction, draining), chemical processes (stabilization), or by inclusion of resisting elements (reinforcement).

In the past, natural inclusions have most often been used to improve the mechanical properties of soils and structures : straw in clayey soils to develop construction material ; palms or branches at the base of structures founded on soft soils and so on. It is also interesting to mention the reinforcement of natural slopes provided by plant roots in the nature.

During the past twenty years, the important development of Reinforced Earth, and the concept of reinforced soil as a construction material, introduced by its inventor H. Vidal, in the sixties, have contributed to the birth of a new area of soil improvement : soil reinforcement. This area is based on a generalization of the "reinforced soil" concept, including various techniques, for slopes and embankments, retaining walls, foundations, dams...

The concept of soil reinforcement is based on the existence of a strong interaction between the soil and the inclusion. The most common interaction is friction, but passive pressure may also be mobilized. Frictional interaction requires good mechanical properties of the soil, particularly in terms of friction angle and drainage ; granular soils are best adapted for this kind of reinforcement. Passive pressure is generally associated with anchors, which needs more sophisticated inclusions. It can be mobilized in all types of soils, including saturated fine-grained soils, where frictional properties are poor.

The number, type, and arrangement of the inclusions may be quite variable. Depending on the type of the inclusion, two extreme cases may be considered (Schlosser et al, 1983).

1) a "uniform inclusion", for which the soil-reinforcement interaction can develop at any point along the inclusion. In this case, a relatively high and uniform density of reinforcements will result in a new composite material called "Reinforced Soil". The behaviour of the reinforced soil mass can be investigated through laboratory testing considering a representative sample of the new composite material. This concept is illustrated in figure 1.

P. M. Jarrett and A. McGown (eds.), The Application of Polymeric Reinforcement in Soil Retaining Structures, 3–65.

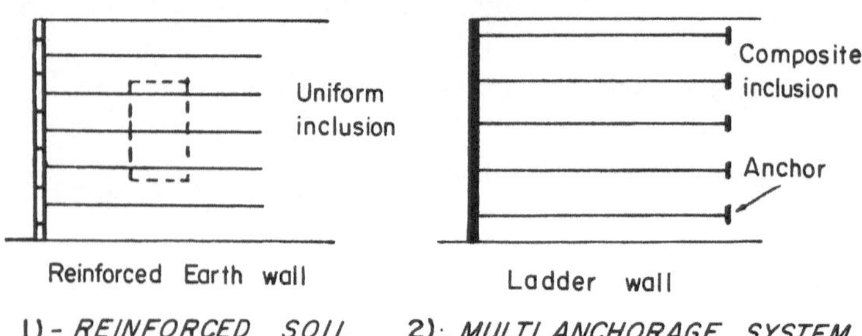

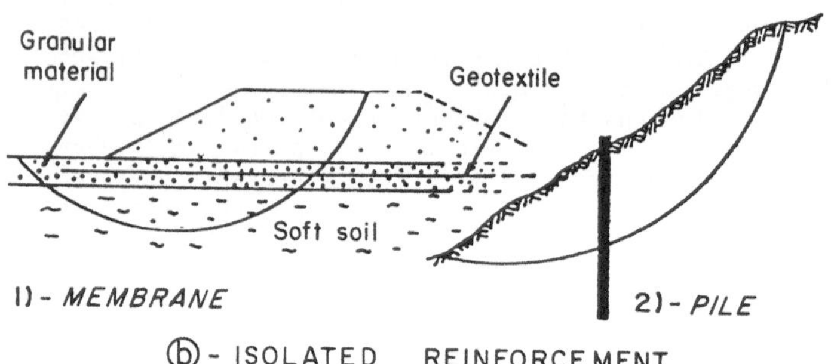

FIGURE 1. Types of soil reinforcement systems (Schlosser et al., 1983)

2) a "composite inclusion", which consists of an inclusion reinforced at particular locations where the soil-reinforcement interaction is concentrated. Generally, as for multi-anchorage systems, these points are located at the ends of the inclusions.

These considerations lead to the classification of soil reinforcement systems presented in Table 1.

Type of reinforcement / Density of reinforcement	uniform	composite
periodical	reinforced soil	multi-anchorages systems
isolated	membranes piles	anchorages

Table I - Classification of soil reinforcement systems

At present time, the development of polymer reinforcements, which are essentially two dimensionnal (geotextiles, geogrids, geomembranes) could lead to a modification of this classification, and to a distinction between linear and two dimensionnal reinforcement.

For Reinforced Earth strips, various materials were initially used (Schlosser, 1977) : fiberglass in a polyester resin, Tergal, or passivable metals such as aluminum or stainless steel. Finally, metal, and more particularly galvanized steel, was elected because of low deformability, cost and ease of setting.

In the early seventies, the use of polymers started to develop, mainly through geotextiles. Progressively, designers were attracted by the great variety of available products. They also considered polymers as an alternative, with respect to metal corrosion problems.

Numerous papers have been published on polymer reinforcement in different geotechnical journals and conference proceedings.

Among the conferences partially or totally devoted to soil improvement, those related to the use of polymers in reinforced soil retaining structures are listed below :

1) Soils and fabrics, Paris, 1977.
2) Reinforced Earth and other composite soil techniques, Edinburgh 1977.
3) Earth Reinforcement, ASCE, Pittsburgh 1978.
4) International Conference on Soil Reinforcement, Paris, 1977.
5) 2nd International Conference on Geotextiles, Las Vegas, 1982.
6) 8th ECSMFE, Helsinki, 1983. General Report on Soil Reinforcement.
7) Symposium on Polymer Grid Reinforcement in Civil Engineering, London, 1984.
8) 11th ICSMFE, San Francisco, 1985 : General Report on Geotechnical Construction.
9) 3rd International Conference on Geotextiles, Vienna 1986.

A journal exclusively dealing with the subject, entitled "Geotextiles and Geomembranes", is now regularly published.

This paper will review the development of reinforced soils systems for retaining walls, with a special attention to the use of polymeric materials such as geotextiles, geogrids and fibers. The history and development of soil reinforcement systems for retaining walls are first presented, after which the main types of existing polymer reinforced soil walls are described. The properties of polymers in relation to the behaviour and the design of reinforced soil walls are briefly considered. Finally, advantages and drawbacks of using polymers in such walls are discussed.

2. HISTORY AND DEVELOPMENTS

2.1 Coyne's ladder wall

In 1929, André Coyne patented in Paris a multi-anchorage system to be used for the construction of retaining walls and especially quaywalls, dykes and so on. The idea of using such a system was the principle of constructing a wall by successive horizontal elements, formed by a light facing element linked to continuous or discrete anchors with ties. In this system, the ratio between the total height of the wall, and the length of the anchor ties was approximately 2.5. Figure 2a shows a schematic view of such a structure as specified in the patent, and Figure 2b a cross section of an actual quaywall, 200 m long, built in Brest harbour in 1928, which was the first major application of the system. Despite some studies performed on reduced scale models (Coyne, 1945), the real mechanism of the ladder wall was not fully explained by Coyne. He indicated that the structure formed by the facing, the tie-rods, the back-fill material located between this facing and the anchorages may apparently behave as a solid, sustaining small deformations, as compared to the displacements of the wall. Coyne (1945) describes some applications of this patent : the work presented in figure 2b is a dyke submitted to tides, which may even be submerged by big waves ; internal seepage may then occur, with 1,50 or 2 m hydraulic gradients, and the structure behaves as a dam ; good rocky material was used as a fill material between the ties ; this structure supported 0,50 m settlements without any problems, because of good articulations between the facings. Coyne also mentions a 200 m long quay wall in Brest, and the construction of 10 to 20 m high retaining structures, such as the sidewalls of an overflow in Marèges, a 10 m high coffer-dam at the same location, and a 14 m high dam on the river Laurenti (Pyrénées) ; in the case of dams, the ladder wall may constitute either the up-stream or the down-stream facing. In all cases, good rocky material is placed in contact with the ties.

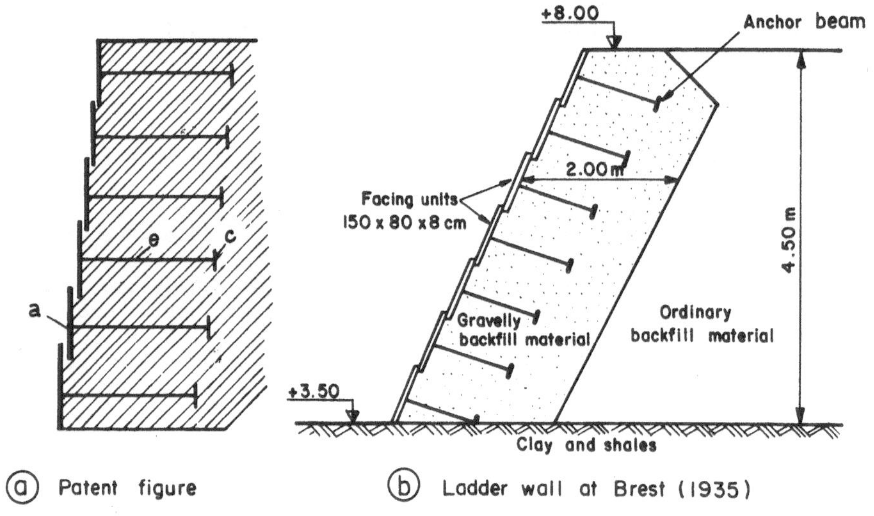

FIGURE 2. Ladder wall system invented by Coyne (1929)

Apparently, the first developments of this technique were stopped during World War II, and, in spite of Coyne's wishes, it was not sufficiently used later during the reconstruction of the country.

After a long period of time during which the Ladder Wall system was not used, several systems related to this type of reinforcement have been proposed since the starting point of Reinforced Earth development. They differ from each other with respect to the type of anchorage and facing used. Anchored Earth (Murray and Irwin, 1981) and Microanchorages (Costa Nunes, 1978) use frictional anchorages, while American Geotech and other systems use passive pressure beams or plate anchors (Figure 3).

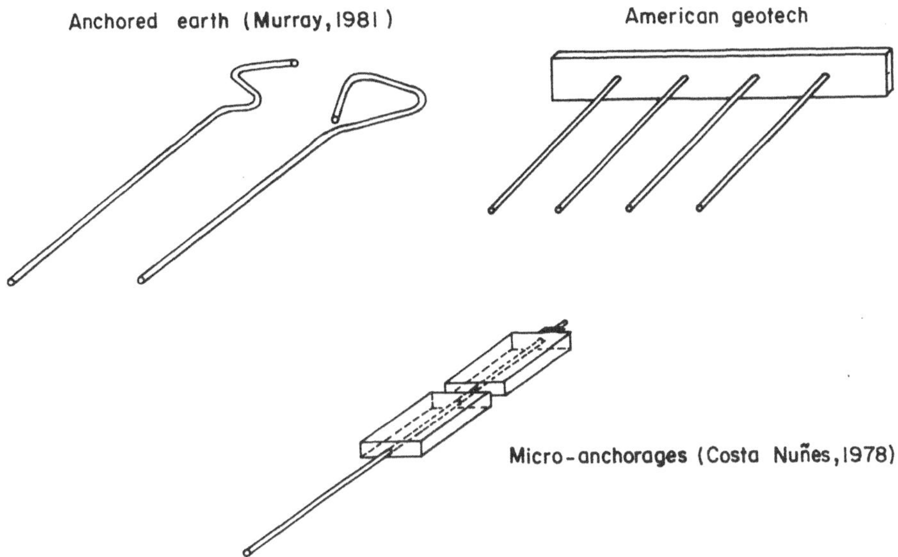

FIGURE 3. Reinforcements in recently developed ladder type walls

It is interesting to mention a mixed system developed by Fukuoka et al. (1982), in which the facing used is made of fabric attached to vertical columns and the anchors are concrete plates. The type of behavioral mechanism involved in this system was demonstrated with a full-scale experiment : the displacement (rotation) of the rigid columns is sufficient to reach the value of the active earth pressure on the facing, whereas the pressures on the anchor plates remain equal to the K_o state of stress.

Chabar et al. (1983) described the construction of a 21 m high dam, built according to the classical Coyne's ladder wall system.

More recently a multi-anchorage system, called Actimur (1984) has been proposed in France. It combines a vertical sheet-pile facing and horizontal tie-rods with vertical metallic anchor discs.

2.2 Reinforced Earth

The invention of Reinforced Earth by Henri Vidal in 1963 and the rapid development of this new technique at the end of the 60's, has been the starting point of reinforcement systems, especially dealing with soil retaining techniques where the soil reinforcement is periodical and where the soil-reinforcement interaction acts all along the reinforcement. These systems have been denominated reinforced soils by Schlosser et al. (1983).

Henri Vidal considered Reinforced Earth as a new composite material and consequently introduced the very interesting concept of reinforced soil material, which has prooved to be general, realistic and efficient.

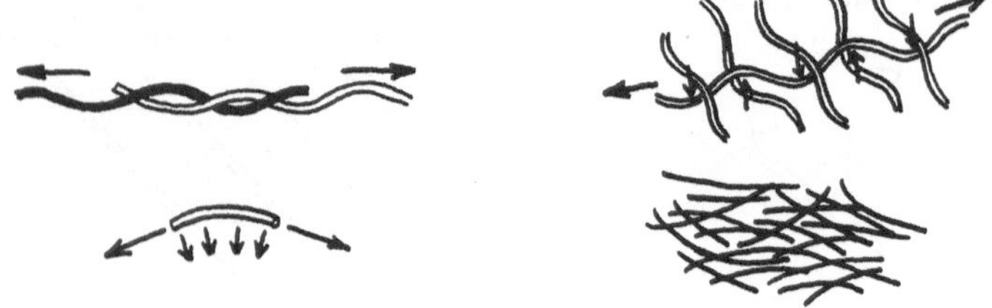

(a) Combination of reinforcements alone

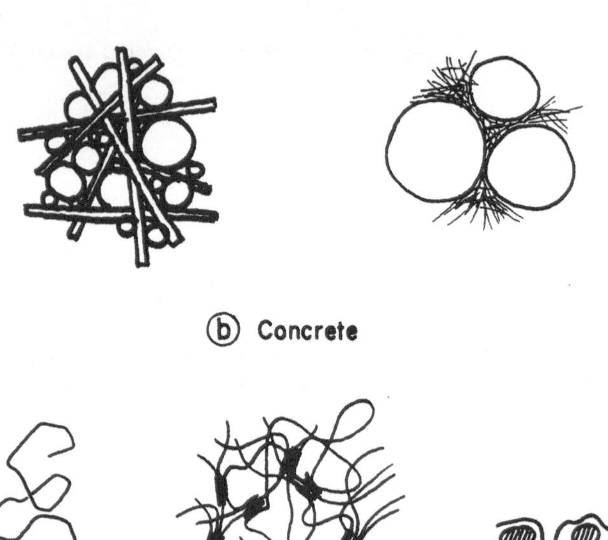

(c) Macro molecules and human body materials

FIGURE 4. Cohesive materials as combinations of "grains" and reinforcements (H. Vidal, 1966)

It must be noticed that, in his first paper (1966), H. Vidal developed a large theory, presenting the different manners of producing a cohesive material using independant grains and reinforcements. As indicated in figure 4 he dealt first with the texture of reinforcements made with fibers (nonwoven, wooven, etc...) and explained the behaviour of several materials (wood, paper, clay, concrete and finally human body materials) by associations of "grains" and reinforcements interacting through frictional forces.

An experimental study of the behaviour of the Reinforced Earth material was performed at the Laboratoire Central des Ponts et Chaussées (Schlosser and Long, 1972) by testing samples of sand reinforced with horizontal and regularly spaced aluminium foil discs in the triaxial apparatus. It was shown that two failure modes can develop in such reinforced sand samples : failure by slippage of the reinforcement, and failure by reinforcement breakage. The yield line in the (σ_1, σ_3) principal stresses axis is presented in figure 5 : at low confining pressures, failure occurs by slippage, leading to a curved yield line passing through the origin ; at higher confining pressure, this failure line is a straight line which proves that the reinforced sand behaves as a cohesive material having the same friction angle as the original sand and an anisotropic pseudo-cohesion due to the reinforcements. This pseudo-cohesion is very rapidly mobilized at low axial deformations, since the reinforcements behave in a rigid way compared to the relatively deformable sand. Tensile stress measurements using strain gauges show that the maximum stress value in the discs is obtained at points located at a distance from the center approximately equal to two-third of the radius. The inclined failure plane developing when the sample fails by "reinforcement breakage", indicates that a bifurcation phenomenon occurs in the development of the failure surfaces.

A very interesting theoretical contribution to the subject is due to Bassett and Last (1978). These authors considered that the mechanism of tensile reinforcement involves anisotropic restraint of the soil deformations in the direction of the reinforcements. Then they used Roscoe's failure criteria for sands, based on zero extension concept, to demonstrate that the presence of the reinforcements leads to a rotation of the principal directions of the deformations tensor. They showed (figure 6) that since in a Reinforced Earth wall, the direction of the reinforcement must be aligned with the zero extension direction, the failure surface must be vertical to comply with the assumption of suppressed dilation rate (it is assumed that the Reinforced Earth material exhibits a zero dilation angle). In other words, it can be said that, due to the soil-reinforcement interaction, the presence of the reinforcements in a soil mass greatly modify the strain and stress patterns. Moreover this is consistent with the development of cracks along a cylindrical surface in reinforced sand samples in the triaxial apparatus.

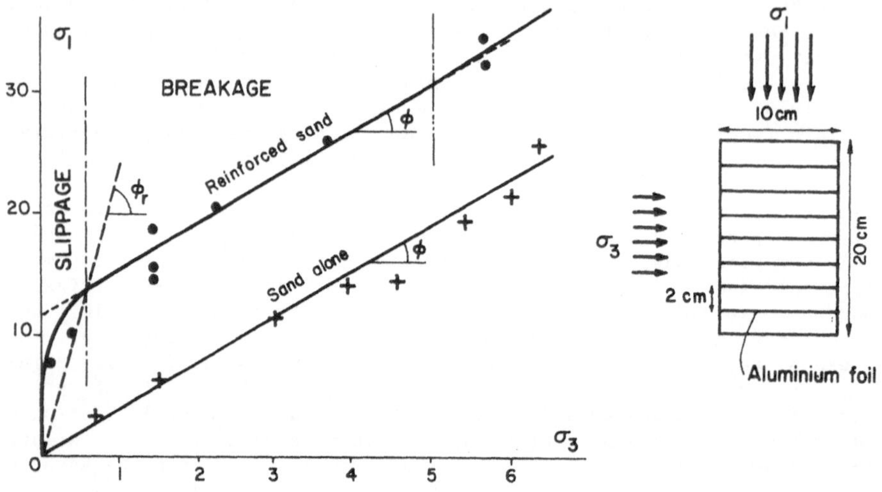

ⓐ Reinforced Earth material failure line (Schlosser and Long,1972)

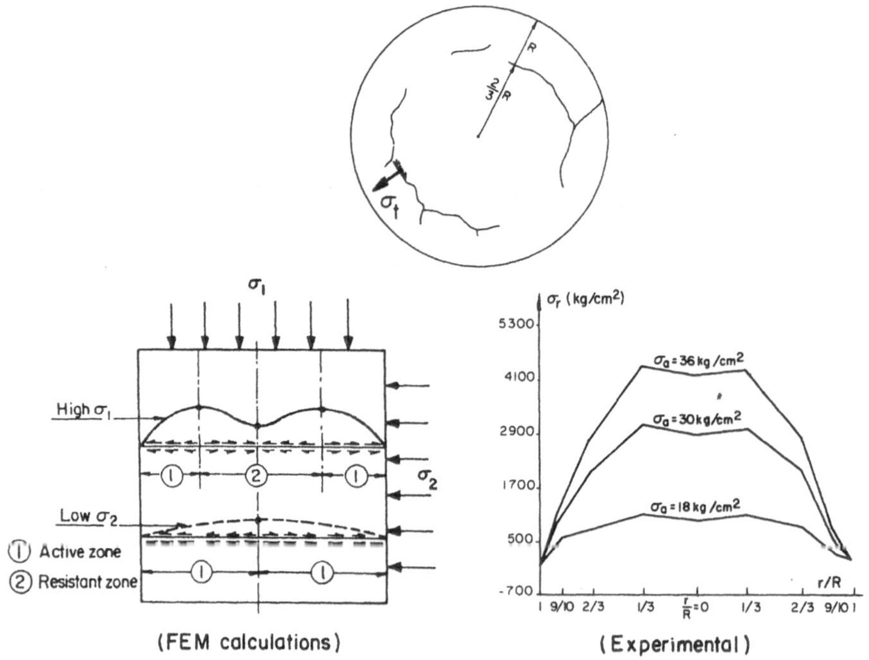

ⓑ Tensile stresses distribution and crack locations in the discs

FIGURE 5. Behaviour of the reinforced earth material at the triaxial apparatus

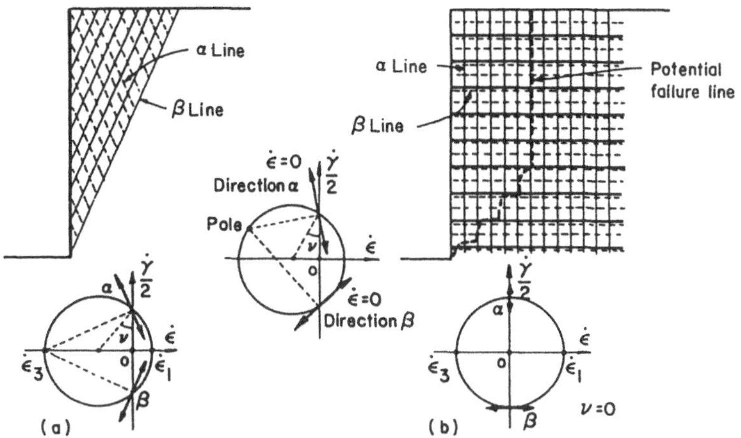

FIGURE 6. Influence of the reinforcement on the potential failure lines (Basset and Last, 1978)

Based on the above principles, Reinforced Earth consists essentially of the following components : 1) a granular backfill material, 2) linear reinforcements, generally strips 3) a facing made of pre-cast elements attached to the strips. The two major components are the granular backfill and the strips ; the purpose of the facing is only to retain locally the backfill between two horizontal reinforcement layers. Figure 7 shows a schematic view of a Reinforced Earth wall, showing these three components.

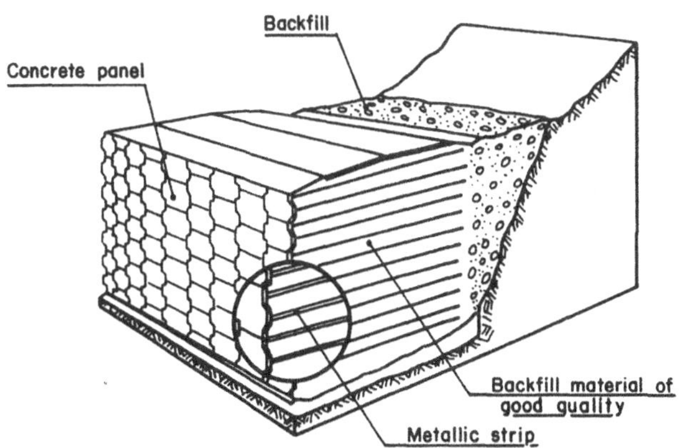

FIGURE 7. Schematic view of a reinforced earth wall

There has been a great improvement in the technological development of the Reinforced Earth technique, with respect to all three components, i.e. the facing, the strips and the backfill material. Initially the facing was made of U-shaped elements, 33 cm high, the weight of each being light enough to enable an easy handling. The strips were completely smooth, generally 60 mm wide and 3 mm thick. The backfill was a good granular material with less than 15 % in weight of grains smaller than 80 μm (n° 200 sieve). Considering the purpose of the present paper, it is interesting to note that H. Vidal first planned to use plastic strips and plastic facing elements in order to avoid corrosion problems, as indicated in his first paper (1966). He was rapidly able to produce industrially in 1967 facing elements and strips made of fiber-glass coated with polyester resin. It will be seen further why the use of such a material was stopped and replaced by metals.

Three events have marked the Reinforced Earth technological development. Firstly, the choice of galvanized steel for strips and facing, after a first tentative with polyester coated fiber-glass and stainless steel and aluminum used for some years in France. These two metals were at that time (and even now) considered to be particularly efficient against corrosion, even when embedded in soils. They are theoretically protected by a thin layer of undestructible oxyde on their surface. However, it must be now accepted 15 years later that this is absolutely not true and that these metals may be corroded in some cases more drastically and rapidly than galvanized steel, due to a special phenomenon involving an accelerated corrosion rate.

The second event has been the development in 1971 of a typical cruciform panel for the facing. This type of facing enables architectural possibilities, curved facings and it is now worldwide representative of the Reinforced Earth development.

In 1975, the Reinforced Earth Company patented the ribbed strip. This new technological aspect was directly issued from research on the soil-reinforcement frictional interaction. As indicated later by Schlosser et Elias (1978), the main phenomenon in this 3-dimensional friction mechanism is the restrained dilatancy effect (figure 8). The consequently apparent friction coefficient is much influenced by the volume of the sheared soil zone around the strip. Ribs increase this volume and thus largely increase the μ^* value.

After 20 years, the Reinforced Earth major development appears to be related to following features : 1) R.E. behaves satisfactorily even in various critical situations (large differential settlements, movements in the foundation soil, seismic event, etc...). 2) R.E. cost is competitive and generally low compared to other solutions. 3) R.E. wall facings are attractive and aesthetic.

For the time being, the only problem is related to the special corrosion of the stainless steel and aluminum strips embedded in walls built in France 10 to 15 years ago.

2.3 Geotextiles

The use of geotextiles in earthworks for reinforcement and separation at the base of an embankment on soft soil started approximately at the same period as the Reinforced Earth early development. In fact, the first paper dealing with such an application has been published in 1969 (Vautrain and Puig).

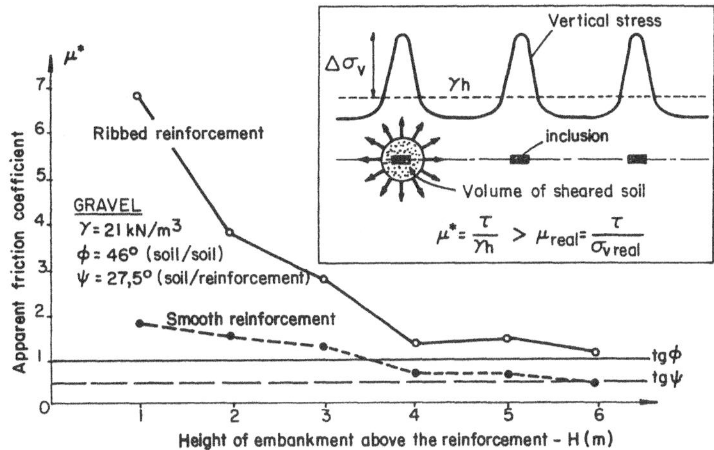

(a) Restrained dilatancy effect mechanism

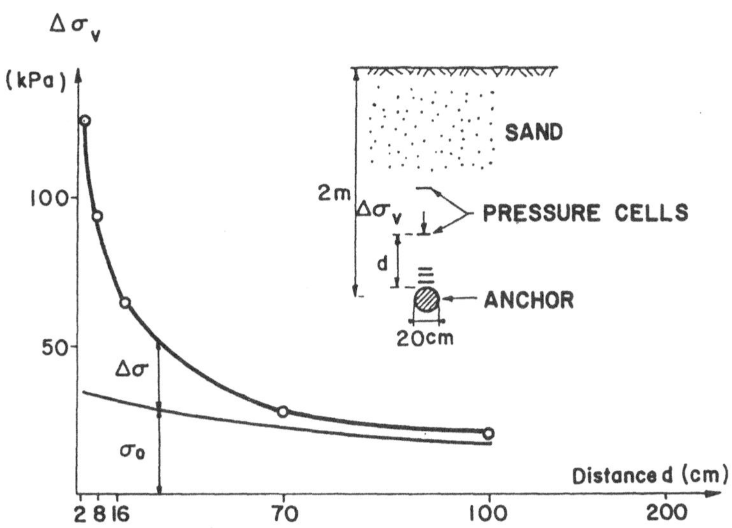

(b) Normal pressures measurements around a tensioned inclusion (Plumelle, 1984)

FIGURE 8. Restrained dilatancy effect on soil-linear inclusion friction

Since this time the application of geotextiles to roadways, embankments and slopes has intensively increased. According to Giroud and Carroll (1983), the largest quantity of geotextiles is now utilized for roadway construction, principally temporary and construction roads.

The first application of geotextiles to multilayered soil-fabric retaining systems was done in 1971 (Puig et al, 1977). It was an experimental wall using a nonwoven fabric (Bidim) and a very poor backfill material (wet clayey and sensitive soil). The wall was 4 m high and was founded on a very compressible soil (peat layer, 3 m thick).

Since this first application, geotextiles have been used for retaining walls, and for earth dams (Kern, 1977). They present interesting features : low cost, drainage, possibility of using poor backfill material. However their utilization has been rather limited until now, probably because of their deformability (particularly in the case of unwoven geotextiles) and to the relatively inaesthetic appearance of the facing.

2.4 Grids

The first reinforced soil retaining structure using grids as reinforcements was constructed in 1974 on Interstate Route 5, near Dunsmuir, California (FORSYTH, 1978). One year before in 1973, the California Transportation Laboratory developed a large direct shear device in order to test the pull out resistance of different reinforcement systems (smooth strips, ribbed strips, bars, bar mats). The purpose was to find a reinforcement system which could enable the use of granular backfill material containing a large percentage of fine-grained material. It was found that the best system was the grid or bar mat, which provides a relatively linear reinforcement, wisthstanding large pull-out forces, thanks to the passive thrust mobilized against the transversal bars. However, compared with the pure frictional interaction reinforcements, i.e. strips, these bar mats require large displacements, 5 cm and more, to fully mobilize the pull-out resistance (CHANG et al., 1977).

Figure 9a shows the welded bar mat tested and used by Caltrans in the construction of the first "mechanically stabilized embankment" at Dunsmuir. This structure was approximately 120 m long and consisted of two walls, 6 m high. The vertical facings were made from rectangular and long precast concrete elements 3.75 m X 0.6 m X 0.2 m in size (figure 8c) . The bar mats, which were 1.2 m wide and 3 to 4.5 m in length, were attached to the facing elements by inversion of the two-bar yoke through precast holes in the facing panels. These two prethreaded bars were bolted into position, inducing some interesting prestressing of the reinforcement. The construction was very similar to Reinforced Earth.

As indicated by Forsyth (1978), it was "anticipated that the bar- mat mode of reinforcement would have significant economic advantage in certain areas of the state of California where high quality backfill material is not readily available".

This argument was considered and put forward by VSL when promoting in 1980 an equivalent retaining system, called Retained Earth, in which the same type of welded bar-mat was used.

As shown by Schlosser et al. (1983, 1985), the bar-mat interaction mechanism is complex and involves both friction along the longitudinal bars and passive thrust again the transversal bars. For small soil-reinforcement displacements (< 0.5 cm) there is initially a mobilization of the friction along the longitudinal bars. For larger displace-

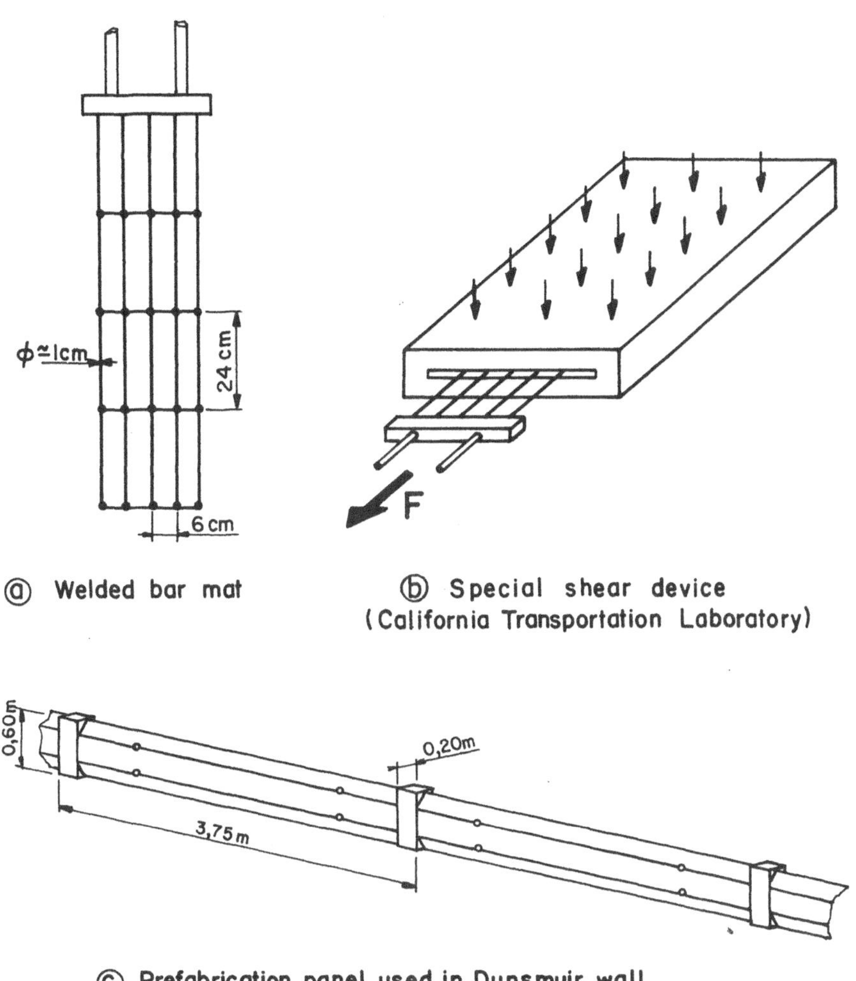

FIGURE 9. Type of welded bar mat and panel used by California transportation (Caltrans)

ments there is a mobilization of the passive pressure on the transverse bars and the stress-displacement curve keeps increasing even for displacements greater than 10 cm.

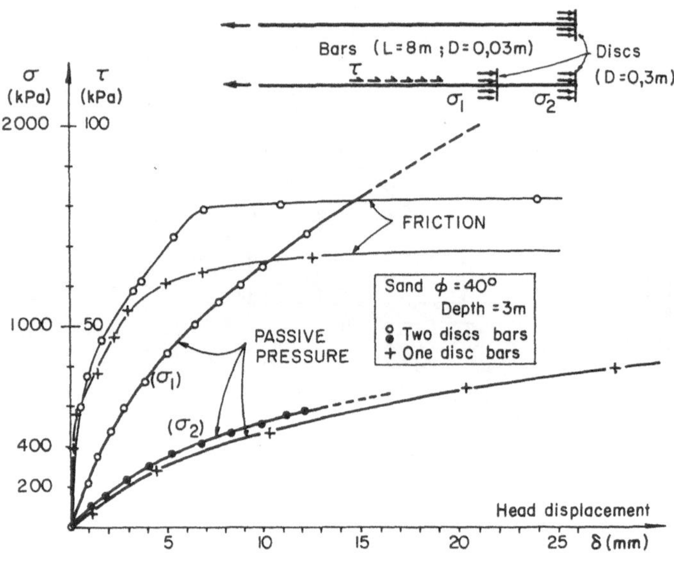

ⓐ Respective mobilization of friction and passive pressure on bars equipped with anchoring discs (after Morbois and Long, 1984)

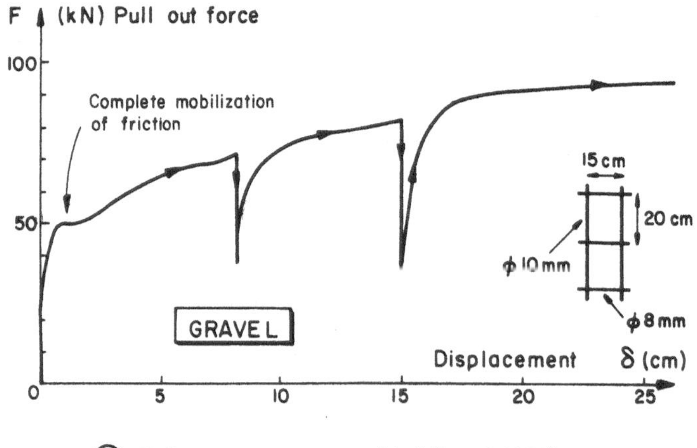

ⓑ Pull out test on a grid (Bacot, 1981)

FIGURE 10. Mechanism of bar pull-out resistance

Figure 10 shows two typical results about this phenomenon : 1) the respective mobilization of friction and passive pressure on bars equipped with anchoring discs ; 2) the pull-out force-displacement curve on a grid.

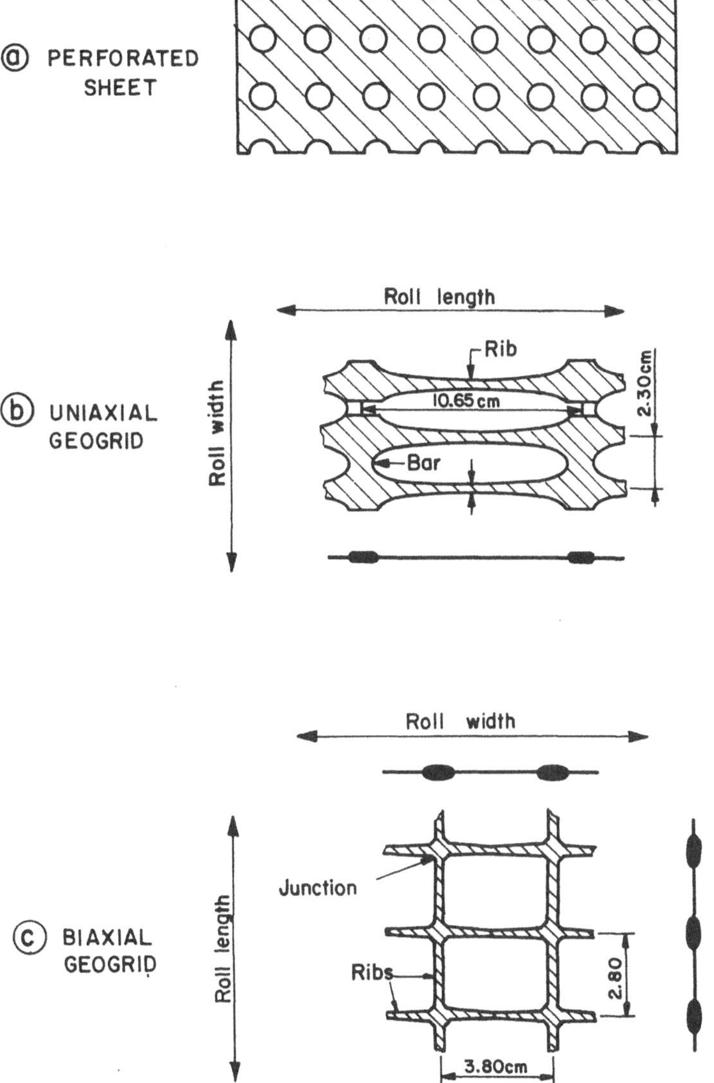

FIGURE 11. Typical geometry of perforated sheet and uniaxial and biaxial geogrids (Bonaparte et al., 1984)

Because of this mechanism, bar-mats are more resistant in pull-out than frictional reinforcements (bars, strips) only for large displacements (5 to 10 cm). If such lateral displacements values are allowable for the structure, it appears that bars-mat reinforcements permit to use poor quality backfill material with a large percentage of fine-grained soil in a retaining system. However, further research in this field is still required to specify suitable soils and what may be the best mat geometry.

In the early 80's, Netlon manufactured and developed a plastic grid product called Tensar. This material consists of a high strength, oriented polymer grid structure obtained from punched and stretched polymer sheets. The rapid development of this product, used in a variety of soil reinforcement applications (embankment reinforcement, retaining walls, rafts, repairs of slope failure, gabions), led to a new type of tridimensional reinforcements called geogrids. A geogrid has a small opening size (about a few centimeters) compared with a bar-mat (10 cm and more), but compared with non woven geotextiles, it exhibits a larger deformation modulus and tensile resistance (figure 11).

Properties and applications of geogrids will be further discussed. With respect to soil-grid friction, there is some similarity with the soil-bar mat interaction. However, the mechanism involves a new phenomenon in coarse granular soils, resulting from interlocked soil particles within the grid apertures, which act as an anchor for the transverse ribs of the grid. Forsyth and Bieber (1984) have compared a plastic grid (Tensar) (mesh size of a few centimeters) with a metallic bar-mat (20 cm opening size), both having identical surface areas. The force-displacement curves obtained in pull-out tests for a normal stress equal to 34.5 kPa are proportional with a ratio of about 3, because the passive pressure effect observed on transverse elements is lower for the Tensar (Figure 12). The type of soil used was decomposed granite ($\phi' \cong 35°$).

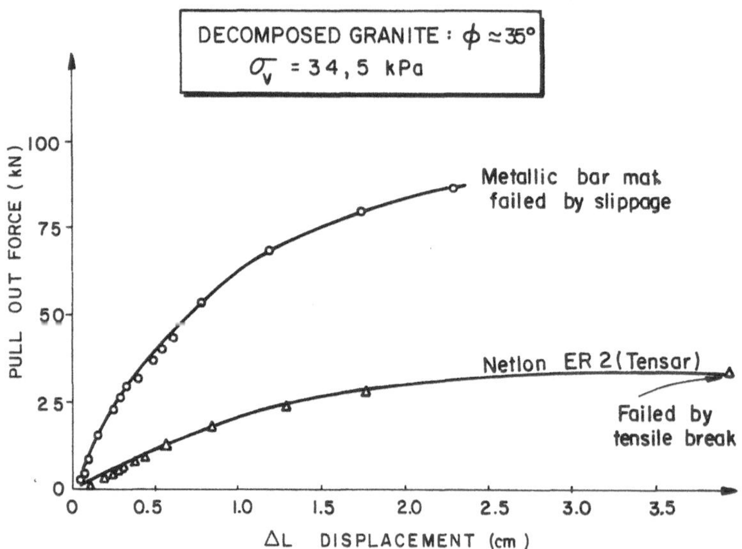

FIGURE 12. Pull-out force/displacement curves for a metallic bar mat and a plastic grid

However, comparing rapid direct shear tests on clay samples transversally reinforced by a metallic or plastic grid, Jewell and Jones (1981), and Ingold (1983) found no difference between plastic and metallic grids. The reinforcement effect on the undrained shear resistance was the following: 12 % increase with smooth steel sheet, 37 % increase with corrugated steel sheet, 42 % increase with steel grid and 44 % increase with the geogrid.

It must be noted that there is a great difference between the two types of tests : pull-out test and Jewell's direct shear on transversally reinforced samples. The latest is more representative of the friction phenomenum close to the potential failure surface, for instance in a reinforced soil retaining wall. However when using long reinforcements beyond this surface, as generally recommended for a design, the first type of test appears more adequate, according to the authors opinion (figure 13).

According to Bonaparte et al. (1984) high density polyethylene or polypropylene are suitable for soil reinforcement because of their in-ground durability and resistance to chemical, as well as micro-organisms attack. Generally speaking, durability is one of the most important problems, because reinforced soil structures are alternatives to classical reinforced concrete structures and they must therefore present an equivalent service life. Durability of polymeric materials will be treated in a subsequent section.

At present time, geogrids have largely been used in embankment reinforcements, rafts, gabions and corrections of landslides, but only a small amount of retaining walls have been built. It seems that like for geotextiles, the problem related to the facing still need to be solved : inaesthetic aspect and erection difficulties in geogrid facings,geogrid attachment to the panels in prefabricated facings.

2.5 Three-dimensional reinforcement : Texsol and fibers

It is technically very difficult to obtain in-situ soil reinforcement with flexible inclusions oriented in all directions as it is the case with the natural in-situ reinforcement provided by roots. However, the development of geotextiles and synthetic fibers has raised interest for three-dimensional reinforcement of soil fills, a process in which the soil is mixed with small inclusions (fibers, small plates) or continuous filaments (Texsol). The materials obtained in this manner, in which many particles are bound to each-other by the reinforcements, have a structure resembling concrete, rather than one or two dimensional materials. Their behaviour mechanism is complex and has not been sufficiently studied until now. Despite their interesting potential use, applications are still limited.

The principle of the patented Texsol consists in placing within the mass of a granular soil one or several continuous filaments (Leflaive et al., 1983), in order to obtain a three-dimensional random mixture of filaments and soil particles. The inclusion is a polyester filament, with a diameter 0.1 mm and a tensile strength of about 10 N. The placement technique involves a simultaneous projection of sand, water and filament, as indicated in figure 14. In spite of a low percentage by weight of the filament (0.1 to 0.2 %), the total length is very important (about 20 cm per cm^3 of material), which explains the good mechanical behaviour at failure.

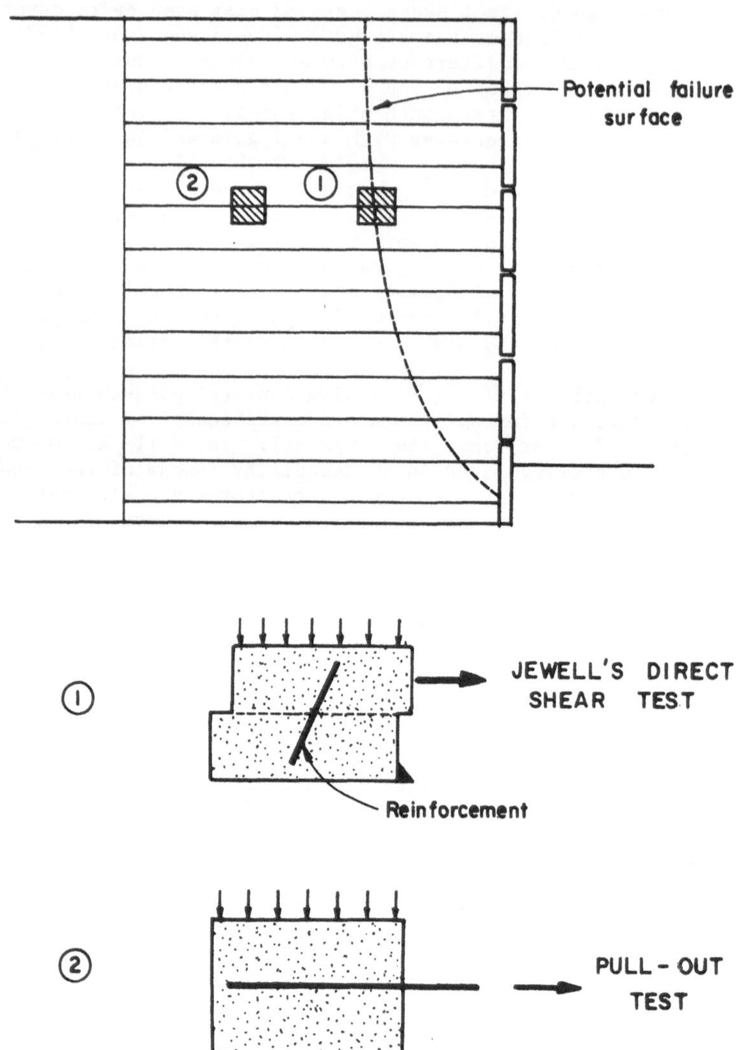

FIGURE 13. Differences between Jewell's direct shear test and pull out test

Figure 14 shows the first job carried out with this material (Leflaive et al., 1983). The problem was to repair a slide in a cliff of chalk and to restore a path at the top of the cliff. About 300 m^3 of Texsol were placed, at a slope angle of 60° and with difficult access conditions.

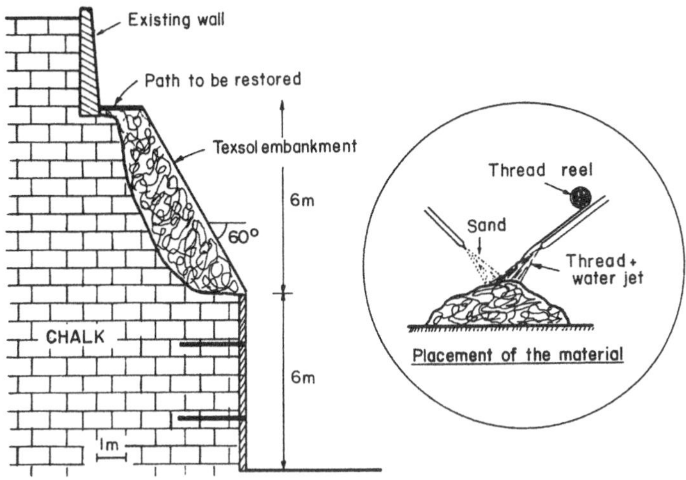

FIGURE 14. Texsol embankment of Caudebec-en-Caux and placement of the material (Leflaive, 1983)

At present time, Texsol is essentially used for retaining walls, but some very interesting applications could be found in relation with foundations, slabs or road construction.

The presence of continuous filaments increases the resistance and the ductility of the sand. A real cohesion of about 100 to 250 kPa is created. However, due to the placement by approximately horizontal layers, this cohesion could be relatively anisotropic with a small value on horizontal planes.

In fact, only very few papers have been published on the mechanical behaviour of Texsol. CBR tests, unconfined compression tests, static and dynamic triaxial tests have been carried out (Leflaive et al., 1983), but no stress-strain curves and no details about the preparation of the samples have been given. It is interesting to notice that the relative density of the sand has a main effect on the cohesion value : for a loose sand this value is smaller than 100 kPa, for a dense sand, it ranges from 150 to 250 kPa. But the way this cohesion is mobilized with the axial deformation in a triaxial test, as it has been analysed for sand samples reinforced by discs, remains a fundamental question. As reinforcement by filaments do not increase the deformation modulus, it could be expected that this mobilization would require relatively large deformations.

The inventor of this technique, Leflaive, says that besides the friction interaction between the soil and the filaments, two other phenomena are involved in the reinforcement mechanism with a continuous filament :

1) The curvature effect of the filament, associated with the friction, considerably increases the resistance to sliding.

2) The three-dimensional entanglement effect, coating each volume of soil with a large number of threads.

The curvature effect is partly similar to the inclusive undulation effect studied by Bacot (1981). This author has experimentally and theoretically studied the influence of the undulation when using linear and flexible inclusions. Such an undulation is mostly characteristic of reinforcement of fills (reinforced earth, embankment reinforcement, mat foundations, etc...) for which the inclusions are set in place layer after layer on an embankment surface which is not perfectly plane. The variation of the apparent friction coefficient μ^* as a function of the total angular variation θ_t of the inclusion with respect to the horizontal, is exponential. Bacot considers that this effect could be added to the dilatancy effect, which would explain the observation of high values of μ^*, higher than tan ϕ , even at large depths. This favourable effect of flexibility may however be balanced by an unfavourable effect due to the extensibility of the inclusion, for soils whose residual strength is very different from the peak strength.

Besides Texsol and reinforcement by continuous filaments, a relatively important research has been performed and interesting results have been published on reinforcement with fibers or small plates. Although it does not exhibit even a high resistance as Texsol, and has not yet given rise to many applications, such a type of reinforcement appears to be an interesting alternative, because of the easibility for mixing soil and plates (or fibers) and the better isotropy of the reinforced soil thus obtained.

The failure of a sand reinforced with small high strength inclusions (metallic discs) was studied for the first time by Hausmann (1976) with the triaxial apparatus. The failure obtained occurs by lack of adhesion, and it is characterized by an apparent internal friction angle ϕ_a increasing with the radius of the discs and by a cohesion value limited to 30 kPa when the size of the inclusions remains below 8 mm. These results were confirmed by Verma and Char (1978) who carried out 120 triaxial tests on a sand reinforced with small metallic inclusions having various shapes.

The use of natural or synthetic flexible fibers, having low deformation moduli (Gray, 1978, 1983), or small pieces of geotextiles (Hoare, 1979 ; Mercer et al., 1984), show that the failure mode of the reinforced sand by breakage of the inclusions is never obtained.

Figure 15 shows direct shear tests carried out by Gray (1983) on a sand reinforced with natural fibers according to the procedure devised by Jewell (1980). The fibers increase the shear strength of the soil, but in different ways according to the angle they form with the horizontal. The optimum improvement is obtained for a 60° angle corresponding to the maximum extension line for the sand alone. A critical normal stress defines two distinct zones for the failure curves ; above, the presence of the fibers results in a cohesion but it does not increase the internal friction angle ; below, there is only an increase of the internal friction angle.

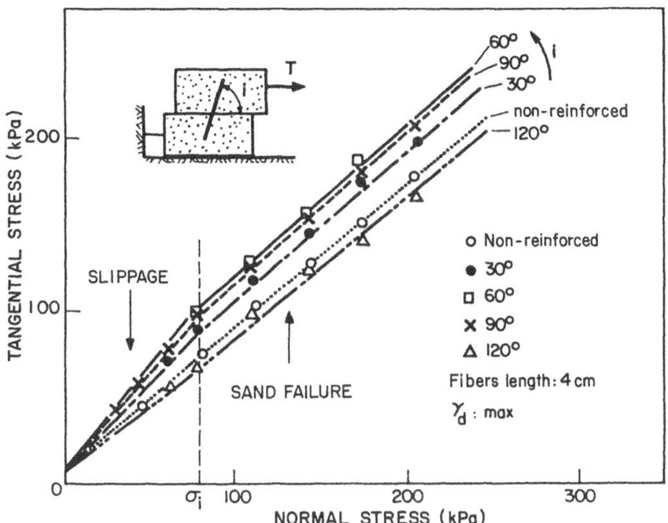

FIGURE 15. Shear strength of fiber reinforced soil (influence of the fiber inclination) (after Gray, 1983)

This phenomenon can be explained as follows : in the first case, the soil fails before the fibers because of the extensibility of the latter; in the second case, there is a failure by lack of adhesion and slipping of the fibers within the sand mass. This failure mechanism is different from the one obtained for rigid inclusions in the triaxial apparatus and for which the critical stress provides a distinction between a slippage type of failure ($\sigma < \sigma_i$) and a failure with reinforcement breakage ($\sigma > \sigma_i$) (Schlosser and Long, 1972).

Thus, it appears that for a soil reinforced with fibers, breakage of the fibers is practically never observed, except may be in the case of very long and inextensible (metallic) fibers. The partial conclusion to be drawn from this is that reinforcement with fibers improves the material ductility and at the same time, it decreases dilatancy (Gray, 1983).

Research is at present time in progress, at Ecole Nationale des Ponts et Chaussées, on sand reinforced with metallic or plastic fibers. Parameters such as fiber length, fiber extensibility, number of fibers per volume unit, relative density of the sand are tested in the triaxial apparatus.

It appears that a similar mechanical behaviour has to be expected for a sand reinforced with plastic grid type square plates, 4 cm on side, constituting very flexible inclusions (Mercer et al., 1984). Up to 0.6 % in weight of inclusions, these plates do not increase the sand void ratio, and there is a 25 to 60 % increase in the maximum deviatoric stress for a percentage of plates by weight of 0.19 %. However, due to a large interlocking effect and entanglement between the inclusions and the soil particles, noticeable shear strength increase occurs even at low values of the confining stress. This type of soil improvement appears to be easy, cheap and effective.

Table II shows cohesion values for reinforced sand, obtained for different types of inclusions (fibers, small plates, continuous filaments), for the same percentage of inclusion by weight (about 0.2 %). This table also indicates the different soil-inclusion interactions involved.

INCLUSIONS	FIBERS 5 cm	FIBERS 15 à 20 cm	GRID TYPE PLATES (4 x 4 cm)	CONTINUOUS FILAMENTS
Cohesion	10 kPa	100 kPa	50 kPa	200 kPa
Types of interaction	Friction Extensibility		Friction Extensibility Entanglement Interlocking	Friction Extensibility Entanglement Curvature effect

Table II - Values of cohesion and soil-inclusion interaction types in sand reinforced with synthetic inclusions

2.6 Another new retaining system : Tervoile

Among the most recently developed retaining systems, it appears interesting to mention Tervoile, which is both a frictional and structural system.

As presented in Figure 16a, the basic structural element is a thin corrugated and U-shaped membrane, (the corrugations being in a plane perpendicular to the cylinder axis). This element is made of plane and bent plates bolted together during construction. Until now, metallic (galvanized corrugated steel) elements are used.

Figure 16b shows basic elements juxtaposed in order to form a retaining structure. The needed interaction (i.e. pressure and frictional shear stress) between the soil and the basic elements of the structure, is provided by the backfill, spread and compacted in successive layers. The structural elements work only in tension (or compression in the vertical direction) : this feature is characteristic of this particular system.

Besides the backfill, the basic elements may either be used as the only structural parts or be coupled with horizontal reinforcing layers made of wiremesh, fabric membranes, etc...

The mechanism of the retaining system is the following (Fig. 16c):

1) In a basic element, the backfill located in the bent part acts as an active zone and applies pressure on the membrane. On the contrary, backfill located between the two plane parts acts as a resistant zone. There is some similarity with the reinforced soil system mechanism, but the vertical location of the frictional elements and their large spacing (2 m) makes the system very different.

2) The system has a relatively important rigidity in vertical planes and can therefore withstand thrust by acting as a beam. This important aspect of the mechanism enables a reduction of the length of the vertical frictional elements.

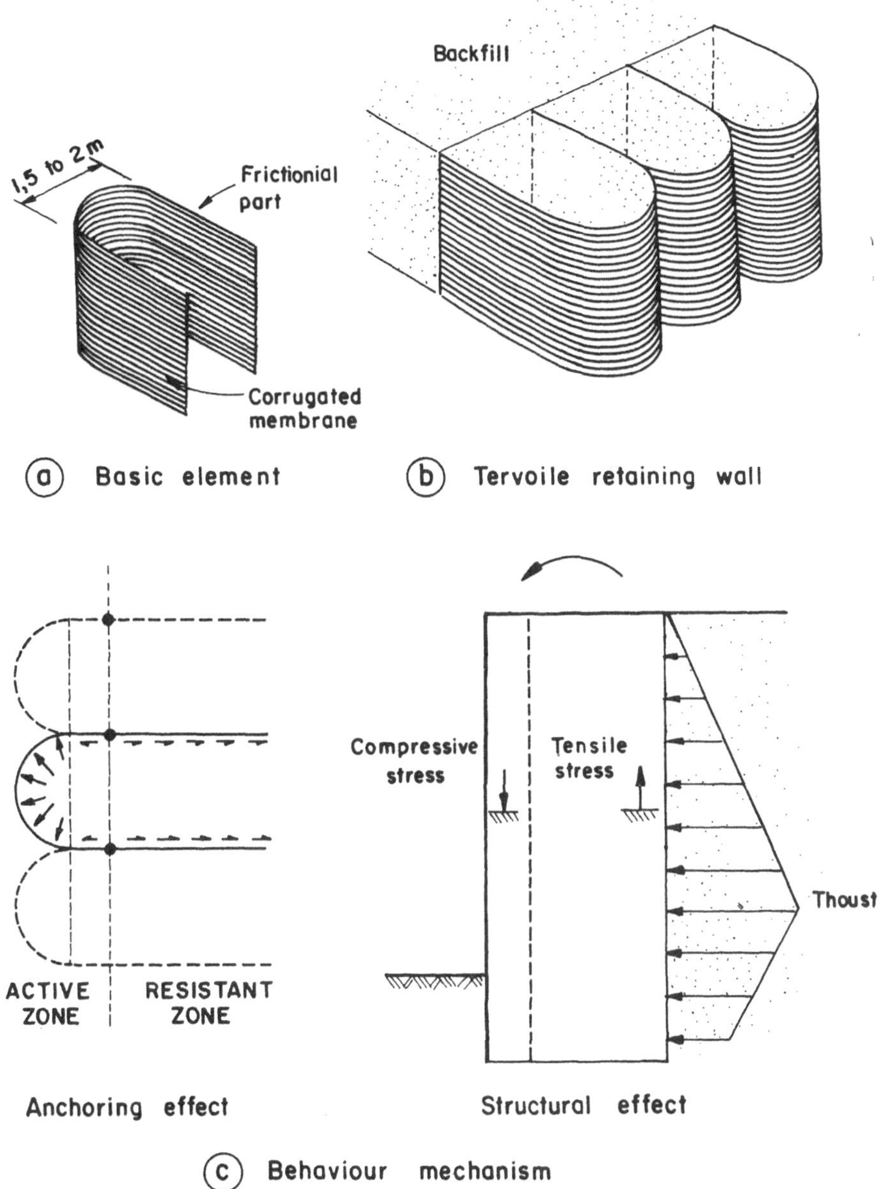

FIGURE 16. Tervoile retaining wall system

According to the inventor, V. Curt, a Canadian engineer, and his french associate, A. Metulesco, this system is flexible and can tolerate large differential settlements.

The advantages of such a retaining system are : 1) low cost, 2) easibility of construction, 3) very light prefabricated elements.

A full scale experiment was performed in 1985 at Sherbrooke in Canada, and in 1986 the Ministry of Transport in Quebec used this technique in road construction at Grandes Piles for a retaining wall, 7 m high to be built along a river.

2.7 Soil nailing

In-situ reinforcement of geotechnical materials by resistant inclusions is a very old process of construction often resorted to in the past. Applications to in-situ retaining walls is more recent, although the first reference in rock bolting is mentioned in 1930. In recent decades, stabilization of rock slopes by passive metallic anchors grouted within the rock mass has been intensively used. The purpose of this technique is essentially to limit the decompression and the opening of preexisting discontinuities by restraining the deformations, and to create a mass in which rock blocks are locked together.

The first application of this technique to soils occurred in 1973 (Rabejact and Toudic, 1975) at Versailles (France), but the concept used was different : the principle of soil nailing consists in obtaining a homogeneous and resistant material by associating soil and inclusions able to withstand tensile forces, but also shearing forces and bending moments. However as indicated on figure 17, it can be considered that soil nailed walls result from both rock slope stabilization and Reinforced Earth. The first soil nailed wall in Versailles was used as a temporary support of the slope in order to build the definitive reinforced concrete wall, despite of difficult conditions (urban area, large railway traffic).

Inclusions in a soil nailed wall are placed approximately horizontal and are mainly subjected to tensile forces. They are generally made of steel or another strong material like for instance, fiber-glass, and are inserted into the soil either by simple driving or by grouting in a predrilled borehole.

Figure 18a shows the different construction steps for a soil nailed wall : successive excavations with limited height (1 to 2 m) are generally stable if the soil exhibits a short term cohesion, which is commonly the case. Then the inclusions are placed. In most cases a facing is required to ensure local stability between the bars ; it can be made of shotcrete or steel mesh.

The behaviour of this type of retaining structure is very similar to that of Reinforced Earth : the high elasticity modulus of the inclusions restrains the displacements of the wall and modifies the shape of the potential failure surface, which is different from the classical Coulomb's wedge. This potential failure surface, which is the locus of the maximum tensile forces in the bars, is parallel to the facing in the upper part of the wall and divides the soil into two zones (Fig. 18b), an active one and a resistant one. Moreover, the equivalent earth pressure distribution related to the maximum tensile forces is quite different from that predicted by the Rankine's theory.

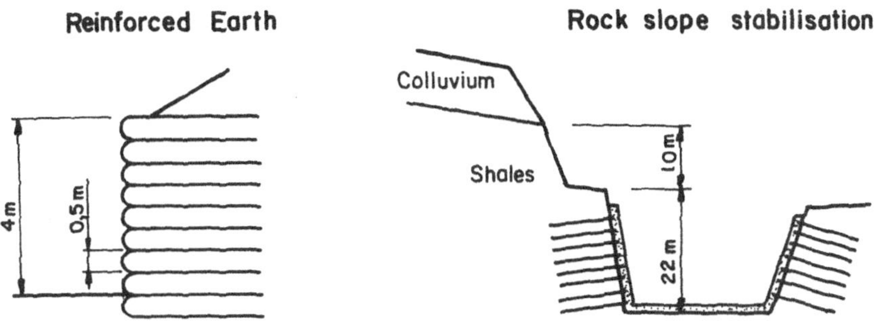

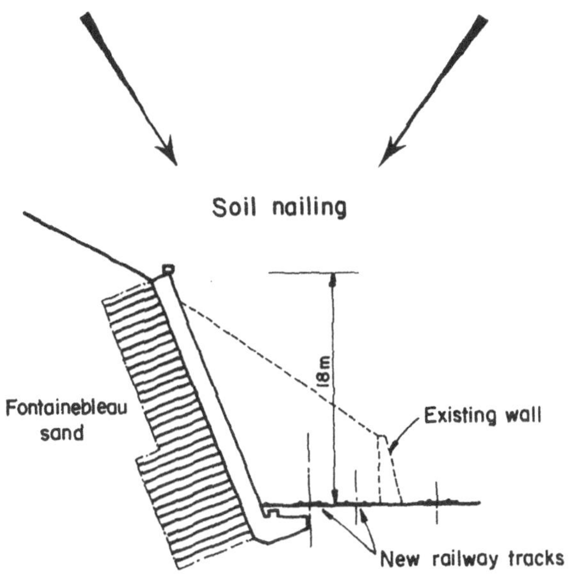

FIGURE 17. First soil nailed wall, reinforced earth wall and rock bolted wall

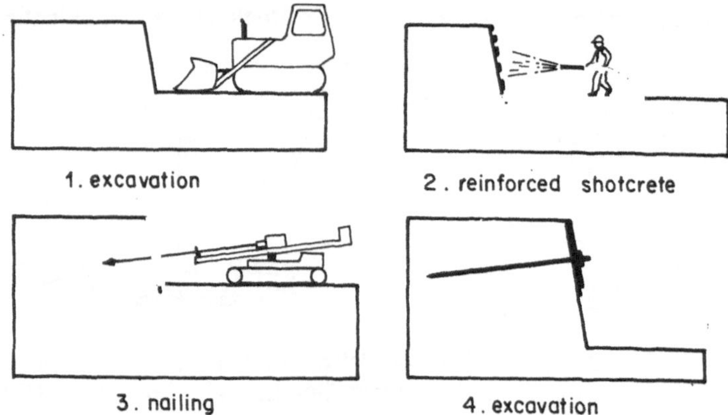

ⓐ SOIL NAILED CONSTRUCTION

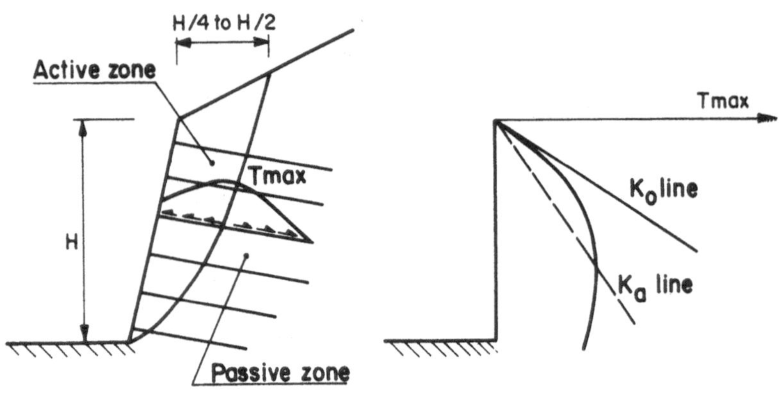

ⓑ SOIL NAILED BEHAVIOUR

FIGURE 18. Construction and mechanical behaviour of a soil nailed wall

However, the behaviour of soil nailed walls is not yet very well understood. In particular, the influence of the construction method (excavation of steps) hasn't yet been investigated. Due to the great interest of the technique, several research programs have been initiated in different countries (Germany, France, USA). The french project CLOUTERRE on soil nailing is a 3 years research program : it mainly consists of 5 to 6 full scale experiments, centrifuge model tests, laboratory and field research on friction, in order to propose reliable design methods including an evaluation of the inclusions durability. The aim of the project is to promote the technique in order to use it for permanent structures. Nail durability is therefore carefully considered and close attention is given to nails specially designed for preventing metallic rod or cable corrosion (figure 19).

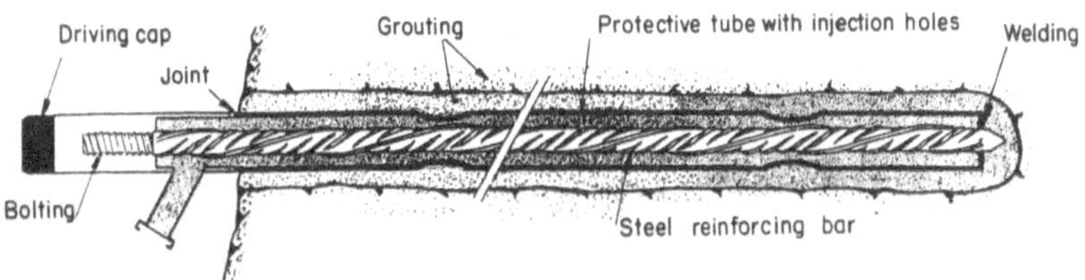

ⓐ "TBHA" NAIL PATENTED AND DEVELOPED BY SOLRENFOR

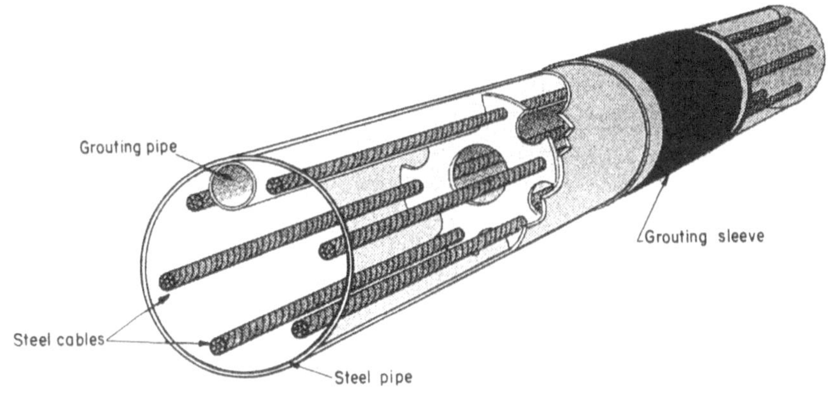

ⓑ PRESTRESSED MULTIREINFORCED NAIL "INTRAPAC" DEVELOPED BY INTRAFOR - COFOR - (France)

FIGURE 19. Special nails developed for preventing corrosion

3. MAIN TYPES OF POLYMER SOIL RETAINING WALLS

3.1 Types of polymer used as reinforcements

Since the beginning of Reinforced Earth development, tentative efforts have been made for using polymeric reinforcements instead of metallic ones. Compared to metals, polymeric materials have large ranges of deformation modulus and tensile strength, and the following polymeric products have been used as reinforcements : geotextile sheets, geogrid sheets, woven geotextile strips, coated fiber strips, rigid plastic strips. Figure 20 shows for instance some mechanical properties of geotextiles and geogrids. Generally speaking, polymeric materials are more deformable and less resistant than metals. Moreover, they exhibit creep behaviour ; nevertheless it is possible to adapt in each retaining system the type of polymeric reinforcement used to according the allowable deformation.

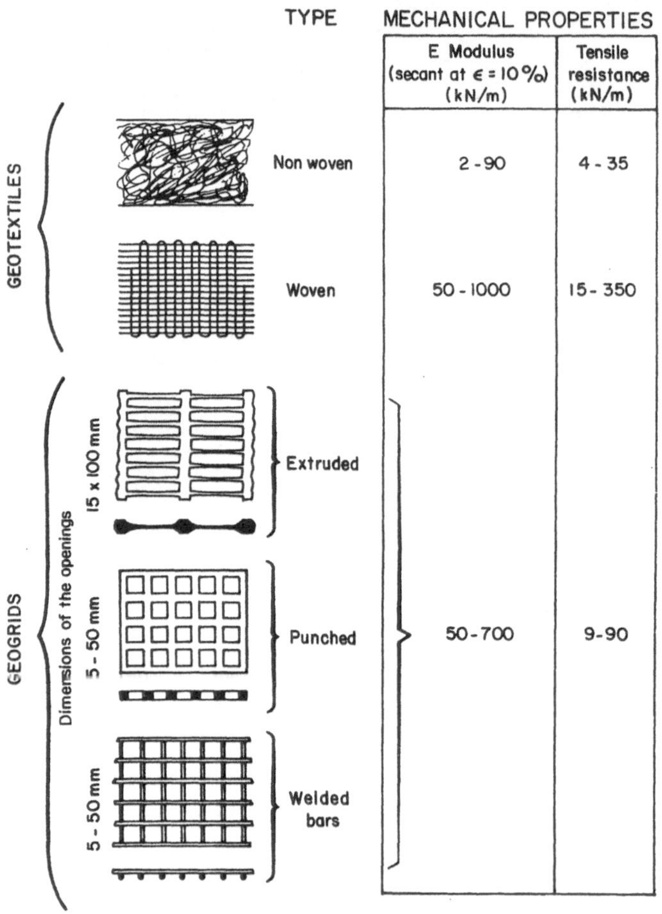

TYPE
MECHANICAL PROPERTIES
E Modulus (secant at ϵ = 10 %) (kN/m)
Tensile resistance (kN/m)
GEOTEXTILES
Non woven
2 - 90
4 - 35
Woven
50 - 1000
15 - 350
GEOGRIDS
Dimensions of the openings
15 x 100 mm
Extruded
5 - 50 mm
Punched
50 - 700
9 - 90
5 - 50 mm
Welded bars

FIGURE 20. Types and mechanical properties of geotextiles and geogrids

Inclusion extensibility greatly influences reinforced soil behaviour. This has been very clearly shown by McGown et al. (1978), who have considered extensible and inextensible inclusions. Besides increasing strength, the principal action of extensible inclusions is to increase soil ductility and decrease or even cancel the softening observed in dense sand behaviour. Inversely, inextensible inclusions mainly increase soil strength and deformation modulus, but they cause the deformation soil modulus to be more brittle. These features are presented in figure 21 and allow the following distinctions to be made :

1 - Reinforcement with ideally inextensible inclusions, mainly represented by Reinforced Earth, for which the reinforcements are generally linear and metallic.

2 - Reinforcement with ideally extensible inclusions, represented by "ply-soil" or "multi-layer soil" (McGown, 1978), for which the reinforcements are generally plane and made of synthetic materials (geotextiles, etc...).

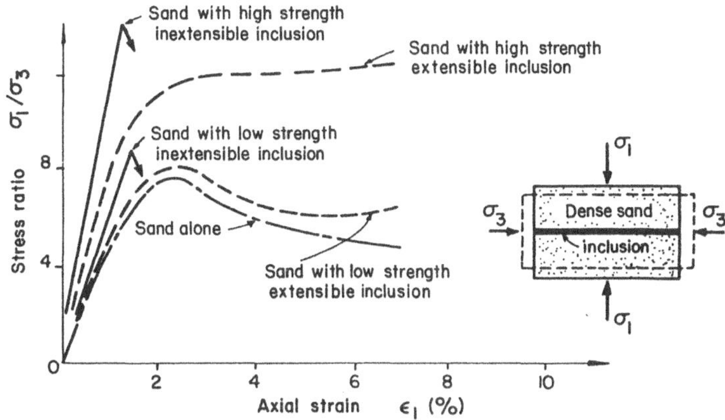

FIGURE 21. Deformability and strength inclusion influence on reinforced dense sand behaviour (Mc Gown et al., 1978)

Table III gives a classification of reinforcement techniques as a function of the relative deformability of the inclusion with respect to the soil, and as a function of the geometrical type of design (one, two, or three dimensional). Initially, and in particular for Reinforced Earth, the reinforcement was made of linear inclusions, and thus it was strongly anisotropic. Geotextiles and grids have led to the development of two-dimensional reinforcement. Tridimensional reinforcement appears theoretically isotropic, but it gives a high deformability to soils reinforced by a continuous filament (Texsol).

Type of rein. / Inclusion	1-D	2-D	3-D
RIGID	Metal strips	Grids (metallic)	Fibers (metallic)
	Rigid plastic strips		
SEMI-DEFORMABLE	Flexible plastic strips	Grids (plastic)	Texsol Fibers (synthetic)
DEFORMABLE		Geotextiles	

Table III - Classification of reinforcement techniques according to the inclusion geometry and to its relative deformability

3.2 Polymer uses in Reinforced Earth Technique

Considering the whole research performed on Reinforced Earth, particularly at the beginning in the 60's, it could appear surprising that nothing would have been done in order to use polymers as reinforcements. In fact, Vidal (1966) planned to use at first polymeric materials : nylon strips, tergal strips and particularly rigid plastic constituted of fiber glass coated with polyester resin. This last material was chosen in 1965 and an important investment was made at that time in order to produce industrially U shaped facing elements and reinforcement strips. More specifically, this material was a fiberglass reinforced plastic, in which strength and stiffness were imparted to easily moulded resins by glass fibers. The individual glass fibers were elastic and as strong as the strongest tensile steels, so that they gave to the composite material a small deformability without creep and a high strength. In 1965, this material had been used for 10 years in underground pipelines and tanks and had proven to behave satisfactory. However, as indicated by Mallinder et al. (1977), some degradation was observed when the material was maintained in wet conditions for long periods of time.

In 1966, a first experimental Reinforced Earth wall using fibreglass reinforced plastic strips and facing units was built, in order to test the construction process and the mechanical behaviour of the wall. Unfortunately, the plastic material was attacked by bacteria and the Reinforced Earth wall was destroyed within 10 months. No biological test was performed on the backfill material after failure and practically no reliable information and expertise about the type of bacteria and the degradability process is now available.

After Mallinder et al. (1977), this failure might have been accidental, since these authors gave biological test results on fiberglass reinforced plastic, indicating that this type of material was not degraded by bacteria (immersion time was however limited to 6 months).

Nevertheless this failure has been the turning point for the use of plastic materials in Reinforced Earth : at the beginning of 1967, VIDAL decided to develop Reinforced Earth with metallic reinforcements.

However, the use of plastics in Reinforced Earth was not yet completely abandoned, since another very interesting attempt was done in

1971 with the construction of the Poitiers wall using Tergal strips. This wall was a temporary structure, 5 m high and 40 m long. The tergal strips were attached to cruciform concrete panels. The calculation of the wall took into account tergal creep and during its service life, the wall behaved satisfactory. However, it appeared during construction that tergal strips had to be slightly prestressed in order to prevent excessive lateral displacements of the facing.

In 1981, ten years after its construction, the wall was dismantled and interesting durability tests were performed on the strip material. As presented further, it was shown that the plastic fibers had been degraded and that the mechanical properties of the plastic strips had decreased.

3.3 Geotextile retaining walls

As previously indicated, the first reinforced soil wall using geotextile was built in 1971. The backfill material used was a wet clayey soil and the foundation soil was poor and compressible (peat).

It appeared from this first encouraging construction that, compared to Reinforced Earth type walls using strip reinforcements and facing units, a geotextile wall presents the following features :

1) It is a very cheap retaining system.

2) Geotextiles enable the use of poor backfill material, because of their large frictional surface area when they are two-dimensionnaly used, and because of their drainage properties.

3) The most important problem is the control of the displacements and the deformations, particularly when using non-woven geotextiles.

4) Building vertical walls is rather tricky.

5) The facing is generally rather bad-looking.

Despite of advantages (1) and (2), such a retaining wall system has been little used until now. Concerning the construction of vertical walls, Jones (1977) has compared different methods which are presented in figure 22. For the first walls built in France (Puig et al., 1977, Kern, 1977), temporary lateral supports were used : embankment or wood plate. It appeared further that a good way of controlling the verticality and the facing deformations was to use posts in addition to the facing geotextile (Jones, 1977 ; Schwantes, 1982). But some other methods are also suitable for controlling lateral displacements and deformations, furthermore leading to aesthetic facings. They are essentially of two types (Figure 23) : 1) use of prefabricated vertical facing units (concrete panels) (Broms, 1977 ; Jones, 1982) ; 2) construction of an inclined facing using another type of reinforcement (grid, Armater, Texsol). Delmas et al. (1986) have published an interesting paper dealing with these different facing construction methods.

The low cost of all types of geotextile reinforced soil walls is thus unfortunately counter-balanced by large deformations and inaesthetic aspect of the facing. In order to keep the cost advantage and to avoid facing deformations, mixed structures have been developed :

1) walls with prefabricated or built in place facing units (concrete panels, gabions, etc ...) and geotextile reinforcements (Broms, 1977 ; Jones, 1982 ; Delmas et al., 1986).

2) Walls with fabric facing and relatively inextensible reinforcements (metallic strips, geogrids) (Schwantes, 1982).

It seems that such combinations could lead to interesting developments in the future.

(a) U.S. Method

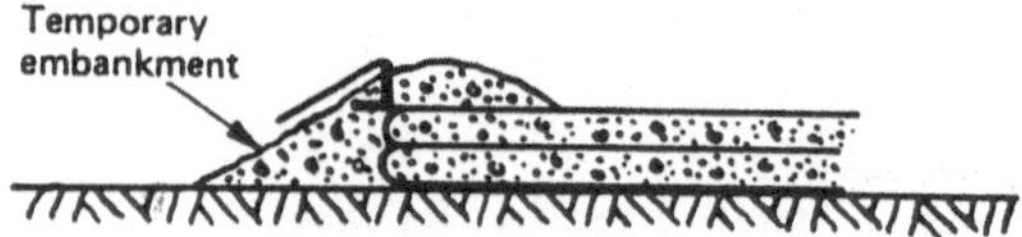

(b) French method

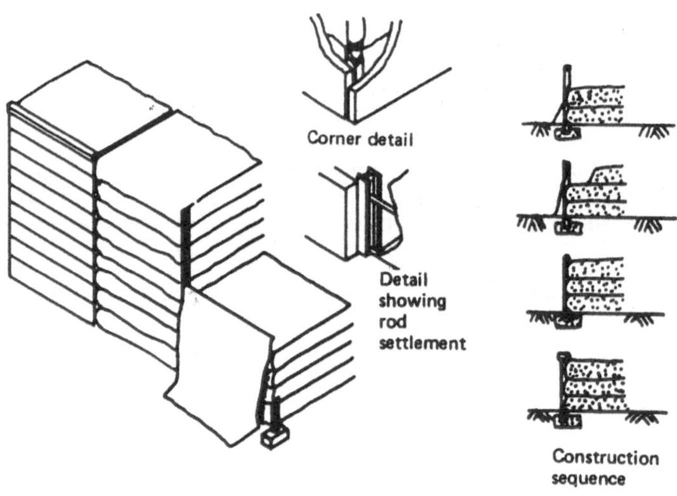

(c) British Yorkshire method

FIGURE 22. Comparison of vertical geotextile wall construction methods (Jones, 1977)

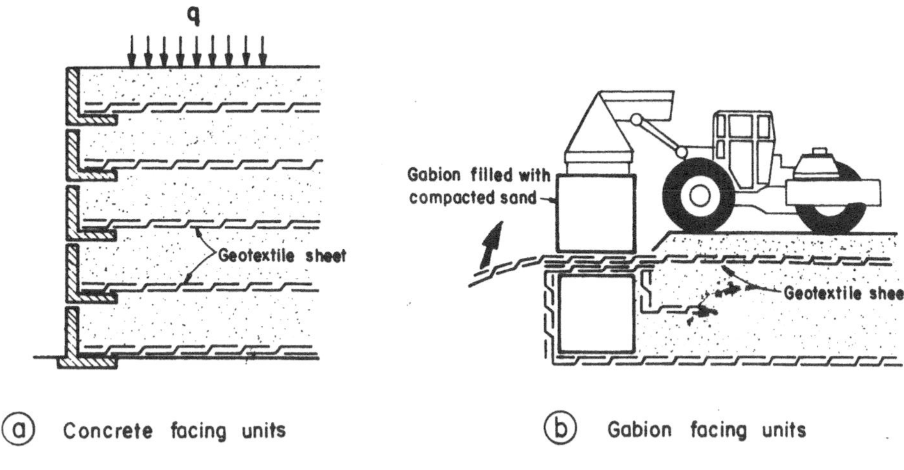

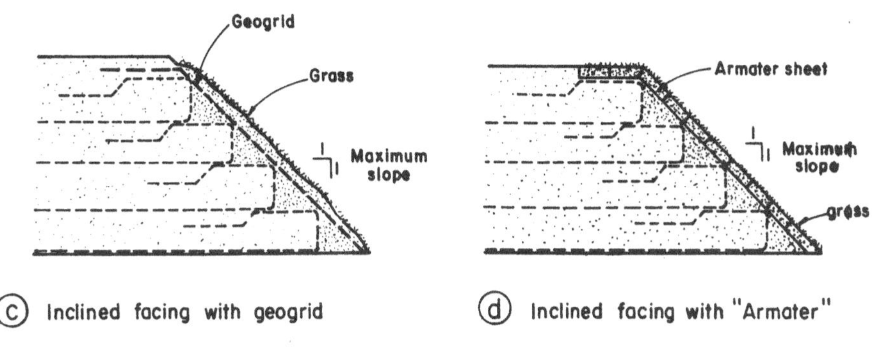

FIGURE 23. Different types of improving geotextile wall facings (Delmas et al., 1986)

Because of their relatively large extensibility, no deformation and tensile stress measurements have been carried out in the geotextile reinforcement sheets of actual walls. However it is now well recognized that the behaviour of a wall reinforced with geotextiles is controlled by the reinforcement deformations and it is therefore different from the behaviour of a wall built with quasi inextensible reinforcements (Reinforced Earth). This fact results from theoretical considerations (McGown, 1978) and from triaxial tests on sand samples reinforced with geotextiles (Blivet, 1979 ; Gray et al., 1982). In particular, although reinforcements with very extensible synthetic fabric increase the ultimate strength, they tend to reduce the overall stiffness of the sand.

Delmas et al. (1986 b) have recently developed a design method for geotextile soil reinforced wall. Contrary to the limit equilibrium methods, this method is based on an evaluation of the deformations and comprises the three following steps for calculating the safety factor along any potential failure surface :

1) Determination of a displacement field compatible with the acceptable wall displacements at the facing and at the top.

2) Determination of the stresses in the geotextile sheets resulting from the above displacements.

3) Checking of the reinforced soil mass equilibrium along potential failure surfaces.

3.4 The York method

The York method for the design of reinforced soil retaining walls has been developed by JONES (1973) on behalf of the Departement of Transportation in U.K. This method is similar to the Reinforced Earth technique despite two small differences :

1) All the prefabricated elements (facing units, strips) can be made of plastic.

2) Reinforcing strips can slide with respect to the facing thanks to the use of vertical poles.

According to JONES (1977), differential settlements can easily be accomodated by the York method, also called the sliding method.

Figure 24 shows the comparison between Reinforced Earth and York method prefabricated elements. In the York method, the facing units are hexagonal membranes, made from glassfiber reinforced cement and bolted together. Strips are either metallic or plastic (fiberglass reinforced plastic as previously described).

Until now, this type of polymer reinforced soil wall has only been developed in U.K. and even in this country a few number of walls have been built. Although nothing has been published yet about the long term behaviour of these structures, it seems that plastic membrane facing units are brittle and could have been damaged. In fact, a relatively long experience in reinforced soil wall construction shows that concrete panel is one of the best and the most resistant facing units. Despite of this, the York method has to be remembered as the first reinforced soil wall totally built with plastic material.

3.5 Geogrids reinforced soil retaining walls

The satisfactory behaviour of the Dunsmuir wall (1974), in which welded metallic bar-mats were used, and subsequent research performed on soil-grid friction have proved that grids were suitable for reinforcing a large range of soils.

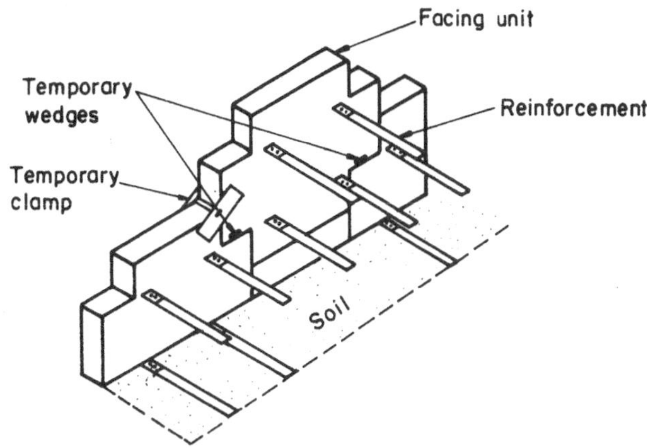

ⓐ Reinforced Earth (telescope method of construction)

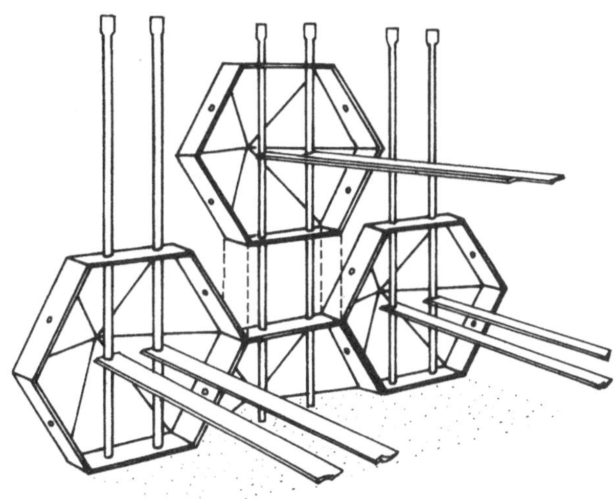

ⓑ York method (sliding method of construction)

FIGURE 24. Comparison between reinforced earth and York method

Furthermore, the relatively large development of geogrids at the beginning of the 80's led to an attempt to use such reinforcements in retaining wall systems. At present time, only a few walls have been constructed with geogrids and, it seems, because of the same facing problems than those encountered with geotextile walls. Considering the interest presented by geogrids in soil reinforcement, the use of prefabricated facing units associated with an easy and efficient attachment system would certainly be an interesting possibility of development.

However, special attention has to be given to creep and durability.

3.6 Websol

This type of soil retaining system has been developed in U.K. by Soil Structures Limited at the end of the 70s and it is very similar to the Reinforced Earth technique.

Paraweb is a composite plastic material consisting of polyester yarns (terylene) coated and protected by a sheet of black polyethylene (alkathene). The fibers give high strength as well as relatively low creep characteristics to the material, and the Alkathene sheet protects it from ultraviolet rays as well as other chemical degradation. Despite its extensibility, Paraweb appears to be an interesting plastic material for soil reinforcement. The Paraweb strips are 2 mm thick and 90 mm wide (figure 25a), and its rupture load ranges between 50 and 100 kN.

Facing elements are T-shaped concrete panels (figure 25b). The reinforcements are formed by a continuous strip which has to be prestressed, because of the extensibility of Paraweb, in order to avoid displacement of the facing (figure 25c).

John et al. (1983) have performed two full-scale experiments on Websol reinforced soil walls. The distribution of maximum tensile forces measured the one corresponding to inextensible reinforcements as steel strips. In the upper part of the wall, it appears to be close to the Ka line, suggesting that the lateral deformations of the structure are large enough to attain the Ka state of stress in the soil. This effect of the extensibility of the reinforcement strips on the maximum tensile forces has since been confirmed by several studies ; for instance, fiberglass reinforced plastic behaves as an inextensible reinforcement material. Reinforcement extensibility has also an influence on the potential failure line (maximum tensile forces line) which is close to Rankine's plane in the case of extensible reinforcements.

3.7 Texsol walls

Since its first use at Caudebec-en-Caux (Leflaive et al., 1983), Texsol has been until now essentially used in retaining wall construction.

Leflaive and Liausu (1986) presented a large project in which Texsol was used for enlarging the french A7 highway. Figure 26a shows a comparison between two possible solutions : a concrete wall and Texsol. Texsol was chosen because of the following requirements :

- Construction had to be achieved very quickly.
- Environment protection was to be considered.
- Traffic had to be maintained.
- Low cost.

It must be noted that the Texsol walls were grassed to provide good environemental conditions and facing protection.

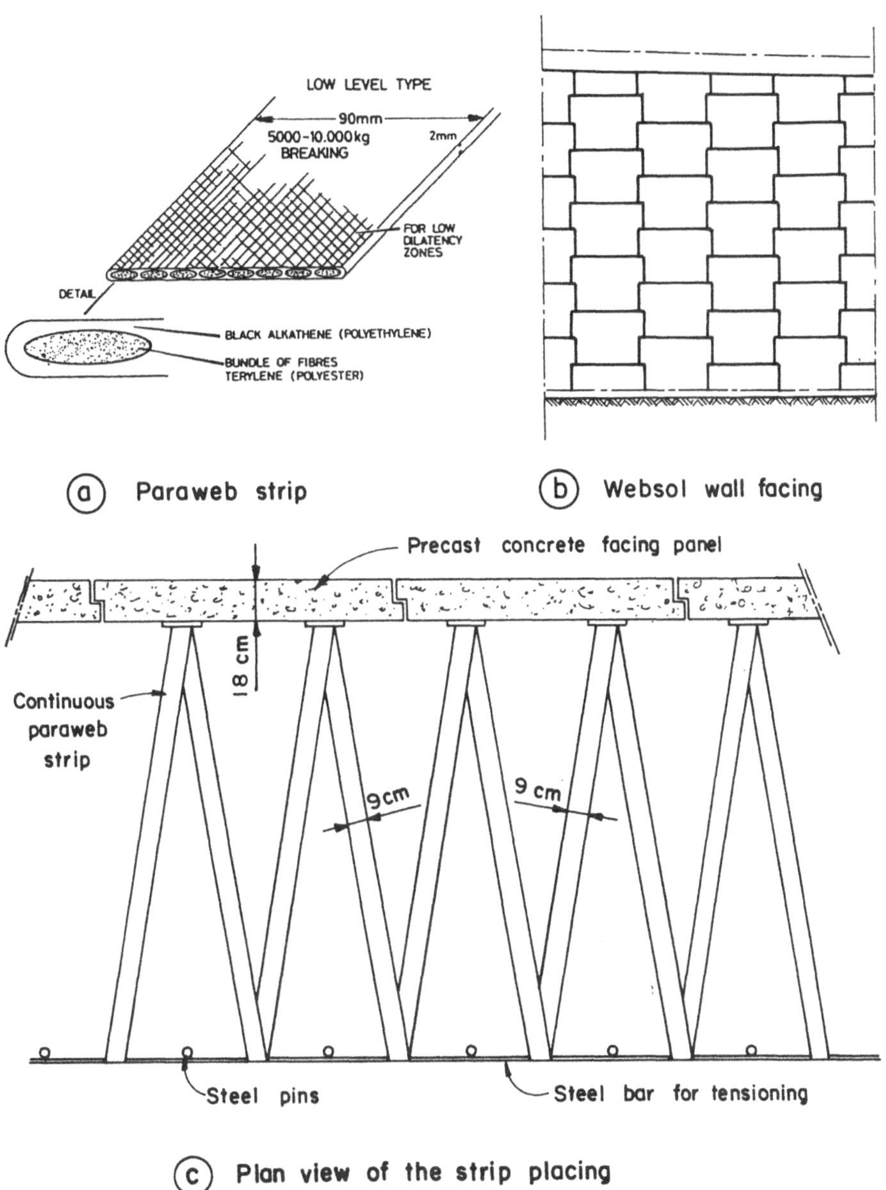

FIGURE 25. Websol retaining wall system

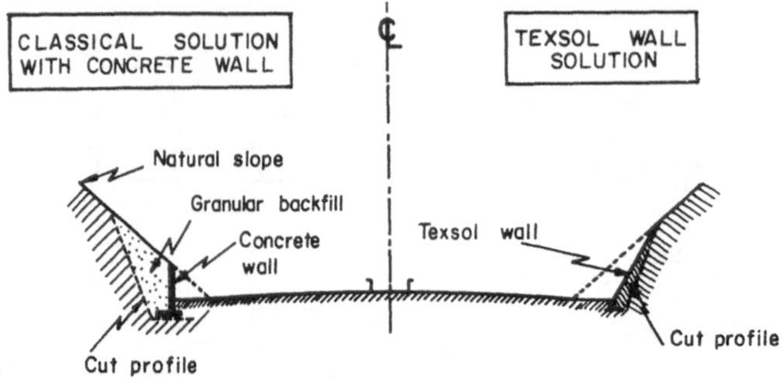

(a) Comparison between classical solution and Texsol solution

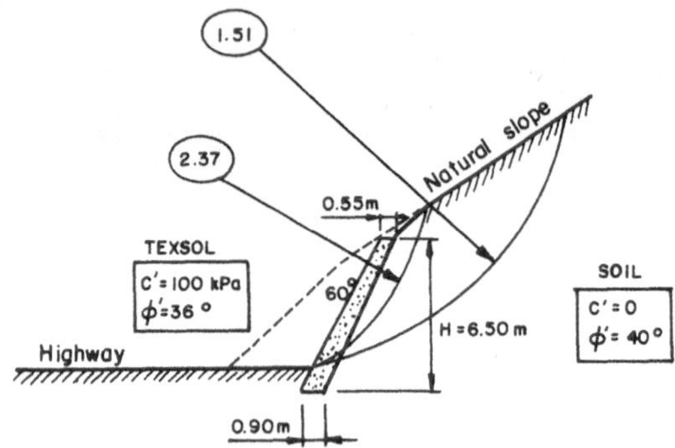

(b) Texsol wall design method

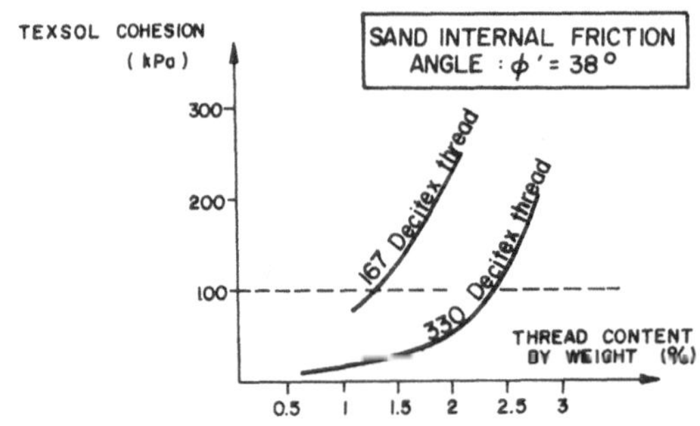

(c) Texsol cohesion value versus thread content by weight

FIGURE 26. Texsol wall solution used for enlarging the french A7 highway (1985-1986)

The following geometry was adopted for the walls :

- External facing inclination : 60° on the horizontal plane.
- 0.5 m base embedment in the foundation soil.
- Height ranging from 3 m to 7 m.
- Width at the top : 0,55 m.
- Width at the base : determined from design calculations in order to have a minimum slope/wall stability of 1.5 .

As mentioned by Leflaive and Liausu, there are two steps in Texsol wall design :

1) Considering the available granular material, triaxial tests have to be carried out on reinforced soil samples in order to determine the influence on Texsol cohesion of parameters such as thread resistance and thread content by weight.

2) Wall stability along circular potential surfaces has to be analysed and wall geometry must be adapted in order to obtain a minimum safety factor of 1.5 .

This wall design must be considered as a first approach and it will certainly need to be further improved in order to account for the two following phenomena related to Texsol behaviour. The first one is the cohesion anisotropy resulting from placement technique used for Texsol : Texsol cohesion determined from triaxial tests may thus not be completely representative of the actual cohesion mobilized along horizontal planes. The second phenomenon is the difference in the at peak deformations observed on the stress-strain curves of soil and Texsol. This difference certainly modifies the failure criterium to be taken into account along the potential failure surface.

Placing Texsol is carried out with a large special machine which builds successive horizontal layers 5 to 10 cm high. Compaction of Texsol is done by a vibrating plate, in the lower part of the wall (up to a height of 1.6 m above the base).

On the A7 highway project 20 000 m^3 of Texsol have been used for the construction of retaining walls and behave satisfactory.

Texsol has promoted the use of three-dimensional soil reinforcement, which certainly will be developed in the future not only for retaining-walls, but in other fields like foundations where such a reinforcement appears to be particularly attractive. Fibers and mesh elements must be considered as potentially interesting reinforcements.

However, it seems that future development of Texsol requires further studies that should be focussed on the following points :

- Mechanical behaviour of the material with respect to the placement method, the deformations, etc ...
- Observations on real structures and for instance on centrifugal models brought to failure, in order to improve the design methods.
- Durability of the material with respect to the facing protection and creep.

4. PROPERTIES OF POLYMERS WITH RESPECT TO THE BEHAVIOUR OF REINFORCED SOIL WALLS

When metallic reinforcements are inserted into the soil, they may be considered as rigid with respect to the soil deformability, and to the magnitude of the induced stresses. On the contrary, polymeric inclusions are characterized by weaker mechanical properties, i.e high extensibility, low tensile strength, associated with long term creep. Furthermore, although not sensitive to electro-chemical corrosion, polymers may also suffer some degradations under soil physico-chemical environment, and durability problems are to be considered.

Several studies have been conducted on strength-strain properties of geotextiles since they have been used in various applications of geotechnical engineering. It has been observed that among all those applications, reinforced soil retaining walls constitute a case in which the magnitude of stresses applied by the soil to the geotextile is the higher. In fact, it is quite difficult to fully understand the load transfer mechanism which occurs between the soil and the textile, and to know the exact stress field the textile is submitted to.

4.1 Extensibility

4.1.1 Polymers properties

Table IV shows some typical mechanical data for the more current polymers used in geotechnical engineering, as compared to steel properties. It should be noticed that, for each polymer, both bulk material and filament properties are given. In fact, filaments are produced by extrusion of the heated polymer mass, and high tenacity yarns, as described in the table, are obtained by controlling the cooling process of the filament, and by stretching it during extrusion. In such a way, polymeric chains have a preferential orientation and an increased anisotropy, thus resulting in much higher (up to ten times) mechanical properties of the material, such as tensile strength or elastic modulus.

Material	Density	Elastic modulus (MPa)		Tensile strength (MPa)		Failure elongation filament (%)
		Bulk material	Filament	Bulk material	Filament	
Steel	7,85	200 000			2340	3
Polyester ...	1,38	2100	13500 18500	60	800	6-20
Polypropylene	0,91	1100	7400	35	380-780	15-25
Polyamide ...	1,1	2400	12500	40-120	600-900	12-26
Polyethylene high density	0,95	800	4300 6500	30	390-700	10-20

Table IV - Typical polymers characteristics, as compared to steel

It may be seen from table IV that the main difference existing between metallic and polymeric materials corresponds to the high extensibility of polymers. This is illustrated by the values of elastic

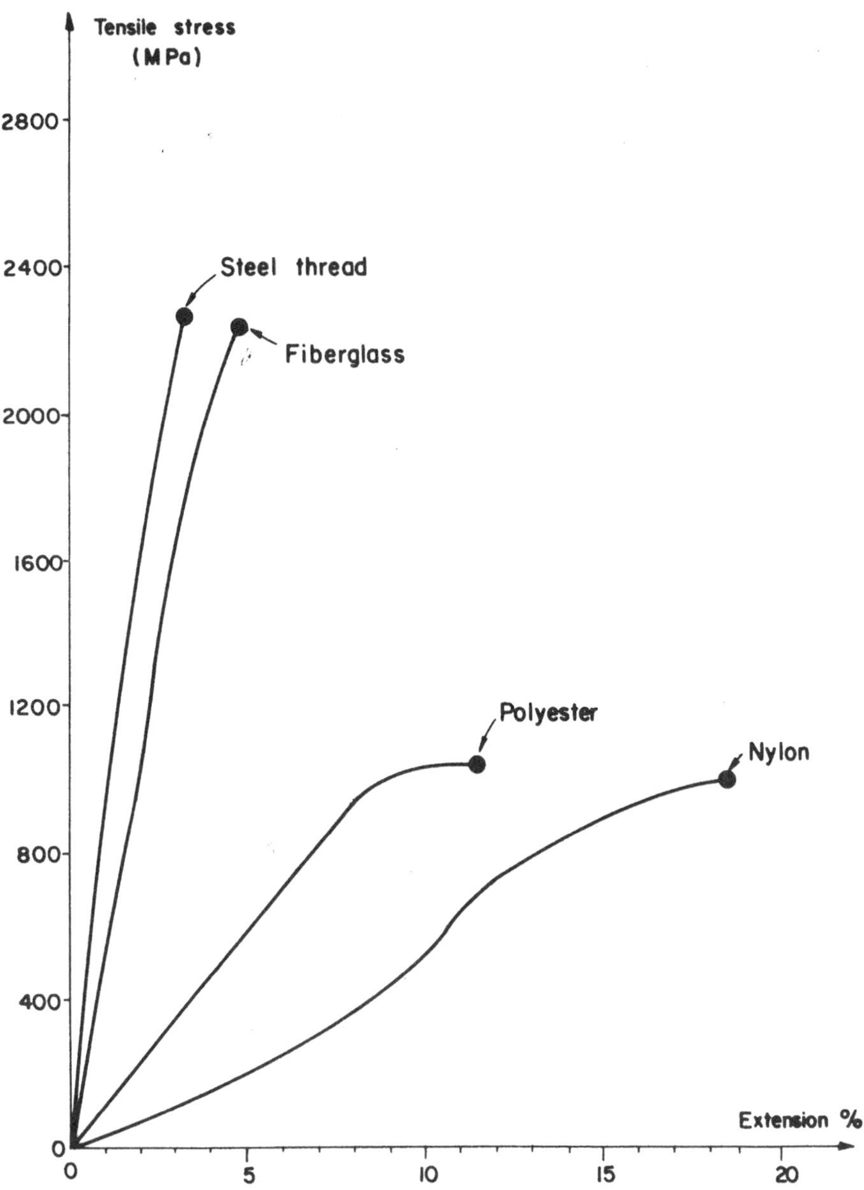

FIGURE 27. Stress-strain behaviour of different material fibers

moduli, which are, for current polymers, from 10 to 30 times lower than for metals ; the tensile strength is also lower (2,5 to 5 times), whereas the deformation at failure is much higher (20-30 % instead of 3 %). When comparing current polymers, it may be seen that polyester and polyamide, which have higher densities (1,38 and 1,1 respectively) have higher stiffness and tensile strength than polypropylene and polyethylene high density (0,91 and 0,95 respectively).

Figure 27 shows tensile test results for steel, fiberglass, polyester and polyamide filaments. Whereas failure occurs, for steel, at a deformation of 3.2 % for a tensile strength of 2340 MPa, polyester and nylon fail for much larger deformation (respectively 11 % and 19 % on the figure), and a tensile strength of about 1000 Mpa.

Baudonnel et al. (1982) took some filaments from current polyester and polypropylene geotextiles, and elongation tests results are shown in figure 28.

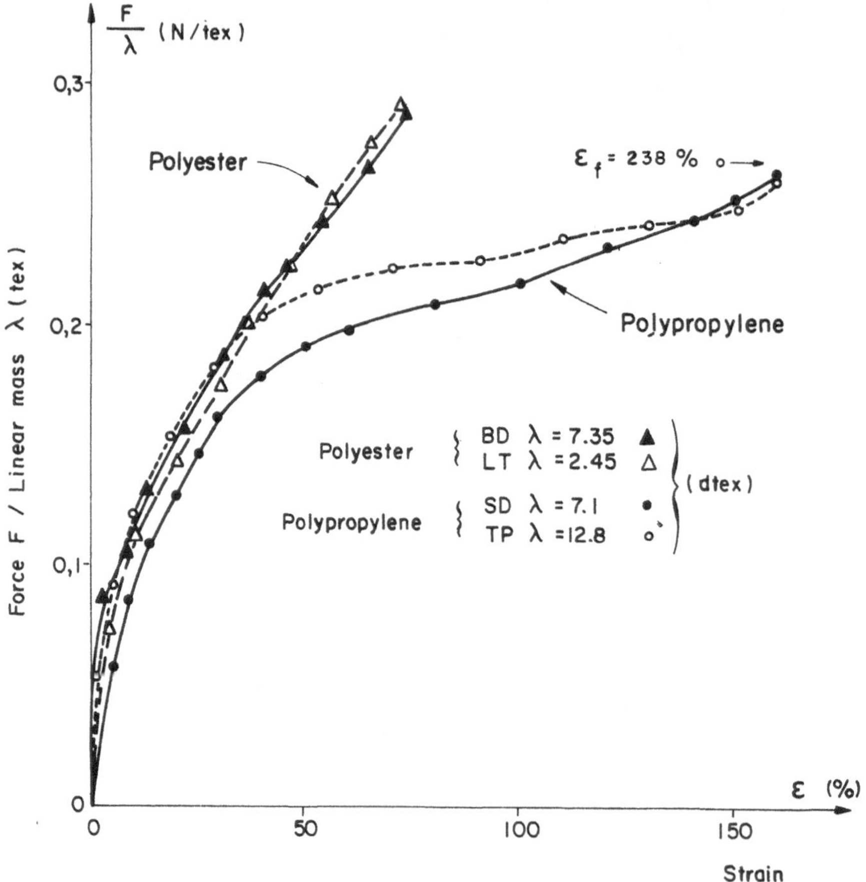

FIGURE 28. Elongation tests on filament extracted from non-woven fabrics (Baudonnel et al., 1987)

The strength value is expressed in terms of tenacity, say the force divided by the linear mass of the yarn, expressed in N/tex. In the case of non-woven fabric extracted filaments, the deformation at failure is in both cases quite higher (71-78 % for polyester, and 155-239 % for polypropylene). The polypropylene curve is composed of two parts having different slopes, the first one being identical to the polyester one. These results illustrate the decrease of the mechanical properties which occur when fibers are processed for fabric construction.

The previous data were related to solids and fibers, but the great variety of polymeric inclusions used in geotechnical engineering may exhibit various properties, which do not only depend on the nature of the polymer, but also on the structure of the inclusion, and on the influence of the soil confinement.

The influence of those parameters has been explicitely evidenced by McGown et al. (1982), who performed load-extension tests on geotextiles confined in soil, with the help of the apparatus presented in figure 29.

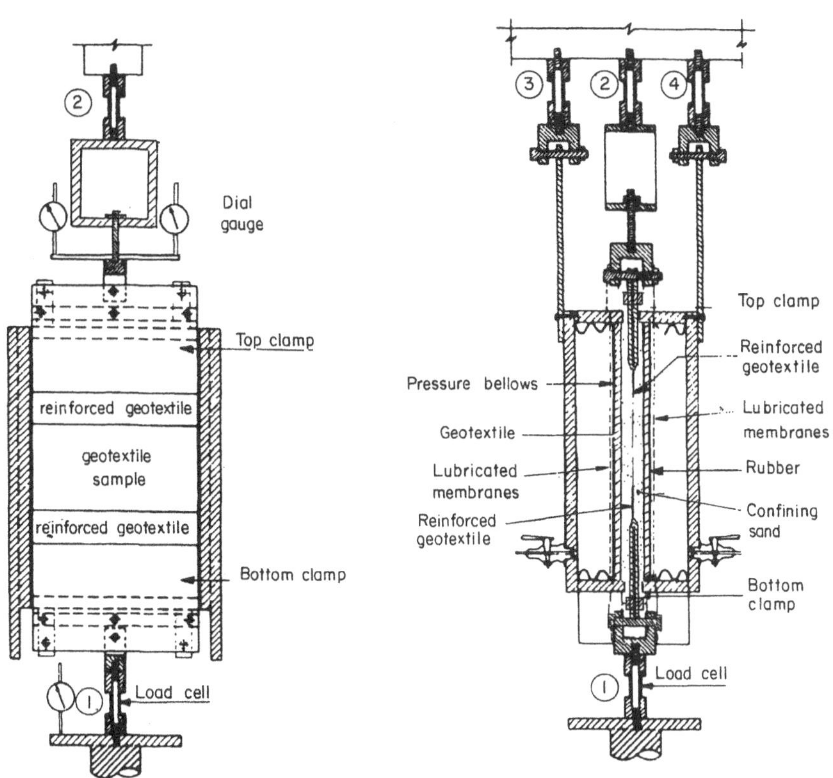

FIGURE 29. In soil tension apparatus (Mc Gown et al., 1982)

In this apparatus, a geotextile is included between two layers of soil (Leighton Buzzard sand, D_{50} = 0,85 mm), and pressure is applied on each side of the fabric by two air-activated rubber pressure bellows. McGown et al. tested four different fabrics : woven, non-woven meltbonded, non-woven needle punched, composite woven and needle punched. The elongation tests were performed at 20°C and at a constant rate of strain of 2 % per minute. Some of the results obtained are presented in figure 29. It may be seen that for the woven polypropylene fabric (Lotrak 16/15, 120 g/m^2), the in-soil confinement has a very little influence. On the contrary, in the case of a polyester non-woven needle punched fabric (Bidim U24, 210 g/m^2), the effect of in-soil confinement is quite important, and a 100 kPa confinement pressure results in a strengthening of the fabric, which corresponds to a higher stiffness, as illustrated by a higher slope of the elongation curve. This phenomenon may be interpretated in the following manner : the fabric elongation does not involve individual filaments, but rather induces a rearrangement of the needle punched filaments, which affects the bonds between the filaments. When soil is in contact with the fabric under a given pressure, it contributes to the stability of the bonds between the filaments, and provides a higher strength to the fabric.

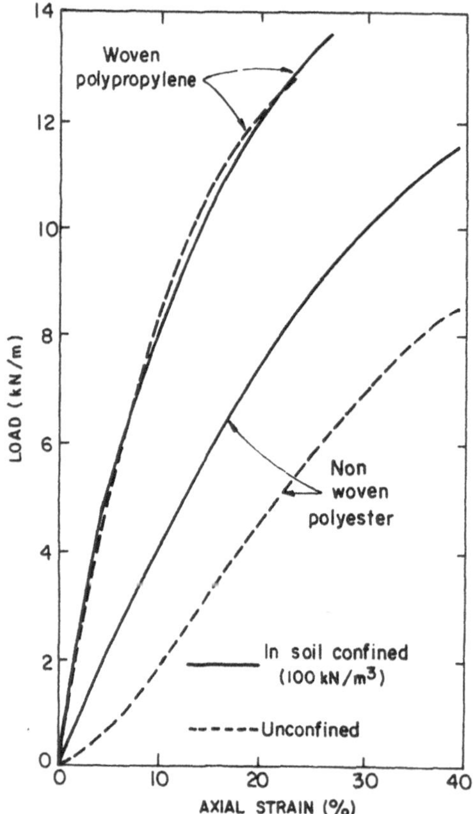

FIGURE 30. Influence of the in-soil confinement and of the fabric structure on elongation properties (Mc Gown et al., 1982)

It is interesting to notice, on figure 30, the predominant influence of the structure of the fabric as compared to the polymer type, since a 120 g/m^2 polypropylene woven fabric is stiffer than a confined 210 g/m^2 non-woven needle punched polyester fabric. For the woven fabric, failure occurs at 23-28 % deformation, whereas for the non-woven material, lower levels of load induce deformations as high as 40 %, without occasioning rupture.

The problem of tensile testing of fabrics is then particularly important for non-woven fabrics, since unconfined testing induce an important retraction of the strip, as shown in figure 31. For this reason, an initially current textile test which consisted of stretching up to failure a 50 mm wide and 200-300 mm long strip has been replaced by some other tests, with wider strips.

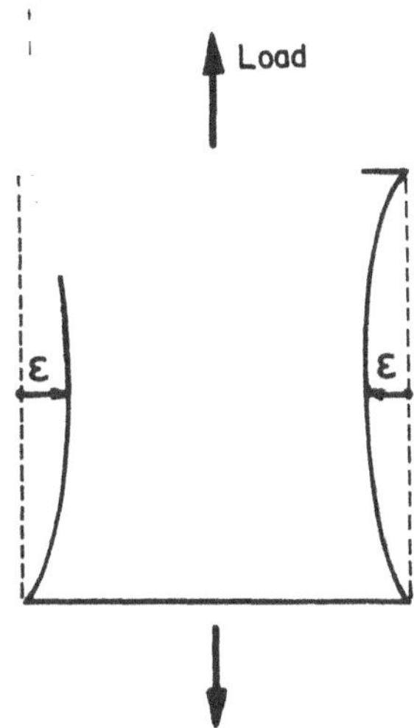

FIGURE 31. Lateral contraction occuring during an elongation test on a non-woven fabric

Presently, agreement does not exist on the best width to be selected for the strip. McGown et al. (1982) have studied the influence of the width on unconfined elongation tests and have compared their results with in-soil 200 mm wide tests results (figure 32). It may be observed that, due to retraction, width has a great influence on strength-strain behaviour, and that strip widths of 50 or 100 mm are definitly too small. It is interesting to notice that the 200 mm test corresponds to the 0 kPa in-soil confined test, whereas the 500 mm test corresponds to the 10 kPa in-soil confined stress.

In fact, present discussions on strip width are related to those two values of 200 and 500 mm, for a length of 100 mm, and argumentation elements have been obtained from special tests, where lateral retraction was avoided. Such special tests include a hydraulic tensile test (Raumann, 1979), biaxial tensile tests (Viergever et al., 1979), cylindrical sleeve test (Paute et Ségouin, 1977), and a special test in which lateral restraint is achieved by means of lightweight wooden brackets in which steel pins have been set (Sissons, 1977). The fabric is pressed on ten of these brackets regularly scattered along the length of the 200 mm wide strip. The pins cross the fabrics and avoid restraint. Shresta and Bell (1982) have used this device for testing several geotextiles. Testing of 200 mm wide strips has been also performed by Murray et al. (1986), and Richard and Scott (1986).

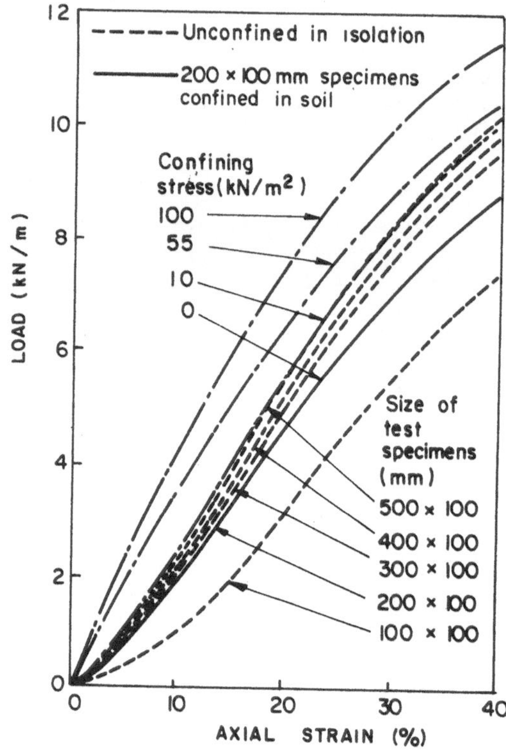

FIGURE 32. Unconfined in-isolation and confined in-soil load-axial strain data for Bidim U24 (Mc Gown et al., 1982)

Leflaive et al. (1982) performed 500 mm wide tests, and proposed a correction corresponding to the lateral contraction. The corrected results they obtained compared favourably with results of cylindrical sleeve test, and in-soil confined tests (McGown et al., 1982). This approach has been adopted by Cazuffi et al. (1986), Leclerq and Prudon (1986). Rowe and Ho (1986) also suggest a 500 mm wide value.

Other parameters have been studied. Figure 33 shows the influence of the strain rate on the maximum tensile strength of some woven fabrics (Rowe and Ho, 1986). In this case, where the constitutive fibers are directly sollicitated by the tensile test, the effect of strain rate is important, and the writers suggest a 2 % rate. Several writers considered the influence of the tensile direction, as compared to warp, weft, or diagonal direction for woven fabrics, and production direction for non-woven fabrics. In the later case, Van Leeuven (1977), Leclerq and

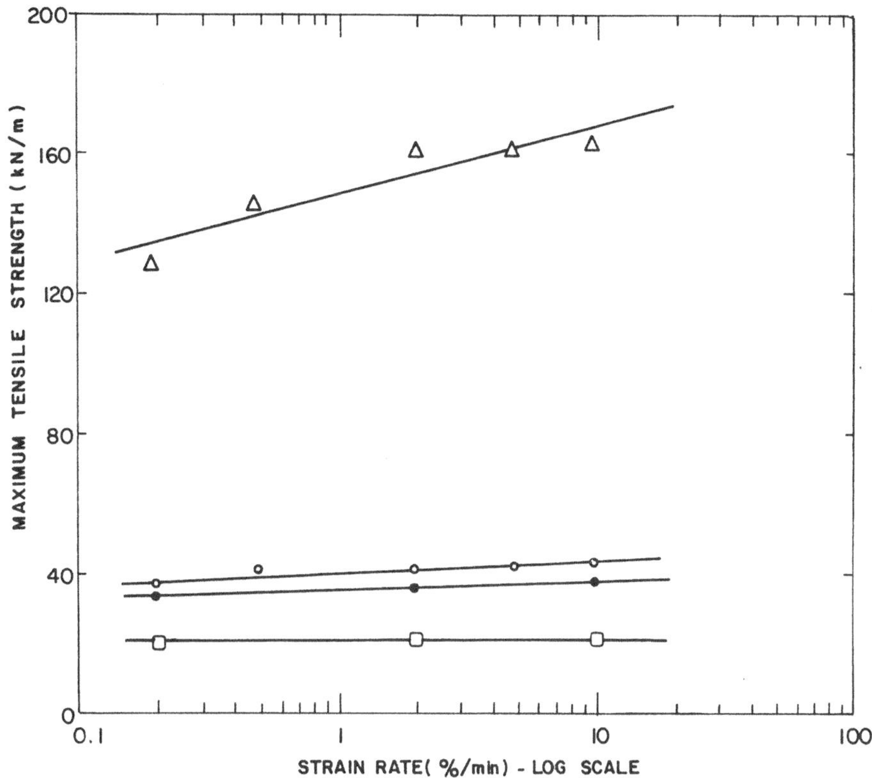

FIGURE 33. Variation in tensile strength of some woven fabrics with strain rate (Rowe and Ho, 1986)

Prudon (1986) observed no variations of the tensile strength, whereas Paute et Ségouin (1977) mention some decrease in the production direction. In the former case, warp and weft directions give generally similar results, whereas a 20-40 % decrease is observed in diagonal directions. However, Rowe and Ho (1986) observed significant variations on some woven fabrics, the warp direction being sometimes stronger.

For non-woven fabrics, and for a given type of polymer, the tensile properties are dependent on the weight per unit area of the fabric, as shown on figure 34 (Paute and Ségouin, 1977), from results of the cylindrical sleeve. The curves show, for various non-woven polyester and polypropylene fabrics, the influence of the value of the weight per unit area, expressed in g/m^2. The mechanical data are the rupture strength (34a), and the deformation modulus E (34b). The increase at rupture strength is fairly linear for all fabrics, whereas the deformation modulus shows, in the case of polyester, a slope increase of about 400 g/m^2. It is interesting to notice that the nature of the polymer has little influence on the mechanical properties of the material. These curves give an idea on non-woven rupture strength, which may vary, according to the weight per unit surface, between 10 and 50 kN/m whereas the deformation modulus is comprised between 30 and 200 kN/m except for the weaker fabric.

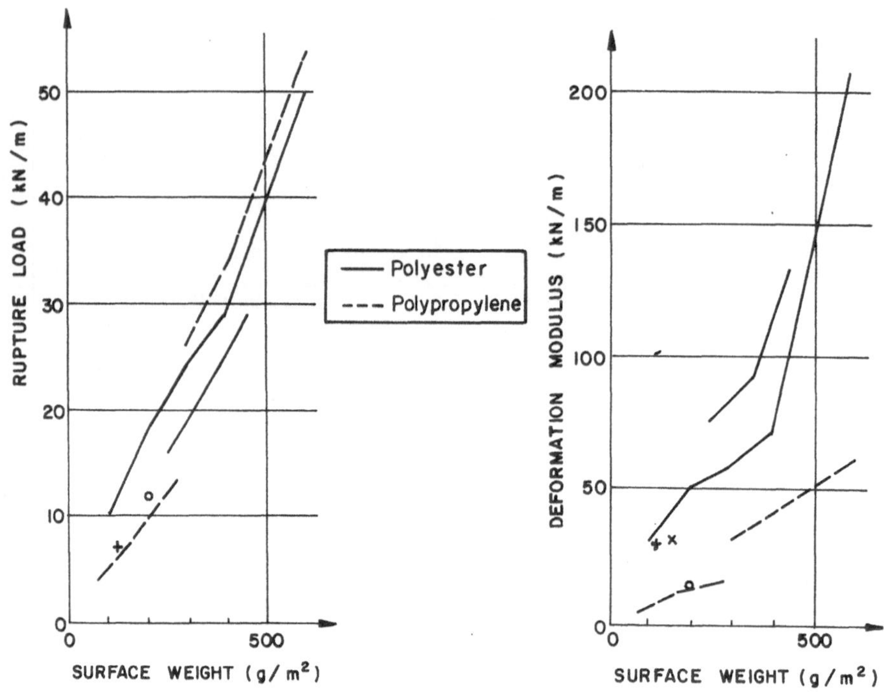

FIGURE 34. Tensile strength and deformation moduli of polyesters and polypropylene fabrics listed on the sleeve-cylinder apparatus (Paute and Segouin, 1977)

For woven fabrics, tensile strength data depends more on the individual filament characteristics. Table V shows data from Roscoe and Ho (1986).

Woven Geotextile	Initial tangent modulus (kN/m)	Tangent modulus from 10 % strain (kN/m)
Geolon 1250 ..	200	800
Permealiner 1195	220	140

Table V - Properties of two woven fabrics (Rowe and Ho, 1986)

For geogrids, the secant modulus at 10 % strain is included between 50 and 700 kN/m, whereas the tensile resistance varies between 9 and 90 kN/m. It must be noted that, as mentionned previously, extruded geogrids have better mechanical properties that punched ones.

The table of figure 20 gives compared values of E Modulus (secant at 10 %) and tensile resistance of the various types of polymers used in reinforcement. It may be seen that, due to stretching of the filaments, woven geotextiles exhibits the better properties.

4.2 Creep behaviour of polymers

In the case of metallic reinforcement, the level of stress induced through soil-structure interaction in retaining structures is quite low as compared to metal tensile strength, and creep of metals is not significant. On the contrary, the tensile strength of polymers is much lower, and creep behaviour has to be seriously considered for assessment of long term stability of retaining structures.

Creep of polymers yarns is a well known phenomenon, and figure 35 presents results of long term elongation tests, on yarns submitted to constant tensile loads (Greenwood and Myles, 1982), during 10 000 hours, i-e 1 year and 50 days. For this duration, and for loads not exceeding 40 % of the rupture load, strain is a linear function of the logarithm of time, as in many other materials like soils, for example. Polyester yarns are characterized by a relatively high instantaneous strain. The values of the instantaneous strain and of the creep rates are in good agreement, for polyester, with previous results presented by Finnigan (1977). As compared to polyester, polypropylene exhibits lower instantaneous strain, but higher creep rates. This creep tendency of polypropylene is well known among textile people. At 60 % of the rupture load, there is an upturn of the curves, which may be characteristic of the rupture phenomenon initiation.

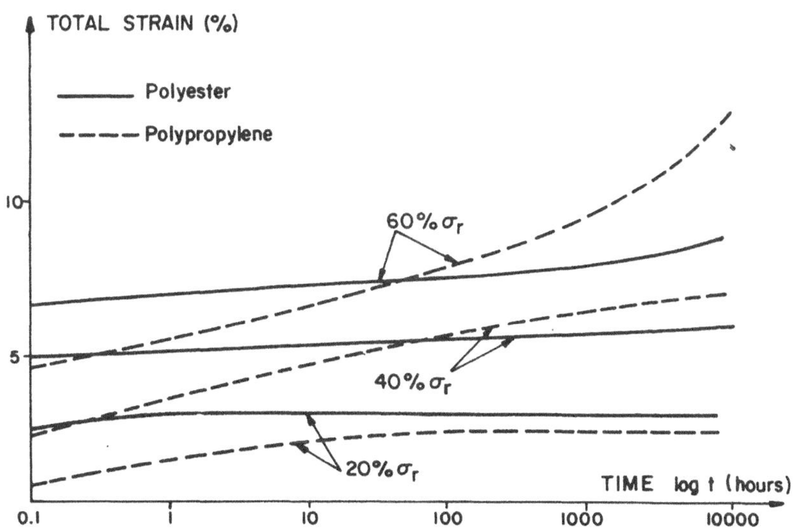

FIGURE 35. Creep of polyester and polypropylene yarns (Greenwood and Myles, 1986)

In fact, long duration tests mentionned by Greenwood and Myles on other polymers yarns (parafilrape) during seven years, showed such upturns, which were initiated after more than 10 000 hours testing. Therefore, linear extrapolation of the curves for long durations may not be realistic, and it may underestimate creep strains. As mentionned by Finnigan (1977), techniques such as heat stretching of the filament (2 % to 10 % at 235°C for 75 seconds) may significantly reduce the creep tendency.

In the case of woven fabrics, data from Van Leeuwen (1977), presented in the discussion session of the Paris Conference (Vol. III, p. 102) are shown in figure 36.

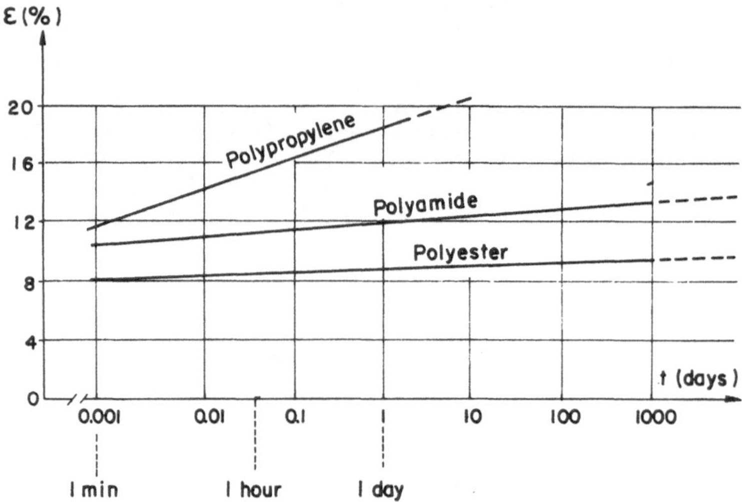

FIGURE 36. Creep of synthetic woven fabrics under prolonged loading (50% of the breaking strength) (Van Leeuwen, 1977)

These data concern polyester, polyamide and polypropylene, for loads equal to 50 % of the rupture strength. As for short term elongation tests described previously, the properties of the constitutive filament has a strong influence on the behaviour of the woven fabric. The best creep behaviour is observed for polyester fabric. Polyamide fabric exhibits a little higher tendency to creep, whereas the creep observed for polypropylene fabric is quite important. However, the author mentions that no heat treatment was performed to improve creep properties of those polymers fabrics. In the case of polyester, table VI shows a comparison between the creep of yarn and woven fabrics, based on strain values corresponding to one decimal logarithm cycle (creep rate).

It may be seen that few differences exist, in terms of creep behaviour, between yarns and woven fabrics. Such a statement has formerly been made by Finnigan (1977).

Such a comparison made for polypropylene woven fabrics shows, according to the results of Van Leeuwen, that a larger increase in creep is observed when passing from yarns to fabric ; creep rates at 50 % of breaking load increase from approximately 1,13 (Greenwood and Myles) to 2,1 (Van Leeuwen). Bell et al. (1982) mention the possibility of improving the polypropylene yarns creep rate under 40 % breaking load, from 1,5 to 0,40 by an adequate treatment. However, for woven fabrics, they only give creep rates at 20 % breaking load (0,40 - 0,73), which does not allow a direct comparison to be made.

Reference	Type of material	% of breaking load	Creep rate (% extension for per $\log_{10}t$ cycle)
Finnigan (1977)	Polyester yarn	41,2	0,192
Finnigan (1977)	Polyester yarn	58,8	0,194
Greenwood and Myles (1986)	Polyester yarn	40	0,125
Greenwood and Myles (1986)	Polyester yarn	60	0,27
Van Leeuwen (1977)	Polyester woven fabric	50	0,2

Table VI - Comparison of creep properties of polyester yarns and woven fabrics

For the SR2 Tensar geogrid, composed of high density polyethylene, data from the isochronous load-strain curves presented by McGown et al. (1984) are reported in strain/logarithm of time diagrams in figure 37.

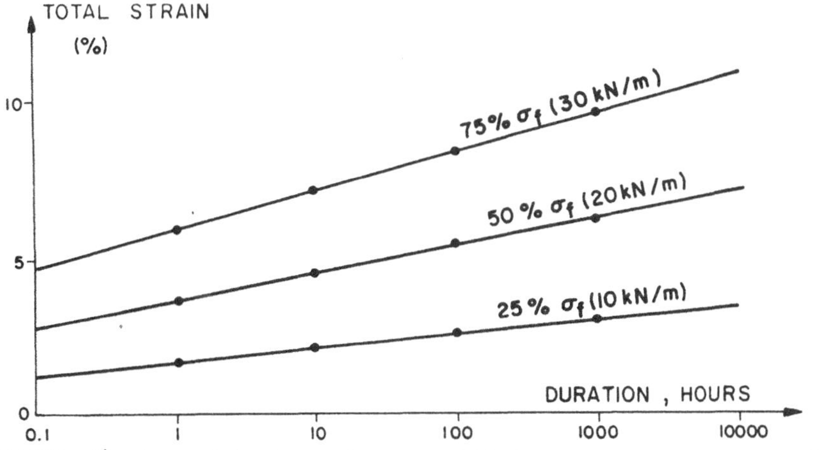

FIGURE 37. Creep behaviour of tensar SR2 (after Mc Gown et al., 1982)

It is interesting to note that the creep rates of those polyethylene grids are in the same order of magnitude as the polypropylene yarns studied by Greenwood and Myles (1986). Unfortunately, such data were not found for the polypropylene SS2 geogrid, which however should exhibit a similar behaviour. In fact, due to the structure of the grids, it seems logical to consider that, as in the case of woven fabrics, geogrids should have a creep behaviour directly related to their composition.

As in the case of elongation tests, problems arose for investigating the creep behaviour of non-woven fabrics, because of the lateral contraction of the elongated strip during non confined tests. For this type of test, predominant influence of the structure of the fabric and of the soil confinement was evidenced, as shown in figure 38 (McGown et al., 1982). The creep rate decrease is particularly evident in the case of Terram 1000 (67 % polypropylene ; 33 % polyethylene melt bonded). For Bidim U24 (polyester, needle punched), the reduction is more efficient in terms of instantaneous strain. Such in-soil creep tests are not yet very numerous, and discussion on the combined effect of polymer composition and fabric construction still has to rely on non-confined tests.

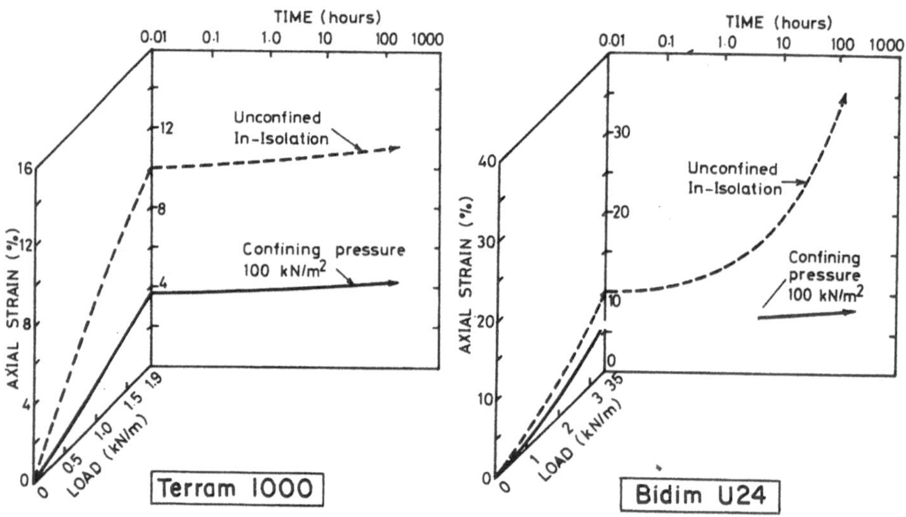

FIGURE 38. Influence of in-soil confinement on creep properties of two non-woven fabrics (Mc Gown et al., 1982)

Table VII shows the results of creep tests which were performed on 200 mm wide strips by Shrestha and Bell (1982). Lowest creep was exhibited by polyester resin bonded NW1, and highest creep by polypropylene staple needle punched NW6. Creep is most sensitive to load levels for the continuous filament polypropylene geotextile NW3 (Heat bonded) and NW5 (Needle punched), which show a 3 to 5 fold increase in creep when the sustained load is doubled. The similar results obtained for these two fabrics indicate an identical effect of heat bonding and needle punching regarding creep behaviour. Staple filaments (NW6) seem quite unfavourable with respect to creep properties. For NW1 and NW6, creep increased by only 1 % to 2 % for increases in the sustained load levels ranging from 40 % to 57 %, and 33 % to 56 % respectively. The creep of polypropylene woven fabrics W4 and C1 is higher than the creep of non-woven polyester fabric NW1, and smaller than the creep of all non-woven polypropylene fabrics.

Fabric	Geotextile construction	Filament	Creep measured in 20 hours			
			Load level (%)	Creep (%)	Load level (%)	Creep (%)
NW1	Non woven resin bonded	Polyester continuous	40	3	57	4
NW3	Non woven heat bonded	Polypropylene continuous	35	5	63	27
NW5	Non woven needle punched	Polypropylene continuous	33	9	57	31
NW6	Non woven needle punched	Polypropylene staple	33	20	56	22
W4	Woven	Polypropylene mono filament	31	11	44	16
C1	Woven, with needle nap	Polypropylene slit film	36	5	55	8

Table VII : Creep properties of various fabrics

Elements on the relative influence of both polymer composition and geotextile construction may be drawn from results of Allen et al. (1982), as shown in figure 39.

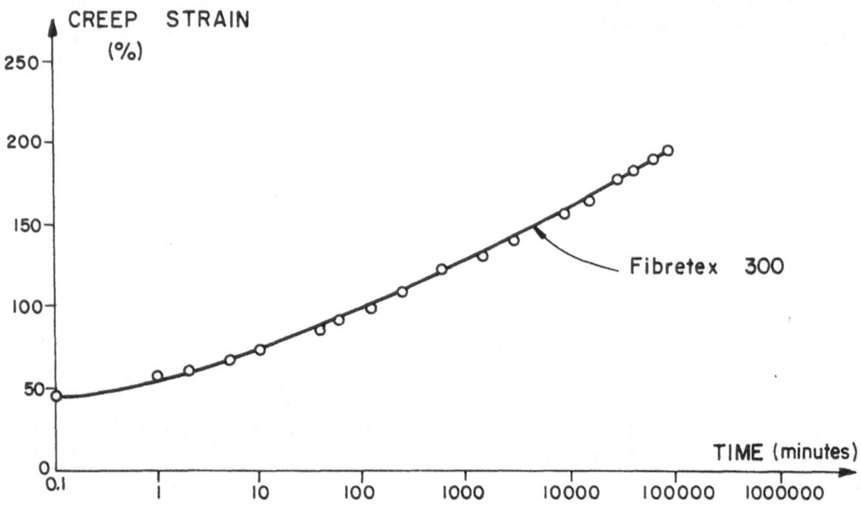

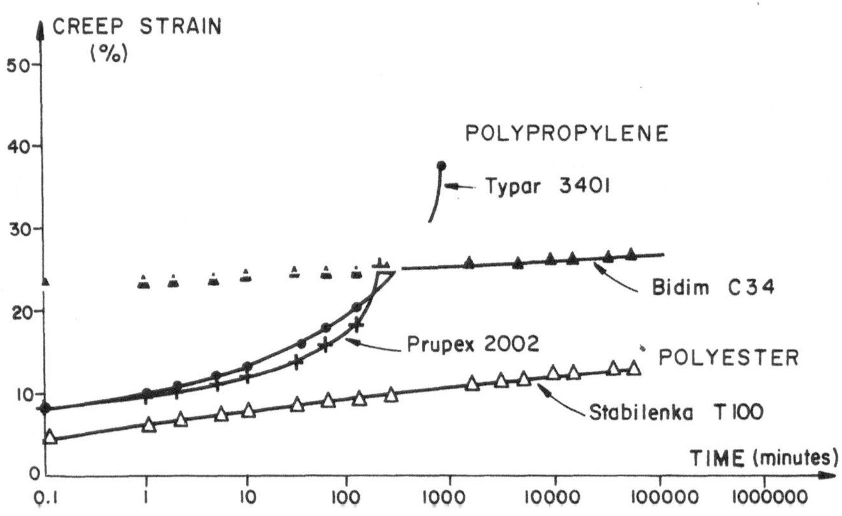

FIGURE 39. Creep properties of various non-woven fabrics (Allen et al., 1982)

This figure presents results of creep tests at 50 % of breaking load, 22°C, on 152 mm wide strips. It may be seen that two of the three polypropylene fabrics (Propex, woven, and Typar, non woven heat bonded) were broken. The third one (Fibretex, non woven, needle punched) exhibits, after 100 000 minutes (69 days), creep strains as high as 200 %. On the contrary, polyester non woven fabrics exhibit low creep strains. It should be noted that the needle punched Bidim C34 shows higher instantaneous strains than the resin bonded Stabilenka T100, whereas the creep rate of the former is much smaller than the latter's one.

In order to check the possibility of using polymers in cold regions, these writers also studied the influence of low temperatures, and they could observe that, whereas polyester presents few variations when lowering the temperature from 22° to -12°C, polypropylene creep properties are radically improved. This positive influence of lowering temperature on creep behaviour of polymers is a well-known phenomenon.

Generally speaking, it may be concluded that creep properties of the different types of geotextile used in the soil reinforcement are highly dependent on the nature of the constituant polymer. Whereas polyester products exhibit satisfactory behaviour, the performances of polypropylene products are poor. Designers must then be careful and impose smooth stress conditions to these reinforcements, with respect to their breaking strength. Further studies are then highly desirable in order to get a precise insight on the consequences of these creep properties on the long term stability of reinforced retaining structures.

4.3 Durability

As mentionned by many writers, one of the reasons why people started considering polymers as a possible material for soil reinforcement was the risks of electro-chemical corrosion encountered when using metallic inclusions.

In a general manner, the problem of long term durability of any type of material submitted to given physico-chemical and mechanical conditions is very much involved, and long term real experiments seem to be the only reliable solution. Thus, adopting a new technology in which durability problems are involved is always dangerous, and may lead to important errors, due to previously unknown phenomena.

For example, in the case of metallic strips, previous medium term experiments developed on stressed stainless steel strips confined in-soil gave excellent results, which led the Reinforced Earth Company to adopt this metal for strips. Nowadays, some of the walls built with this type of strips are presenting some problems due to a rapid and unexpected corrosion, and have to be repaired. On the contrary, galvanized soft steel generally shows excellent behaviour, and the numerous corresponding walls, which were erected 15-20 years ago, still behave satisfactorily.

Paragraph 2.2. of this report described the use of polymers in Reinforced Earth Technique. In the case of the wall of Poitiers, strip samples were taken from the upper part of the wall. According to tensile tests, a 50 % loss of strength, and a 25 % decrease of the rupture strain were estimated. Optical and electron micrographs evidenced an important degradation due to setting (flexion, shearing or crushing actions). The micrographs also revealed a degradation of the strip surface, corresponding to a "dirty" macroscopic aspect, which was attributed to a contamination caused by the soil.

In fact, durability problems of polymers, and ageing phenomena are increased, as in any other material, by stressing. The theme of durability was developed in Session 8B of Las Vegas Conference (4 papers). Unfortunately no special session was dedicated to this subject at Vienna (1986), whereas a Rilem symposium was organized in Paris (1986) on "Long term behaviour of Geotextile". Finally, relatively few papers deal with this important problem.

The sensitivity of polymers to Ultra Violet radiations is well known, and Von Vijk and Stoerzer (1982) give a description of the chemical mechanisms involved, based on the presence of oxygen, of a minimal activation energy, and consists in an autoxydation process which finally results in the formation of inert products. Each type of polymer is most sensitive to a critical wave-length in the 300-370 mm range for polyethylene, polyester and polypropylene. Intensity of radiation greatly influences the rate of ageing, together with temperature and humidity. Tropical conditions are for example critical. However, stabilizers may improve resistance against day light.

Figure 40 (Raumann, 1982) shows the degradation in strength, obtained from grab tests, which occured in a polyester non woven fabric, after a 32 weeks exposition in Florida, Arizona and North Carolina. A considerable degradation of the strength may be observed, depending on the localization of the sample.

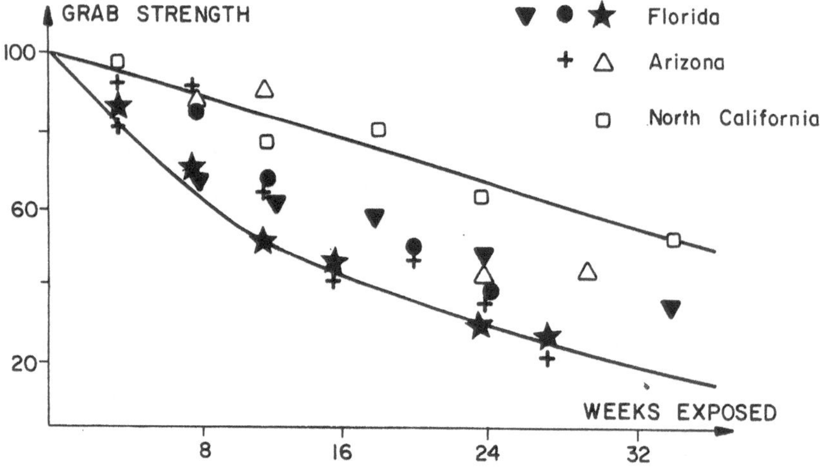

FIGURE 40. Decrease of the grab strength of a non-woven fabrics submitted to light in various parts of the U.S. (Raumann, 1982)

For all these reasons, users of polymers in geotechnical engineering are very cautious, and they never let fabrics exposed to light for long periods. In the case of geotextile walls, protective measures, such as bitumen projection, or non polymer facing, are adopted. Some people consider special the design of facings, allowing periodical replacements.

Other attacks on polymers may be caused by chemical and biological action, and the effects resulting from the burial of stressed polymers within wet soils still need to be well investigated, and require more research.

The a-priori great confidence geotextile users put in polymers may come from the remarkable performances of PVC, since adduction of water under pressure has been based on PVC pipes since 1935, which still behave very satisfactorily. More recent polymers have comparable mechanical properties, and considerable progress has also been made in terms of additive agents and stabilizers. However, these good performances should not hide some reactions polymers may have with soils.

Oxydation of polymers always has to be initiated by another agent, such as U.V. radiation or heat, and this should not occur within a soil mass.

Presence of water may induce some effect, corresponding to absorption of water within the polymer structure. The intensity of this action varies in the same way as permeability, and polyamids are known to be the more reacting polymers to water. The absorption decreases when the cristallinity rate increases. It may only be a physical absorption, which induces a swelling corresponding to a finite decrease of strength. Absorption may however by accompanied by a chemical reaction called hydrolisis, which modifies the structure of the polymer. This reaction is however very slow at current temperature. Polypropylene is known to have a good stability with respect water action.

When a polymer is buried in ground, several factors related to mineral or organic action have to be considered : acid, or alcalis, soluble salts, from iron or magnesium for instance ; organic elements, such as acids which come from bacteriologic destruction of organic matter. Generally, except for polyamids, current polymers have a satisfactory resistance to those agents. Polyester may be more sensitive to alkalis.

Biological action, as coming either from bacteria or fungus may affect polymers. This action is better known for drainage problems, since it corresponds to a biocolmatation caused by microbial produced ferric hydroxide, ferrous sulfide and polysaccharide. However, as mentionned previously in the case of the Reinforced Earth wall built with fiberglass reinforced plastic strips, biological action may be quite agressive, fast and unpredictable. Ionescu et al. (1982) have performed interesting tests by incubating various woven and non woven polyester and polypropylene fabrics in eight different media, including distilled water, iron bacteria, desulfovibrios, levan synthesiying bacteria, sea water, mineral solution, compost and alluvial soil. After 5 and 17 months, they observed some bio-colmatation, but an insignificant change in the tensile strength values. It should be noted however that no stress were applied during those experiments, and that very few results on the chemical and biological behaviour of stressed polymers buried in such media exists. Furthermore, as noted by Mallinder et al. (1977), the 6 months duration laboratory tests performed on fiberglass plastic strips were also satisfactory. Extrapolation of those kind of tests for long term behaviour should then be made very cautiously.

It is generally believed that long term stressing of polymers may induce some superficial microfissurations which may considerably favorize chemical or bio-chemical attacks, through the microdefects developing at the surface. This development of microfissures is much more important for tensile stresses than for compression stresses.

The best way of understanding those problems is to perform full scale experiments, which consist for instance in observing polymers after a long period of true working conditions. An important work, described by Sotton et al. (1982), has been undertaken in this direction by the French Committee of Geotextiles. This study concerns different types of woven and unwoven fabrics, made of polypropylene and polyester filaments. Many observations have been made, and mechanical control tests were performed. Writers insist on the need of conserving reference samples of the geotextiles used, in order to make valid comparisons, particularly in terms of the evolution of the mechanical properties. They observe few variation when the geotextile is not submitted to stress. Exposure to weather variations and geotextile tearing induced large decreases in strength (> 30 %). This may happen when the geotextile stays too long without being covered by a soil layer. For the application to retaining structures, it seems interesting to rather consider, with respect to creep consequences, the studied cases where the geotextiles were not apparently damaged, but had to support a loading. In this perspective, it can be refered to the case histories of Noyalo (non woven needle punched polypropylene) and Thiers (woven polypropylene). The Noyalo fabric appeared to be contaminated by fine grained soils, and suffered a 20-40 % loss of strength. This contamination could be compared to the degradation of the strips of the Poitiers wall. The case of Thiers, which is the only case where the fabric had to sustain a 5 m high embankment, seems more serious, with a 53 % loss strength, attributed by the authors to creep under the considered load.

Such long term applications of load seem to be an important factor of strength reduction, because of creep. This aspect is particularly important in the case of retaining structures, where fabrics are submitted to important loads. For this reason, the problem of long term stability of these structures still has to be discussed regarding polymer properties. In this perspective, the study of the properties of the Bidim fabric of the first geotextile reinforced wall of Rouen (1971) will be extremely interesting. Such a study has been started by the Laboratoire des Ponts et Chaussées (L.P.C.) and the general aspect of the fabric, after 16 years in the ground, is good (Delmas and Gourc, 1987). Mechanical data have now to be precisely studied in order to draw valuable conclusions. On the other hand, the behaviour of the various fabric reinforced soil walls monitored by LPC in France for some years seem at present satisfactory, in terms of external deformations as a function of time.

However, predictions are difficult to make. In fact, the only suitable method to answer the difficult problem of durability is the full scale long term experimentation. Before drawing any valuable conclusions from this kind of approach, relative caution in any assessment concerning long term performance of any material as a reinforcement material seems to be desirable.

GENERAL CONCLUSIONS

Polymers are or can be used in numerous types of retaining systems, ranging from the classical "ladder wall" invented by Coyne to Texsol, recently developed by Leflaive. Of course, each application is different and depends on many factors (cost, aesthetic aspect of the structure, structure deformability, durability, etc ...).

As used in reinforcements, polymers provide an alternative to metals, the latter being so far limited to linear inclusions and grids. However, their deformability is generally higher than metals and it appears to be a disavantage. They are not subject to corrosion, but they can be degraded.

Considering wall construction, the use of deformable polymers generally requires special types of facing in order to built vertical walls.

For reinforced soil walls, in which the frictional soil-inclusion interaction is predominant, the use of polymers has permitted to cover all the range of reinforcement types : 1) linear reinforcement with materials like fiber reinforced plastic ; 2) two-dimensional reinforcement with geotextiles and geogrids ; 3) three-dimensional reinforcement with a continuous thread (Texsol), fibers or mesh elements, but also with combinations of grids like for gabions, rafts, or mattresses.

Future development in the use of polymers for retaining walls construction requires further research in the following areas :

- Reinforced soil mechanical behaviour and durability.
- Design methods taking into account deformations.
- Facing construction techniques.

REFERENCES

1. Bacot J: Contribution à l'étude du frottement entre un matériau souple et un matériau pulvérulent. Thèse de Doctorat es Sciences. INSA de Lyon, 1981.

2. Baudonnel J, Giroud JP, Gourc JP: Etude expérimentale et théorique du comportement en traction des géotextiles non tissés, Proceedings of the 2nd International Conference on Geotextiles, volume 3, pp. 823-828, Las Vegas, 1982.

3. Bassett RH, Last HC: Reinforced Earth below Footings and Embankments. ASCE Symposium on Earth Reinforcement, pp. 222-231, Pittsburgh, April 1978.

4. Bell AL, Green HM, Laverty K: Factors Influencing the Selection of Woven Polypropylene Geotextile for Earth Reinforcement. Proceedings of the 2nd International Conference on Geotextiles, volume 3, pp. 689-694, Las Vegas, 1982.

5. Blivet J.C., Grestin F.: Etude de l'Adhérence entre le Phosphogypse et deux Géotextiles. Coll. Int. Renf. des Sols, Vol. II, Paris, pp. 403-408, 1979.

6. Bonaparte R, Kamel MI, Dixon JH: Use of geotextiles in soil reinforcement. Proceedings of the 1984 annual meeting of the Transportation Research Board, Washington DC, January 1984.

7. Bonazzi D., Colombet G.: Réajustement et Entretien des Ancrages de Talus. Proc. of the Int. Conf. on In-Situ Soil and Rock Reinforcement, Ecole Nationale des Ponts et Chaussées, Paris 9-11 Octobre 1984, pp. 225-230, 1984.

8. Cazzuffi D, Venezia S, Rinaldi H, Zocca A: The Mechanical Properties of Geotextiles : Italian Standard and Interlaboratory Test Comparison. Proceedings of the 3rd International Conference on Geotextiles, volume 3, pp. 879-884, Vienna, 1986.

9. Chabard J.P., Pardieu P., Guerber P., Bertrand J.: Novel Reinforced Fill Dam. Proc. 8th ECSMFE, Elsinki, Session 5, 1983.

10. Chang JC, Hannon JB, Forsyth RA: Pull Resistance and Interaction of Earthwork Reinforcement and Soil. Proceedings of the 1977 Meeting of the Transportation Research Board. Washington DC, January 1977.

11. Coyne A.: Murs de Soutènement et Murs de Quai "à Echelle", Le Génie Civil, 1er et 15 Mai, 1945.

12. Delmas P, Berche JC, Gourc JP: Le dimensionnement des ouvrages renforcés par géotextiles: Programme CARTAGE. Bulletin de Liaison des Ponts et Chaussées, n° 142, pp. 33-44, Paris, Mars-Avril 1986.

13. Delmas P, Gourc JP: Le Renforcement par Géotextiles: Recherches et Réalisations. Réunion du Comité Français de Mécanique des Sols, Paris, Avril 1987.

14. Delmas P, Puig J, Schaeffner M: Mise en oeuvre et parement des massifs de soutènement renforcés par des nappes de géotextiles. Bulletin de Liaison des Ponts et Chaussées, n° 143, pp. 65-77, Paris, Mai-Juin 1986.

15. Finnigan JA: The Creep Behaviour of High Tenacity Yarns and Fabrics Used in Civil Engineering. Proceedings of the International Conference on the Use of Fabrics in Geotechnics, volume 2, pp. 305-310, Paris, 1977.

16. Forsyth RA: Alternative Earth Reinforcements. Proceedings of the ASCE Symposium on Earth Reinforcement, pp. 358-370, Pittsburgh, April 1978.

17. Forsyth RA, Bieber DA: La Honda Slope Repair with Geogrid Reinforcement. Proceedings of the Symposium on Polymer Grid Reinforcement, pp. 54-57, London, 1984.

18. Fukuoka M, Imamora Y: Fabric retaining walls. Proceedings of the 2nd International Conference on Geotextiles, Volume 3, pp. 575-580, Las Vegas, 1982.

19. Gasnier R, Plumelle C: Etude expérimentale en vraie grandeur de tirants d'ancrage. International Conference on in-situ Soil and Rock Reinforcement, Ecole Nationale des Ponts et Chaussées, pp. 333-339, Paris, 1984.

20. Giroud JP, Carroll Jr: Geotextile Products. Geotechnical Fabrics Report, volume 1, n°1, IFAI, pp.12-15, St Paul, Minnesota, 1983.

21. Gray DH: Role of woody vegetation in reinforcing soils and stabilizing slopes, Symposium on soil reinforcement and stabilizing techniques, pp. 253-306, Sydney, 1978.

22. Gray DH, Athanasopoulos G, Ohashi H: Internal-External Fabric Reinforcement of Sand. Proceedings of the 2nd International Conference of Geotextiles, volume 3, pp. 611-616, Las Vegas, 1982.

23. Gray DH, Ohashi H: Mechanics of fiber reinforcement in sand, ASCE, Journal of the Geotechnical Engineering Division, (109) GT3, pp. 335-353, 1983.

24. Greenwood JH, Myles B: Creep and Stress Relaxation of Geotextiles. Proceedings of the 3rd International Conference on Geotextiles, volume 3, pp. 821-826, Vienna, 1986.

25. Hausmann MR: Strength of Reinforced Soil. Australian Road Research Board, Proceedings, (8), 1-8, Session 13, 1976.

26. Hoare DJ: Laboratory study of granular soils reinforced with randomly oriented discrete fibers. Proceedings of the International Conference on Soil Reinforcement (1), pp. 47-52, Paris, 1979.

27. Ingold TS: A laboratory investigation of grid reinforcements in clay. Geotechnical Testing Journal, GTODJ, volume 6, n° 3, September 1983.

28. Ionescu A, Kiss S, Dragan-Bularda M, Rodulescu D, Kolozsi E, Pintea H, Crisan R: Methods used for Testing the Bio-Colmatation and Degradation of Geotextiles Manufactured in Romania. Proceedings of the 2nd International Conference on Geotextiles, volume 3, pp. 547-552, Las Vegas, 1982.

29. Jewell RA: Some effects of reinforcement on the mechanical behaviour of soil. PhD thesis, University of Cambridge, 1980.

30. Jewell RA, Jones CJFP: Reinforcement of clay soils and waste materials using grids. Proceedings of the 10th International Conference of

Soil Mechanic and Foundation Engineering (3), pp. 701-706, Stockholm, 1981.

31. John N.W.H., Ritson R., Johnson P.B., PETLEY D.J.: Instrumentation of Reinforced Soil Walls. Proceedings of the 8th ECSMFE. Helsinki, May 1983, Vol. 2, pp. 509-512, 1983.

32. Jones CJFP: Practical design considerations. Proceedings of the Symposium on Reinforced Earth and Other Composite Soil Techniques. TRRL and Heriot-Watt University, September 1977.

33. Jones CJFP: Practical Construction Techniques for Retaining Structures using Fabric and Geogrids. Proceedings of the 2nd International Conference on Geotextiles, volume 3, pp. 581-585, Las Vegas, 1982.

34. Kern F: Réalisation d'un barrage en terre avec parement aval vertical au moyen de poches en textile. International Conference on the Use of Fabrics in Geotechnics. Ecole Nationale des Ponts et Chaussées, volume 1, pp. 91-94, Paris, 1977.

35. Leclercq B et Prudon R: Comportement en traction des géotextiles en fonction de l'inclinaison relative de l'axe de production par rapport à l'axe de l'effort, Proceedings of the 3rd International Conference on Geotextiles, volume 3, pp. 751-756, Vienna, 1986.

36. Leflaive E, Paute JL et Ségouin M: La Mesure des Caractéristiques de Traction en vue des Applications Pratiques, Proceedings of the 2nd International Conference on Geotextiles, volume 3, pp. 733-738, Las Vegas, 1982.

37. Leflaive E, Khay M, Blivet JC: Un nouveau matériau : le Texsol. Bulletin de Liaison du Laboratoire des Ponts et Chaussées, n° 125, pp. 105-114, Paris, Mai-Juin 1983.

38. Leflaive E, Liausu PH: Le Renforcement des Sols par Fils Continus. Proceedings of the 3rd International Conference on Geotextiles, volume 2, pp. 523-529, Vienna, 1986.

39. Mallinder F.P.: The Use of FRP as Reinforcing Elements in Reinforced Soil Systems. Proceedings of the Symposium on Reinforced Earth and other Composite Soil Technique. TRRL and Heriot-Watt University, Edinburgh, 1977.

40. McGown A, Andrawes KZ, Al Hasani MM: Effect of Inclusion Properties on the Behaviour of Sands. Geotechnique (28), 3, pp. 327-346, 1978.

41. McGown A, Andrawes KZ, Kabir MH: Load-extension Testing of Geotextiles Confined in-soil, Proceedings of the 2nd International Conference on Geotextiles, volume 3, pp. 793-798, Las Vegas, 1982.

42. McGown A, Andrawes KZ, Yeo KC: The Load-Strain-Time Behaviour of Tensar Geogrids. Proceedings of the Conference on Polymer Grid Reinforcement, pp. 11-17, London, 1984.

43. Mercer FD, Andrawes KZ, McGown A, Hytiris N: A new method of soil stabilization. Proceedings Symposium on Polymer Grid Reinforcement in Civil Engineering. Paper 8-1, London, 22-23, March 1984.

44. Morbois A., Long N.T.: Etude du procédé Actimur. Rapport de Recherche. Laboratoire Central des Ponts et Chaussées, Paris, 1984.

45. Murray RT, Irwin MJ. A Preliminary Study of TRRL Anchored Earth. TRRL supplementary report 674, 1981.

46. Murray RT, McGown A, Andrawes KZ, Swan D: Testing Joints in Geotextiles and Geogrids, 3rd International Conference on Geotextiles, volume 3, pp. 731-736, Vienna, 1986.

47. Paute JL, Segouin M: Détermination des caractéristiques de résistance et de déformabilité des textiles par dilatation d'un manchon

cyclindrique, International Conference on the Use of Fabrics in Geotechnics, volume 4, pp. 293-298, Paris 1977.

48. Puig J, Blivet JC, Pasquet P: Remblai armé avec un textile synthétique. International Conference on the Use of Fabrics in Geotechnics. Ecole Nationale des Ponts et Chaussées, volume 1, pp. 85-9, Paris, 1977.

49. Rabejac S, Toudic P: Construction d'un mur de soutènement entre Versailles-Chantiers et Versailles-Matelots. Revue Générale des Chemins de Fer, 93ème année, 1975.

50. Raumann C: A Hydraulic Tensile Test with Zero Transverse Strain for Geotechnical Fabrics. Geotechnical Testing Journal, volume 2, n°3, pp. 69-76, June 1979.

51. Richards EA, Scott JD: Stress-strain Properties of Geotextiles. Proceedings of the 3rd International Conference on Geotextiles, volume 1, pp. 873-878, Vienna, 1986.

52. Rowe RK, Ho SK: Determination of Geotextile Stress-strain Characteristics Using a Wide Strip Test. 3rd International Conference on Geotextiles, volume 3, pp. 885-890, Vienna, 1986.

53. Schlosser F: Discussion Session. International Conference on the Use of Fabrics in Geotechnics, volume 3, pp. 36-37, Paris, 1977.

54. Schlosser F, Long NT: Comportement de la Terre Armée dans les Ouvrages de Soutènement. Proceedings of the 5th European Conference on Soil Mechanics and Foundation Engineering, volume 1, pp. 299-306, Madrid, 1972.

55. Schlosser F, Elias V: Friction in Reinforced Earth. Proceedings of the ASCE Symposium on Earth Reinforcement, pp. 735-764, Pittsburgh, April 1978.

56. Schlosser F, Jacobsen HM, Juran I: Soil Reinforcement. General Report. Speciality Session 5, Proceedings of the 8th European Conference on Soil Mechanics and Foundation Engineering (3), pp. 1159-1180, Helsinki, 1983.

57. Schlosser F, Magnan JP, Holtz RD: Geotechnical Engineered Construction. Theme Lecture 5, Proceedings of the 11th International Conference on Soil Mechanics and Foundation Engineering (1), pp. 211-254, San Francisco, 1985.

58. Schwantes ED, JR: Recent Experience with Fabric-faced Retaining Walls, Proceedings of the 2nd International Conference on Geotextiles, volume 3, pp. 605-609, 1982.

59. Shresta SC, Bell JR: A Wide Strip Tensile Test of Geotextiles, Proceedings of the 2nd International Conference on Geotextiles, volume 3, pp. 739-744, Las Vegas, 1982.

60. Sissons CR: Strength testing of fabrics for use in civil engineering. International Conference on the Use of fabrics in Geotechnics, volume 2, pp. 287-292, Paris, 1977.

61. Sotton M, Leclercq B, Paute JL, Fayoux D: Quelques Eléments de Réponse au Problème de la Durabilité des Géotextiles. Proceedings of the 2nd International Conference on Geotextiles, volume 3, pp. 553-558, Las Vegas, 1982.

62. Van Leeuwen JH: New methods of determining the Stress-strain Behaviour of Woven and Non Woven Fabrics in the Laboratory and in Practice. International Conference on the Use of Fabrics in Geotechnics, volume IV, pp. 299-304, Paris, 1977.

63. Van Vijk W, Stoerzer M: UV Stability of Polypropylene. Proceedings of the 3rd International Conference on Geotextiles, volume 3, pp. 851-856, Vienna, 1986.

64. Vautrain J, Puig J: Expérimentation du Bidim ; Remblai Expérimental de Caen. Bulletin de Liaison des Laboratoires Routiers des Ponts et Chaussées, n° 41, Paris, Novembre 1969.
65. Verma BP, Char ANR: Triaxial Tests on Reinforced Sand. Proceedings of the Symposium on Soil Reinforcing and Stabilizing Techniques, pp. 29-39, Sydney, 1978.
66. Vidal H: La terre armée (un nouveau matériau pour les travaux publics). Annales de l'ITBTP, n° 223-224, Juillet-Aout 1966.
67. Viergever MA, De Feijter JW, Mouw KAG: Biaxial Tensile Strength and Resistance to Cone Penetration of Membranes. International Conference on the Use of Fabrics in Geotechnics, volume II, pp. 311-316, Paris, 1977.

ACKNOWLEDGEMENTS

The authors acknowledge Mr J. CANOU for his help in preparing the English version of the text.

Prediction Exercise

INTRODUCTION AND RATIONALE FOR PREDICTION EXERCISE

PETER M. JARRETT

Royal Military College of Canada

Two large scale polymerically reinforced soil walls were built, instrumented and tested at the Royal Military College of Canada. The design of the experiments, which represented working levels of stress, arose as a result of discussions held under a NATO Collaborative Research Grant between Drs. McGown and Andrawes of the University of Strathclyde in Scotland and Drs. Jarrett and Bathurst at the Royal Military College of Canada.

Prior to construction, details of the proposed tests and the materials to be used were sent to potential predictors. They were asked to make Class A (in advance of construction) predictions of aspects of the test behaviour such as the force on and the displacements of the facing elements, the strains in the reinforcement, the earth pressure distribution and a projected failure load. In addition they were asked to provide details of their analytical methods. The understanding from the outset of the exercise was that the predictors would meet to discuss in a friendly, cooperative manner the measured results and the predictions. It was also stated that to allow freedom to experiment analytically and to maintain openess of discussion that there would be no formal attribution of the predictions to their predictors. The only competitive aspect was the prize of a bottle of Teachers Whiskey provided for the closest prediction courtesy of our Scots colleagues.

The following persons submitted predictions. The list is in Alphabetic order only and bears no relationship to the order of discussion of predictions in a later paper:

Rudolph Bonaparte
Danielle Cazzuffi, Pietro Rimoldi and Roberto Mangiavacchi
Philippe Delmas and Jean-Pierre Gourc
Richard Jewell
Colin Jones
Robert Koerner and Manfred Hausmann
Richard Murray
Gregory Richardson
Jost Studer and B. Graf

The organizers most sincerely thank these gentlemen for their considerable efforts, their spirit in accepting the challenge and for the friendly, open and cooperative manner in which their discussions were carried out.

The outcome of the prediction exercise is presented in the following chapter together with certain amplifying material that it is felt increases

P. M. Jarrett and A. McGown (eds.), The Application of Polymeric Reinforcement in Soil Retaining Structures, 69–70.

the overall value of the basic test program. There are two keystone papers for the prediction exercise and three supplementary papers.

Comprehensive details of the construction of the RMC Test Walls and the measured results are provided in the first paper by Bathurst, Wawrychuk and Jarrett. Most of the information provided to predictors is contained in this paper to enable readers to test their own analytical methods against the results. The second paper by Bathurst and Koerner presents details of the methods employed by predictors and a comparison of the results of the predictions to the measured behaviour. It makes fascinating reading!

The three papers that follow contain most valuable supplementary information and analyses that were completed after the workshop.

During demolition of the test walls, samples of the Tensar SR2 reinforcement were recovered and returned to the manufacturer, Netlon Ltd. for examination and testing to assess the material's resistance to damage during construction. The third paper by Bush and Swan reports the results of this work.

It was suggested during the workshop that measurements made on a propped wall, retaining unreinforced soil might help in the interpretation and comprehension of the reinforced soil results. Most especially it could shed light on the effect of side wall friction in the test facility and the actual angle of friction of the soil. A preliminary test of this nature was carried out after the workshop together with further measurements of the side wall friction potential. These results are presented by Bathurst and Benjamin.

The final paper by Jewell represents a very thorough analysis of the Test Walls. It is especially valuable in the context of this chapter for its interpretation of the strength properties of the sand used in the Test Walls and for the analysis of the effects of sidewall friction on the earth pressures in the RMC Test Facility. The paper also acts as a test for and example of the use of the analytical procedures proposed by Jewell in a companion paper in this volume.

In summary, it is believed that this chapter contains a detailed case history of two reinforced soil walls that may be used as a benchmark test for analytical methods as they are developed and it provides a graphic state of the art of the accuracy of the present analytical methods used by the predictors. There is still much to be learned about both soil and polymer behaviour.

LABORATORY INVESTIGATION OF TWO LARGE-SCALE GEOGRID REINFORCED SOIL WALLS

Richard J. Bathurst, William F. Wawrychuk and **Peter M. Jarrett**

Civil Engineering Department
Royal Military College of Canada

1.0 Introduction

1.1 General

Two(2) large-scale model reinforced soil walls were constructed within the soil retaining wall test facility at the Royal Military College, Kingston, Canada.

The vertical reinforced soil walls were 3 m high and were constructed using:

a) an incremental timber panel facing with a polymer geogrid soil reinforcement, and;

b) a propped (tilt-up) timber panel facing with a polymer geogrid soil reinforcement.

Each wall was subjected to sustained surcharge loadings up to 50 kPa following construction.

1.2 Objectives

The construction and performance monitoring of the two reinforced soil structures was undertaken to provide a focal point for discussion at the NATO Advanced Research Workshop entitled: *Application of Polymeric Reinforcement in Soil Retaining Structures* held at the Royal Military College of Canada in June 1987.

The workshop was broadly directed to review the state-of-the-art in polymer reinforced soil retaining structures and to critically assess the relationships between design methodologies, construction practices and the measured performance of these systems.

In order to stimulate discussion, invited experts were given details of the proposed reinforced soil wall configurations *before* construction (Bathurst and Jarrett, 1986) and asked to make *Class A* predictions of the behaviour of the two reinforced soil structures at various elapsed times after initial construction and after surcharging pressures were applied to the structures. The results of the prediction exercise were presented informally and were the centre of discussion for two workshop sessions.

This paper describes the construction of the two test walls and the measured test results. A summary of the performance predictions offered by Workshop participants is reported by Bathurst and Koerner in these proceedings.

2.0 RMC Retaining Wall Test Facility

2.1 General

The RMC Retaining Wall Test Facility was conceived and constructed to provide a general purpose large-scale test apparatus to examine a variety of reinforced soil wall systems. The test facility is located within the Dolphin Structures Laboratory of the Civil Engineering Department at RMC. The principal structural components of the test facility were completed in November 1985.

P. M. Jarrett and A. McGown (eds.), The Application of Polymeric Reinforcement in Soil Retaining Structures, 71–125.

2.2 Description of Test Facility

An overview of the RMC Retaining Wall Test Facility is given on Plate 1. The principal structural components of the facility are six rigid heavily reinforced concrete counterfort cantilever wall modules which are shown on Figure 1. These modules are arranged to laterally confine a block of soil up to 6.0 m long by 3.6 m high by about 2.4 m wide. Translational stability of the reinforced concrete segments is provided by 32 mm dia. anchor rods/bolts extending through bolt holes to structural anchors in the laboratory floor. Additional lateral stability is provided by six rectangular hollow structural sections bolted across the top of wall segments at the location of the module counterforts.

The back of the soil block between concrete wall segments is confined by a system of braced vertical posts and lagging boards. Reinforced soil wall facings are exposed at the front edge of the test facility.

Surcharging of test configurations is carried out by inflating reinforced air bags confined between the concrete walls and a timber ceiling. The timber ceiling is restrained in turn by the hollow structural steel sections as shown on Figure 2. The current surcharging arrangement allows a vertical pressure equivalent to 3 m of fill to be applied to the soil surface between the concrete modules. The inside walls of the structure are faced with a composite plywood/clear plexiglas/polyethylene sheeting which acts to reduce sidewall friction. In addition, the plywood has been drilled out at regular spacings to allow the soil fill and reinforcing elements to be observed directly during and after construction of test models.

Additional details of the test facility have been reported by Lescoutre (1986) and Wawrychuk (1987).

3.0 Test Configurations

3.1 General

Schematics of the reinforced soil walls tested are shown on Figures 3 and 4. Both constructions comprised a high density polyethylene TENSAR SR2 Geogrid-reinforced sand soil. Reinforcement layers were attached to the plywood bulkhead facing panels shown on the figures. The principal differences between the structures were the front panel configurations and details of construction sequence. The incremental facing panel wall was constructed in four stages with each panel externally supported only until the sand fill behind the panel was placed and compacted. The facing panel supports on the propped wall construction, however, were released only after the full height of sand fill had been placed behind the facing units. Both model walls were constructed with a central instrumented section nominally 1 m wide and two 0.7 m wide edge sections. This construction was adopted to reduce edge-effects on the performance of the monitored central reinforcement strip and facing unit(s). A foam rubber void filler was placed along all panel edges in order to prevent the panels from binding during outward movements. In addition, for the incremental wall test, a layer of compressible foam rubber was placed at each horizontal panel to panel interface to assist in panel levelling during construction of the wall.

A principal objective of the current investigation was to examine the behaviour of these reinforced soil wall systems under *working stress* conditions. Therefore, the dimensions of the trial sections were based essentially on the design methods contained in the U.K. Department of Transport's technical memorandum *Reinforced Earth Retaining Walls and Bridge Abutments for Embankments, BE 3/78* with one exception, the grid lengths were limited to 3 metres instead of 5 metres. Preliminary trial tests at RMC with *wrap-around* geogrid facings have shown that for the same number of reinforcing layers (e.g. four) and a total height of fill equal to 3.25 m, 3 metre reinforcement lengths are more than adequate for stability (Bathurst, 1987).

Plate 1 Overview of RMC Retaining Wall Test Facility

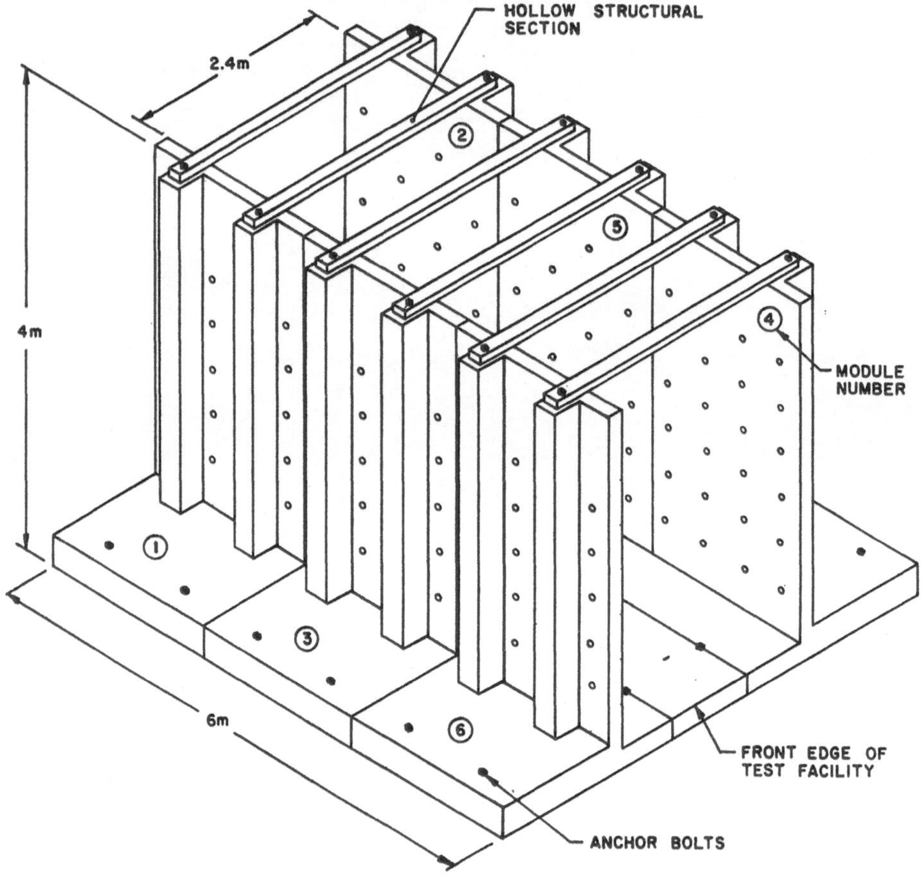

Figure 1 RMC Retaining Wall Test Facility

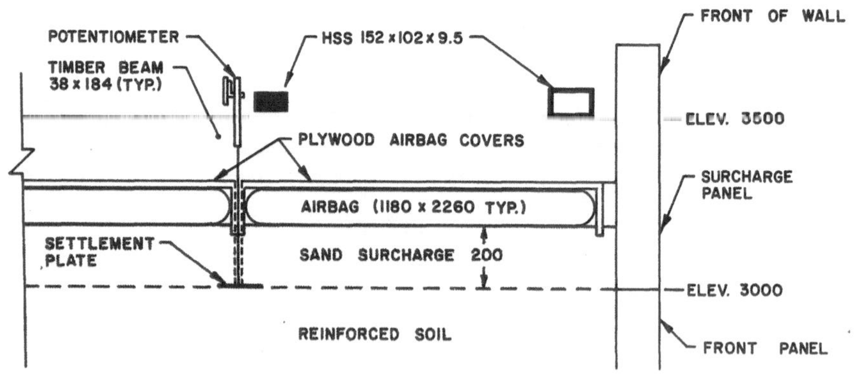

Figure 2 Air Bag Surcharging System

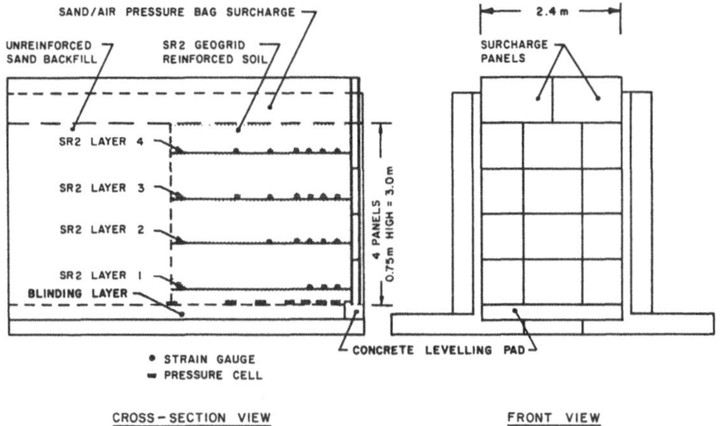

Figure 3 General Arrangement for Incremental Facing Wall Construction

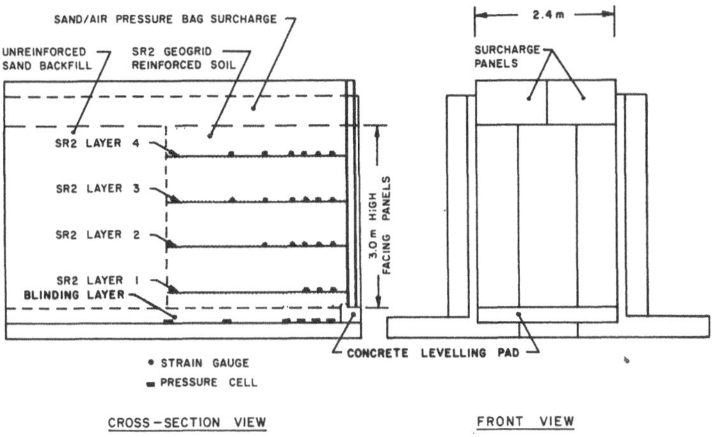

Figure 4 General Arrangement for Propped Facing Wall Construction

Details of wall design and construction were arrived at through collaboration between the authors and Dr. A. McGown, Dr. K. Andrawes and D. Varney at the University of Strathclyde, Scotland. Construction details are considered realistic for actual field structures although some compromises in construction were made to make the performance of these structures easier to predict and monitor. For example, facing panels are not dovetailed or staggered and the facings were arranged without an initial batter in the vertical direction. Another example is the panel/reinforcement connections. The number of bolts per grid was limited to four in order to ensure adequate sensitivity of the bolts as load cells.

3.2 Materials

3.2.1 Sand

The soil used was a washed sand with some gravel. This material is composed of sub-angular to angular quartz and feldspar particles. The grain size distribution for this material is given on Figure 5. Less than 0.3% of all particles by weight are less than 0.075 microns.

The results of direct shear tests and triaxial compression tests carried out at RMC are given in Appendix A. A summary of test results is given on Table 1. The second column of friction angles shown on the table are values representing linear failure envelopes passing through the origin of a $\tau - \sigma_n$ plot. Additional direct shear tests on the RMC sand are reported by Jewell (1987b) as part of these proceedings and confirm (secant) peak friction angles interpreted from the RMC direct shear tests.

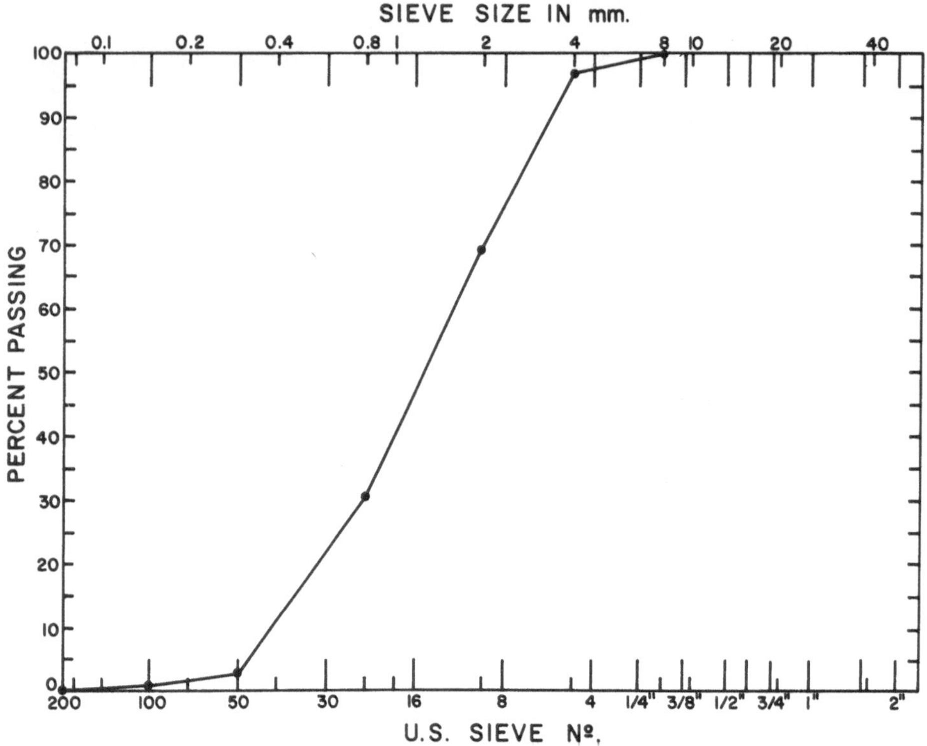

Figure 5 Sand Fill Grain-Size Distribution

3.2.2 Sidewall Friction

A preliminary, large-scale shear box test (1 m by 1 m in plan view) gave a fully mobilized friction angle of 20° for the sand/sidewall interface. The results of a more recent and comprehensive set of direct shear box tests reported in this proceedings by Bathurst and Benjamin shows that 20° may be an upper-bound value. A value of 15° is likely a better estimate of the sidewall friction angle at normal stress levels considered to act against the test facility sidewalls.

Table 1

Results of Shear Strength Tests on Sand Backfill

Type of Test	Dry Soil Density ρ_d (kg/m^3)	Friction Angle (degrees)
Direct Shear Box (6cm x 6cm x 3.7cm high)	1690	42* 44**
Direct Shear Box (6cm x 6cm x 3.7cm high)	1780	43* 48**
Direct Shear Box (6cm x 6cm x 3.7cm high)	1860	(peak) 46* 53** (constant volume) 40* 44**
Consolidated-Drained (Standard) Triaxial Test	1680	41**

* as provided to predictors in *Bulletin 1*, (Bathurst and Jarrett, 1986)

** based on a linear failure envelope with c=0

3.2.3 Geogrid Reinforcement

Four layers of reinforcement comprising 3 m long strips of high density polyethylene TENSAR SR2 Geogrid were used in test configurations. The mechanical properties of samples taken from the *actual* rolls of reinforcement were determined from tests carried out by Netlon Ltd. prior to wall construction. These results are presented in Appendix B.

4.0 Construction Methods and Surcharge Loading

4.1 General

The construction sequence adopted for both walls was similar with the exception of the temporary waling support and release sequence.

Prior to construction of both test configurations, a 250 mm thick blinding layer was placed and compacted behind the concrete floor levelling pad shown on Figures 3 and 4. The timber facing panels were seated on the concrete levelling pad and initially restrained between the concrete lip shown on the figures and a wooden waling located in front of the wall at the same elevation. Subsequently, all sand backfill was placed and compacted in 125 mm lifts covering the full length of the test facility. A vibrating plate tamper was the principal means of compaction although a hand-held Kango hammer with tamper attachment was employed to compact sand in the corners formed by the facing panels and the test facility walls.

Prior to placing fill over any strip of reinforcement, a light pretensioning load of about 0.4 kN/m was applied to the geogrid to ensure that the reinforcement was free of warps. The pretensioning was applied using a bar threaded through the free end of each strip of reinforcement and attached in turn by a cable to a system of weights at the back of the test facility. The pretensioning was released after the grid was covered by 0.5 m of compacted sand.

4.2 Incremental Panel Facing Reinforced Soil Wall Test

In this construction a total of twelve 0.75 m high timber bulkhead facing panels were used to construct a reinforced soil wall 3.0 m high. Temporary lateral support was provided to each row of panels by a pair of horizontal timber walings braced in turn by a system of wedging plates bolted to the front of the RMC test facility. Each panel row support was released after the sand backfill had been placed and compacted to the elevation of the top of the facing panel. The purpose of this form of construction was to progressively mobilize the inherent tensile capacity of the geogrid reinforcement as the height of the composite structure was increased.

The construction sequence is illustrated on Figure 6. Also identified on the figure are critical construction events. A frontal view showing the incremental panel wall facing units at the end of construction is given on Plate 2. Following construction, the incremental facing wall was subjected to a series of surcharge loadings using a sand surcharge or sand surcharge/pressurized airbag combination. The surcharging schedule is given in Table 2. The design surcharge of 12 kPa was sustained for a period in excess of 1000 hrs in order to observe creep behaviour under working (i.e design) loads. Later, a maximum surcharge load of 50 kPa was applied for 500 hrs to observe further creep deformations in the composite system.

Table 2

Surcharging Schedule for Incremental Panel Wall

Event	Elapsed time (hrs)	Duration (hrs)
wall construction commences	0	225
12 kPa surcharge applied	225	1668
3.3 kPa surcharge during airbag construction	1893	938
12 kPa surcharge applied	2831	24
30 kPa surcharge applied	2855	123
40 kPa surcharge applied	2978	101
50 kPa surcharge applied	3079	500
30 kPa surcharge applied	3579	22
12 kPa surcharge applied	3601	22
3.3 kPa surcharge applied	3623	19
excavation commences	3642	76
excavation complete	3718	-

4.3 Propped Panel Facing Reinforced Soil Wall Test

This construction comprised three(3) 3.0 m high timber bulkhead facing panels. The facing panels were temporarily supported at the wall base, and at 1.0 m, 1.75 m and 2.50 m above the base of the wall by the same system of horizontal timber walings described earlier. The walings were released simultaneously only after the sand backfill had been placed and compacted to the full panel height of 3 m. The

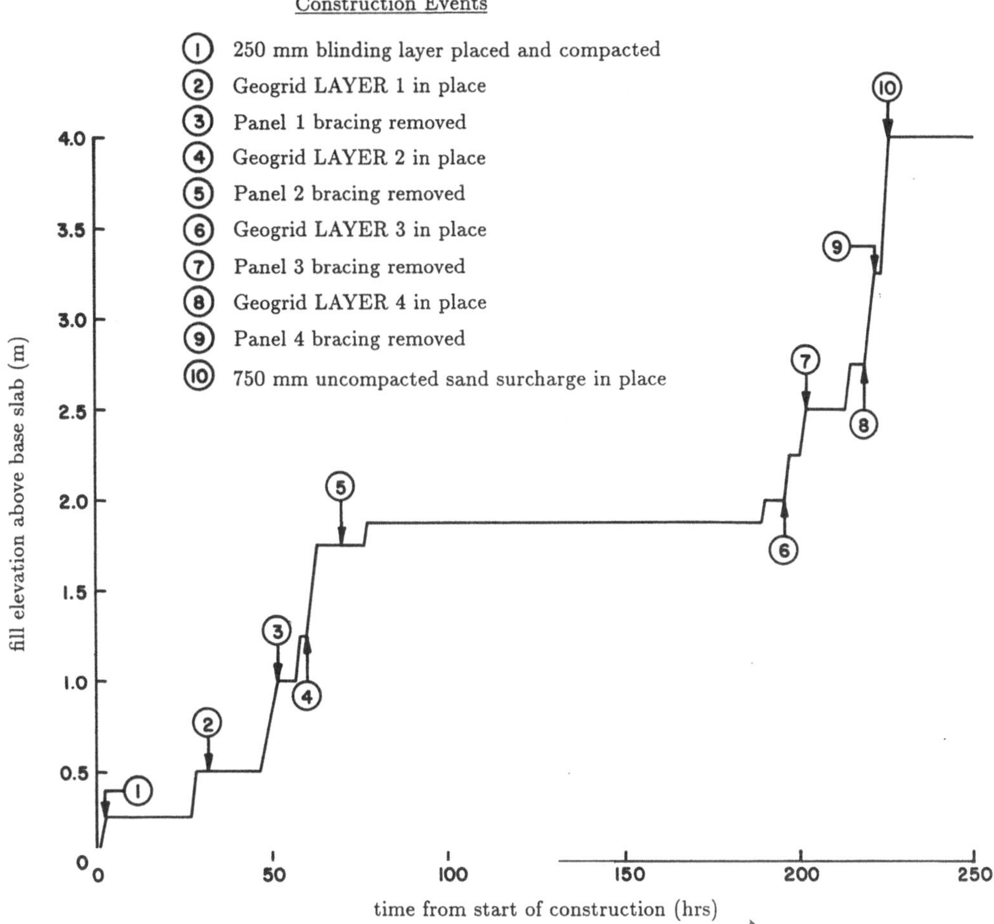

Figure 6 Incremental Panel Wall Construction Sequence

construction sequence for this wall including critical construction events is given on Figure 7. In a similar manner to the incremental panel wall, the propped wall was surcharged using the airbag system installed at the top of the RMC test facility. The surcharging schedule is given in Table 3. In this test the maximum surcharge pressure of 50 kPa was sustained for 1000 hrs.

4.4 Sand Fill Compaction Results

Figures 8a and 8b show sand fill density and moisture content profiles as measured during construction and at wall excavation using a Troxler nuclear density meter. The average bulk density for the two walls was typically 1.79 Mg/m^3 (dry density 1.74 Mg/m^3) at the end of construction. The sand backfill was placed at a relative dry density of about 40% and a moisture content of about 3%. The figures also show that some densification of the soil occurred as a result of soil self-weight and surcharging. As a result, the relative dry density is about 50% at the end of the tests.

Plate 2 Front View of Incremental Panel Wall Test

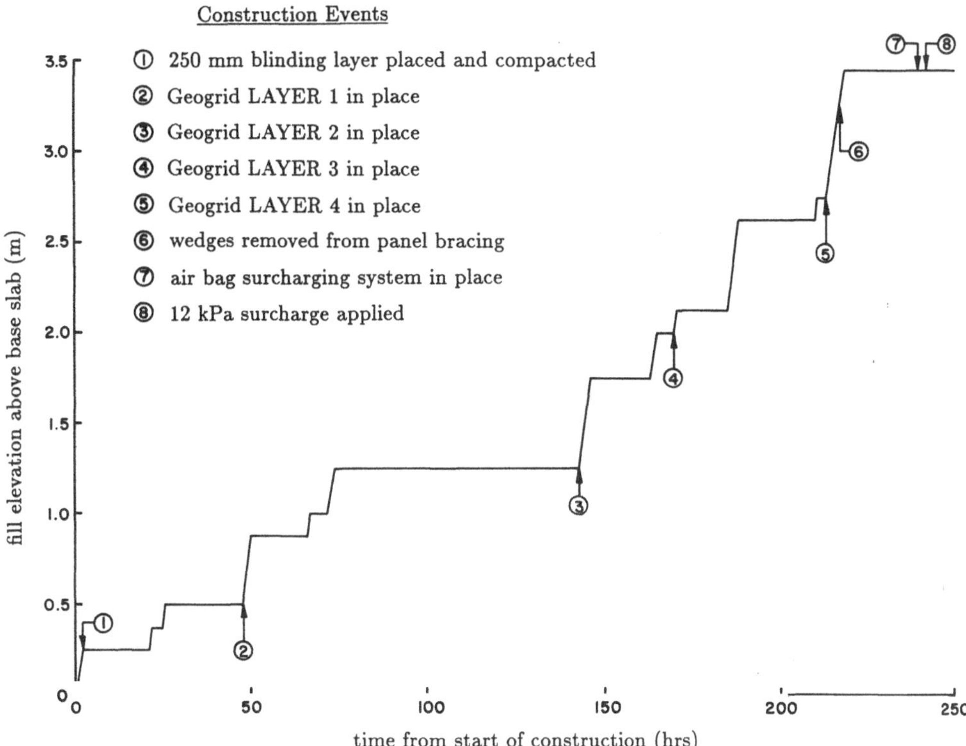

Figure 7 Propped Panel Wall Construction Sequence

Table 3

Surcharging Schedule for Propped Panel Wall

Event	Elapsed time (hrs)	Duration (hrs)
wall construction up to removal of bracing	0	217
air bag installation	217	25
12 kPa surcharge applied	242	430
30 kPa surcharge applied	672	100
40 kPa surcharge applied	772	110
50 kPa surcharge applied	882	1000
30 kPa surcharge applied	1882	16.5
12 kPa surcharge applied	1899	46
3.3 kPa surcharge applied	1945	43
excavation commences	1988	52
excavation complete	2240	-

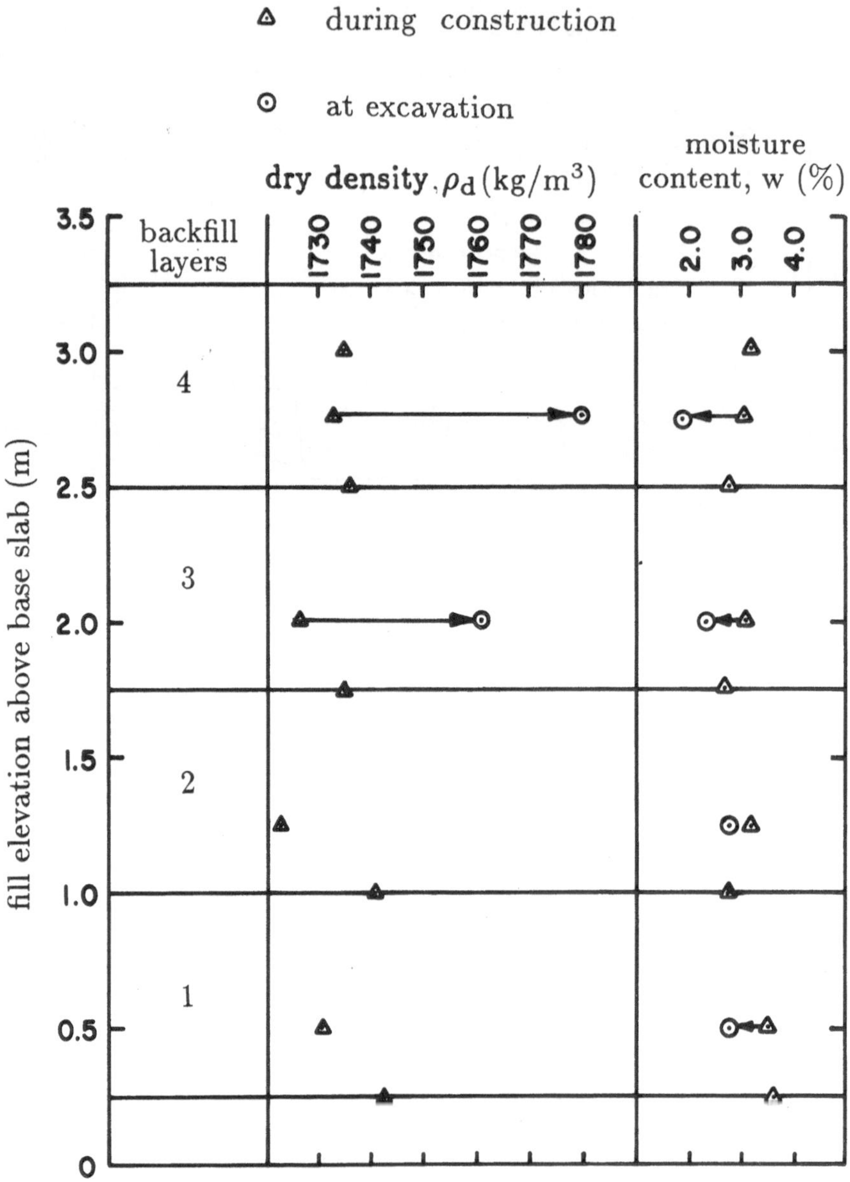

a) Incremental Facing Wall Construction

Figure 8 Sand Fill Density and Moisture Content Profiles

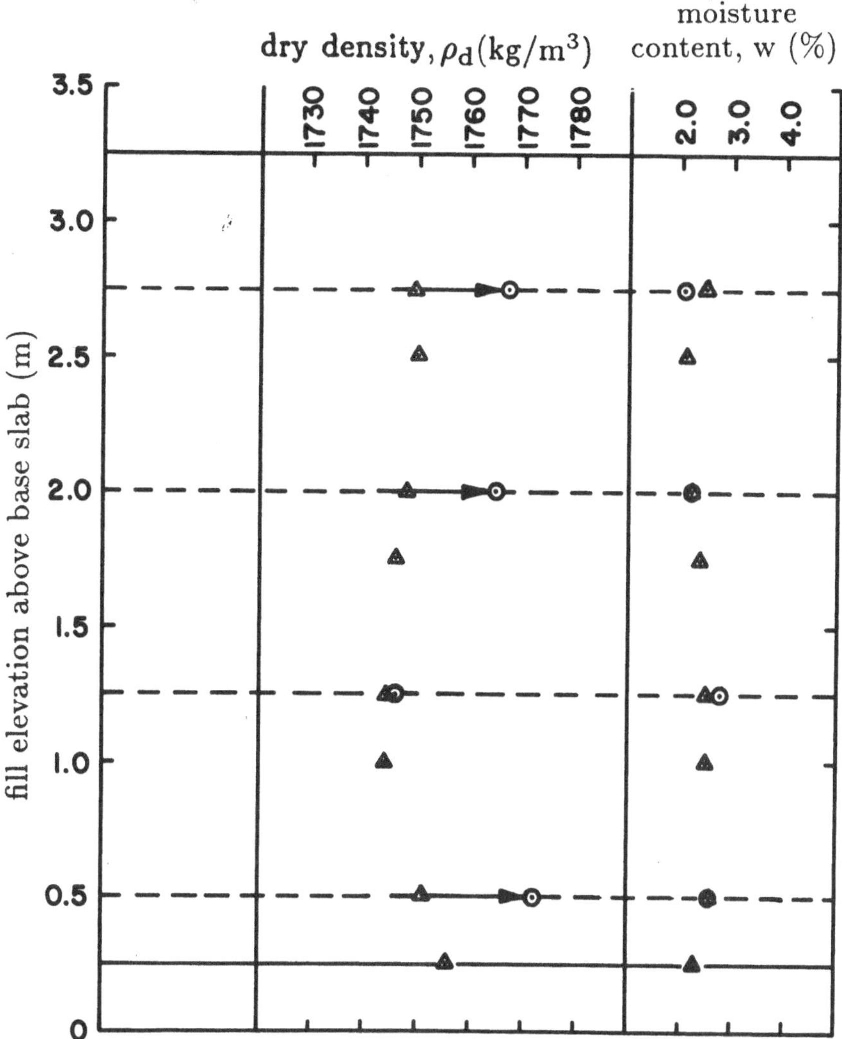

b) Propped Facing Wall Construction

Figure 8 (cont'd) Sand Fill Density and Moisture Content Profiles

5.0 Instrumentation

5.1 General

Each wall was instrumented to record the following:

a) horizontal movement of facing panels,

b) longitudinal displacements and strain in the geogrid reinforcement,

c) load at the panel/geogrid connections,

d) distribution of vertical earth pressure below the reinforced block of soil,

e) vertical settlements at the top of surcharge and,

f) temperature in fill.

The instrumentation used to record the data identified above was installed as construction proceeded and was monitored for the duration of testing including wall excavation.

5.2 Horizontal Movement of Facing Panels

In each test, horizontal movements of the central facing panels were monitored by an array of electrical displacement measuring devices (potentiometer type).

5.3 Longitudinal Displacements and Strains in the Geogrid

5.3.1 General

The displacement of the reinforcement in the longitudinal direction of the model walls was monitored by extensometers (i.e. by attaching tensioned steel wires to selected locations on the mesh). Movement of the thin steel wires was recorded by electronic displacement transducers mounted at the back of the test facility. The wires were protected from the granular fill by passing them through stiff plastic tubes. In the incremental wall construction, two displacement monitoring points per reinforcement layer were used. In the propped wall test up to six displacement monitoring points were used for each strip.

In addition, strain gauges were attached at selected mid-rib locations along the length of each central reinforcement strip. After several years of experimentation with different techniques of geogrid strain-gauging, the authors have found that high-strain, foil-type gauges manufactured by Showa Measuring Instruments Co., Ltd. perform satisfactorily. These strain gauges (together with proper rib surface preparation and a two-part RTC epoxy adhesive) are effective at measuring small levels of strain in the polymer reinforcement while the extensometers are effective at determining large strains after the strain gauges have failed.

5.3.2 Strain Gauge Calibration

Experience with a variety of TENSAR geogrid reinforcement products has shown that, in general, strain distribution is not uniform along any reinforcement rib. This non-uniformity is a consequence of grid geometry and variable material moduli. In order to make comparable strain measurements between strain gauge readings and strains deduced from displacements recorded over several grid apertures it was necessary to correlate strain gauge readings against gross average strains in the grid. In this paper, strains measured over a gauge length consisting of one or more apertures are called *grid strains* and strains measured with a gauge mounted at the mid-point on a rib are called *rib strains*. It can also be appreciated that grid strain is a more meaningful parameter since it facilitates comparison of mechanical response between different sheet reinforcement types and is more easily implemented in analytical models.

Recent investigations at RMC show that the relationship between grid and rib strain can vary with geogrid type (e.g. Jarrett and Bathurst, 1987). In the current investigation the correlation between strain gauge readings giving rib strain and grid strain was determined from load-controlled in-isolation tests carried out on sections of SR2 geogrid. A series of increasing static loads was applied rapidly to test specimens and each load increment was sustained for 24 hrs while strain gauges and grid strains were recorded electronically. The results of this test program are summarized on Figure 9. For SR2 material in the longitudinal direction, the data shows that *grid* strains and *rib* strains are essentially equivalent up to about 1.8% grid strain. The strain gauges in this test were arranged over two rows of ribs. Nevertheless, there was some deviation about the average strain gauge reading at any given grid strain. This variability is thought to be due to the sensitivity of gauge response to small differences in gauge positioning, rib dimensions and non-uniform load distribution between ribs.

5.4 Load at Panel/Geogrid Connections

The facing panel/geogrid connections for the incremental panel wall were constructed by clamping the geogrid between a horizontal steel-reinforced wood batten and the back of the facing panels. Each batten was secured to the timber plywood facing panels by four aluminium bolts passing through the panel.

For the propped wall construction the connection detail was modified so that the geogrid reinforcement and panel bolts were at the same elevation. This configuration proved to be a more secure method of attaching the geogrid to the facing panels.

Each of the four aluminium bolts in the panel/geogrid connections was used as a load cell by attaching four strain gauges to a reduced cross-section of the bolt. The strain gauges were monitored to determine (horizontal) panel/reinforcement connection loads at each reinforcement layer.

5.5 Distribution of Vertical Earth Pressures

In both tests a series of earth pressures cells was located at the bottom of the test facility to record the magnitude and distribution of vertical overburden pressures below the reinforced block of soil. The location of these devices is shown on Figures 3 and 4.

For the incremental wall construction, a series of 130 mm diameter aluminium diaphragm-type pressure cells with an aspect ratio of 0.1 were manufactured in-house. These cells proved to be unsatisfactory. For the propped wall, larger (230 mm diameter) custom-built GEOKON earth pressure cells were used. These devices are 13 mm thick (aspect ratio 0.06) and are of the cavity diaphragm-type with the cavity filled with an incompressible liquid. Fluid pressures in equilibrium with vertical earth pressures are measured by pressure transducers. These devices appeared to give a more accurate measure of earth pressures.

5.6 Additional Instrumentation and Data Acquisition

Additional instrumentation included three displacement transducers which were installed to measure vertical deformations at the top of the reinforced soil. Lastly, temperature gauges were buried in the soil. This instrumentation was included to ensure that measured ambient soil temperatures were close to 20°C corresponding to the standard laboratory temperature for in-isolation tests carried out at the Netlon laboratories and at RMC on SR2 reinforcement samples.

The instrumentation described in the previous sections was monitored using a Hewlett Packard 3497A data acquisition system with an HP Vectra AT micro-computer as the controller. The system allowed us to efficiently monitor up to 300 electrical devices and provided fully-integrated data processing.

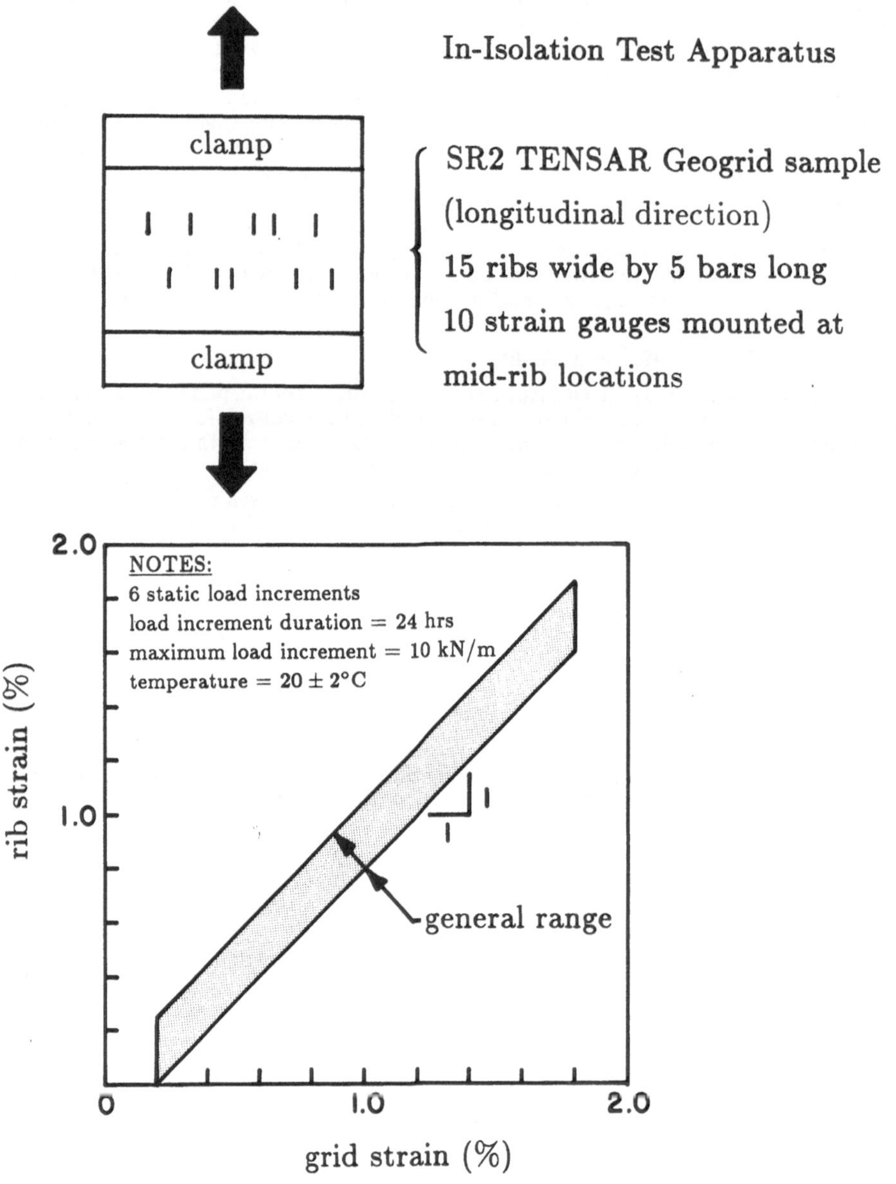

Figure 9 Calibration of Strain Gauges to Tensar SR2 Grid Strain using Static Load Increment In-isolation Tensile Test

6.0 Test Results

6.1 General

The following sections summarize the results of measurements taken over the course of each test including wall construction, surcharging, and wall excavation.

6.2 Incremental Panel Facing Wall

6.2.1 Horizontal Panel Movements

Figure 10 shows panel movements recorded by the four displacement potentiometers attached to the second panel of the incremental wall. The results of similar movements recorded for all panels have been combined to produce Figure 11.

Figure 11 shows that the position of the top of the wall at the the end of the surcharging program was about 40 mm from the initial position of the bottom panel. However, about 14 mm of this movement was accumulated prior to placement of the top panel. From Figure 10 it can be seen that the largest outward movements were recorded as the surcharging loads were applied. However, at constant surcharge loads, time-dependent outward panel movements were also apparent. For example, about 2 mm of outward movement in the top panel can be attributed to creep in the reinforced wall model between 100 hrs and 500 hrs of the 50 kPa surcharging increment (Figure 11). Figure 11 also illustrates that progressive rotation of each facing unit about its base occurred over the course of testing and that net rotation of the full wall height about the toe of the structure was evident.

6.2.2 Geogrid Displacements

The incremental wall model was instrumented with two extensometers attached to each reinforcement layer. The horizontal displacements recorded by these devices at selected times during and after wall construction are given on Figure 12. Also shown on the figure are outward panel movements recorded at the end of the 50 kPa surcharging increment. The figure shows that horizontal outward translation measured at the geogrid 160 mm behind the panel facings was less than the magnitude of outward panel movements recorded at the reinforcement elevations. However, the difference can be attributed to strain in the grid, some play in the extensometers and compliance in the panel/geogrid connections and the panels themselves. Nevertheless, qualitative features of the geogrid movements measured immediately behind the panel units are similar to those reported for the panels themselves on Figure 11. In particular, horizontal deformations increase with height of reinforcement in the composite section and with the magnitude of surcharge loading.

The data on Figure 12 indicates that there was significant horizontal translation of the grid at distances up to a monitored distance of 1260 mm behind each panel. Furthermore, the figure shows that, between 160 mm and 1260 mm behind each panel, grid displacements were progressively attenuated indicating that tensioning of the grid took place with increasing surcharge level.

6.2.3 Geogrid Strains

Selected strain gauge readings with time for the incremental panel wall test are given on Figure 13. Some variability in strain gauge readings is observed for gauges located at nominally equivalent distances behind the facing panels on the same reinforcement strip. This variability is likely due to non-uniform load distribution across the reinforcement width, local variability in sand/geogrid interlock and small deviations in details of strain gauge application. Nevertheless, all gauges on a given row showed similar trends and grid strain distribution in the longitudinal direction of the reinforcement is considered to be well-represented by the *average* strain gauge reading at each row.

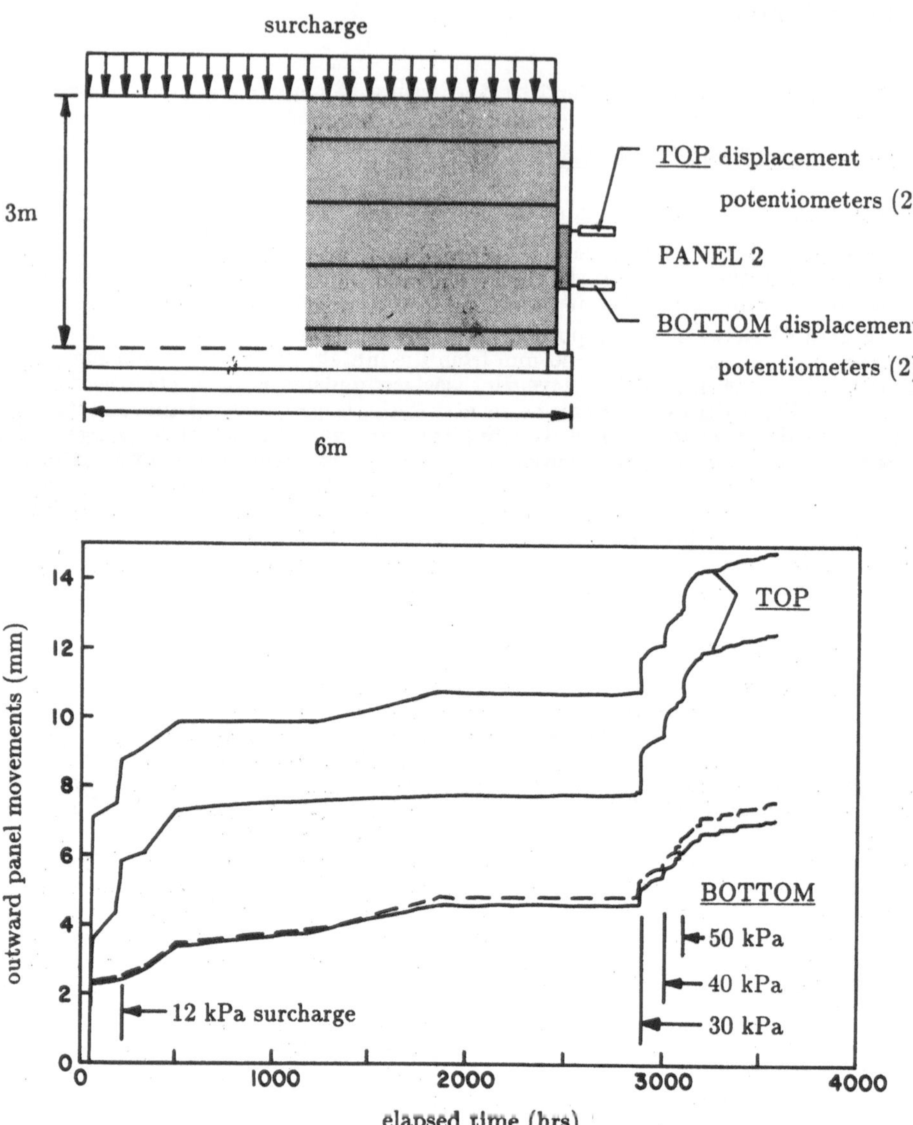

Figure 10 Panel 2 Movement versus Elapsed Time (Incremental Panel Wall Test)

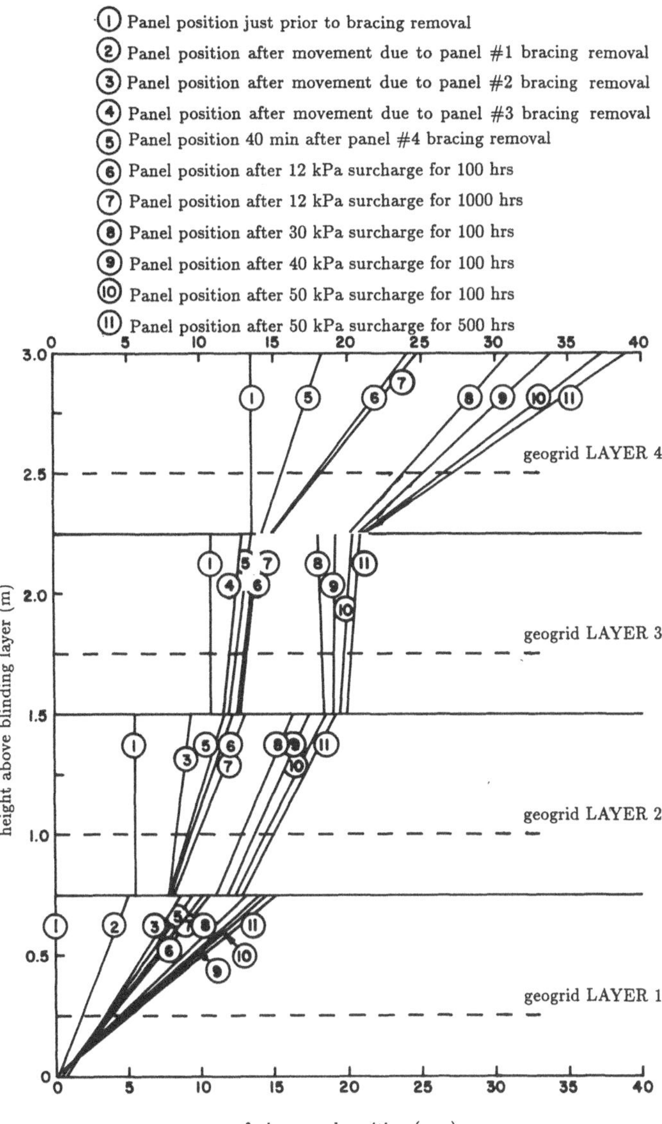

(1) Panel position just prior to bracing removal
(2) Panel position after movement due to panel #1 bracing removal
(3) Panel position after movement due to panel #2 bracing removal
(4) Panel position after movement due to panel #3 bracing removal
(5) Panel position 40 min after panel #4 bracing removal
(6) Panel position after 12 kPa surcharge for 100 hrs
(7) Panel position after 12 kPa surcharge for 1000 hrs
(8) Panel position after 30 kPa surcharge for 100 hrs
(9) Panel position after 40 kPa surcharge for 100 hrs
(10) Panel position after 50 kPa surcharge for 100 hrs
(11) Panel position after 50 kPa surcharge for 500 hrs
geogrid LAYER 4
geogrid LAYER 3
geogrid LAYER 2
geogrid LAYER 1
height above blinding layer (m)
facing panel position (mm)

Figure 11 Summary of Panel Positions (Incremental Panel Wall Test)

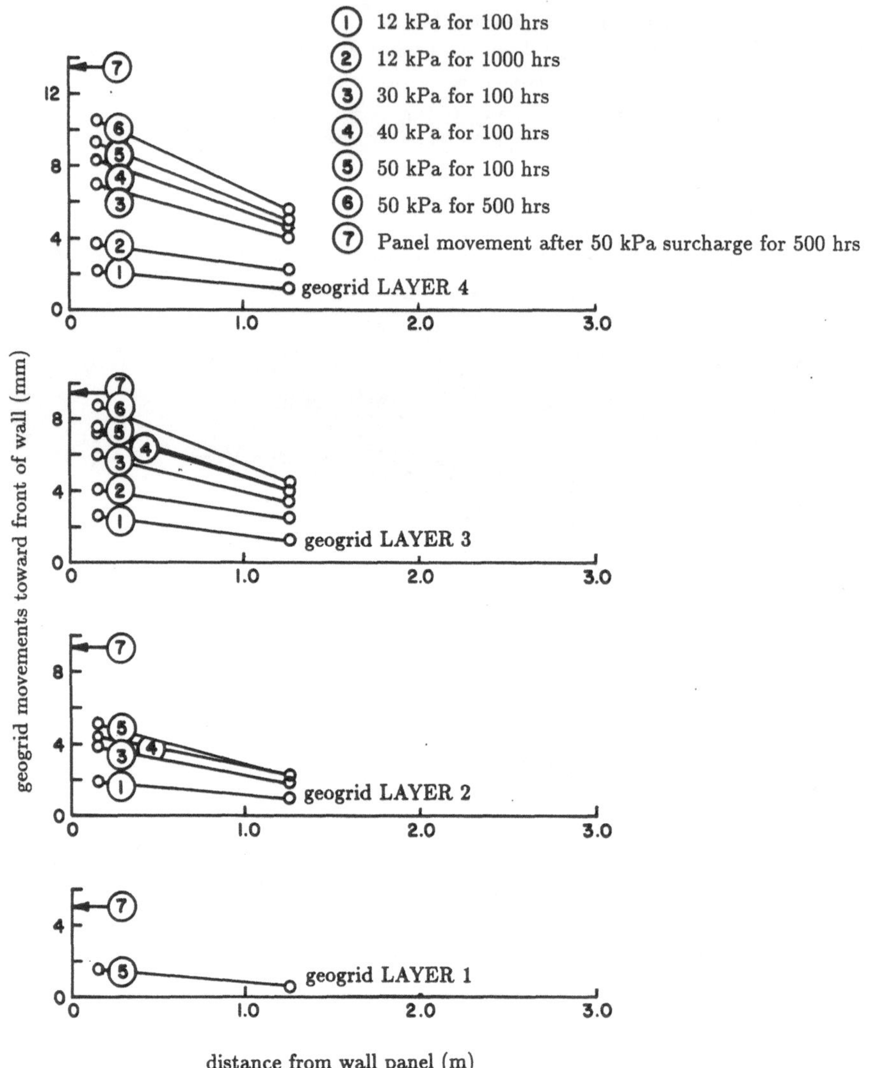

Figure 12 Summary of Geogrid Movements (Incremental Panel Wall Test)

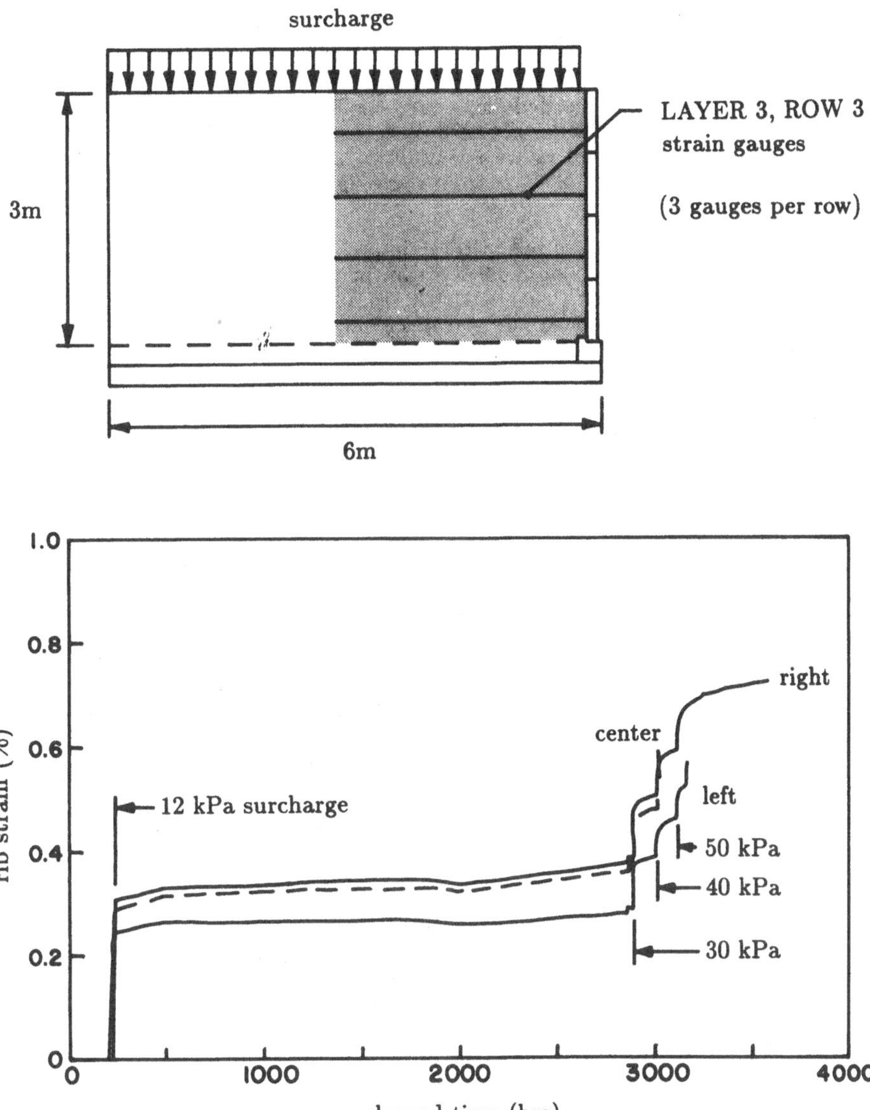

Figure 13 Typical Strain Gauge Response versus Elapsed Time (Incremental Panel Wall Test)

Figure 13 also shows that sudden changes in strain gauge response match critical events in wall construction and the application of increased surcharge loads. It is also apparent from the figure that the polymer reinforcement continued to deform under constant surcharge loading.

The *average* rib strains recorded along the length of each reinforcement layer are summarized on Figure 14. The results of in-isolation tests described earlier indicate that *rib* strains shown on the figures are essentially equivalent to *grid* strains. A number of important observations can be made from these figures:

a) Significant strains were recorded in the reinforcement during construction of the wall. The progressive development of strain in the reinforcement due to the incremental wall construction can be noted from the data. The maximum strain in the reinforcement layers at the end of construction decreases with increasing reinforcement elevation. In layer 1, more than 50% of the total strain recorded at the front of the reinforcement occurred *during* construction.

b) Strain in the reinforcement is observed to increase with surcharge load level and elapsed time under constant surcharge load.

c) In reinforcement layers 3 and 4 there is a marked peak in the longitudinal distribution of strain at about 450 mm behind the facing panels. A similar trend could not be established for layer 2 due to premature gauge failure.

Figure 15 shows a comparison of grid strains calculated from the extensometers with grid strains inferred from rib-mounted strain gauges. The data shows that the values from the displacement devices were never less than 70% of the average strain gauge values taken over the same gauge length. The lower values can be attributed to compliances within the extensometers. The relative values gave us confidence that the reinforcement strains inferred from the strain gauges mounted directly on the reinforcement were reasonable and that strains calculated from the extensometers were *lower-bound* estimates of grid strains.

6.2.4 Load in Panel/Geogrid Connections

Strain gauges mounted on the bolts anchoring the panel/geogrid connections at the back of each facing panel were used to calculate horizontal load in these connections at the end of the test (i.e. 50 kPa surcharge for 500 hrs). The results of these measurements are plotted on Figure 16. Also shown on the figure are estimated connection loads based on strains recorded in the first row of strain gauges. To calculate the loads in the grid, *isochronous* load-strain-time data supplied for the actual reinforcement material was used (see Appendix B). The method used to equate grid strain readings to load has been reported by McGown et al. (1984) and Yeo (1985). Geogrid forces calculated in this manner are considered to give *upper-bound* values on the connection loads. Calculations using both methods to estimate connection loads give a range of values between 0.5 and 4 kN/m at the end of the test with a 50 kPa surcharge. If the maximum connection loads from the range of values shown on the figure are considered then, the data indicates that the largest connection loads occur in the top and bottom panels.

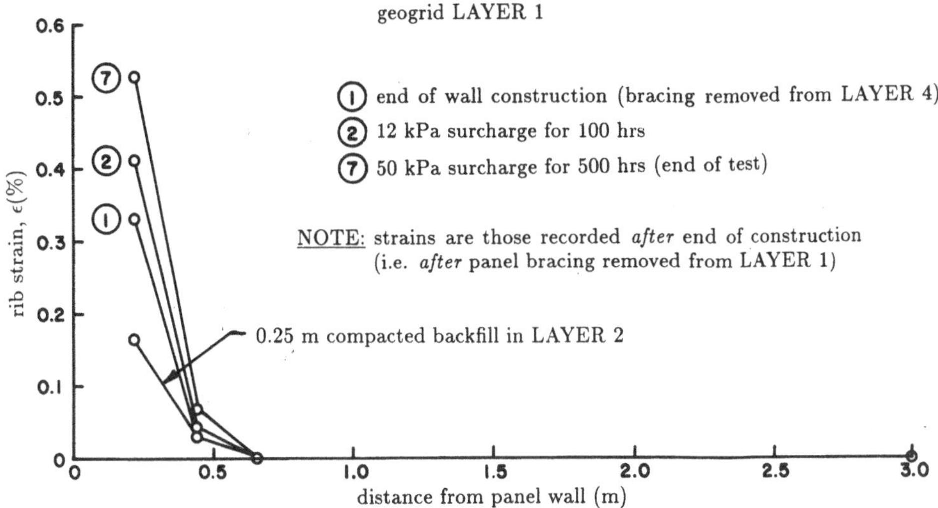

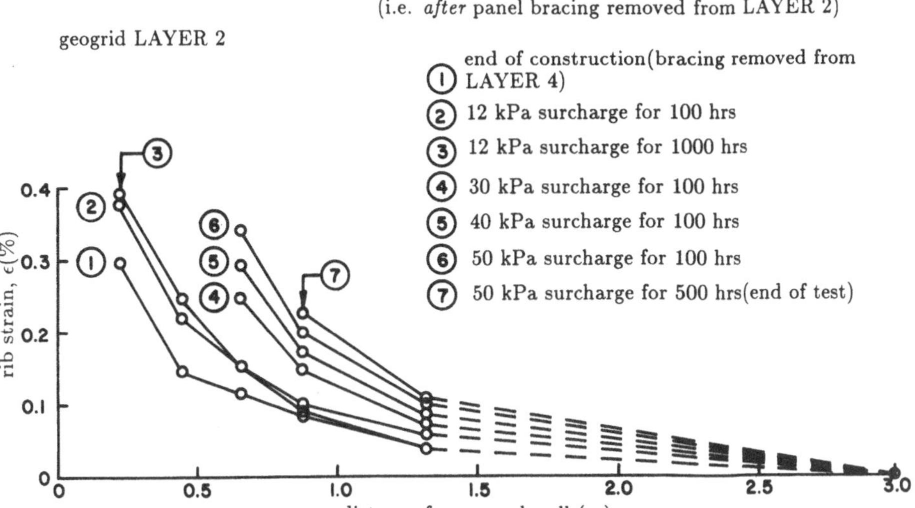

Figure 14 **Summary of Strain Gauge Readings (Incremental Panel Wall Test)**

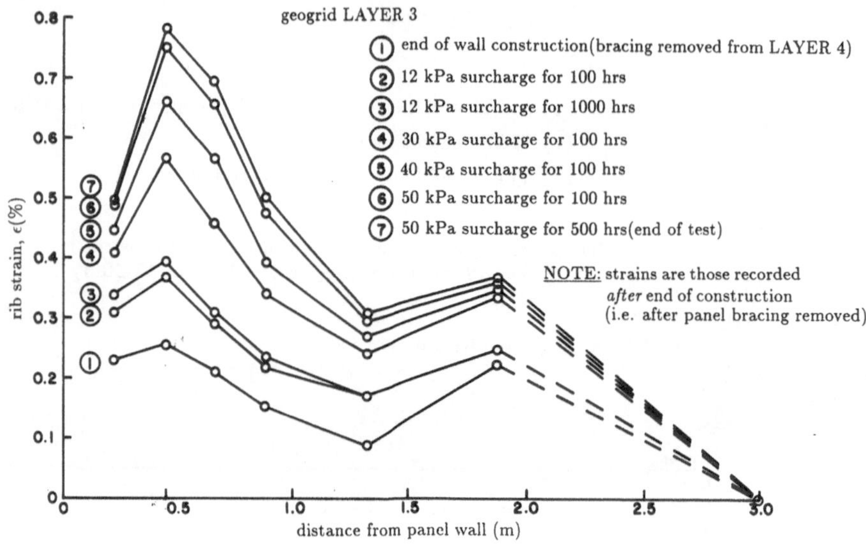

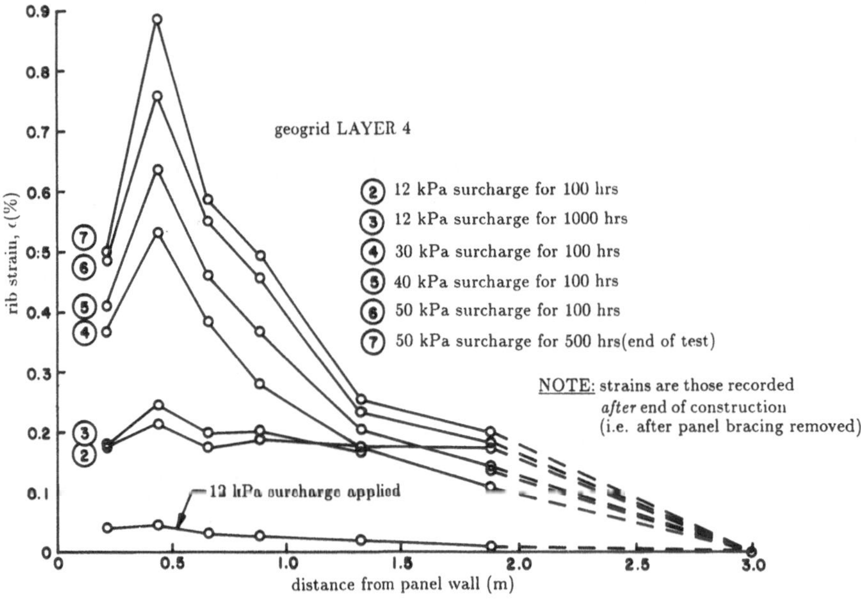

Figure 14 (Cont'd) Summary of Strain Gauge Readings

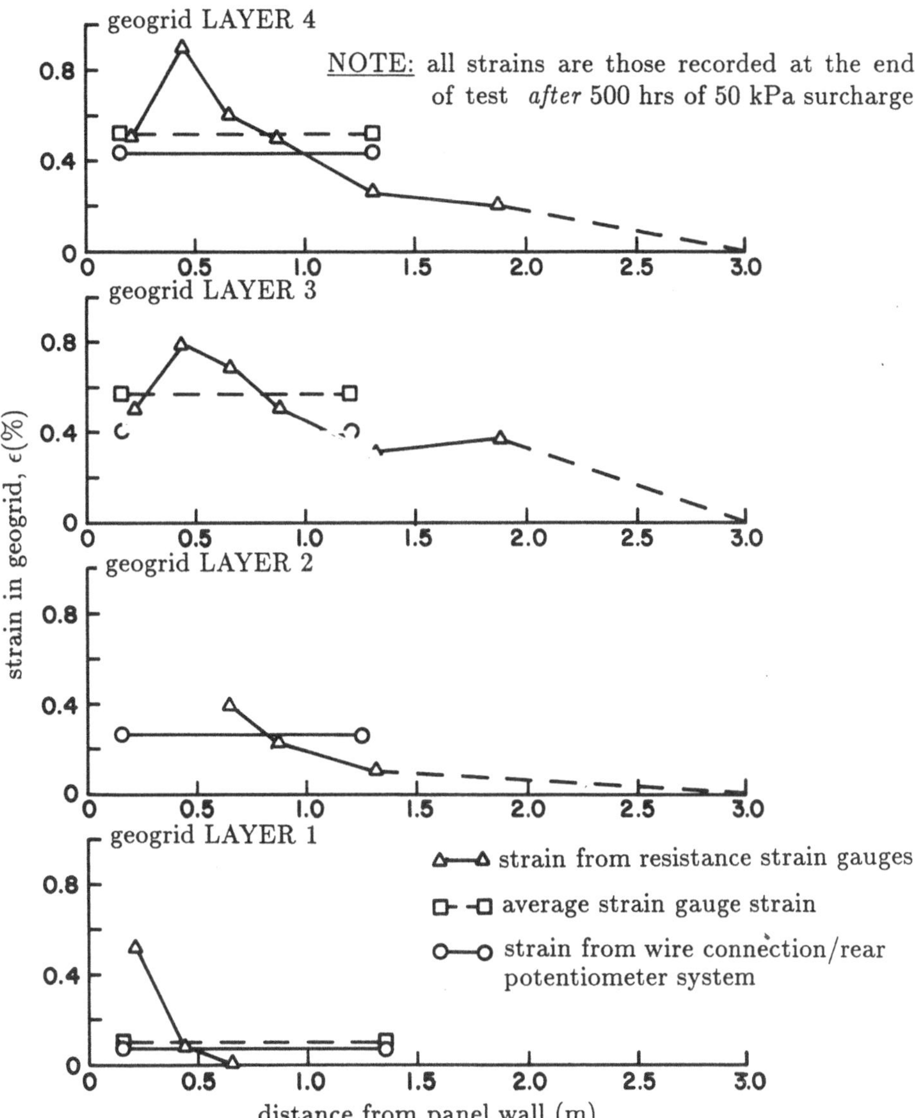

Figure 15 Comparison of Grid Strains from Strain Gauge Readings and Steel Wire Displacement Devices (Incremental Panel Wall Test)

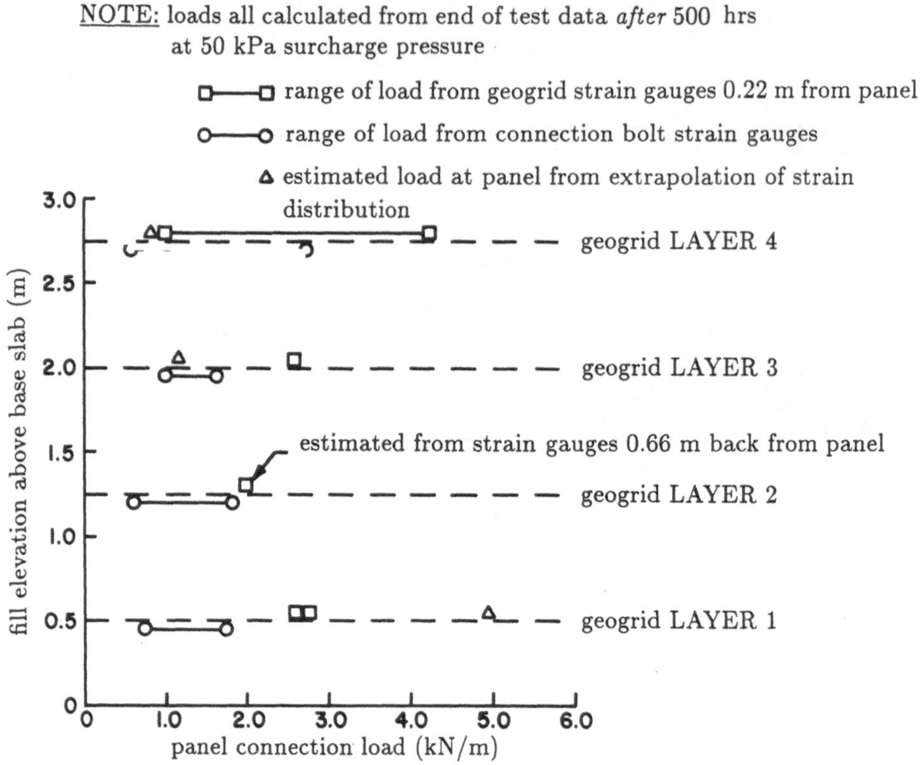

NOTE: loads all calculated from end of test data *after* 500 hrs at 50 kPa surcharge pressure

range of load from geogrid strain gauges 0.22 m from panel

range of load from connection bolt strain gauges

estimated load at panel from extrapolation of strain distribution

geogrid LAYER 4

geogrid LAYER 3

estimated from strain gauges 0.66 m back from panel

geogrid LAYER 2

geogrid LAYER 1

fill elevation above base slab (m)

panel connection load (kN/m)

Figure 16 Geogrid/Panel Connection Loads from Incremental Panel Wall Test

6.3 Propped Panel Facing Wall

6.3.1 Horizontal Panel Movements

Outward panel movements with time are plotted on Figure 17. These movements are referenced to the end of construction corresponding to the panel geometry just prior to bracing removal. Similar to the results of the incremental wall, the propped panel wall showed large outward movements at the application of each surcharge load increment and significant deformation under constant surcharge loading.

Figure 18 shows outward panel movements recorded at selected times during surcharging of the propped panel wall. The data shows a progressive rotation of the facing unit about the toe with a total outward movement at the top of the wall of about 13 mm at the end of the test. As in the incremental panel wall test, the data shows significant wall movements under constant surcharging load increments.

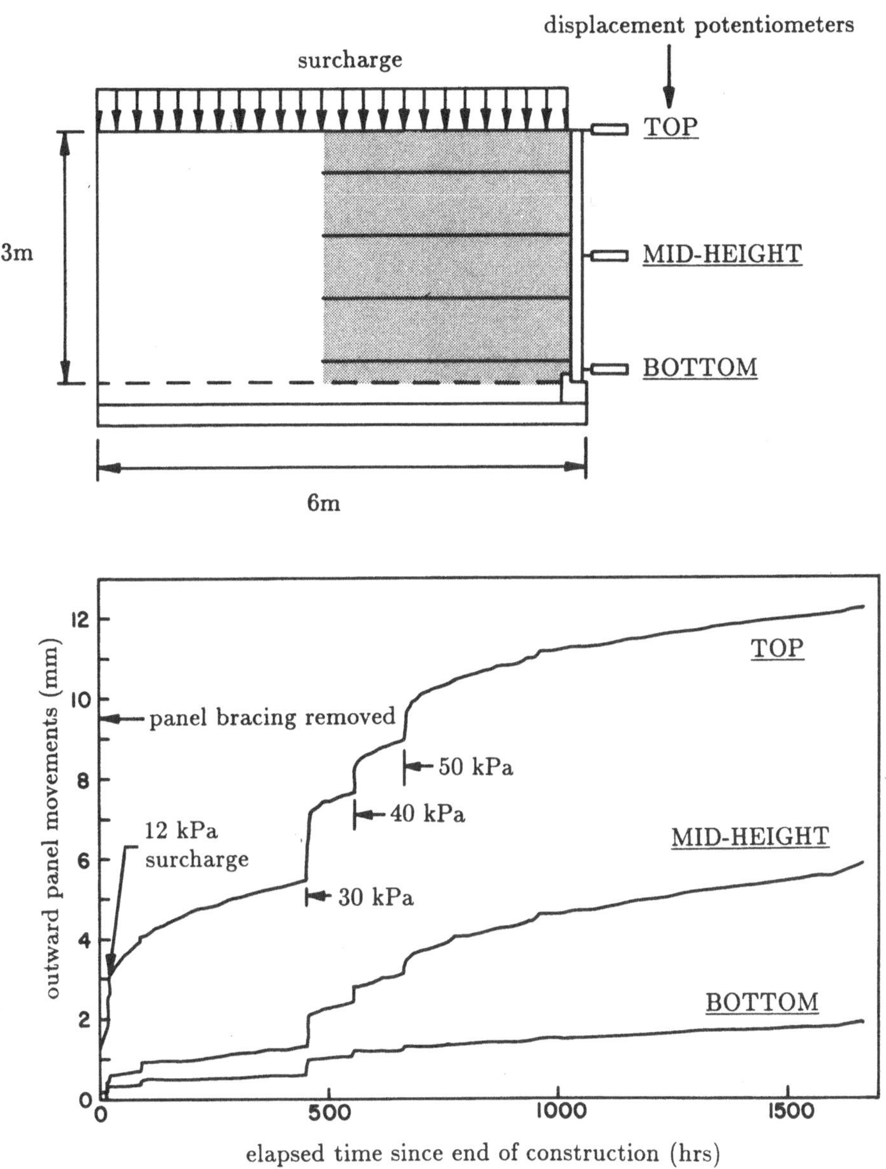

Figure 17 Panel Movement versus Elapsed Time (Propped Panel Wall Test)

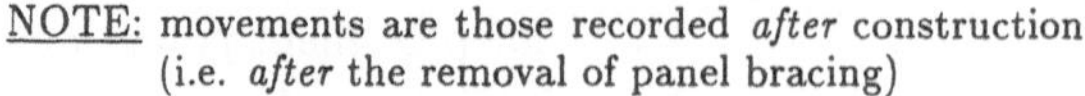

NOTE: movements are those recorded *after* construction
(i.e. *after* the removal of panel bracing)

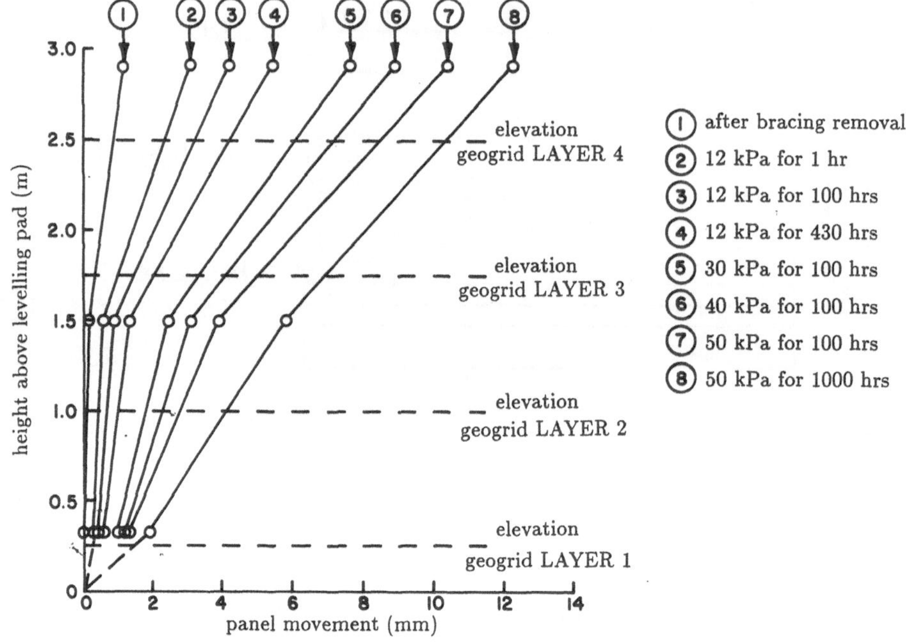

Figure 18 Summary of Panel Movements (Propped Panel Wall Test)

6.3.2 Geogrid Displacements

The propped panel wall was instrumented with up to six extensometers attached to each reinforcement layer. The horizontal displacements recorded by these devices are given on Figure 19. Also shown on the figure are outward panel movements recorded at the end of the 50 kPa surcharging increment.

Figure 19 shows that there was essentially uniform displacement of the grid layers after the initial 100 hrs of the 12 kPa surcharging increment. For example, in the top layer this movement was about 2 mm towards the wall which is somewhat less than the 3.3 mm outward movement recorded at the panel facing opposite this reinforcement layer (Figure 18). This observation suggests that there was some initial slip of the grid as soil/grid interlock was established during the initial loading increment. At higher surcharge loads the increased slope of the displacement profiles shows that tensile straining in the reinforcement layers and load transfer between the grid and soil was occurring.

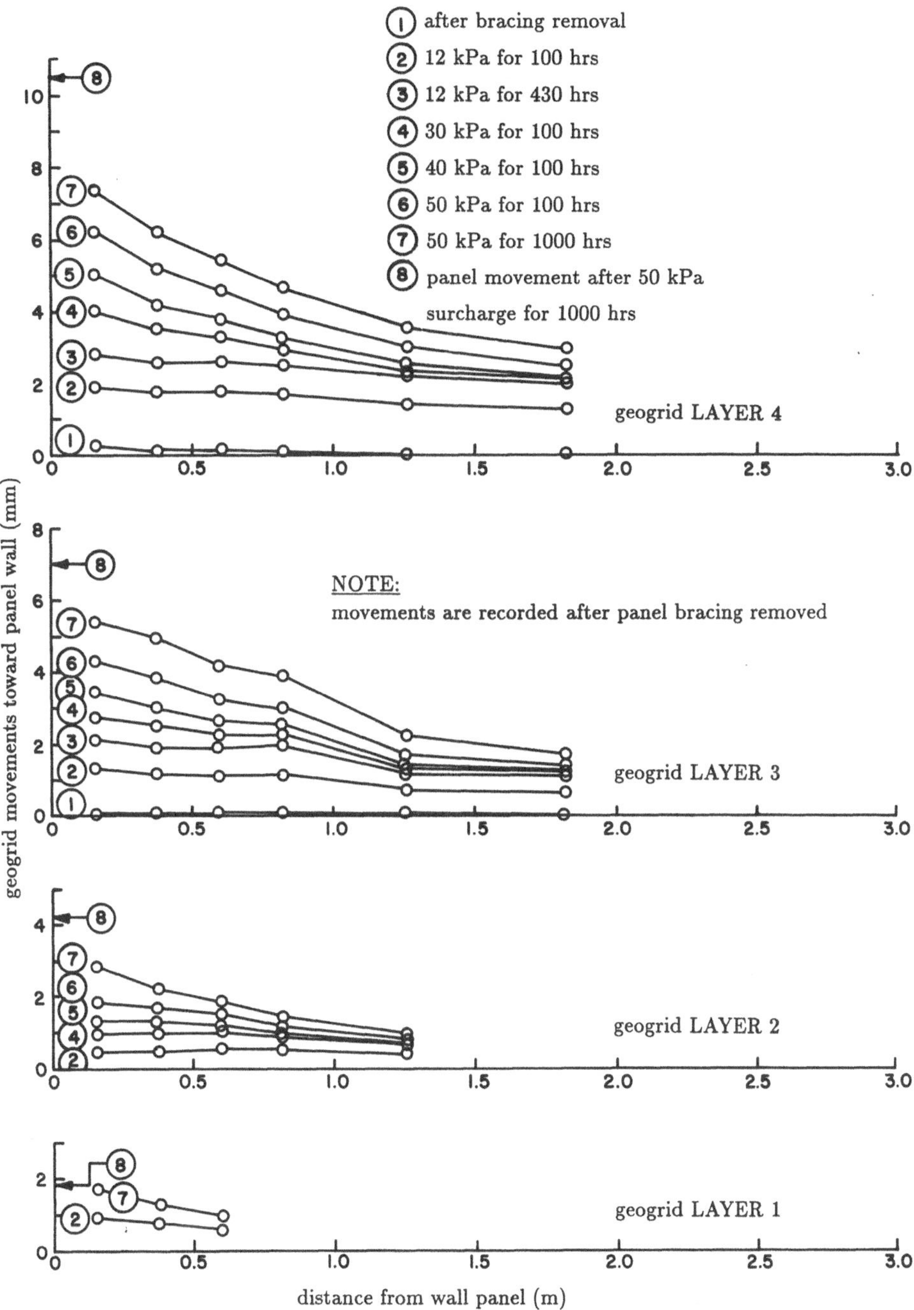

Figure 19 Summary of Geogrid Movements (Propped Panel Wall Test)

6.3.3 Geogrid Strains

Selected strain gauge readings with time for the propped wall test are given on Figure 20. Similar to observations made for the incremental wall test, the strain gauges showed increased tension in the grid as surcharge load increments were applied and they also recorded time-dependent strain under constant surcharge loadings.

The distribution of rib strains at selected times in the propped wall test are plotted on Figure 21. A number of important observations can be made from Figures 20 and 21:

a) The magnitude of strain recorded at a given distance behind the facing can be seen to increase with time and surcharge load level.

b) The propagation of strain into the reinforcement can be seen to increase with time and surcharge load level.

c) In geogrid layers 1, 2 and 3 there is a distinct change in strain gradient at about 0.45 to 0.65 m behind the wall.

The strain distributions for the propped wall show qualitative features in the immediate vicinity of the wall facing which are different from those recorded for the incremental wall. In particular, layer 4 shows a progressive increase in geogrid strain as the distance to the panel facing decreases. At the same location in the incremental wall there were reduced strain values. Layers 2 and 3 show qualitative features that fall between the strain profiles recorded for the incremental wall and layer 4 in the propped wall. The difference in strain response is considered to be due to the relative vertical compliance in the two wall constructions. In the incremental wall test a horizontal layer of compressible foam was placed between panel units. After fill compaction and surcharging, it was noted that the foam filler compressed leading to an overall *shortening* of the full wall height by roughly 10 mm. This degree of freedom was not available in the propped panel wall which was constructed as a single facing unit. Consequently, the fill moved down with respect to the panel unit during fill compaction and surcharging and in the process generated additional tensile loading in the reinforcement close to the geogrid/panel connections. The relative soil/panel movement can be expected to decrease with distance below the top of the wall and this accounts for the observation that additional geogrid loading in the vicinity of the panel connections appears to diminish with lower reinforcement elevation.

Grid strains were also calculated from the results of extensometers (i.e. from the gradients plotted on Figure 19). The results of these calculations are given on Figure 22. The figure shows that these calculated strains under-register the grid strains inferred from the rib-mounted strain gauges. Nevertheless, both sets of data show generally consistent trends with the extensometers tending to give lower-bound estimates of grid strain. Again, the discrepancy is thought to be due to mechanical compliances in the steel wire systems.

6.3.4 Load in Panel/Geogrid Connections

A significant number of strain gauges attached to the panel connection bolts were observed to have failed when the propped wall was excavated. The poor performance of the strain gauged bolts did not allow panel/geogrid loads to be calculated with any confidence using this method of instrumentation. Nevertheless, an *upper-bound* estimate of panel loads can be determined from the strains recorded in the grid at strain gauge locations closest to the facing unit. The results of this exercise are given on Figure 23 and show that the load in the panel connections increases with reinforcement elevation.

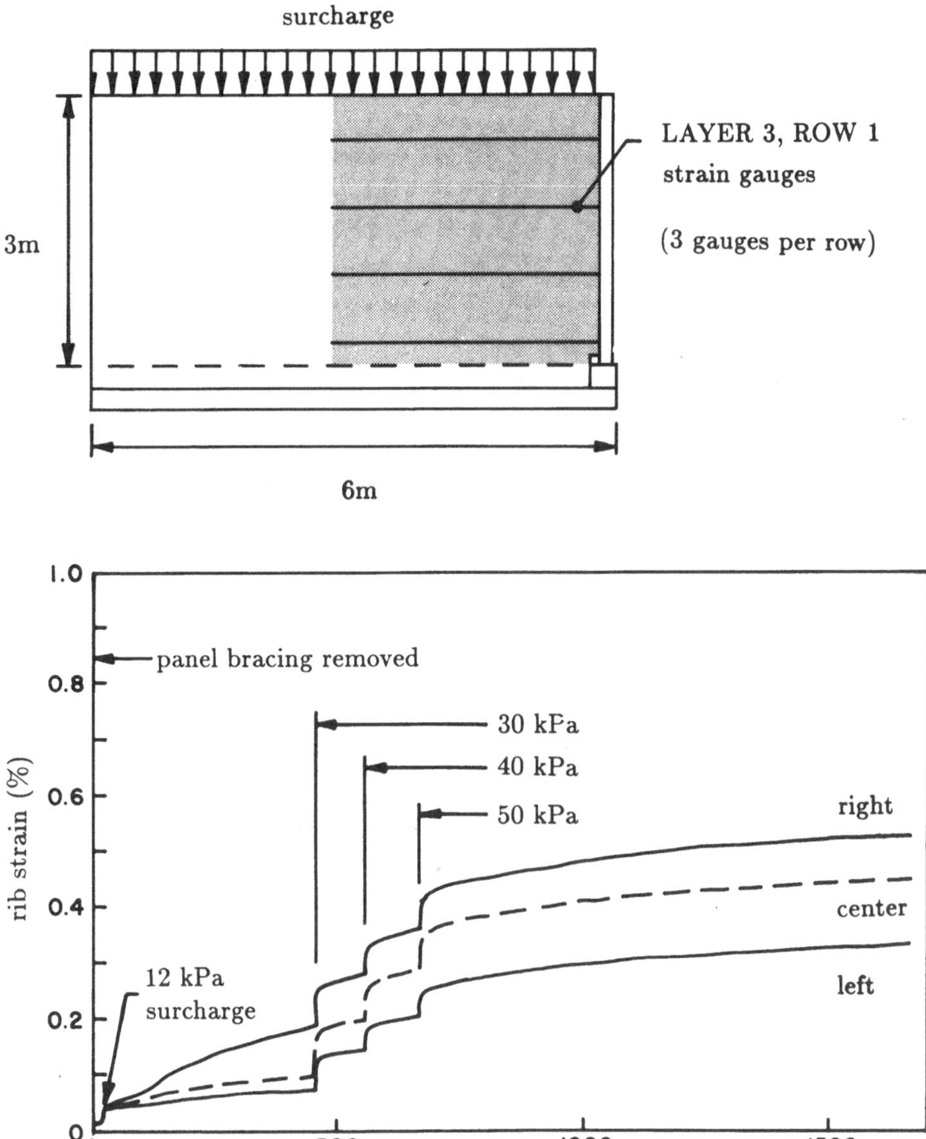

Figure 20 Typical Strain Gauge Response versus Elapsed Time (Propped Panel Wall Test)

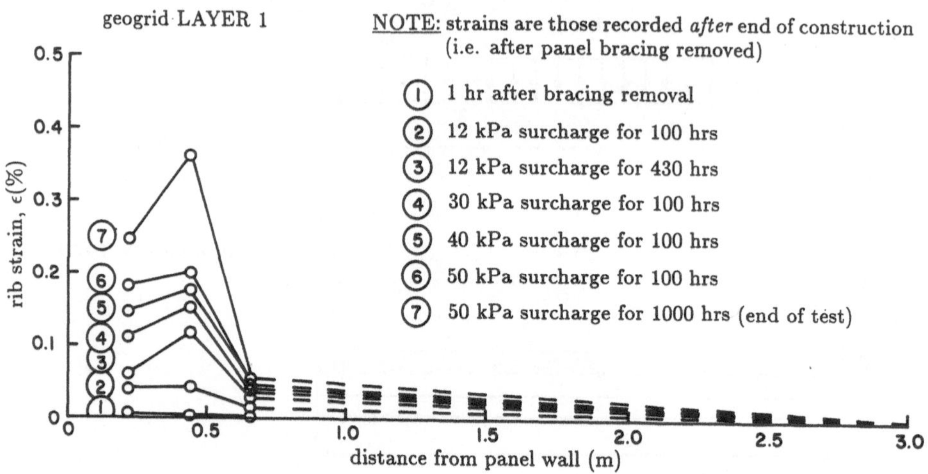

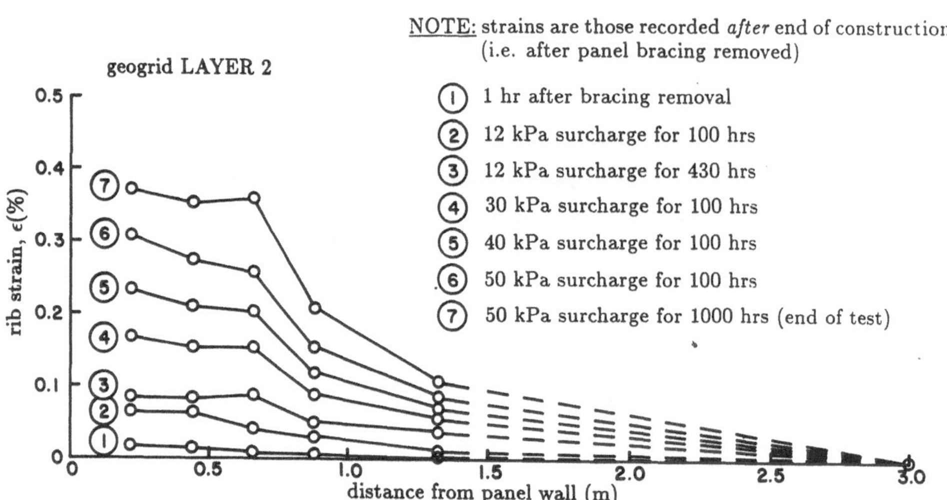

Figure 21 Summary of Strain Gauge Readings (Propped Panel Wall Test)

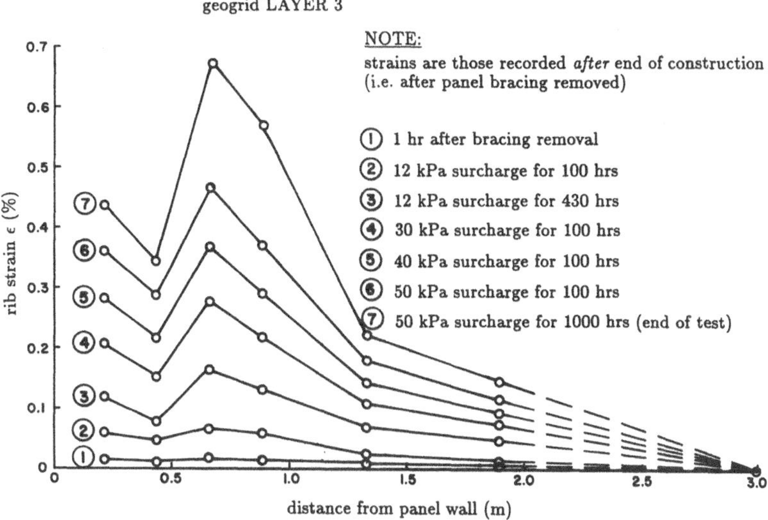

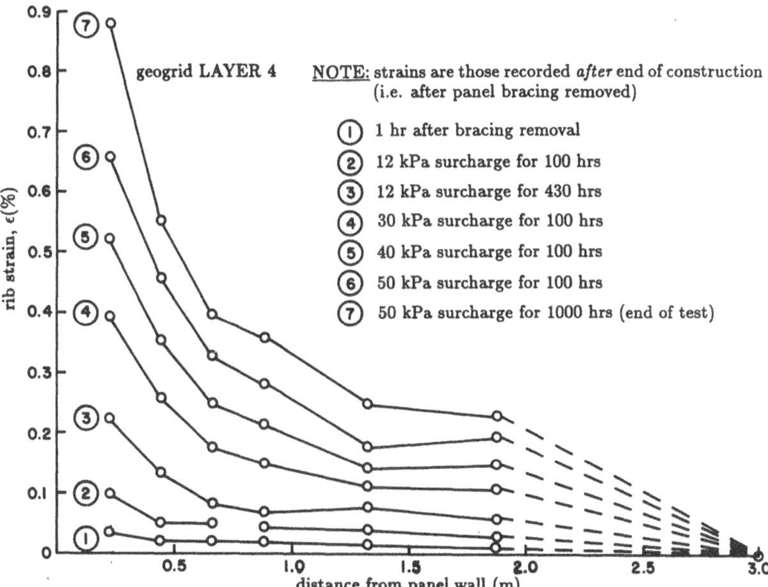

Figure 21 (Cont'd) Summary of Strain Gauge Readings (Propped Panel Wall Test)

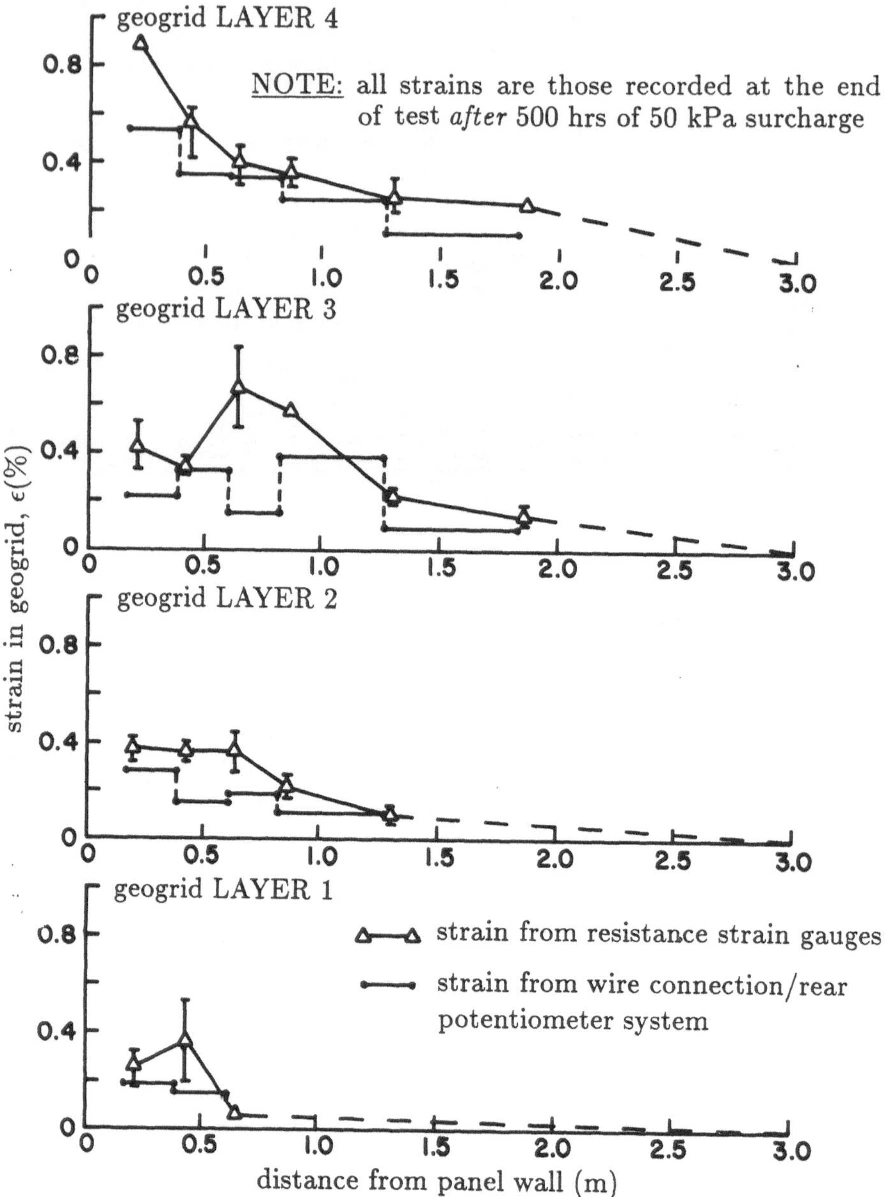

Figure 22 Comparison of Grid Strains from Strain Gauge Readings and Steel Wire Displacement Devices (Propped Panel Wall Test)

NOTE: loads all calculated from end of test data *after* 1000 hrs of 50 kPa surcharge pressure

□—□ range of load from geogrid strain gauges 0.22 m from panel

o estimated load at panel from extrapolation of strain distribution

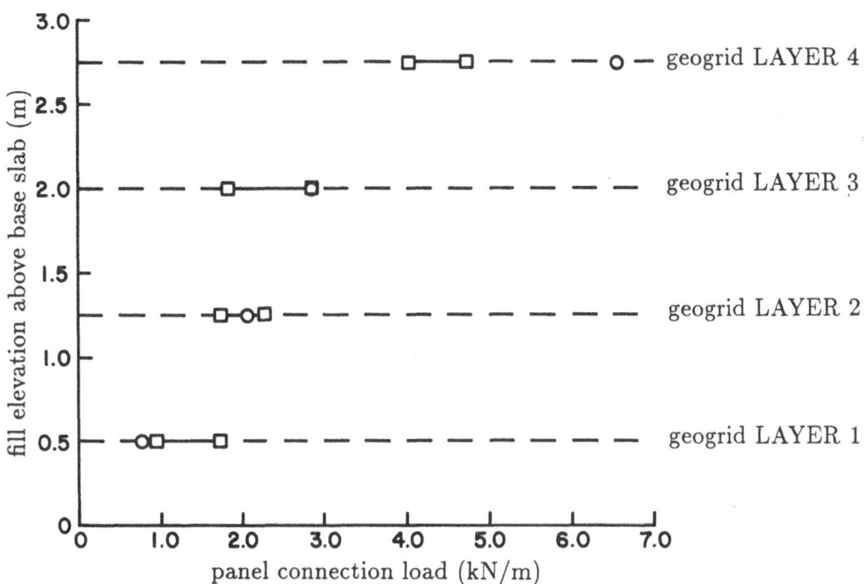

Figure 23 Geogrid/Panel Connection Loads from Propped Panel Wall Test

6.4 Additional Results from Both Tests

The results of earth pressure cell measurements taken at the bottom of the reinforced soil block in the incremental wall test were disappointing due to poor cell sensitivity. The cell response problem was overcome in the subsequent propped wall test by using different earth pressure cells with a larger diameter. The results of earth pressure readings taken during the propped panel wall test at the end of construction and at the end of surcharging are given on Figure 24. The data shown was arrived at by calibrating each cell response *insitu* against the initial 2 m depth of fill placed during wall construction.

In general, all pressure cells at the base of the test facility recorded vertical earth pressures in the reinforced propped wall test that were below the value calculated from soil self-weight and surcharge. Portions of this under-registration may be attributed to pressure cell response, sidewall friction in the test facility and facing/backfill interaction for those cells closer to the front of the test facility. The influence of sidewall friction was discussed at length during the NATO Workshop. As a result of these discussions, a large-scale *unreinforced* propped wall test was undertaken to examine the contribution of sidewall friction and facing friction to stability of soil within the RMC Test Facility. The results of this investigation are reported by Bathurst and Benjamin in these proceedings and show that about 10 to 12% of vertical earth pressure under-registration at the soil base may be due to sidewall friction. Futhermore, the unreinforced test showed that the effective friction angle between the rigid facing panel and backfill with the geogrid/panel connections in place is about 33°.

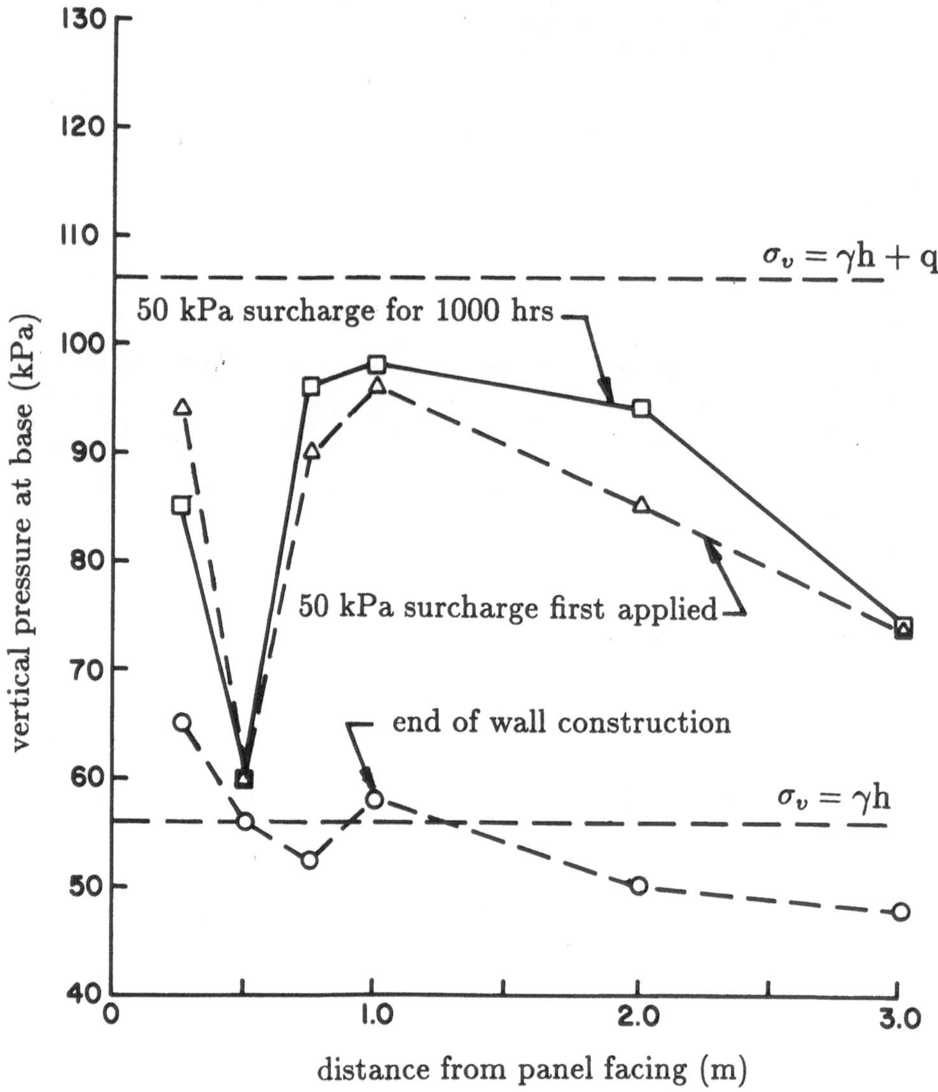

Figure 24 Vertical Pressure Distributions from Propped Panel Wall Test

The results of temperature measurements taken within the reinforced soil mass showed that the fill temperature never varied from $21 \pm 2°C$. As a result, the mechanical properties of the SR2 geogrid material established from laboratory in-isolation tests at 20°C are considered representative of the *insitu* condition.

Finally, it should be noted that an attempt was made to monitor vertical settlement of the reinforced sand backfill using displacement devices mounted at the top of the test facility. However, the installation of these devices proved difficult and no useful data was recovered.

7.0 Discussion of Test Results

7.1 General

The purpose of this paper is to report the results of two large-scale reinforced soil walls constructed at RMC. The trial walls provided a starting point for discussions related to the general theme of the Workshop (i.e. application of polymeric reinforcement in retaining structures). This section discusses the RMC test facility and qualitative features of the model tests which can assist the reader to interpret the test results.

7.2 Wall Stability

Both test walls proved to be stable during construction and during sustained surcharging of up to 50 kPa magnitude. While time-dependent panel movements were still underway at the completion of each test, the rate of deformation was diminishing with time. The *grid* strains inferred from strain gauges showed that the SR2 was well within the 10% strain *performance limit* recommended for this material (McGown et al., 1984). Even if the maximum recorded logarithmic rate of strain at the end of each test was to persist over an additional three logarithmic time cycles (i.e. 120 years) the reinforcement layers would still have adequate tensile capacity under the 50 kPa surcharge. It is clear from the comments made above that the laboratory walls were subject to *working load* conditions rather than a condition close to the ultimate surcharge capacity of the composite systems.

Wall stability calculations based on *limiting equilibrium* methods may not be appropriate to determine working tensile loads in the reinforcement layers. Having said this it is of interest to note that the locus of maximum geogrid strains in layers 3 and 4 from the incremental panel wall test suggests that (if surcharge loads could be increased) a near-vertical failure surface through the top half of the composite soil mass might be generated about 0.45 m back from the wall. This feature is qualitatively similar to critical failure surfaces assumed in *coherent gravity methods* of wall design (e.g. Tensar Corp., 1986) and by Juran and Schlosser (1978) for reinforced soil walls. A similar locus of maximum geogrid strains is masked in the propped panel wall test by the generation of additional geogrid strains close to the panel/geogrid connections resulting from the relative vertical movement between the sand and the panels during fill placement and surcharging. While some features of the coherent gravity method of design are observed in the data from the current study, both walls showed outward rotation of the walls about the toe at working loads rather than rotation about the top which is often assumed in limiting equilibrium-based coherent gravity methods of design. The results of these tests suggest that the condition of bottom restraint in these walls is critical to the performance of these structures. In the current study, the facing unit/levelling pad interface was sufficiently rough to offer an essentially fully restrained horizontal and vertical support. A different deformation response would be anticipated for walls constructed with a less restrained horizontal degree of freedom at this boundary.

The RMC retaining wall test facility was constructed to act as a (near) plane-strain apparatus. However, despite friction-reducing sidewall construction and separation of the reinforced soil blocks into three sections, the sidewalls will contribute to the overall stability of the model walls. The magnitude of this contribution is, however, difficult to quantify based on measured results from the trial walls. In the companion paper by Bathurst and Benjamin, the results of an *unreinforced* test using the same sand backfill and propped wall construction are reported. The purpose of this test was to determine the contribution of sidewall friction and facing/sand interaction to the stability of the soil within the retaining wall test facility. The results of analyses using measured boundary forces on the unreinforced wall and results of direct shear box tests on the sidewall/sand interface suggests that for the propped reinforced wall approximately 14% of the resistance to active earth force may be due to sidewall friction (assuming that limiting equilibrium conditions are applicable to the reinforced test).

The sand backfill employed in these large-scale tests has a high shear strength. The results of direct shear tests show that the material has a peak friction angle between 42° and 53° and possibly as high as 55 or 56° if plane-strain conditions are considered (Jewell, 1987b). In addition, it was observed during excavation of earlier *wrap-around* constructions that the residual moisture content in the sand gave the material an apparent cohesion great enough to allow a cut face one meter high to stand unsupported. The inherently high shear strength of the sand backfill is thought to explain in part why relatively low tensile loads were generated in the reinforcing elements.

7.3 Summary of Measured Results

As part of the Workshop, participants were asked to make *Class A* predictions of the behaviour of the incremental and propped panel walls reported in this paper. A summary of measured results is appended to this paper (Appendix C).

8.0 Conclusions

Two(2) large-scale model reinforced soil walls were constructed within the soil retaining wall test facility at RMC. The walls comprised an incremental panel wall construction and a single propped panel wall system. Both walls were constructed with a Tensar SR2 geogrid as the sand backfill reinforcement. Each wall was subjected to sustained surcharge loading up to 50 kPa following construction.

The following points summarize some of the important observations made during the construction and surcharging of these structures.

1) Both large-scale model walls proved to be stable during construction and surcharging. The low levels of strain developed in the reinforcement layers and decreasing rates of panel movement suggest that both walls would have remained stable for many years under sustained 50 kPa surcharging.

2) The relative vertical compressibility of the facing units in the two tests influenced the generation of geogrid strains in the vicinity of the wall panels. In the incremental wall construction, the panel units were better able to settle with the retained sand backfill owing to a layer of compressible foam placed at each horizontal joint between panels. In the propped panel wall this degree of freedom was not present and additional tensile geogrid straining was recorded in the vicinity of the geogrid/panel connections as the backfill settled under the action of fill self-weight and surcharging. These observations clearly have important implications to the design of these structures. Specifically, details of wall type and construction can lead to additional geogrid loads which are not accounted for directly in current design methodologies.

3) The history of panel deformation and strain development in the reinforcing layers was sensitive to the construction technique employed. In the incremental wall, significant tensile strains were developed during construction. In addition, larger outward panel movements were recorded for the incremental panel wall than for the propped wall test.

4) Estimated geogrid/panel connection loads were very low in these large-scale models. In the incremental wall construction, the trend in geogrid strains close to the panels suggests that earth pressures on the facing units are very small due to the reinforcing effect of the geogrid layers.

5) The distribution of vertical earth pressures below the reinforced soil block in the propped wall test may be markedly reduced close to the wall facing indicating that significant backfill/facing interaction occurs. A portion of the under-registration recorded by pressure cells further from the rigid wall facing is considered to be due to sidewall friction.

Acknowledgements

The authors would like to acknowledge the collaboration of Drs. A. McGown and K. Andrawes of the University of Strathclyde, Scotland in the conceptual development and pursuit of this testing program. Appreciation is also extended to D. Varney and D. Swan, graduate students at the University of Strathclyde for assisting in the construction of the walls. Funding to support this liaison between RMC and Strathclyde was provided through a NATO Collaborative Research Grant. The authors would also like to thank Netlon U.K. for carrying out the in-isolation tests used to determine the load-strain-time data for the SR2 reinforcement and for the provision of the material. Additional support was provided by members of the Civil Engineering Dept. at RMC including Messrs. J. Bell, D. Wawrychuk, S. Prunster and J. DiPietrantonio. The in-isolation tests for strain gauge calibrations were performed by Capt. G. Richardson, graduate student at RMC. Financial support for construction of the test facility and reinforced soil wall models was provided through the ARP program and the Chief of Construction and Properties, DND (Canada). Finally the authors would like to thank Dr. G.W.E. Milligan for his many fruitful discussions while on a sabbatical visit from Oxford University.

References

1. BATHURST, R.J. and JARRETT, P.M. (1986)
Class A Prediction Exercise for Reinforced Earth Walls,
Bulletin No.1 for NATO Advanced Research Workshop, Application of Polymeric Reinforcement in Soil Retaining Structures
Departments of Civil Engineering RMC and University of Strathclyde

2. BATHURST, R.J. (1987)
Large-Scale Reinforced-Earth Wall Tests at RMC
Report to Chief of Construction and Properties
Dept. National Defence (Canada)
Department of Civil Engineering RMC

3. BATHURST, R.J. and BENJAMIN, D.J. (1987)
Preliminary Assessment of Sidewall Friction on Large-Scale Wall Models in the RMC Test Facility
Application of Polymeric Reinforcement in Soil Retaining Structures, Nato Advanced Research Workshop,
Royal Military College of Canada, June 1987

4. BATHURST, R.J. and KOERNER, R.M. (1987)
Results of Class A Predictions for the RMC Large-Scale Reinforced Soil Wall Trials
Application of Polymeric Reinforcement in Soil Retaining Structures, Nato Advanced Research Workshop,
Royal Military College of Canada, June 1987

5. DEPARTMENT OF TRANSPORT, U.K. (1978)
Reinforced Earth Retaining Walls and Bridge Abutments for Embankments, *Technical Memorandum BE3/78*

6. JARRETT, P.M. and BATHURST, R.J. (1987)
Strain Development in Anchorage Zones
Proc. of Geosynthetics '87 Conference, New Orleans

7. JEWELL, R.A. (1987b)
Analysis and Predicted Behaviour for the RMC Trial Wall
Application of Polymeric Reinforcement in Soil Retaining Structures, Nato Advanced Research Workshop,
Royal Military College of Canada, June 1987

8. JURAN, I. and SCHLOSSER, F. (1978)
Theoretical Analysis of Failure in Reinforced Earth Structures
Proc. Symp. on Earth Reinforcement, ASCE, Pittsburgh

9. LESCOUTRE, S.R. (1986)
The Development of a Large-Scale Test Facility for Reinforced Soil Retaining Walls
M.Eng. thesis, Royal Military College of Canada, Kingston

10. McGOWN, A., ANDRAWES, K., YEO, K. and DUBOIS, D. (1984)
The Load-Strain-Time Behaviour of Tensar Geogrids
Symp. on Polymer Grid Reinforcement in Civil Engineering
Paper No. 1.2, London

11. TENSAR CORP. (1986)
Guidelines for the Design of Tensar Geogrid Reinforced Soil Retaining Walls

12. WAWRYCHUK, W.F. (1987)
Two Geogrid Reinforced Soil Retaining Walls
M.Eng. thesis, Royal Military College of Canada, Kingston

13. YEO, K.C. (1985)
The Behaviour of Polymeric Grids used for Soil Reinforcement
Ph.D. thesis, University of Strathclyde, Scotland

Appendix A

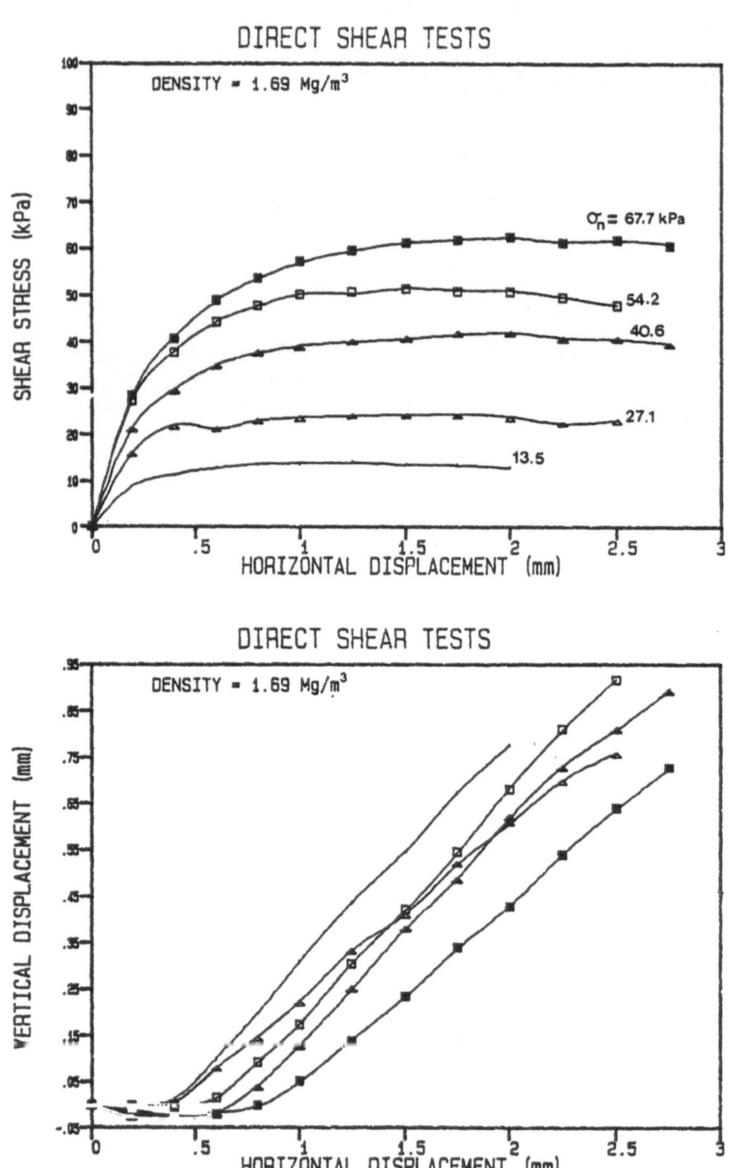

Figure A.1a Load-Deformation of Sand Backfill Material from Direct Shear Box Tests ($\rho_d = 1.69\text{Mg/m}^3$)

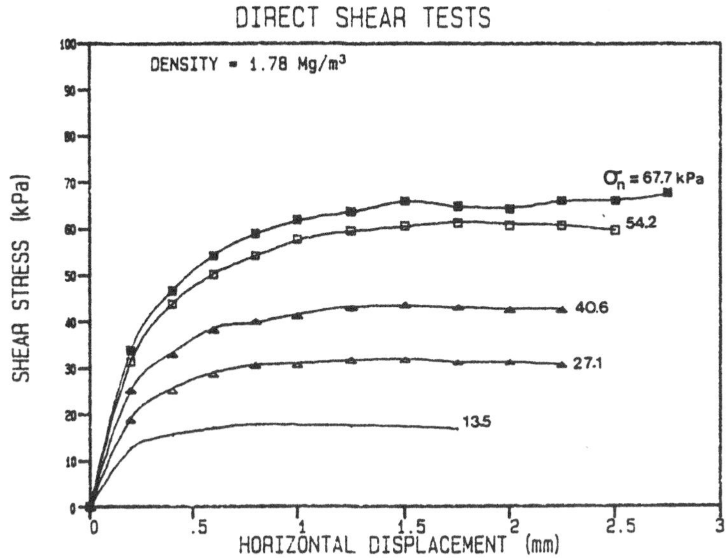

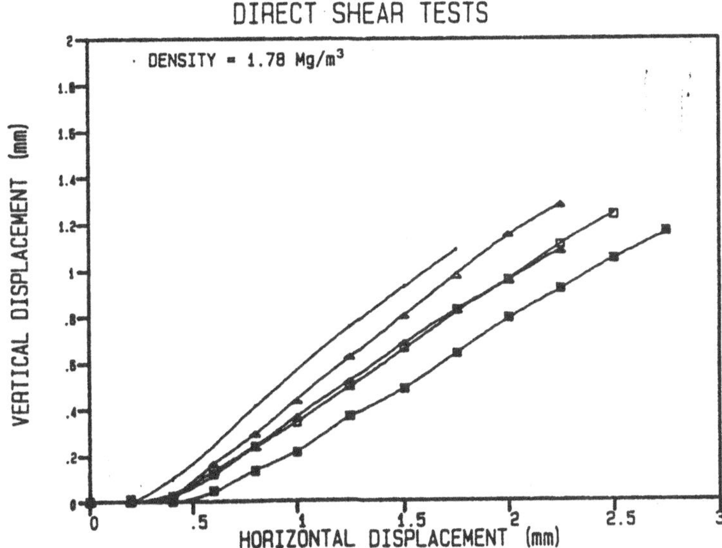

Figure A.1b Load-Deformation of Sand Backfill Material from Direct Shear Box Tests ($\rho_d = 1.78\text{Mg/m}^3$)

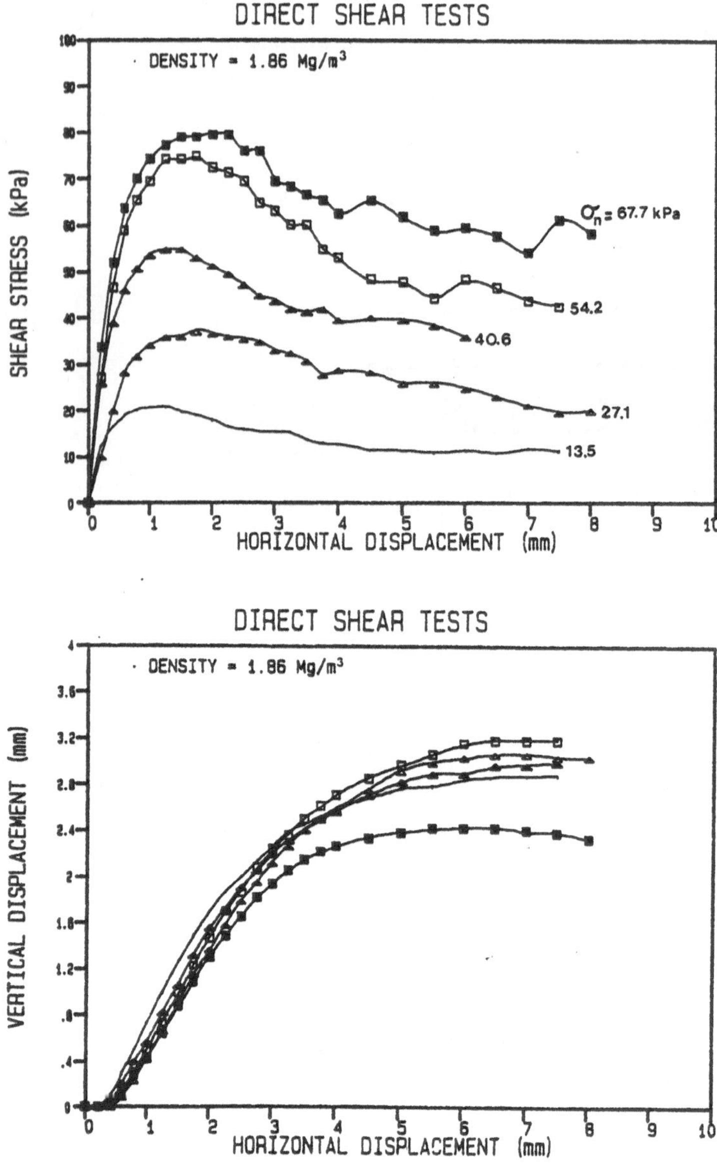

Figure A.1c Load-Deformation of Sand Backfill Material from Direct Shear Box Tests ($\rho_d = 1.86\text{Mg/m}^3$)

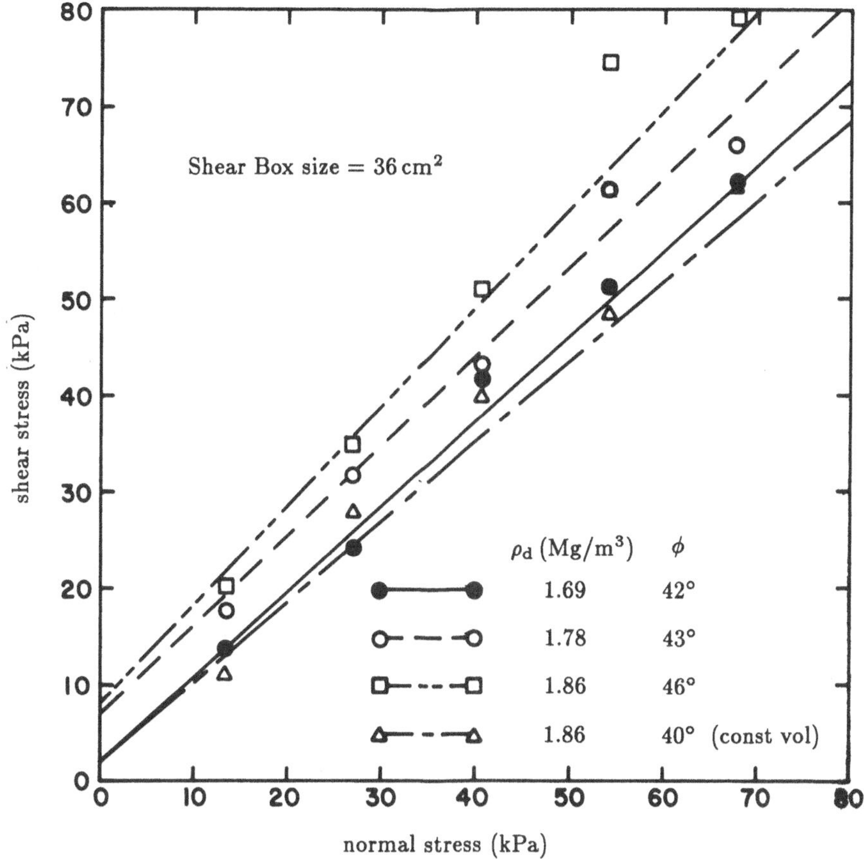

Figure A.2 Mohr-Coulomb Failure Envelope for Sand Backfill Material from Direct Shear Box Tests

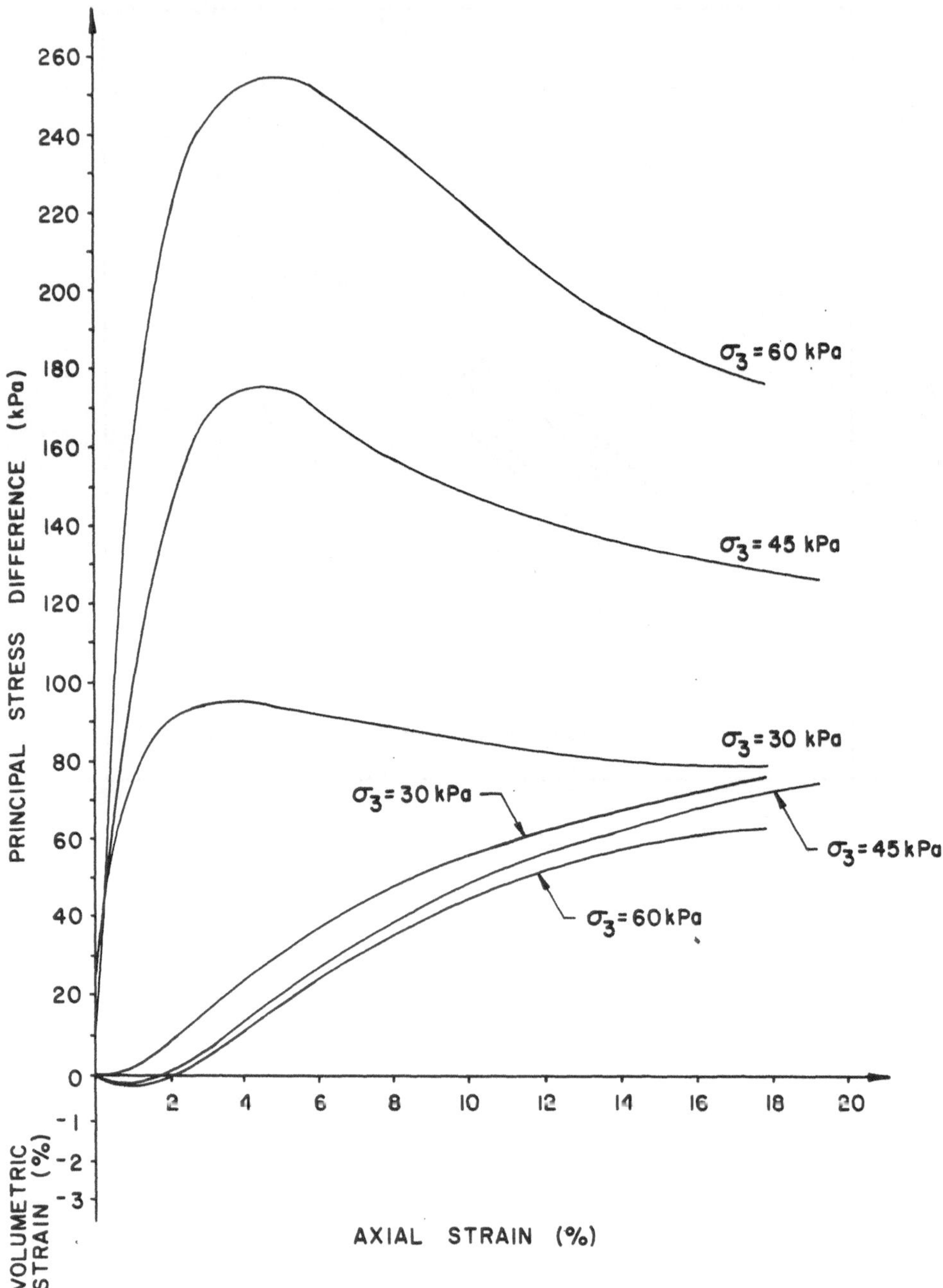

Figure A.3 Stress-Strain Behaviour of Sand Backfill Material from Standard Consolidated-Drained Triaxial Compression Tests ($\rho_d = 1.68\text{Mg/m}^3$)

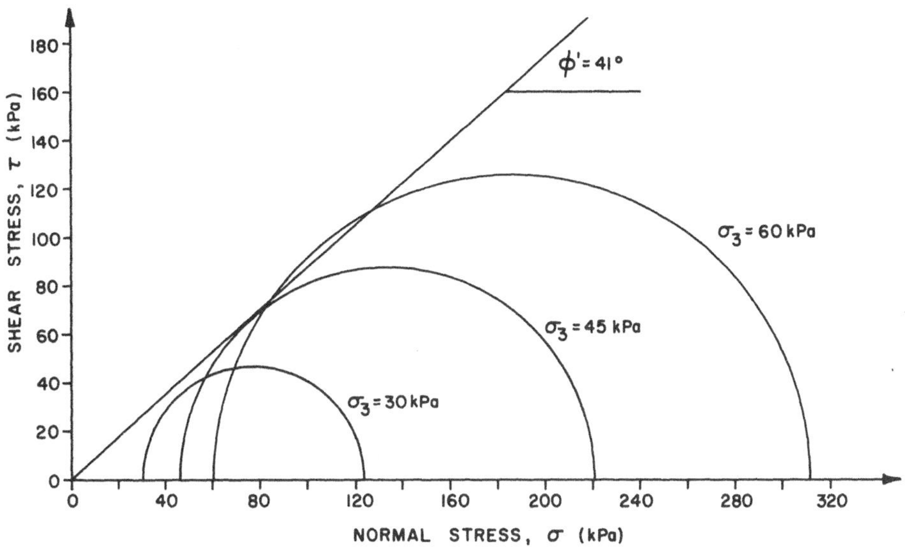

Figure A.4 Mohr-Coulomb Failure Envelope for Sand Backfill Material from Standard Consolidated-Drained Triaxial Compression Tests ($\rho_d = 1.68 \mathrm{Mg/m^3}$)

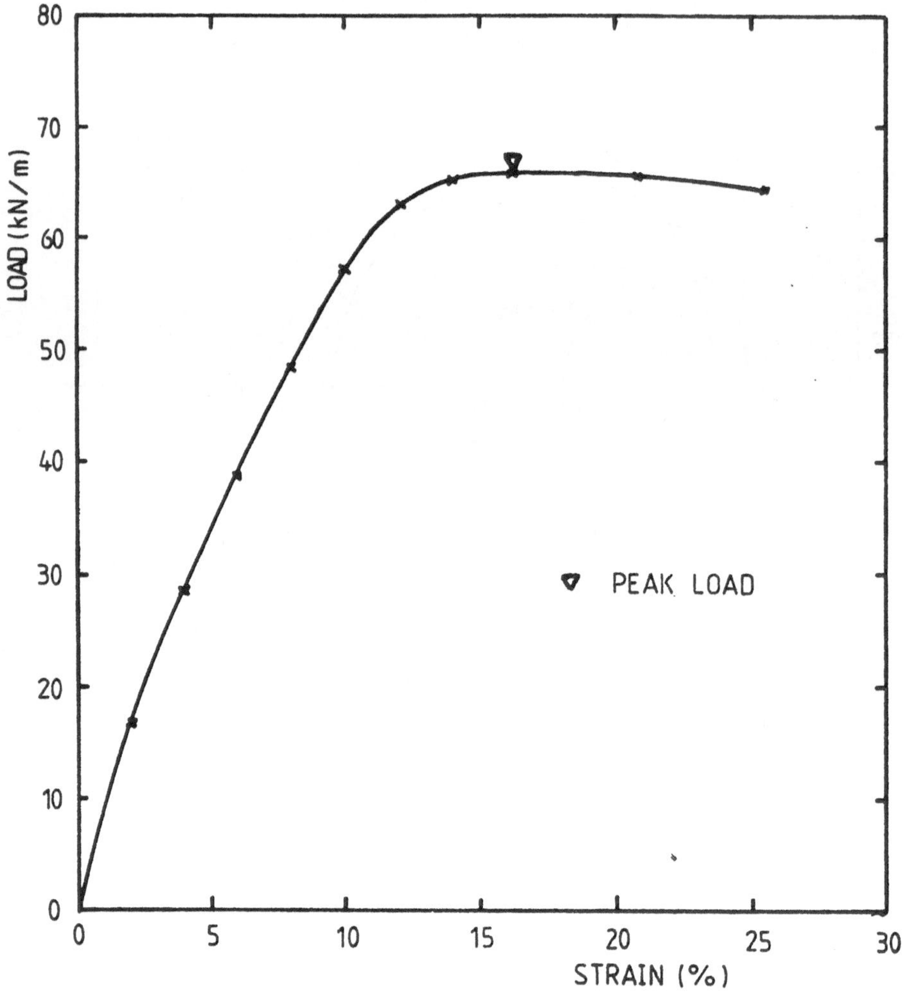

LOAD (kN/m)

PEAK LOAD

STRAIN (%)

Figure B.1

Constant Rate of Strain Test, Load-Extension for TENSAR SR2 at 20°C

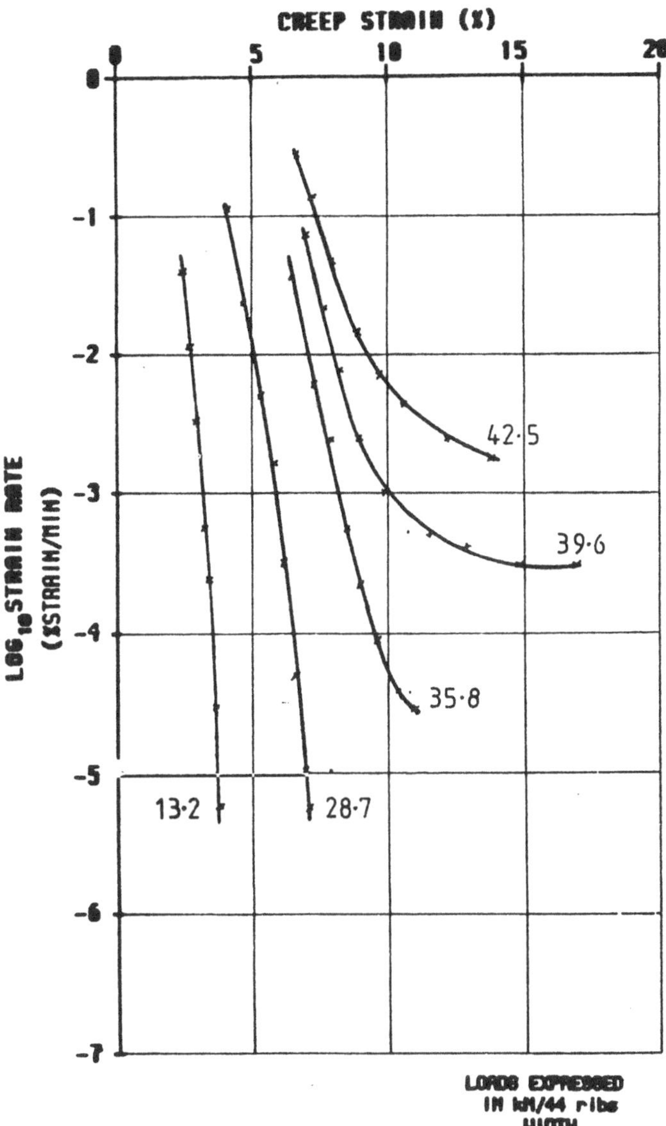

TEST TEMPERATURE : 20°±1°C

Figure B.2 Sherby-Dorn Plot for TENSAR SR2 at 20°C

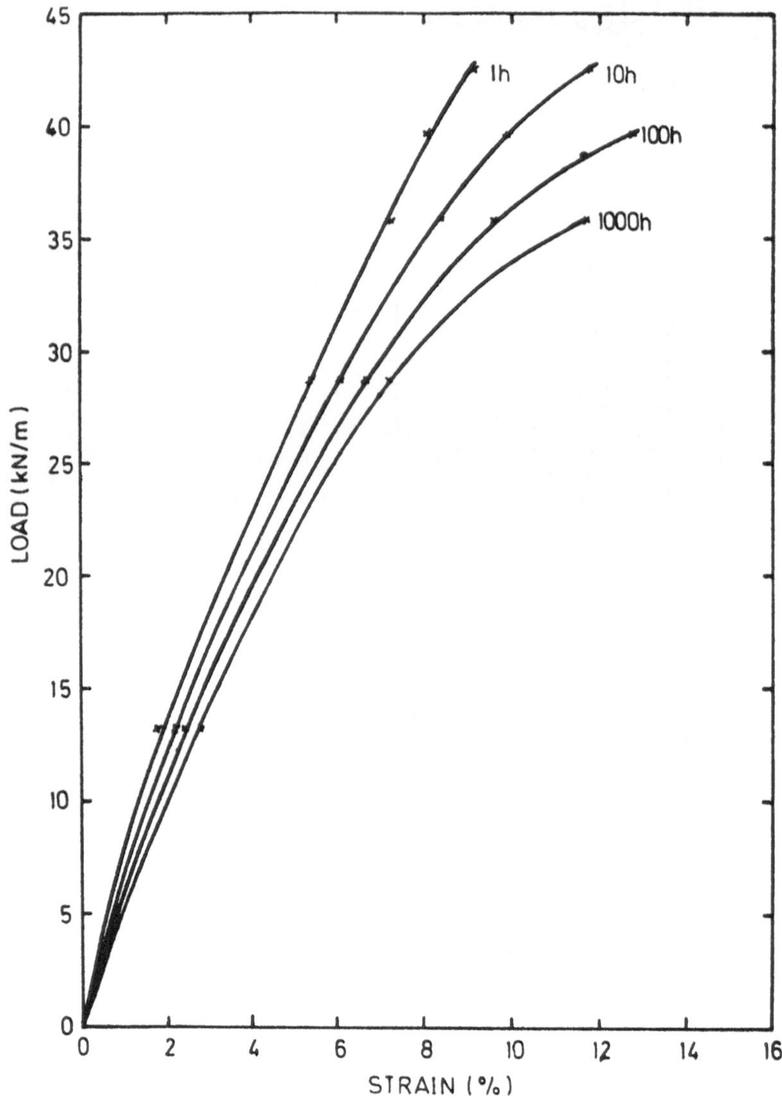

Figure B.3 Isochronous Load-Strain Curves for TENSAR SR2 at 20°C

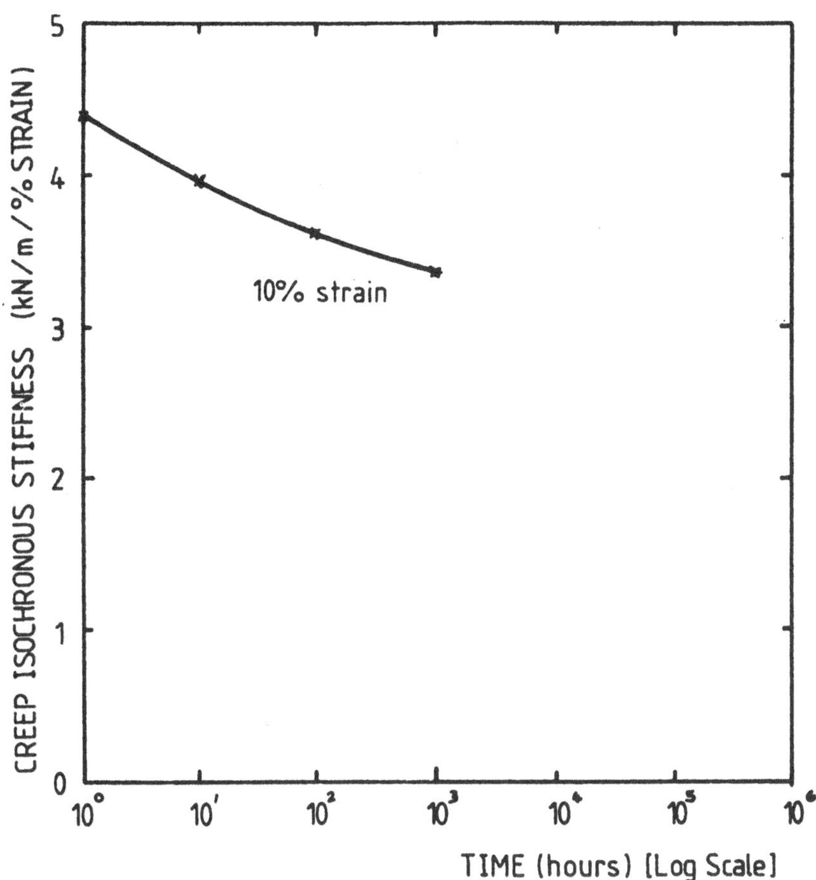

Figure B.4 Creep Isochronous Stiffness versus Log Time Relationship for TENSAR SR2 at 20°C

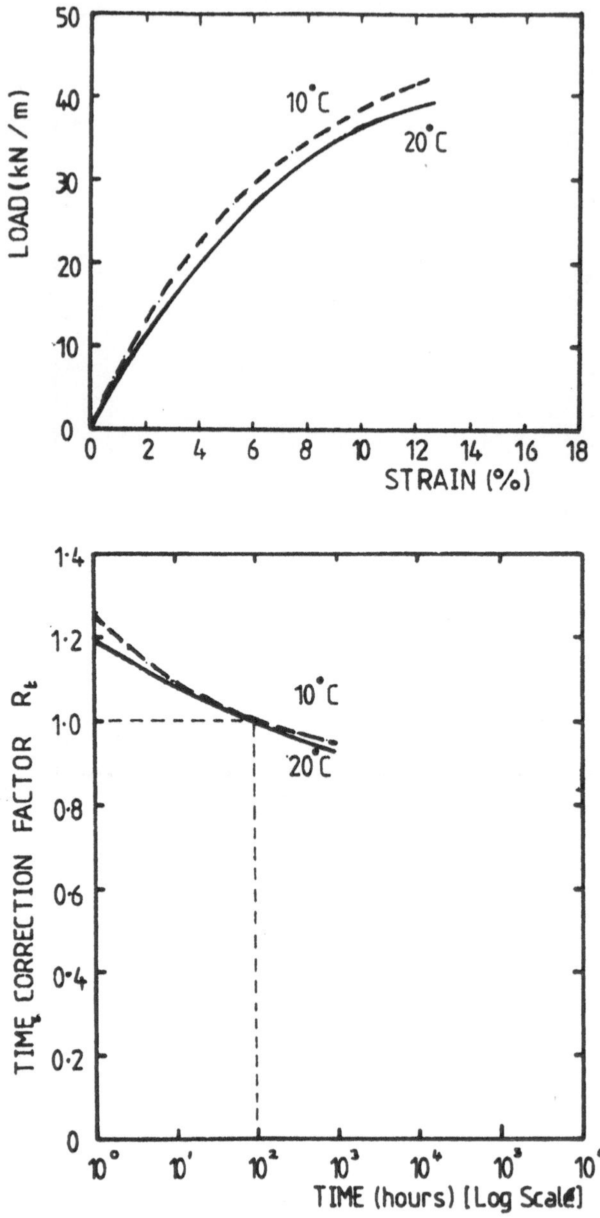

Figure B.5 100 Hour Isochronous Curve and Time Correction Factor versus Log Time Relationship for TENSAR SR2

APPENDIX C

TABLE C.1

Horizontal Panel Positions at Reinforcement Elevations (Incremental Panel Wall Test)*

surcharge= duration=	0kPa ** (mm)	12kPa 100hrs (mm)	12kPa 1000hrs (mm)	12kPa 1600hrs (mm)	30kPa 100hrs (mm)	40kPa 100hrs (mm)	50kPa 100hrs (mm)	50kPa 500hrs (mm)
layer 1	0	4	4	4	4	5	5	5
layer 2	6	9	10	11	13	14	14	15
layer 3	11	13	13	16	18	19	20	20
layer 4	14	18	18	21	24	25	26	27

* refer to Figure 3 for wall geometry
** panel positions at time of installation, post-construction movements can be calculated by subtracting first column of numbers from subsequent measurements

TABLE C.2

Horizontal Panel Movement at Reinforcement Elevations (Propped Panel Wall Test)*

surcharge= duration=	12kPa 100hrs (mm)	12kPa 430hrs (mm)	30kPa 100hrs (mm)	40kPa 100hrs (mm)	50kPa 100hrs (mm)	50kPa 100hrs (mm)
layer 1	1	1	1	1	1	2
layer 2	1	1	2	2	3	4
layer 3	2	2	3	4	5	7
layer 4	3	4	6	7	8	10

* refer to Figure 4 for wall geometry

TABLE C.3

Strain in Reinforcement (Incremental Panel Wall Test)*

		surcharge duration strain (%)						
layer	location** (mm)	12kPa 100hrs	12kPa 1000hrs	12kPa 1600hrs	30kPa 100hrs	40kPa 100hrs	50kPa 100hrs	50kPa 500hrs
1	220	0.41	0.42	0.42	0.47	0.50	0.52	0.52
	445	0.07	0.04	0.04	0.06	0.07	0.08	0.07
	675	0.03	0	0	0.01	0.01	0.01	0.01
2	220	0.38	0.39	-	-	-	-	-
	445	0.22	0.25	0.26	-	-	-	-
	675	0.16	0.15	0.15	0.25	0.29	0.34	-
	900	0.10	0.09	0.09	0.15	0.17	0.20	0.23
	1350	0.06	0.04	0.04	0.07	0.09	0.10	0.11
3	220	0.31	0.34	0.33	0.41	0.45	0.49	0.49
	445	0.37	0.39	0.39	0.57	0.66	0.75	0.79
	675	0.29	0.31	0.31	0.46	0.56	0.65	0.70
	900	0.22	0.24	0.24	0.34	0.39	0.47	0.50
	1350	0.17	0.17	0.17	0.24	0.27	0.29	0.31
	1900	0.25	0.25	0.25	0.33	0.35	0.36	0.37
4	220	0.18	0.18	0.17	0.37	0.41	0.49	0.50
	445	0.22	0.25	0.25	0.53	0.64	0.76	0.89
	675	0.18	0.20	0.20	0.38	0.46	0.55	0.58
	900	0.19	0.20	0.21	0.28	0.37	0.46	0.49
	1350	0.18	0.17	0.17	0.17	0.20	0.23	0.25
	1900	0.17	0.14	0.13	0.11	0.14	0.18	0.20

* strains recorded <u>after</u> each panel support released
** refer to Figure 3 for wall geometry and strain gauge locations
\- denotes gauge failure

TABLE C.4

Strain in Reinforcement (Propped Panel Wall Test)

		surcharge duration strain (%)					
layer	location* (mm)	12kPa 100hrs	12kPa 430hrs	30kPa 100hrs	40kPa 100hrs	50kPa 100hrs	50kPa 1000hrs
1	220	0.04	0.06	0.11	0.15	0.18	0.25
	445	0.04	0.12	0.15	0.18	0.20	0.36
	675	0.01	0.03	0.03	0.04	0.04	0.06
2	220	0.07	0.08	0.17	0.23	0.31	0.37
	445	0.06	0.08	0.15	0.21	0.28	0.35
	675	0.04	0.09	0.15	0.20	0.26	0.36
	900	0.03	0.05	0.09	0.12	0.15	0.21
	1350	0.01	0.04	0.06	0.07	0.09	0.11
3	220	0.06	0.12	0.21	0.28	0.36	0.44
	445	0.05	0.08	0.15	0.22	0.29	0.35
	675	0.07	0.16	0.28	0.37	0.47	0.67
	900	0.06	0.13	0.22	0.29	0.37	0.57
	1350	0.02	0.07	0.11	0.14	0.18	0.22
	1900	0.01	0.05	0.07	0.10	0.12	0.15
4	220	0.10	0.22	0.39	0.52	0.66	0.88
	445	0.05	0.13	0.26	0.35	0.46	0.55
	675	0.05	0.08	0.18	0.25	0.33	0.40
	900	0.04	0.07	0.15	0.21	0.28	0.36
	1350	0.04	0.08	0.11	0.14	0.18	0.25
	1900	0.03	0.06	0.11	0.15	0.19	0.23

* refer to Figure 4 for wall geometry and strain gauge locations

RESULTS OF CLASS A PREDICTIONS FOR THE RMC REINFORCED SOIL WALL TRIALS

RICHARD J. BATHURST, Royal Military College of Canada

ROBERT M. KOERNER, Drexel University, U.S.A.

1. INTRODUCTION

The NATO Advanced Workshop on Application of Polymeric Reinforcement in Soil Retaining Structures was undertaken to critically assess the state-of-the-art in polymeric reinforced soil retaining structures. In order to establish common ground for discussions, a number of participants were asked to make detailed Class A predictions of the performance of two large-scale model polymeric reinforced soil walls constructed within the RMC Retaining Wall Test Facility.

A description of the two tests and the test results are presented in the companion paper in this volume by Bathurst, Wawrychuk and Jarrett.

Participants were issued with details of the proposed walls "in advance" of construction (Bathurst and Jarrett, 1986). The data included:

a) description of the RMC Retaining Wall Test Facility;

b) proposed construction sequence for each wall;

c) construction drawings for proposed walls;

d) material properties for the retained soil;

e) load-strain-time properties of the polymeric reinforcement; and

f) surcharge loading schedule.

Participants were asked to predict the performance of the two walls at various elapsed times after initial construction and after surcharging loads had been applied to the top of the retained soil. Predictions were to include the deformations and strains incurred as a result of the two different construction techniques employed and time-dependent behaviour of the polymeric reinforcement.

The following predictions were requested at the times shown in Table 1:

a) horizontal panel movements at the reinforcement elevations;

b) strain in the reinforcement;

c) total load in the panel/reinforcement connections; and

P. M. Jarrett and A. McGown (eds.), The Application of Polymeric Reinforcement in Soil Retaining Structures, 127–171.

d) vertical pressure along the base of the reinforced soil mass.

Surcharge Load (kPa)	Duration (hrs)
12	100
12	1000
30	100
40	100
50	100
50	1000

Table 1
Schedule of Prediction Times for Proposed RMC Trial Walls

In addition, the predictors were asked to estimate the magnitude of the uniformly distributed surcharge load required to initiate wall failure.

Along with each set of predictions the predictors were asked to submit a brief description of the method of analysis that they adopted to arrive at their estimates. These submissions form a large part of the current paper.

In keeping with the informal spirit of the workshop and to encourage as many predictions as possible it was decided not to attribute predictions to their authors in the published proceedings.

2. RMC TEST WALLS

The large-scale reinforced soil test walls comprised:

a) an incremental panel construction; and

b) a propped (or tilt-up) facing construction.

The general arrangement for the two test wall configurations is given on Figures 1 and 2. Both walls used a medium to coarse grained sand as the soil backfill and an SR2 Tensar "Geogrid" as the reinforcement. The facing panels were constructed as timber bulkheads.

3. METHODS OF ANALYSIS

A total of 10 predictors submitted estimates of wall performance. In some cases, only one of the walls was attempted and in others only a portion of the prediction exercise was attempted. In several cases the submissions were made by a team of workers. It should be noted that the predictors have a wide variety of experience. Some are researchers at universities and government laboratories, others are practicing geotechnical engineers and many are both. Each predictor or team adopted a different approach to arrive at their best estimate of wall performance. In light of these comments it is fair to say that the predictions represent a wide range of current strategies that can be employed to estimate the performance of polymeric reinforced soil walls.

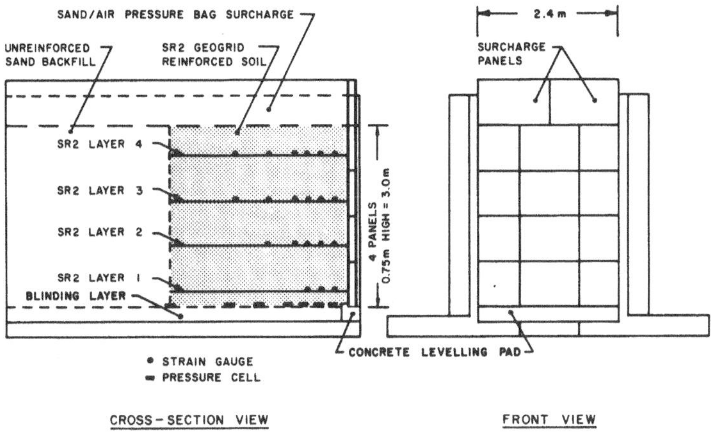

Figure 1 General Arrangement for Incremental Wall Test

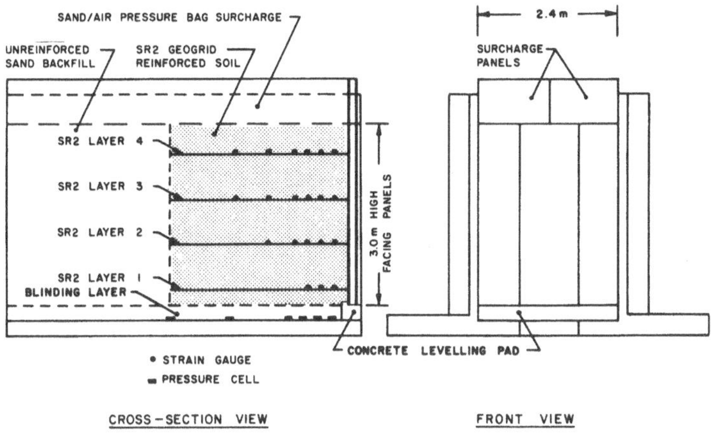

Figure 2 General Arrangement for Propped (Tilt Up) Wall Test

4. PREDICTION METHODOLOGIES

4.1 General

The following descriptions are design method summaries submitted by the predictors prior to the Workshop in June 1987. These overviews can be used to compare essential features of the models employed by the participants.

4.2 Methodology Summaries

PREDICTOR A

A1. PHILOSOPHY OF METHOD

The predictions are based upon a mathematical simulation of the construction and loading procedures used to build the two walls. The mathematical soil model used is based upon the hyperbolic equation proposed by Konder (1963) and described by Duncan and Chang (1970) in which:

$$(\sigma_1 - \sigma_3) = \frac{\varepsilon}{a + b\varepsilon} \qquad \text{(A1)}$$

where σ_1 and σ_3 = major and minor principal stresses.

and ε = axial strain

a and b = constants

By expressing the parameters a and b in terms of the initial tangent modulus value, the compressive strength equation (A1) may be expressed as:

$$(\sigma_1 - \sigma_3) = \frac{\varepsilon}{\frac{1}{E_i} + \frac{\varepsilon R_f}{(\sigma_1 - \sigma_3)_f}} \qquad \text{(A2)}$$

where $(\sigma_1 - \sigma_3)_f$ = compressive strength or stress difference at failure

E_i = initial tangent modulus

R_f = failure ratio defined by:

$$(\sigma_1 - \sigma_3)_f = R_f(\sigma_1 - \sigma_3)_{ult} \qquad \text{(A3)}$$

and $(\sigma_1 - \sigma_3)_{ult}$ = the asymptotic value of stress difference

A2. MODELLING OF SYSTEM

Simulations of the construction and loading for the two walls are shown on Figures A1 and A2.

A3. PARAMETERS USED IN ANALYSIS

A3.1 Fill

The fill was assumed to be a non-linear elastic material. No unloading-reloading was assumed, and Poisson's ratio was taken as constant (ν = 0.35). Stress-strain behaviour of the sand backfill was taken from standard Consolidated-Drained Triaxial Compression Tests provided.

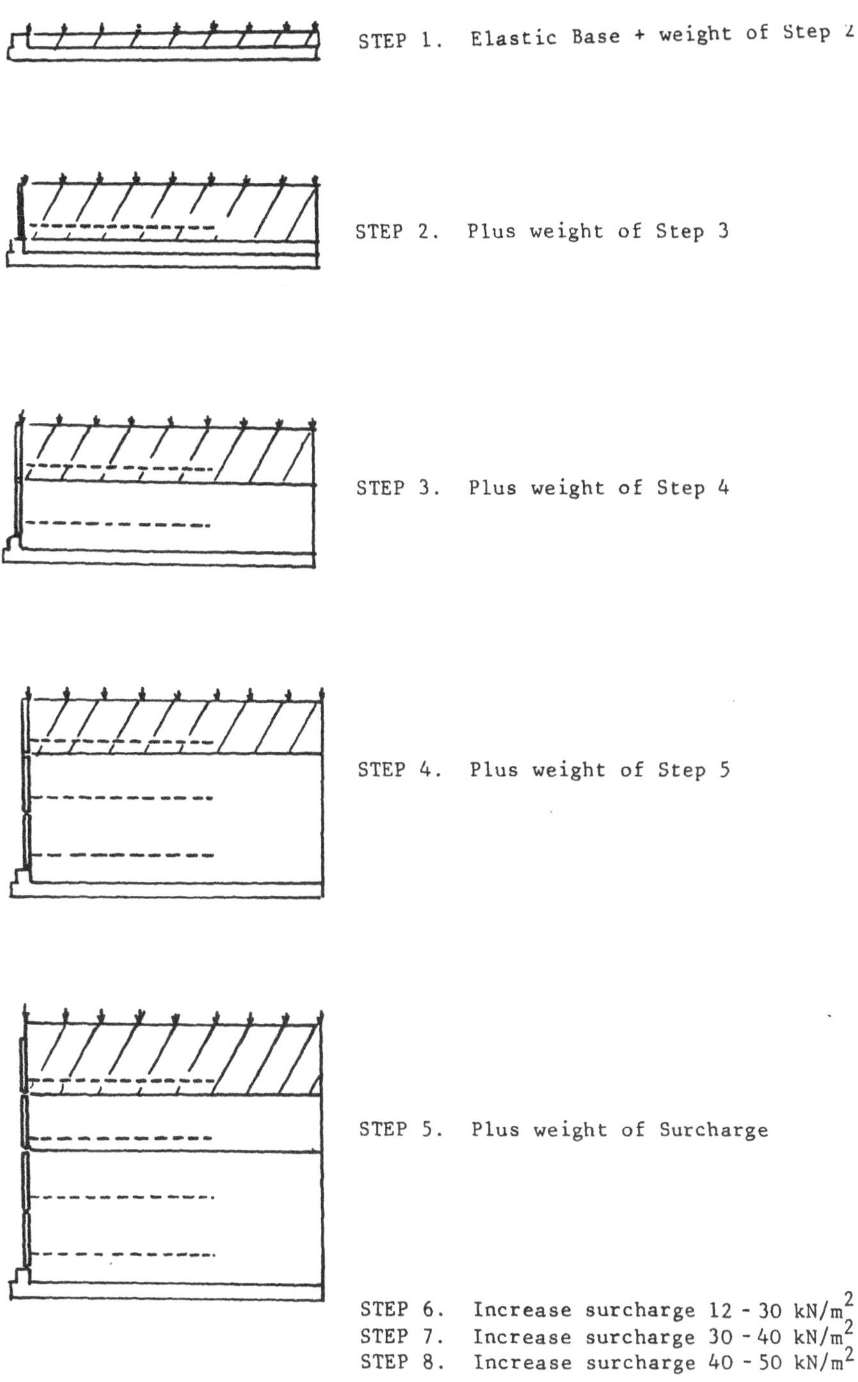

Figure A1 Simulation of Construction and Loading of Incremental Wall

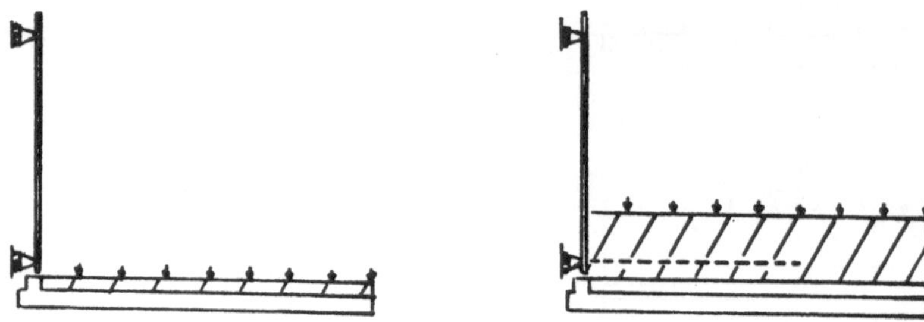

STEP 1. Elastic Base + weight of Step 2 STEP 2. Plus weight of Step 3

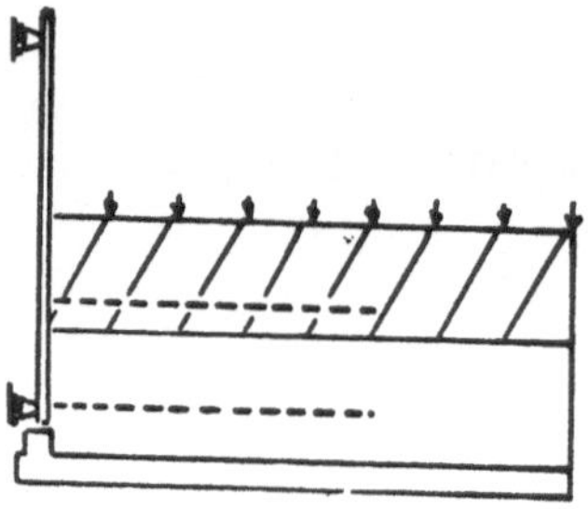

STEP 3. Plus weight of Step 4

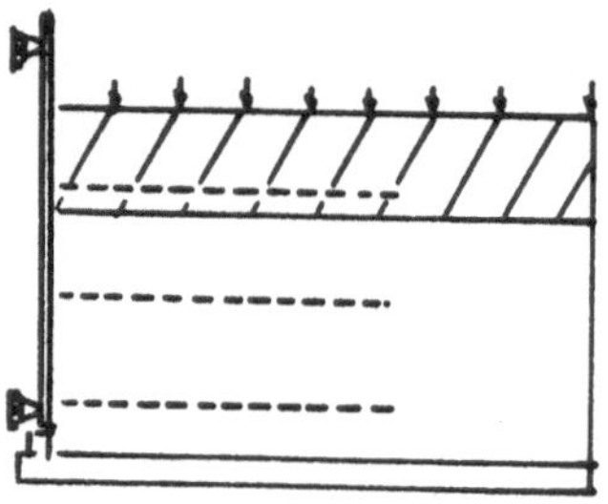

STEP 4. Plus weight of Step 5

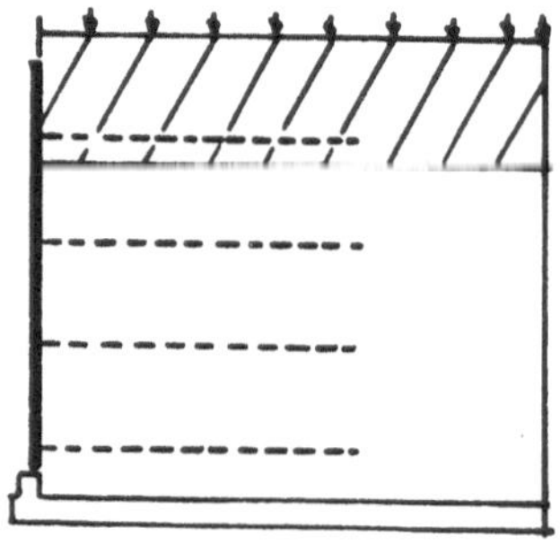

STEP 5. Plus weight of surcharge

STEP 6. Increase surcharge 12 - 30 kN/m^2
STEP 7. Increase surcharge 30 - 40 kN/m^2
STEP 8. Increase surcharge 40 - 50 kN/m^2

Figure A2 Simulation of Construction and Loading of Propped Wall

A3.2 Compressive Stress

A compaction stress of 3.4 kN/m² was assumed for the reinforced soil fill. This was determined by the use of the empirical formula:

$$\sigma_h' = \sqrt{(\frac{2P\gamma}{\pi})}$$

where P = effective weight of the vibrating tamper.

A3.3 Reinforcement

An isochronous stiffness of 3.6 kN/m was assumed for the initial 100 hour analysis. A reduction of 10% in stiffness was assumed for the 1000 hour condition.

A3.4 Facing

The facing was assumed to have the same stiffness as the concrete footing and to be linear elastic.

$$E_{concrete} = 0.24 \times 10^8 \text{ kN/m}^2$$

$$\nu_{concrete} = 0.17$$

A3.5 Fill Immediately Behind Facing

The layer of fill immediately behind the facing was assumed to be elastic and to have a low modulus but high Poisson's ratio.

$$E = 0.1 \times 10^3 \text{ kN/m}^2$$

$$\nu = 0.49$$

A3.6 Fill Beneath Reinforced Soil Block

The layer of fill beneath the reinforced soil block was assumed to be elastic.

$$E = 0.35 \times 10^5 \text{ kN/m}^2$$

$$\nu = 0.35$$

A3.7 Overburden/Surcharge

The overburden was assumed to be elastic.

$$E = 0.1 \times 10^4 \text{ kN/m}^2$$

$$\nu = 0.49$$

A4. COMMENTS ON PREDICTION

A4.1 General

Negative movements resulting from numerical simulations represent movement away from the fill. Positive values represent movement into the fill. The Horizontal Panel movements are very susceptible to the quality of the facing/reinforcement connection. This prediction assumes complete fixity. In reality a large degree of slackness can be expected, in which case the movement of the facing will be away from the fill (i.e. this prediction is expected to fail at this point).

A4.2 Propped Panel Wall

In this trial wall, the connection of the reinforcement to the facing makes no allowance for vertical movements. Hence, this structure does not follow one of the recognized methods of construction (concertina, telescope or sliding). The case can be classified as a 'special' propped wall. Some level of increased stress at the reinforcement/facing connections can be expected.

REFERENCES

1. Duncan, J.M. and Chang, Chin-Yung (1970). "Nonlinear Analysis of Stress and Strain in Soils". Proc. ASCE SM5, September.

2. Konder, R.L. and Zelasko, J.S. (1963). "A Hyperbolic Stress Strain Formulation for Sands". Proc. 2nd Pan-American Conf. on Soil Mech. and Founds. Brazil. Vol. 1, pp. 289-324.

PREDICTOR B

B1. PREDICTION METHODOLOGY

B1.1 Sidewall Influence

Bransby and Smith (1975) report up to 14% decrease in the active earth pressure coefficient due to sidewall friction in model tests. This was considered negligible in wall prediction calculations and therefore neglected.

B1.2 Calculation of Force in Reinforcement

A simple Rankine model was used to estimate lateral earth pressures on the wall. Tributary wall areas were used to calculate reinforcement forces in each layer. The earth pressure distribution was based on the assumption of K_a in the lower third and K_o in the upper two thirds of the wall height. This method was based on Reinforced Earth Company data and personal measurements that indicate that forces in the bottom layers of reinforced walls are always much less than predicted.

B1.3 Distribution of Reinforcement Tension

The maximum reinforcement tension was assumed to act at the Rankine failure plane. A linear reduction of tensile force from a maximum at the failure plane to zero at the embedded end of the reinforcement was assumed. Constant tension from the failure plane to the front of the wall was assumed for all layers except the top layer where the tension at the facing was assumed to be 50% of the maximum value. This approach is based on experience with Reinforced Earth Company walls.

B1.4 Wall Displacement

Wall deformations were calculated using reinforcement stress distributions calculated above together with a PL/AE approach. The embedded end of the reinforcement was considered to be fixed. The modulus of the reinforcement was taken from the initial tangent of standard Index tensile tests curves for SR2 geogrid. Creep in the reinforcement was not considered.

B1.5 Base Pressure

A Meyerhof distribution was assumed for the calculation of base vertical pressures.

REFERENCE

1. Bransby, P.L. and Smith, I.A.A. (1975)
Side Friction in Model Retaining-Wall Experiments, Journal of the Geotechnical Engineering Division, ASCE, GT7.

PREDICTOR C

C1. METHOD OF ANALYSIS

C1.1 Step 1

The vertical stress distribution at the base of the wall and other levels was evaluated by considering moment equilibrium about the centre of the reinforced fill (refer to Figure C1).

Although mobilised friction (ϕ_m') was considered to be acting at the rear of the reinforced fill, at this stage the friction at the facing was ignored. In performing the calculations, side friction was assumed to reduce the effective density of the fill by about 5% and a unit weight of fill of 17.5 kN/m³ was used throughout.

The resulting equation obtained was as follows:

$$\sigma_f, \sigma_b = \gamma z \left[1 \pm 3K_a \frac{z}{L}\left(m \frac{z}{L} \cos\phi_m' - \frac{1}{2} \sin\phi_m'\right)\right]$$

$$+ q\left[1 \pm 3K_a \frac{z}{L}\left(\frac{z}{L} \cos\phi_m' - \sin\phi_m'\right)\right] \qquad \text{(C1)}$$

The stress at the front of the wall (σ_f) now requires to be modified to take account of friction at the facing.

C1.2 Step 2

Close to the facing it is assumed that friction will provide some support to the soil and reduce the vertical stresses in this region. The friction effect is assumed to vary linearly from a maximum at the face to zero at a distance x (refer to Figure C2).

Considering the equilibrium conditions over a layer of thickness dz produces the following differential equation:

$$\frac{d\sigma_w}{dz} = \frac{2k \tan \phi_w}{x} (\gamma z - \sigma_w) \qquad \text{(C2)}$$

where the subscript w refers to conditions at the facing. Solution of the above equation with the appropriate boundary conditions gives:

$$\sigma_w = \frac{\gamma x}{2 \tan \phi_w K}\left[\left(1 - \frac{2K \tan \phi_w q}{\gamma x}\right) e^{\frac{-2K \tan \phi_w z}{x}} - 1\right] + \gamma z$$

where K = lateral pressure coefficient

σ_w = vertical stress reduction at the facing

The vertical stresses determined in Step 1 were reduced on the basis of the above equation.

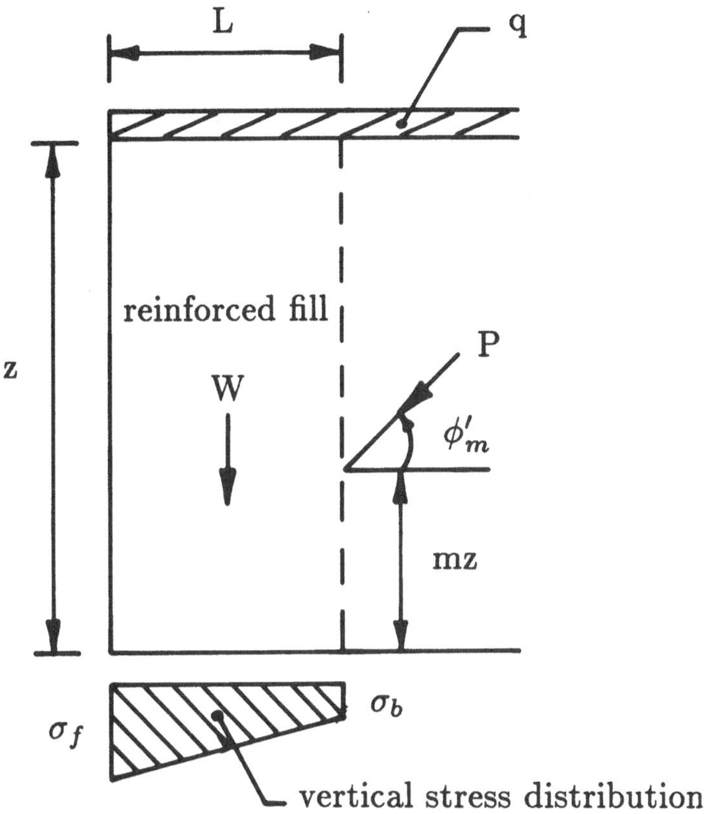

Figure C1 Vertical Stress Distribution

C1.3 Step 3

To evaluate the tension distribution and the strains in the reinforcement, the horizontal equilibrium of an element of reinforced soil was assessed (refer to Figure C3) and the resulting differential equation obtained:

$$\frac{dT}{dx} = VK \frac{d\sigma_v}{dx} + 2\sigma_v \tan \phi_u \tag{C3}$$

In the above equation both the earth pressure coefficient (K) and the interface friction coefficient ($\tan \phi_\mu = \mu_w$) were assumed to be functions of x.

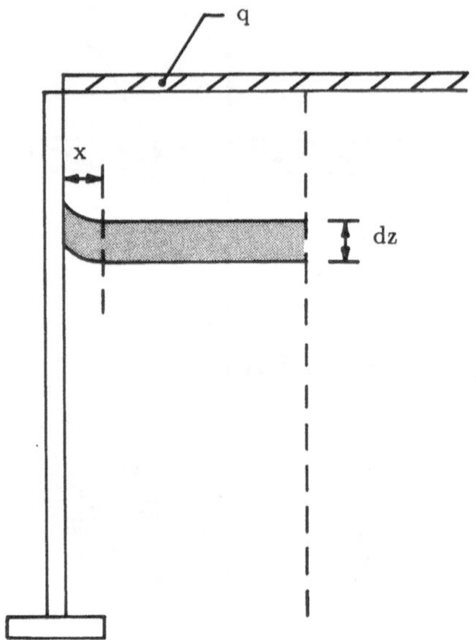

Figure C2 Soil/Facing Friction

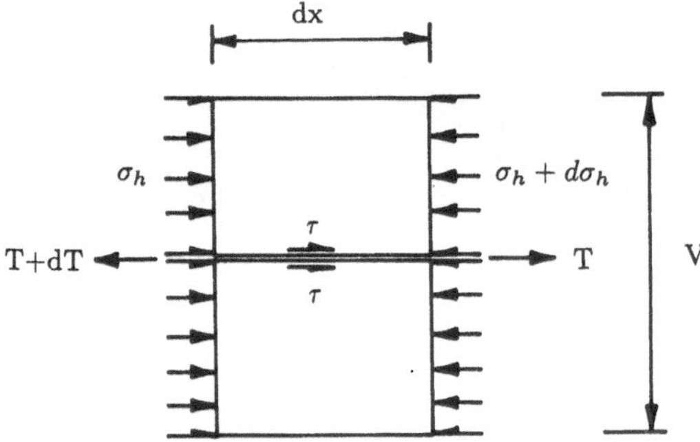

Figure C3 Horizontal Equilibrium in Soil/Reinforcement Element

Solution of the above differential equation with the appropriate boundary conditions produced the following equation:

$$T = \sigma_v \left[\alpha_1 x + \alpha_2 x^2 + \alpha_3 x^3 + \alpha_4 x^4 + \alpha_5 x^5 \right] + \sigma_h V \qquad \text{(C4)}$$

where $\alpha_1 = 2K_w\mu_w - \beta_1$

$\alpha_2 = \beta_1\mu_w + \beta_3 K_w - \beta_2$

$\alpha_3 = 2/3\ (\mu_w\beta_2 + \beta_1\beta_3 + \beta_4 K_w)$

$\alpha_4 = 1/2\ (\beta_2\beta_3 + \beta_1\beta_4)$

$\alpha_5 = 2/5\ \beta_2\beta_4$

σ_h = horizontal stress at face

σ_v = vertical stress at distance x, as given in Step 1

V = vertical spacing of reinforcement

$\beta_1 = [K_a - K_o\ m^2/L^2 + K_w(m^2/L^2 - 1)]/m(1 - m/L)$

$\beta_2 = (K_a - K_w - \beta_1 m)/m^2$

β_3 and $\beta_4 = f(\alpha_1, \alpha_2, \alpha_3, \alpha_4, \alpha_5, \beta_1, \beta_2)$

C1.4 Step 4

In determining the tension distribution, the horizontal stress at the facing is required. The precise value of horizontal stress is influenced by the mode of wall movement.

As a first step assume that the total horizontal force is given by the Coulomb equation, i.e.

$$F = 1/2K\gamma z^2 \qquad \text{(C5)}$$

In the above expression it is further assumed that the amount of soil friction mobilized will be determined by the type and magnitude of the movement. For rotation about the base (propped wall) the same friction angle will be developed at all depths. For incremental construction, movement conforms more closely to rotation about the top. The mobilised friction angle will thus increase with depth and may be approximated by:

$$\phi_m = \phi_o + \eta z \qquad \text{(C6)}$$

where ϕ_o = friction angle at the surface

η = friction angle gradient with depth

To obtain the lateral stress, the expression for total force is differentiated with respect to z, i.e.

$$\frac{dF}{dZ} = K\gamma z + \frac{1}{2}\gamma z^2 \frac{d}{dz}(K) = \sigma_h \qquad \text{(C7)}$$

Now using the simple active earth pressure coefficient expression:

$$K = \frac{1 - \sin\psi}{1 + \sin\psi} \qquad \text{(C8)}$$

where $\psi = \phi_o + \eta z$

Therefore the lateral stress for rotation about the top is obtained from the following equation:

$$\sigma_h(\text{top}) = \gamma z \left[K - \frac{z}{2}\left(\frac{\eta \cos \psi}{1 + \sin \psi}\right)(1 + K)\right] \tag{C9}$$

While for rotation about the base the conventional expression is used:

$$\sigma_h(\text{base}) = K_a \gamma z \tag{C10}$$

$$\text{where } K_a = \frac{1 - \sin \phi_m}{1 + \sin \phi_m}$$

In calculating the horizontal stress directly at the facing, the vertical stress is reduced on the basis of the previous equation developed for friction at the facing.

For both situations the maximum mobilised angle of friction was assumed to be 30°. This friction angle was used generally throughout all the calculations as it provided better strain compatibility with the soil.

C1.5 Step 5

The strain values at the required locations were calculated from the tension distributions obtained according to Step 3 and isochronous data for the reinforcement material. Over a range of 0 - 20 kN the curves for 100 hrs and 1000 hrs are approximately linear and stiffness values were determined accordingly.

The total displacements were calculated by integrating the tension distribution curves over the effective length of reinforcement and dividing the result by the appropriate stiffness.

C1.6 Step 6

Taking a connection strength as 90% of the short-term strength, the failure load was estimated by plotting load at connection versus surcharge stress and extrapolating to determine where the curve intersected the connection strength. It is appreciated that as failure approaches the behaviour will be very plastic and the approach adopted is likely to be conservative.

PREDICTOR D

D1. PHILOSOPHY OF METHOD

For the design of reinforcement layers (length, tensile forces) limit states corresponding to internal and external stability are assumed. For details cf. (2). Creep is prevented by restriction of the strains corresponding to the predicted tensile forces. Empirical estimates are used for predicting the displacements and strain-distributions in the reinforcement.

D2. MODELLING OF SYSTEM

D2.1 Horizontal panel movements

Horizontal displacements at the top of the wall are assumed to be equal to some percentage: d_e of the height of the wall:

d_e = 1.5%; incremental panel wall

d_e = 1%; propped panel wall

and to vary linearly with height. The given values for d_e come from observations from other large-scale tests. For the propped panel wall a greater stiffness, with respect to the incremental panel wall, was assumed, reflected by d_e = 1% and d_e = 1.5%. The assumed amount and distribution of displacements is supported by experimental data cf. (2), (3). Following the arguments of (4), creep displacements can be neglected (experience from granular backfill and limit-analysis based design).

D2.2 Strain in reinforcement

Within our design concept, the strain-distributions for the reinforcement layers cannot be predicted. We are restricted to the following statements:

- The strain distribution can be expected (qualitatively) as shown in Figure D1.

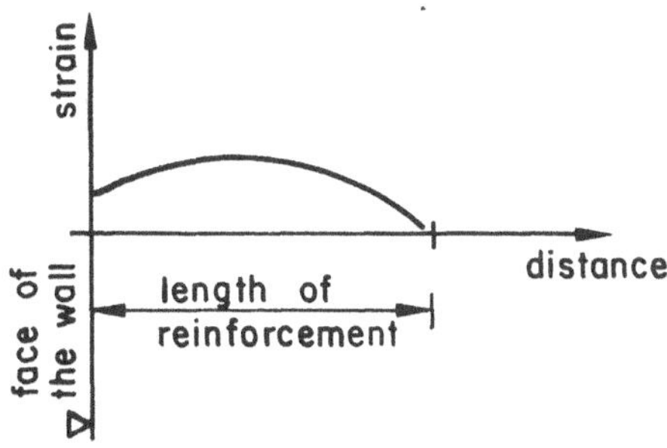

Figure D1 Strain in a Reinforcement Layer

- The maximum strain can be expected for an intermediate reinforcement layer.

- The maximum tensile force (due to internal stability) can be calculated as follows:

$$Z = 0.9\ \gamma \cdot h \cdot \Delta h \cdot K \qquad (D1)$$

where γ = unit weight of soil

h = height of the wall

Δh = spacing between reinforcement layers

K = coefficient of earth-pressure

whereas K_a was assumed for the incremental panel wall and $K = 0.5\ (K_a + K_o)$ for the propped panel wall.

Surcharge loads have been treated as follows:

$$Z_p = p \cdot \Delta h \cdot K \qquad (D2)$$

where p = surcharge load

The maximum strain (due to internal stability), therefore, corresponds to the following tensile force:

$$Z_m = Z + Z_p \qquad (D3)$$

D2.3 Total load at panel/reinforcement connection

For calculation of the connector forces, the earth-pressure distribution from Figure D2 was assumed cf. (2).

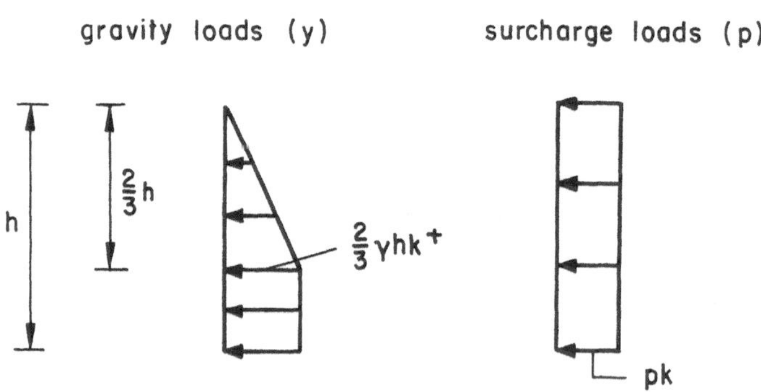

Figure D2 Earth-Pressure Distribution

For the incremental panel wall $K = K_a$ was assumed and, because of the greater stiffness, $K = 0.5\ (K_a + K_o)$ for the propped panel wall. The assumed earth pressure distribution is supported by experimental data cf. (2).

D2.4 Pressure along base of reinforced block of soil

Due to local base failure, the vertical pressure, in particular in the vicinity of the face of a wall, can be essentially lower than the overburden pressure. This phenomena can be expected to be very pronounced for a weak soil, underlying a reinforced block of soil, and to be less pronounced for a concrete floor, underlaying a wall (RMC-test facility). Our prediction was established for a wall resting on a soil half-space and is based on experiences reported by others from large-scale tests cf. (2), (1).

The following assumptions have been used for predicting the earth-pressure distributions:

- At the face of the wall the lateral and vertical pressures are assumed to be correlated via the earth-pressure coefficient (cf. Figure D2).

- The mean-vertical pressure, defined for the length of the reinforcement layer, is equal to the overburden pressure.

D2.5 Surcharge pressure required to produce failure

Because of the great reinforcement length ℓ: $\ell = h$ with $\ell = 0.4\ h$ is usually sufficient for external stability cf. (2), no external failure was expected. This statement is also supported by experiences from large-scale tests reported by (3).

Back-calculation of surcharge loads due to internal stability (failure of reinforcement or connectors) leads to a very high value:

$$p = 1234\ kN/m^2$$

i.e. before a loss of internal stability can be expected, probably base failure under the surcharge load occurs cf. (3).

D3. PARAMETERS USED IN ANALYSIS

From Bulletin No. 1, Class A Prediction Exercise for (RMC) Reinforced Soil Walls.

REFERENCES

1. Carroll, R.G. and Richardson, G.N.: Geosynthetic reinforced retaining walls. Proceedings of the Third International Conference on Geotextiles, Vienna, Austria, 1986.

2. Gudehus, G. and Schwing, E.: Standsicherheit Kunststoffbewehrter Erdbauwerke an Geländesprüngen. Proceedings of the Baugrundtagung, Nürnberg, West Germany, 1986.

3. Wichter, L., Risseuw, P. and Guy, G.: Grossversuch zum Tragverhalten einer Steilwand aus Gewebe und Mergel. Proceedings of the Third International Conference on Geotextiles, Vienna, Austria, 1986.

4. Yamanouchi, T., Fukuda, N. and Ikegami, M.: Design and techniques of steep reinforced embankments without edge supportings. Third International Conference on Geotextiles, Vienna, Austria, 1986.

PREDICTOR E

E1. INTRODUCTION

This prediction utilized classical earth pressure concepts as generally practiced in the design of reinforced earth walls using metallic strips. The Geotechnical Engineering literature is abundant with such design information and the performance of these walls has been excellent. To the authors' knowledge there have been no failures nor cases of excessive deformation using such design procedures.

However, even within such a design framework there are certain design modifications and assumptions which become necessary. Our thinking on how we arrived at these various decisions will be outlined in the text to follow.

Lastly, it was recognized that limit equilibrium represents a failure condition but was used to predict pre-failure situations. Thus it was tempting to include an arbitrary reduction value on the predicted values. This temptation was resisted on the following grounds:

(a) We could not decide on what value to use.

(b) We did not know how to reduce the factor as failure approached.

(c) We did not know if the value would change for different parts of the prediction request.

(d) There would be no way to back-calculate to the "correct" answers.

E2. EARTH PRESSURES

E2.1 Options

Vertical pressure: Uniform, trapezoidal or Meyerhof (Figure E1)
Horizontal pressure = K times vertical pressure
K-value: K_a, K_o or higher; constant or varying with depth (Figure E2)
Used in final prediction: Maximum earth pressure related to overburden+ surcharge by K_o for top three grid levels and by $K_o/2$ for lowest grid level (Figure E3).

E3. ALLOCATION OF FORCES TO GRIDS

For the incremental (Element-) wall the grids are assumed to take on the resultant of the earth pressure per face element (0.75m times 1 m).

For the single (rigid) panel wall, the earth pressure allocated to each grid is determined by lines drawn midway between grids and/or the horizontal boundaries of the backfill.

These earth pressure allocations are illustrated in Figure E4.

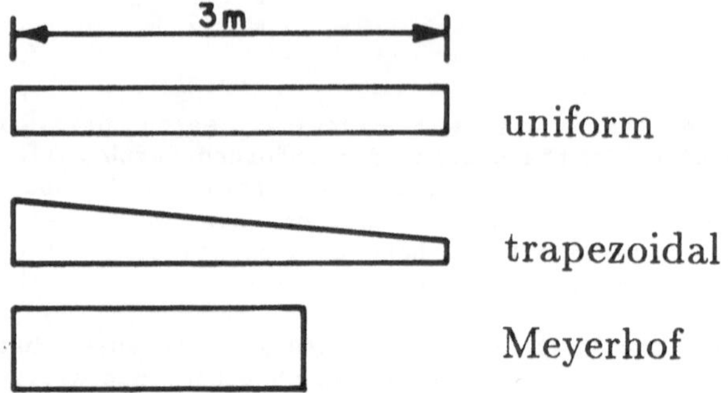

Figure E1 Vertical Stress Distribution Options

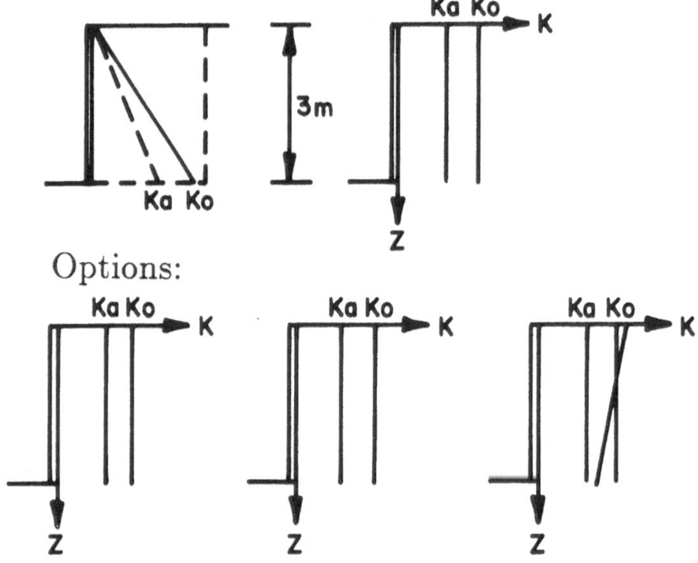

Figure E2 Horizontal Earth Pressure Distribution Options

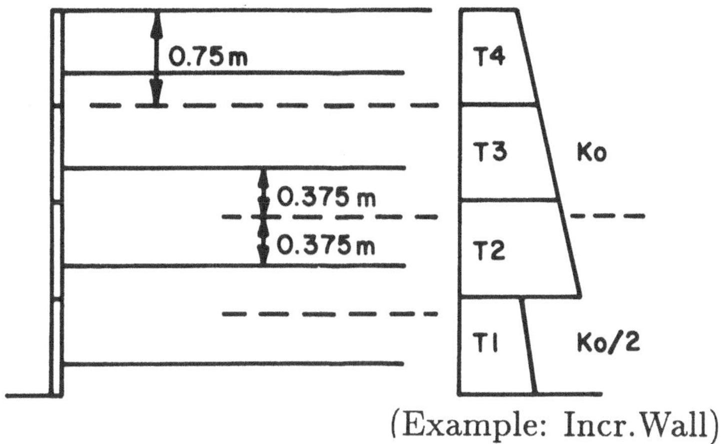

Figure E3 Earth Pressure used in Calculations

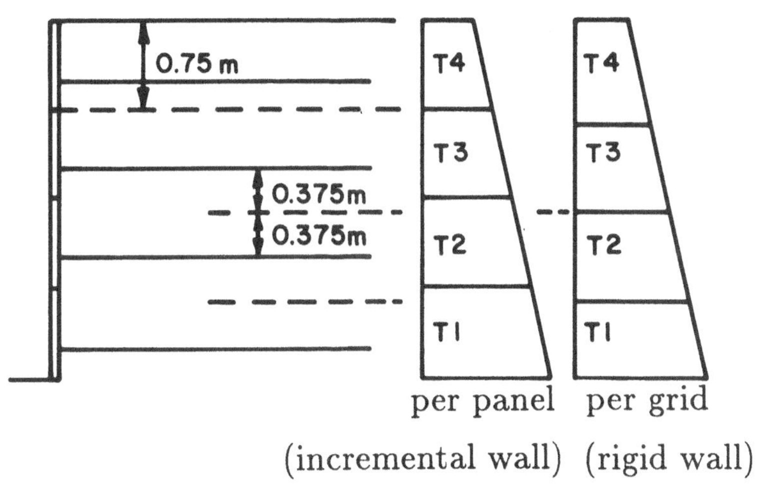

Figure E4 Allocation of Earth Pressure to Grids

E4. DISTRIBUTION OF SHEAR STRESSES ON AND TENSION IN THE GRIDS

Options considered are shown in Figure E5.

Some initial calculations were done assuming uniform distribution of shear stresses and the resulting triangular distribution of tension. These conditions would resemble those attained during an ideal pullout test at the point of failure.

For the prediction, tension T was assumed to have a polynomial distribution with T_{max} at a distance of 1m (=1/3 of total length of grid) and $T_{max}/2$ at the face while the curvature of the distribution is reversed in the last third of the grid to reach zero at the end. The corresponding distribution of shear stresses is triangular, with a stress reversal at the 1m point (Figure E6).

E4.1 Mathematically

With x = distance from wall face and T_{max} = Resultant of earth pressure

0-2m: $T = (0.5 + x - 0.5x^2)\ T_{max}$ and $\tau = (1-x)\ T_{max}$

$\tau_{max} = (+/-)\ T_{max}$

$T_{avg} = (11/18)\ T_{max} = 0.61\ T_{max}$

At face: $T_o = T_{max}/2$

2-3m: $T = (4.5 - 3x + 0.5x^2)\ T_{max}$

E5. REDISTRIBUTION OF FORCES FOR RIGID WALL

Resultants of earth pressure on each section of the rigid panel are redistributed so that they vary linearly with height.

E5.1 Procedure

(a) Calculate resultant forces F_i (Total Force) on each part of the panel.

(b) Calculate sum of all 4 forces: $\sum F_i = F_R$.

(c) Calculate sum of all overturning moments due to F_i:$M = \sum F_i h_i$.

(d) Calculate parameters s and a:

$$s = \frac{F_R\left[\frac{\sum h_i}{4} - h_r\right]}{\sum h_i^2 - \left[\frac{\sum h_i \sum h_i}{4}\right]} \qquad \text{(E1)}$$

$$a = \frac{F_R + s\sum h_i}{4} \qquad \text{(E2)}$$

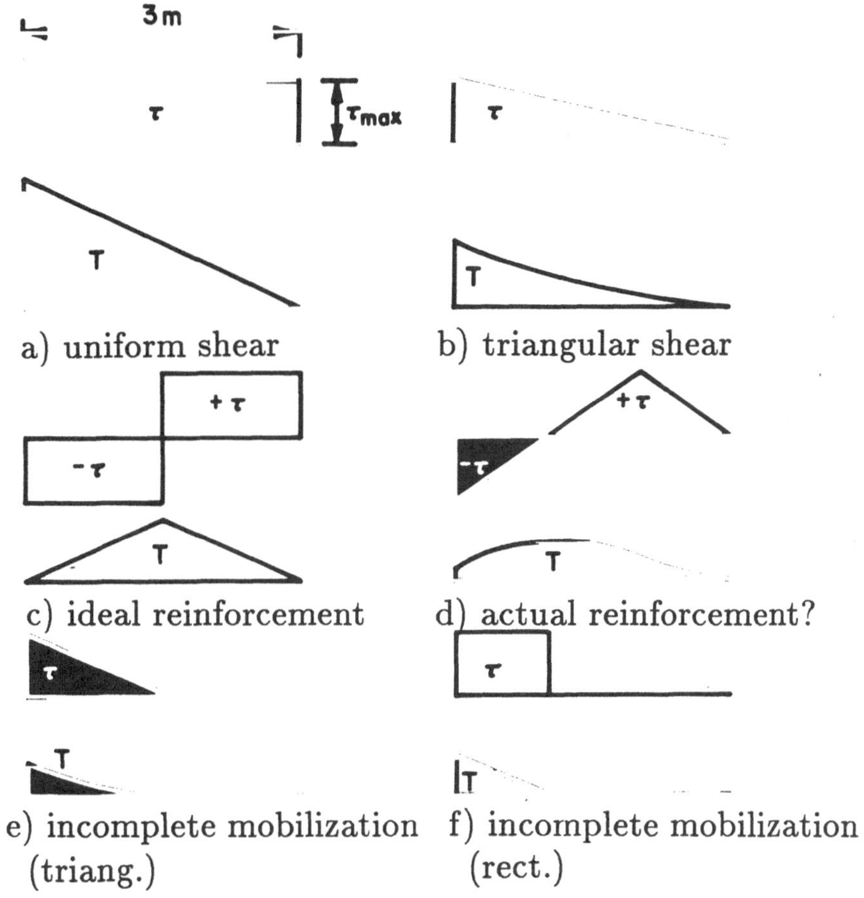

3m
τ
τmax
τ
T
T
a) uniform shear
b) triangular shear
+τ
-τ
+τ
-τ
T
T
c) ideal reinforcement
d) actual reinforcement?
τ
τ
T
T
e) incomplete mobilization (triang.)
f) incomplete mobilization (rect.)

Figure E5 Various Possible Distribution of Tension and Shear Stresses

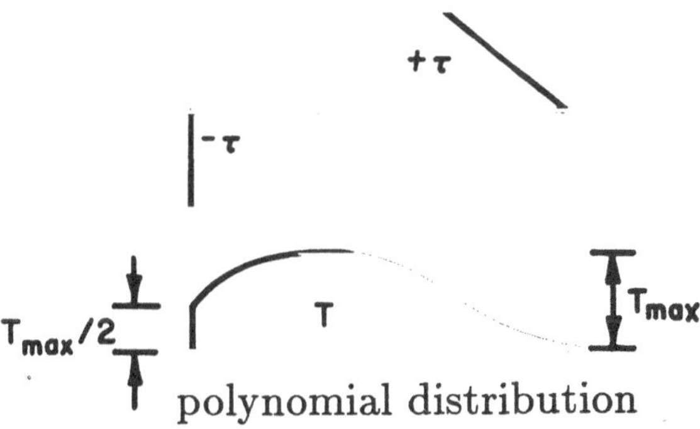

+τ
-τ
Tmax/2
T
Tmax
polynomial distribution
Tmax = resultant of earth pressure per panel

Figure E6 Distribution of Shear and Tension as Assumed in Prediction

(e) Calculate redistributed forces as follows:

$$F_i = a - sh_i \qquad \text{(E3)}$$

(f) Calculate strains based on $F_{avg} = 0.61$ F. If the modulus is constant for all stress levels, then the deformation should also be constant.

E6. SPECIAL CONSIDERATION OF BASE PRESSURE

A check showed that trapezoidal distribution would result in negative pressures at the back of the reinforced soil. A Meyerhof distribution was considered unsuitable for a prediction.

Chosen was an arbitrary distribution which resembles what is generally measured below Reinforced Earth structures, although it may still tend to be conservative as far as maximum values are concerned (Figure E7).

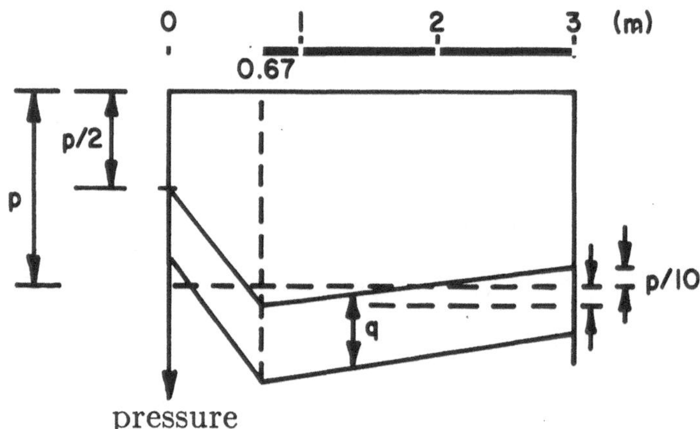

p = initial overburden stress (including 12 kPa surcharge)

q = additional surcharge

Figure E7 Assumed Distribution of Base Pressure

E7. STRAINS AND DEFORMATIONS

Initial wall deformations (incl. 12 kPa at 0 hours) were calculated as the cumulative deformations due to earth pressure for T_{avg}.

Additional deformations were added for each stress increase at a particular level (non-cumulative).

Modulus chosen was the secant modulus for the appropriate stress level and creep time. For each stress level the appropriate curve was selected (no correction for "accumulative" creep).

PREDICTOR F

F1. PHILOSOPHY OF METHOD

This work is aimed to present a model for the prediction of the behaviour of a geogrid reinforced earth wall (forces and strains in the inclusions, horizontal movements of the wall face, pressure under the reinforced soil block). In particular, the model developed for the present prediction refers to the Incremental Facing Reinforced Soil Test, built at the Royal Military College, Kingston, Ontario.

The predictors have chosen to develop a rather simple method, in order to obtain satisfactory results, carrying out the calculations with a small computer and in a short time. The use of a simple model involved some simplifying hypotheses, based only on the experience achieved from previous field investigations on geogrid reinforced earth walls.

F2. PARAMETERS USED IN ANALYSIS

F2.1 Soil

On the basis of the results of Direct Shear Box tests and according to the dry density of the backfill sand (1790 kg/m³ = 17.56 kN/m³) an internal friction angle of 43° was assumed.

F2.2 Reinforcement - Inclusion

The parameters for the Reinforcement - Inclusion behaviour were obtained from the Load - Extension data and from Isochronous Load-Strain curves data.

The Isochronous Curves were schematized by means of the parabolic equation:

$$\varepsilon(F,t) = a(t) \cdot F^2 + b(t) \cdot F \qquad \text{(F1)}$$

where the coefficients a(t) and b(t) were obtained from the system:

$$\begin{vmatrix} F_A^2 & F_A \\ F_B^2 & F_B \end{vmatrix} \cdot \begin{vmatrix} a(t) \\ b(t) \end{vmatrix} = \begin{vmatrix} \varepsilon_A(t) \\ \varepsilon_B(t) \end{vmatrix} \qquad \text{(F2)}$$

The input data (F_A,ε_A), (F_B,ε_B) were obtained from the Isochronous Curves corresponding to F_A = 10 kN/m and F_B = 30 kN/m.

F2.3 Soil - Inclusion Interaction

It was assumed that no relative movement between the geogrid and soil occurred in the considered range of forces.

F2.4 Surcharge

A uniform surcharge applied according to the schedule in Table 1 was assumed.

F3. MODELLING OF SYSTEM

F3.1 Maximum Forces in the Inclusions

The evaluation of the forces induced in the geogrids was done by giving an "influence zone" to each reinforcing layer (geogrid). The load applied to each layer was calculated as the sum of the Rankine active pressure and of the pressure due to the surcharge, as shown in Figure F1. The active pressure coefficient K_a was assumed to be constant and equal to: $\tan^2(45°-43°/2) = 0.189$.

In order to simplify the calculations, the zero reference for the vertical axis was not positioned at the crest of the wall, but elevated by a quantity equal to (W_s/y). Therefore the formula used was:

$$P_i(W_s) = 1/2 \cdot y \cdot K_a \cdot (V_i^2 - V_{i+1}^2) = 1/2 \cdot y \cdot K_a \cdot A_i(W_s) \qquad \text{(F3)}$$

where: $V_5 = W_s/y$ (see Figure F1)

$$A_i(W_s) = (V_i^2 - V_{i+1}^2)$$

V_i = as shown on Figure F1

The total force per unit width was calculated as:

$$T_{tot} = 1/2 \cdot y \cdot K_a \cdot (H+W_s/y)^2 - 1/2 \cdot y \cdot K_a \cdot (W_s/y)^2 = \sum_i P_i(W_s) \qquad \text{(F4)}$$

F3.2 Distribution of the Tensile Forces in each Inclusion

The calculation of the tensile force distribution in each geogrid layer was based on the following assumptions:

- the tensile force versus the distance from the wall face was represented by a curve, approximated by two straight lines, exhibiting a maximum at a distance equal to x_i* from the wall face (see Figure F2);

- the maximum value of the tensile force in the geogrid (F_{imax}) was assumed to be equal to the above calculated $P_i(W_s)$;

- the position of F_{imax} was given by a line defining the "locus of maximum tensile force"; in this case it was assumed to be at a slope of 20°, based on the results given by YAMANOUCHI et al. (1986) and BERG et al. (1986), as shown in Figure F2;

- the tensile force distribution in each inclusion ($F_i(x)$ diagram) was assumed to have the same slope on both sides of the maximum: that is, the tensile load transfer by friction with the adjacent soil was assumed to be of the same intensity on both sides of the peak force; the remaining force at the wall face was the load at panel/reinforcement connection.

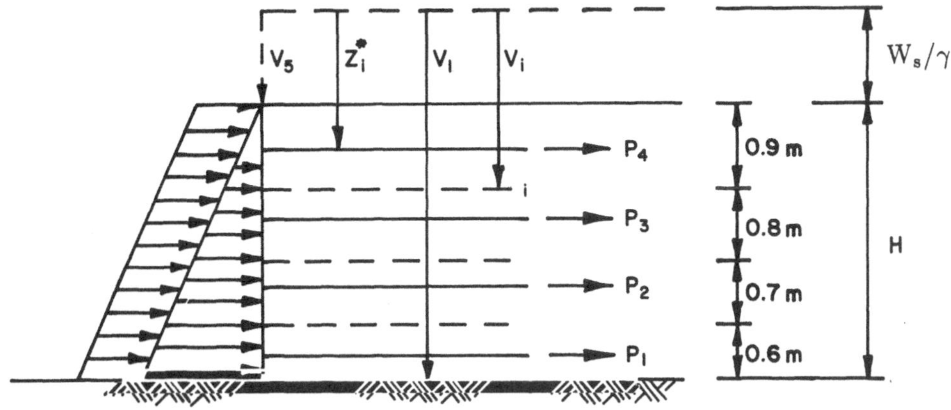

Figure F1 Scheme for the Calculation of the Forces in the Inclusions

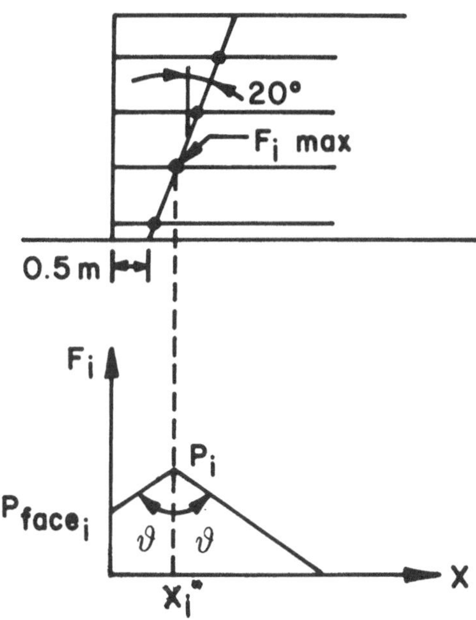

Figure F2 Tensile Force Distribution

Therefore the formulas used for the calculation of the tensile forces in each inclusion and at the face were:

$$F_i(x) = P_i - (dF/dx)_i : |x - x_i^*| \tag{F5}$$

where: $$(dF/dx)_i = P_i/(L - x_i^*) \tag{F6}$$

$$P_{face\ i} = P_i - (dF/dx)_i : x_i^* \tag{F7}$$

F3.3 <u>Strains in the Inclusions</u>

Isochronous Load-Strain Curves were used for the calculation of strains in the geogrids due to creep, according to the parabolic scheme presented in F2.2.

Following the qualitative pattern of strain versus time and surcharge shown in Figure F3 and taking into account the elastic strain (from load-extension diagram at 0 h) and the creep strain (from isochronous load-strain curves), the values of the strain in each geogrid layer at each distance from the face and at specified elapsed times, were given by:

$$\varepsilon_i(Ws_k,\ x,t) = \{\varepsilon[F_i(Ws_k,\ x,t),\ t=0]\} + \sum_{i}^{k-1}{}_j\ \Delta\varepsilon_j \tag{F8}$$

where: $$F_i(Ws,\ x,t) \simeq F_i(Ws_k,\ x) \tag{F9}$$

$$\sum_{1}^{k-1}{}_j\ \Delta\varepsilon_j = \sum_{1}^{k-1}{}_j\ \{\varepsilon[F_i(Ws,\ x),\Delta t_j] - \varepsilon[F_i(Ws_j,\ x),t=0]\} \cdot C_R(F_i,\Delta t_j) \tag{F10}$$

with: $\varepsilon_i(Ws_k,x,t)$ = strain in the geogrid layer "i", due to the surcharge Ws_k

$F_i(Ws_k,x,t)$ = tensile force in the geogrid layer "i", due to the surcharge Ws_k

$\Delta\varepsilon_j$ = increment of strain due to creep, during surcharge Ws_j

Δt_j = period of application of the surcharge Ws_j

$C_R(F_i,\Delta t_j)$ = In-Soil Creep Factor.

Another hypothesis (eq. F9) was that the tensile force in the geogrids was constant for the entire period of application of each surcharge.

The In-Soil Creep Factor takes into account the difference in creep behaviour between laboratory and field conditions: without any useful information on this matter, it was taken as a constant equal to 0.85.

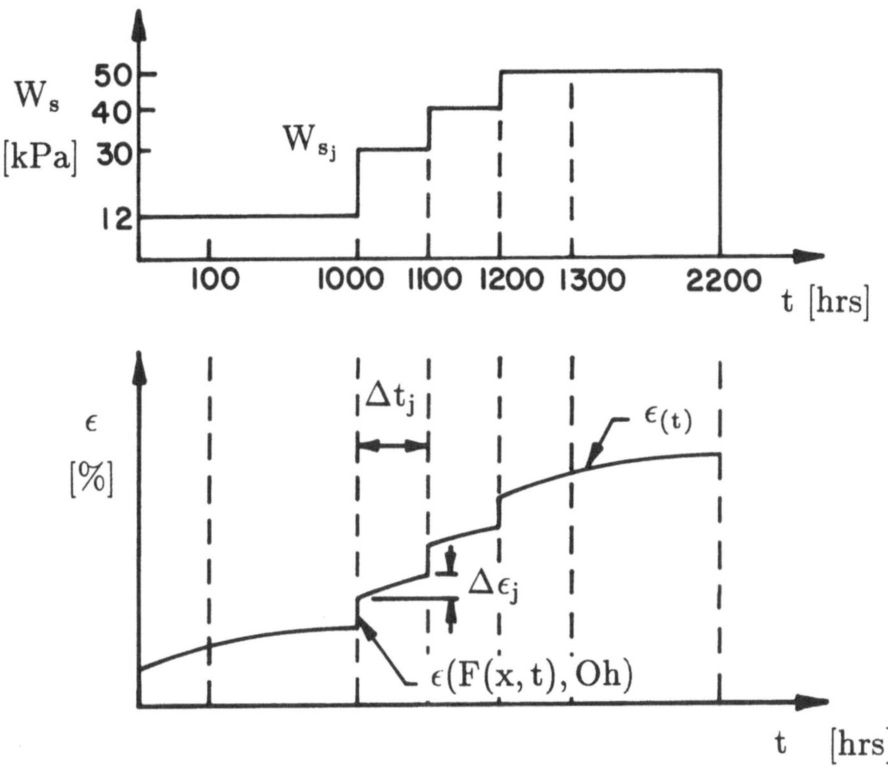

Figure F3 Qualitative Pattern of Strain versus Time and Surcharge

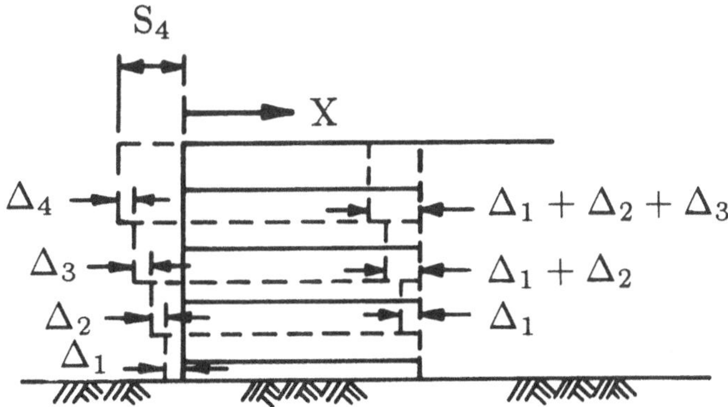

Figure F4 Scheme for the Calculation of the Horizontal Wall Movements

F3.4 Horizontal movements of the wall face

To evaluate the horizontal movements of the wall face, the following hypothesis was used: The reinforced block acts as 4 boxes superimposed one over the other and each of them represented by one of the "influence zones". The horizontal pressure which increases with depth under the crest causes the lowest box to move the other boxes with it. The procedure is repeated for each subsequent box. The horizontal movement of each box was assumed to be equal to the total elongation of each geogrid, calculated as an integration of the strains:

$$\Delta_i(W_s,t) = \int_o^L \varepsilon_i(W_s,x,t)dx \simeq \sum_{1j}^{N} \varepsilon_i(W_s,x_j,t) \cdot L_j \qquad \text{(F11)}$$

where: $\varepsilon_i(W_s,x,t)$ = strain in the geogrid layer "i"

Δ_i = as shown on Figure F4

Assuming the above mentioned mechanism, the horizontal movement of each "influence zone" was given by:

$$S_i(W_s,t) = \sum_{1h}^{i} \cdot \sum_{1j}^{N} \varepsilon_h(W_s,x_j,t) \cdot L_j \qquad \text{(F12)}$$

F3.5 Pressure under the reinforced soil block

Considering that the surcharge was uniformly distributed over the whole crest area, it was assumed that no variation of the pressure at the base of the reinforced block versus the considered position occurred. So, the base pressure was assumed to be a function of the surcharge W_s and of the dry density of the backfill sand:

$$r(x) = W_s + \gamma \cdot H \qquad \text{(F13)}$$

F4. CONCLUSIONS

At the moment, the most important assumptions to be improved are: the actual active pressure distribution; the position of the "locus of maximum tensile force" line; the distribution of the tensile forces in each inclusion, and the definition of the "In-Soil Creep Factor".

REFERENCES

1. Berg, R.R., Bonaparte, R., Anderson, R.P. and Chouery, V.E. (1986). "Design, Construction and Performance of two Geogrid Reinforced Soil Retaining Walls". Proceedings III International Conference on Geotextiles, Vienna, Austria.

2. Yamanouchi, T., Fukuda, N. and Ikegami, M. (1986). "Design and Techniques of Steep Reinforced Embankments without Edge Supportings". Proceedings III International Conference on Geotextiles, Vienna, Austria.

PREDICTORS G, H, I, J

These predictors did not submit prediction methodologies in the manner requested. It should be noted, however, that predictors G, H, I and J adopted design strategies that can be broadly classified as limiting equilibrium approaches.

5. COMPARISON OF PREDICTED VERSUS MEASURED RESULTS

5.1 General

The types of predictions requested have been described earlier in the paper. It should be noted that predictions were submitted for wall performance at 100 hrs and up to 1000 hrs after each surcharge load increment was applied. As a result there is a large amount of data available to base comparisons on. In the interests of brevity, only selected data is compared to illustrate the success or failure of the analytical methods employed. In general, predicted and measured test results have been restricted to predictions corresponding to the 12 kPa surcharging increment after 100 hrs and the 50 kPa surcharge loading increment after 1000 hrs.

6. INCREMENTAL PANEL WALL

The surcharge loading schedule actually applied to the incremental wall test is presented in the companion paper by Bathurst et al. The schedule differs from the proposed schedule originally given to the predictors in that the initial 12 kPa surcharging increment was left on longer than 1000 hrs and the last loading increment of 50 kPa was terminated after 500 hrs (instead of 1000 hrs). Nevertheless, very little movement in the system was recorded in the reinforced wall over the extended 12 kPa loading increment. Similarly, after 500 hrs of the last 50 kPa loading increment, deformation rates were essentially zero and no difference in measurements at 500 hrs or 1000 hrs would have been anticipated.

6.1 Horizontal Panel Deformations

Horizontal panel deformations at selected elapsed times are given on Figure 3a and 3b. The measured data has been presented in two ways: The curve with the larger values represents "total" horizontal deformations referenced to the initial location of the toe of the bottom facing panel unit at the time of construction. In other words these deformations are the sum total of deformations (or total wall out-of-alignment) incurred during construction and during sustained surcharging. The other curve represents only post-construction measurements. Only predictor A considered the contribution of construction-induced deformations in his calculations. Unfortunately, his finite element model predicted values of deformation indicating inward wall movements not observed in the model. All other predictors calculated deformations with respect to the end of construction panel alignments. However, the measured data curves illustrate that system deformations which occurred during construction are significant. Even at the end of the test after sustained surcharging, fully 50% of the top panel movement was due to movements generated during construction. The inability of analytical methods to accomodate construction-induced deformations represents an important shortcoming. Most esti-

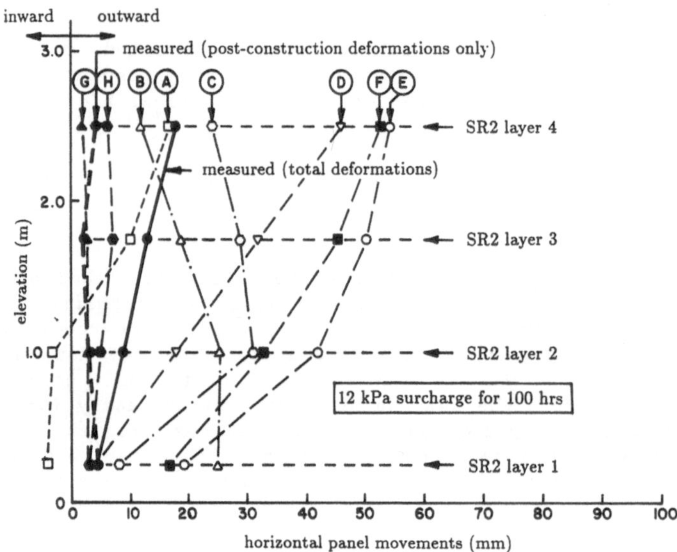

a) Deformations after 12 kPa surcharge for 100 hrs

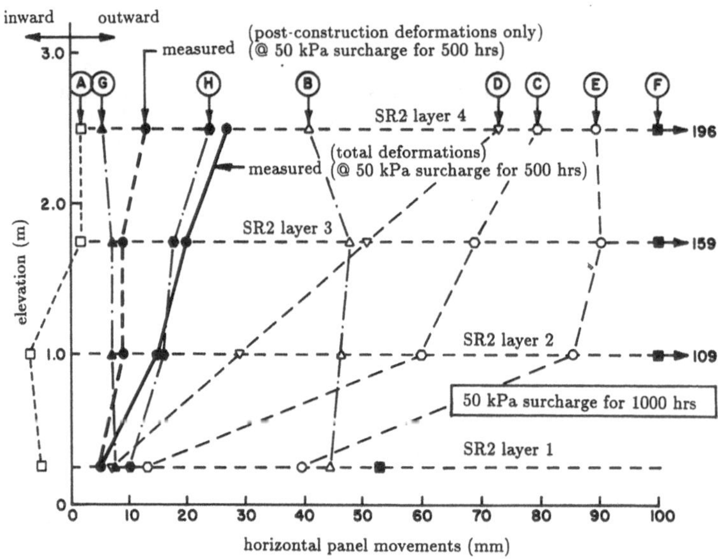

b) Deformations after 50 kPa surcharge for 1000 hrs

Figure 3 Panel Deformations (Incremental Wall Test)

mates of deformations were significantly higher than measured values despite neglecting construction deformations. The limiting equilibrium based methods employed by predictors G and H for post-construction deformations generated values which can be considered in the correct range.

6.2 Strain in Reinforcement

The strain in the geogrid reinforcement at selected times is given on Figure 4a and 4b. Similar to the comments made concerning horizontal panel movements, the measured strains include strain accumulated as a result of construction activities. Again only predictor A considered construction-induced strains in his prediction. All others referenced strain levels to the end of construction condition. All measured strains were less than 1% and the companion paper by Bathurst et al. notes that in layer 1, for instance, more than 50% of the recorded strain at the end of the test (i.e. 50 kPa surcharge for 500 hrs) occurred during construction. With the exception of predictor A, all analytical methods predicted values of strain which are in excess of the strains which could be associated with post-construction surcharging activities. The limiting equilibrium based methods considered to be least in error can be ascribed to predictors G and H.

6.3 Load in Panel/Grid Connections

Predicted and measured connection loads at the end of the test are given on Figure 5. Of all the predictions offered, only the finite-element model employed by predictor A gave values which were consistent with the measured values. Other predictions typically over-predicted the test results.

6.4 Vertical Pressure Along Base of Reinforced Soil

Predicted vertical pressures at the end of the test are given on Figure 6. No reliable physical measurements of vertical earth pressures were obtained from this test. It is interesting to note that predictions A,C,D and E gave similar trends in estimated values. In addition, these predictions can be seen to straddle estimates based on the simplified assumption of uniform vertical earth pressure according to $\gamma h+q$.

6.5 Predicted Failure Surcharge Loads

The RMC test facility at the time of the workshop was only capable of generating surcharge loadings up to 50 kPa. It was recognized by the

Predictor	Estimated Surcharge Load to Initiate Failure (kPa)
A	+200
B	350
C	130
D	1234
E	110
F	365

Table 2 Predicted Values of Surcharge Load to Initiate Failure of the Incremental Panel Wall Trial

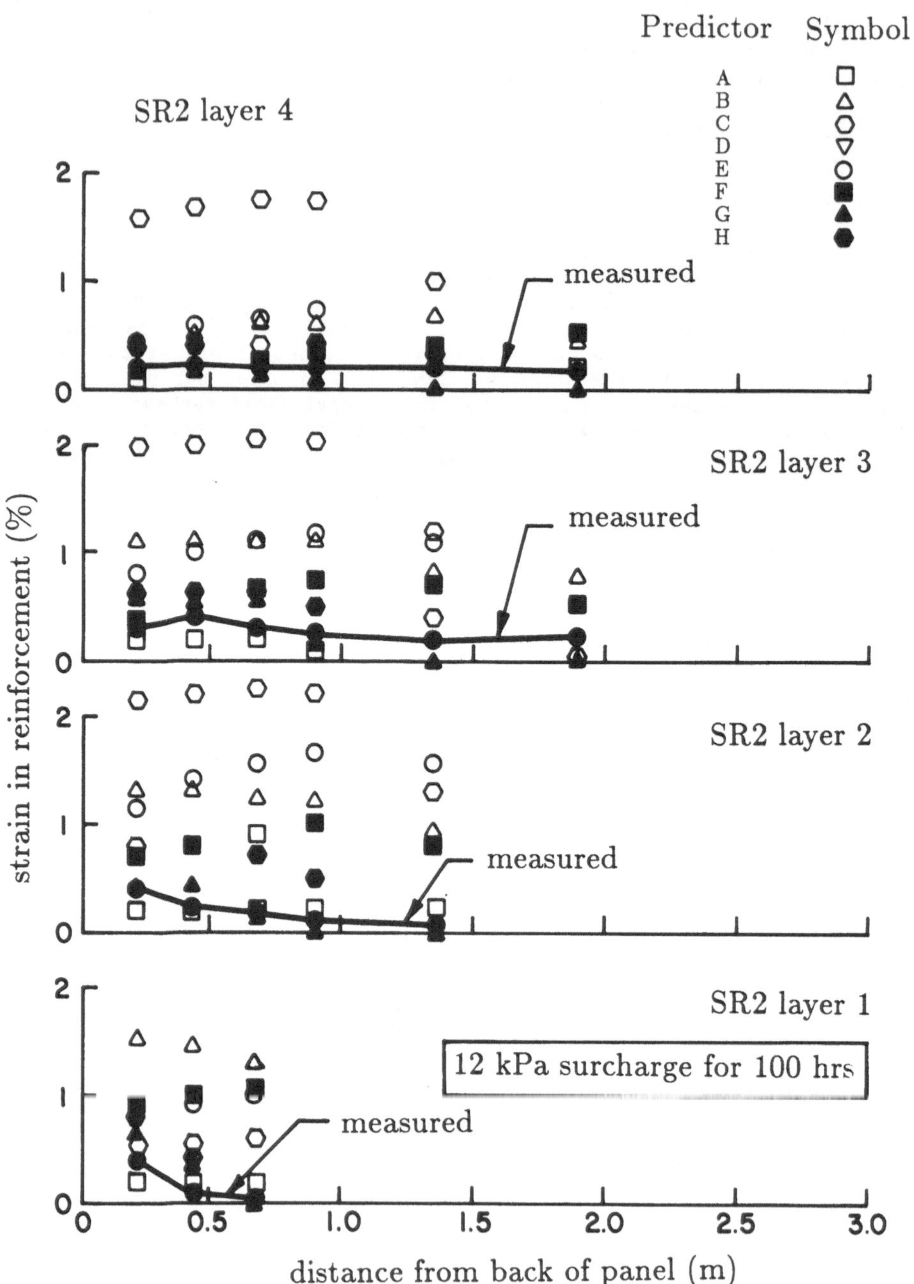

Figure 4a Strain in Reinforcement after 12 kPa Surcharge for 100 hrs (Incremental Wall Test)

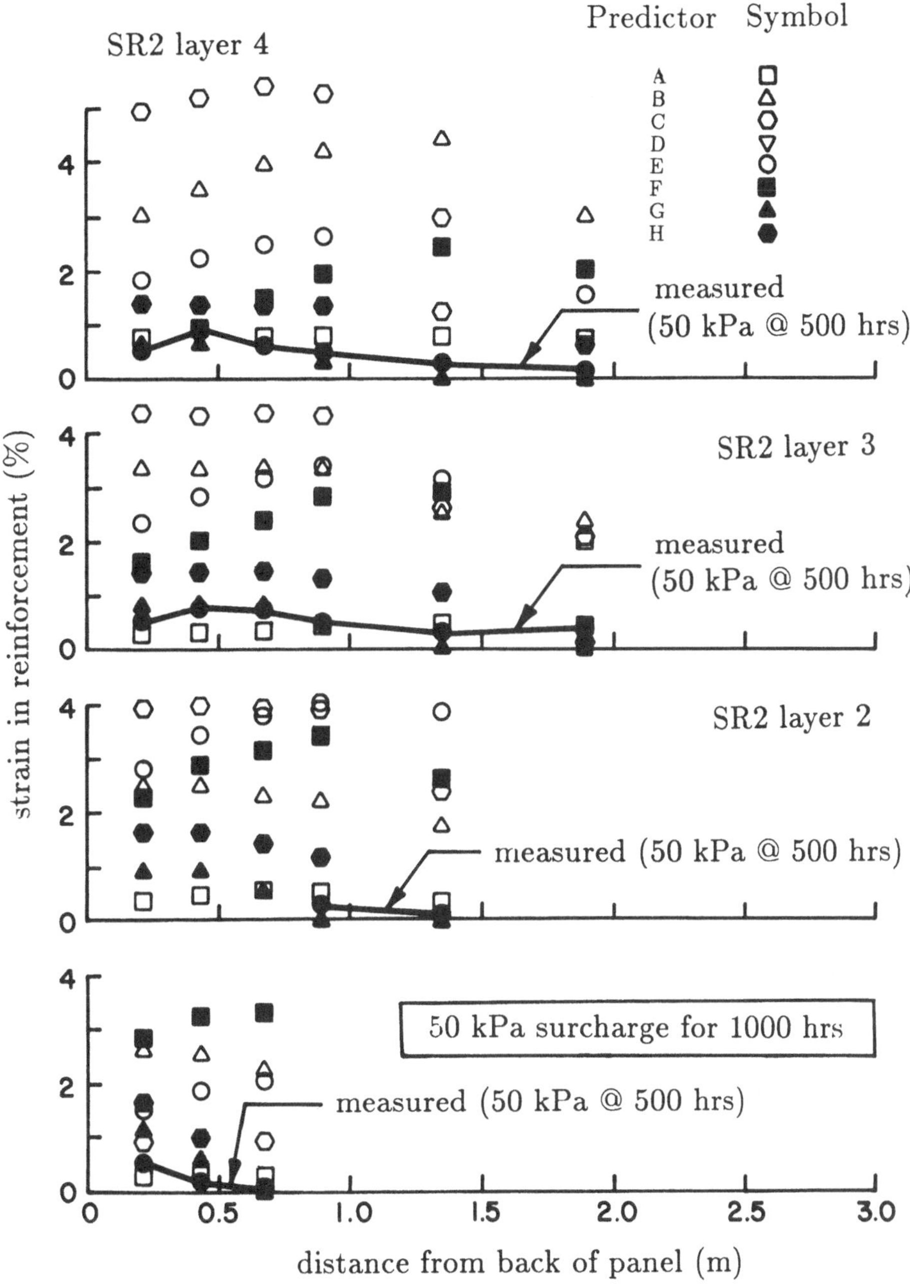

Figure 4b Strain in Reinforcement after 50 kPa Surcharge for 1000 hrs (Incremental Wall Test)

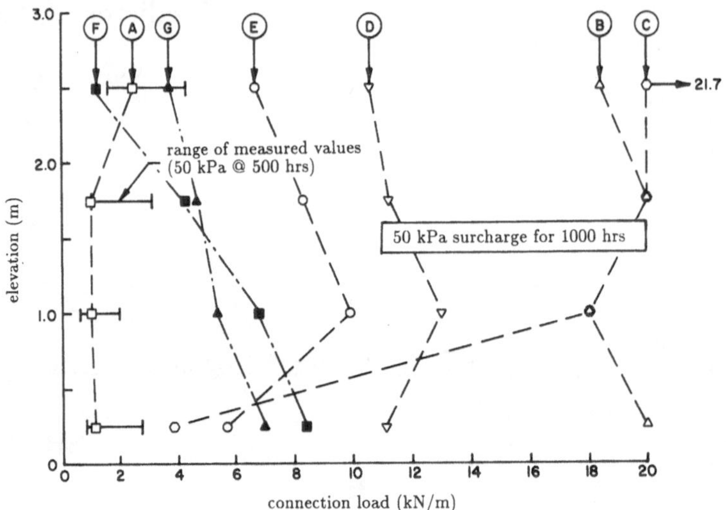

Figure 5 Connection Loads (Incremental Wall Tests)

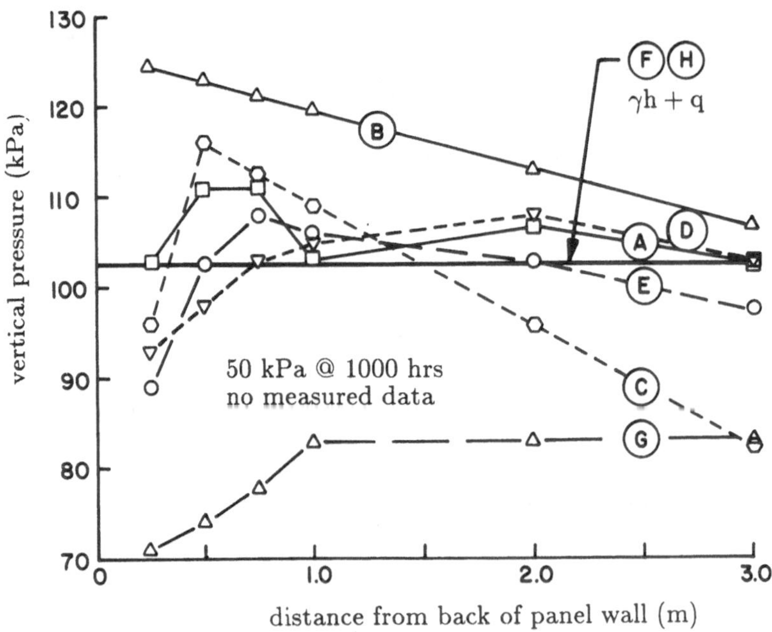

Figure 6 Vertical Pressure along Base of Reinforced Soil (Incremental Wall Test)

researchers at RMC (and by the predictors) that this load level would be insufficient to fail the trial walls. Nevertheless, the predictors were invited to submit their estimates of the critical surcharge loading required to initiate failure. The predicted values fall over a wide range as shown on Table 2.

7. PROPPED PANEL WALL

The number of predictions offered for aspects of the propped wall behaviour was significantly less than the response for the incremental wall. This observation was a surprise to the workshop organizers because the propped wall behaviour is a relatively easier problem since the facing unit was restricted from moving until after the backfill had been placed to the full wall height.

7.1 Horizontal Panel Movement

The predicted and measured wall movements at selected times are given on Figure 7a and 7b. All methods resulted in deformations which were in excess of those measured. Nevertheless, all predicted near-rigid rotation of the facing panel about the toe. Predictor A gave the smallest estimates for predicted wall deformations and hence closest to those actually observed.

7.2 Strain in Reinforcement

The strain in the geogrid reinforcement at selected times is given on Figures 8a and 8b. With the exception of predictor A, all participants significantly over-estimated measured strains in the reinforcement which never exceeded 1%. However, no analytical method could account for the observed increased strain gradient in the vicinity of the top reinforcement connection considered to be the result of relative downward movement of the sand backfill behind the wall.

7.3 Load in Panel/Geogrid Connections

Predicted and estimated values of panel/geogrid connection loads are given on Figure 9.

Predictors A,I and J estimated low to non-existent connection loads for the propped panel wall at the end of the test. The measured values are also low and comparison of estimates shows that predictor A using a finite element model and predictors I and J using limiting equilibrium methods were "qualitatively" correct.

7.4 Vertical Pressures at the Base of the Reinforced Soil

Predicted and measured values of vertical earth pressure are plotted on Figure 10 for the end of test condition. Essentially all predictions and the measured values of vertical earth pressure showed the same trends with relatively high earth pressures recorded 1 m back from the wall and reduced values toward the wall and below the free end of the reinforced block of soil. The discrepancy between measured and predicted values is likely due to loss in base pressure due to sidewall friction and (possibly) some under-registration of cell response due to cell/soil interaction.

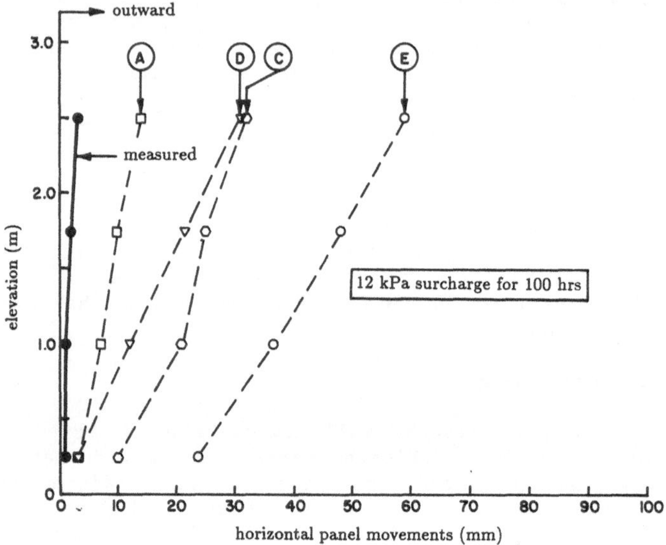

a) Deformations after 12 kPa surcharge for 100 hrs

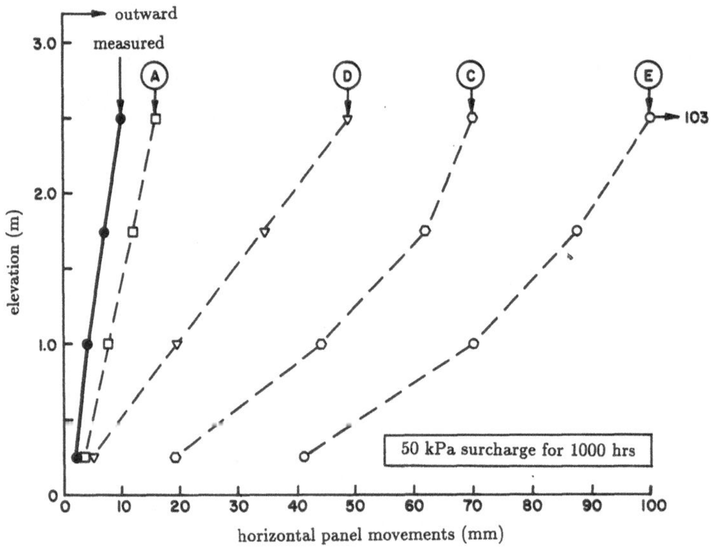

b) Deformations after 50 kPa surcharge for 1000 hrs

Figure 7 Panel Deformations (Propped Wall Test)

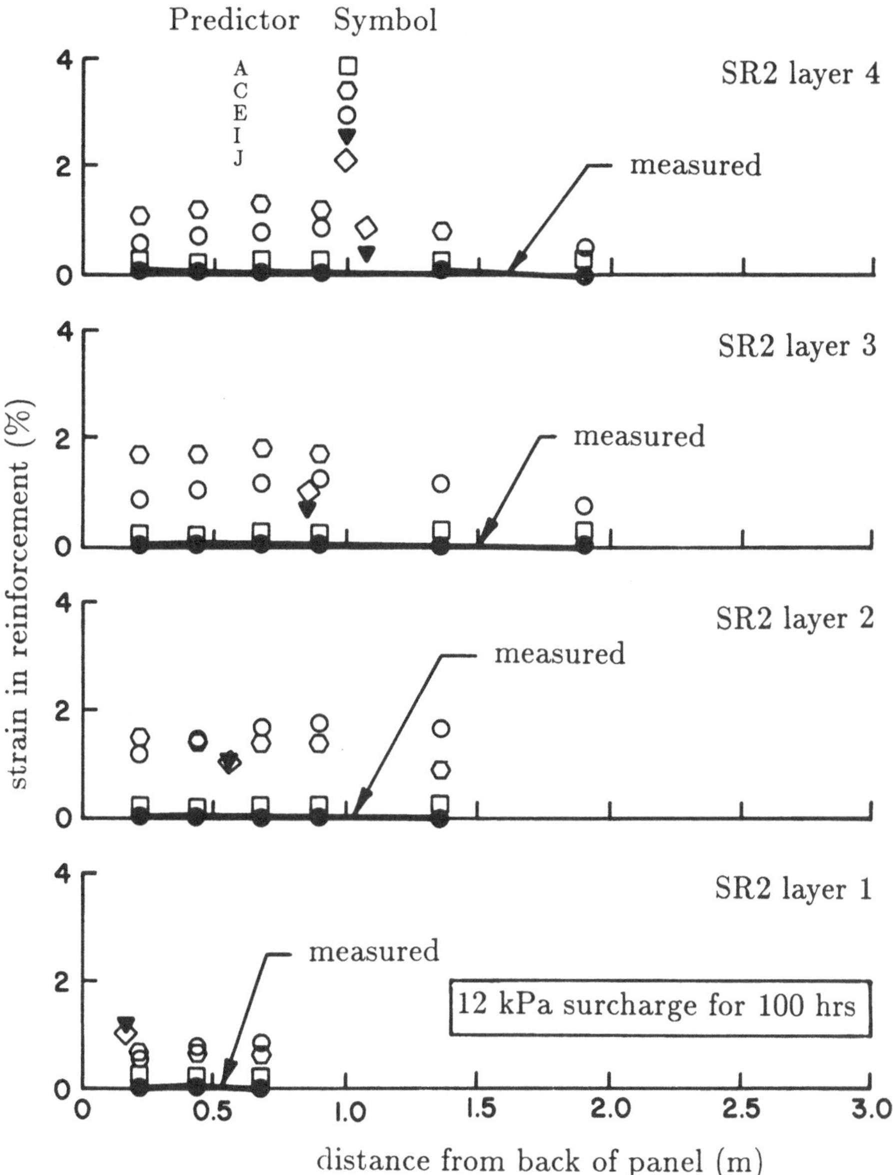

Figure 8a Strain in Reinforcement after 12 kPa Surcharge for 100 hrs (Propped Wall Test)

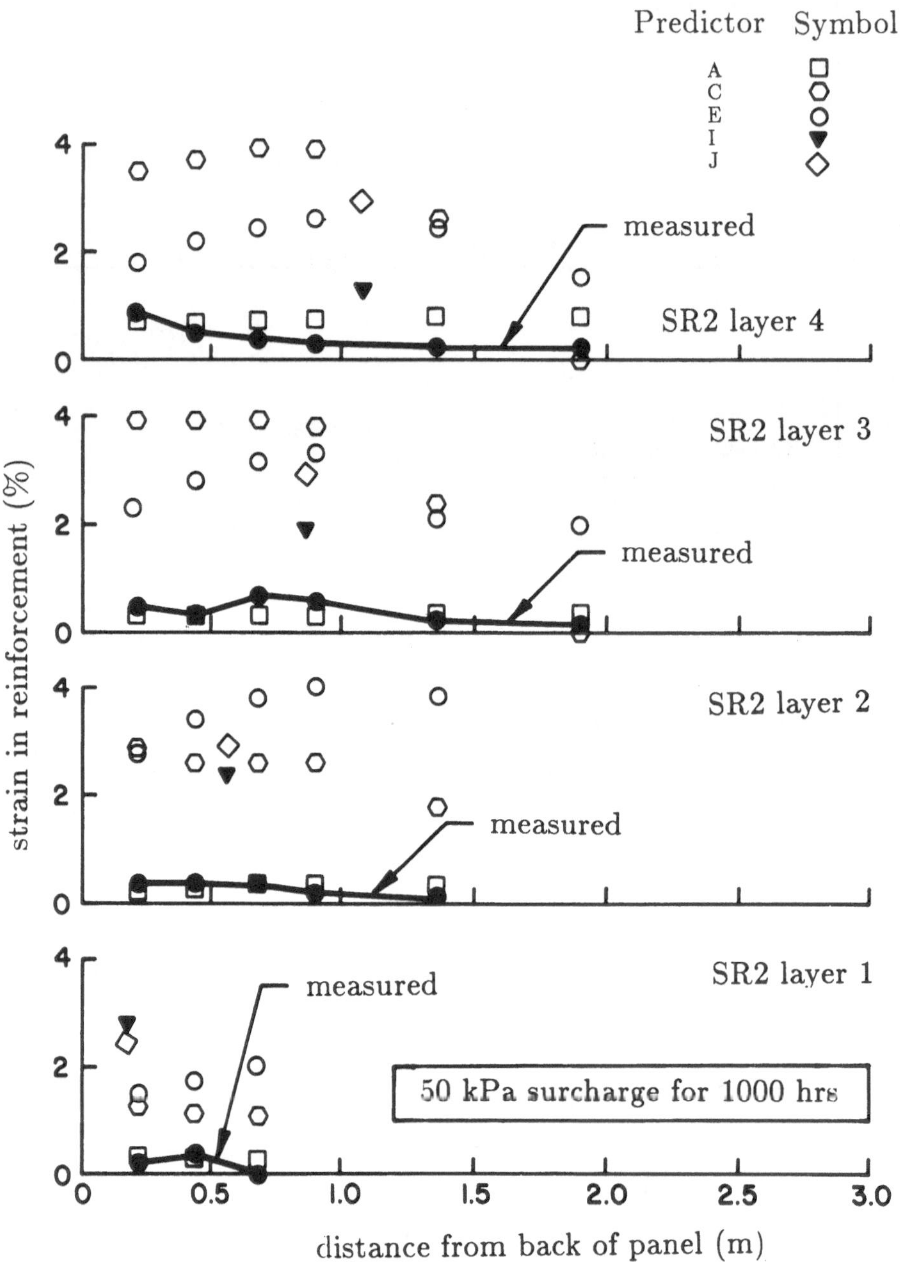

Figure 8b Strain in Reinforcement after 50 kPa Surcharge for 1000 hrs (Propped Wall Test)

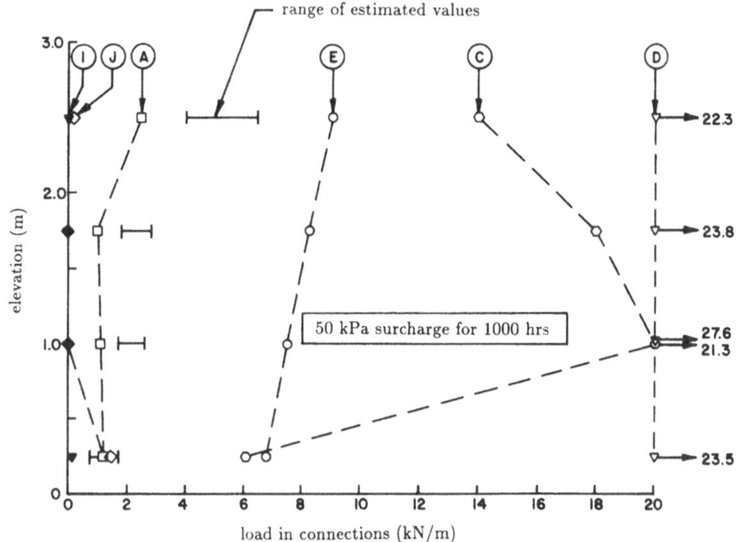

Figure 9 Connection Loads (Propped Wall Test)

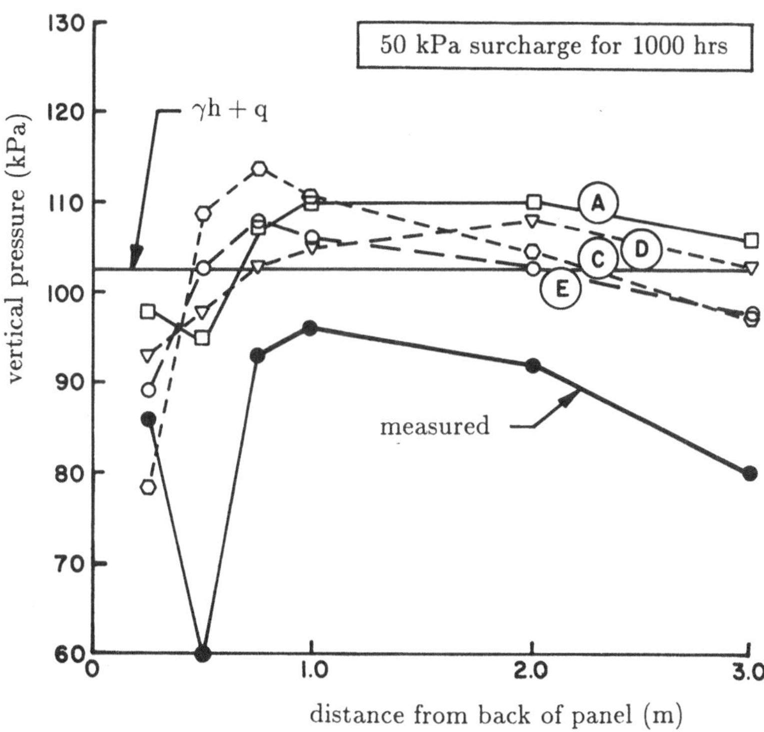

Figure 10 Vertical Pressure along Base of Reinforced Soil (Propped Wall Test)

7.5 <u>Surcharge Loading to Initiate Wall Failure</u>

The estimated magnitude of surcharge loading to initiate failure is given on Table 3.

Predictor	Estimated Surcharge Load to Initiate Failure (kPa)
A	+200
C	101
E	110
F	365
I	540
J	532

Table 3
Predicted Values of Surcharge Load to Initiate Failure of the Propped Panel Wall Trial

Similar to the results of the incremental panel wall predictions, there are a wide range of estimates of the critical failure load to initiate failure.

8. DISCUSSION

At the outset it should be noted that the predictors were faced with a formidable task: Specifically, they were required to model the behaviour of a sand/polymeric material composite. The constituent parts that comprise this composite are themselves difficult materials to model. Furthermore, the synergism between components (interaction, arching, etc.) has yet to be formulated.

With the exception of predictor A the methodologies employed were essentially limiting equilibrium analyses. The use of limiting equilibrium models to predict the behaviour of the trial walls which were at working load conditions is questionable at best and possibly very inaccurate. Nevertheless, the predictors recognized this general deficiency and in most instances attempted to modify their methodologies to accomodate this contradiction. For example: deformations far less than those required for plastic conditions in soil; friction angles below peak values; and earth pressure coefficients greater than K_a.

While most predictors gave geogrid strains that were significantly higher than measured, all strains were less than the 10% performance limit for the inclusions (McGown et al. 1984) confirming that analytical procedures correctly predicted that rupture of the reinforcement would not occur under the surcharges and surcharge durations applied. It can be inferred from the strain data predictions that most analyses predicted not only larger strains but propagation of these elevated strains further from the wall facings than was actually observed.

Based on the Statements of Methodology supplied by the participants as part of the Class A predictions, none of the limiting equilibrium methods were able to differentiate between incremental and propped wall construction and facing type. This is a serious shortcoming since significantly different facing deformation behaviours were recorded for the two retaining wall systems. In particular, the majority of facing movements for the incremental wall occurred during construction.

Limit equilibrium methods may be sensitive to the choice of friction angle employed particularly as the soil was likely at a pre-failure condition. The choice of friction angle is sensitive to the interpretation of direct shear data and further complicated by the (near) plane-strain conditions of the RMC Test Facility. The peak shear friction angles used in limiting equilibrium methods should be increased to account for plane-strain conditions as pointed out by Jewell (1987a, 1987b), Bonaparte and Schmertmann (1987). The unreinforced test reported by Bathurst and Benjamin (1987) in these proceedings confirms that a relatively high soil friction angle may be operative in retaining wall models when these structures are surcharged to failure.

Another shortcoming of all analytical methods was the neglect of facing/sand interaction. The results of the unreinforced propped wall test showed that a facing/sand friction angle of 33 degrees was mobilized. Consequently, a significant portion of vertical soil stress may be transferred to the facing in the reinforced trial walls (although these boundary forces were not monitored). This transfer should be accounted for in calculations that consider (global) reinforced soil block stability.

Facing/sand/reinforcement behaviour was also influenced by the relative vertical movement between the facings and sand backfill in these trial walls. It can also be appreciated that the degree of compaction and support offered to the geogrid in the vicinity of the connection can have an important influence on the strains generated in the extensible reinforcement. These effects are largely related to quality control during construction and consequently difficult to implement within analytical models.

Only two of the predictors considered the influence of sidewall friction directly within calculations. An explanation for this may be that only an estimate of sidewall friction angle was provided. The paper by Bathurst and Benjamin shows that about 14% of the total active earth force may be resisted by sidewall friction. Nevertheless, assuming that this value is representative of the reinforced trial walls, this reduction is not likely to bring the majority of calculated connection loads to values close to measured results. Unfortunately, it is difficult at this point in time to quantify the contribution of sidewall friction to the generation of strain in the reinforcement inclusions. Bonaparte and Schmertmann (1987) suggest that the frictional boundary conditions of the RMC test facility may reduce reinforcement strains by 20 to 30%. Jewell (1987b) suggests reductions up to 50%. Nevertheless, for most of the predictions, even application of these large corrections does not bring calculated values into the range of measured values.

In general, limiting equilibrium methods over-estimated connection loads, facing displacements and strains in the inclusions. However, from an engineering point of view the methods are inherently safe. Perhaps the most important criticism of these methods is that they are excessively safe and hence impose an economic penalty when extensible reinforcement is used in these structures. The finite element approach adopted by predictor A generally gave the most accurate predictions. Nevertheless, the model did fail dramatically at some stages in trial wall testing. Specifically, the model predicted inward facing movements at low surcharge levels. This observation suggests that despite giving relatively good estimates of wall performance at the end of each trial wall, much work remains to be done before FE models (and their "drivers") can be used to predict the full range of behaviour for these structures.

Nevertheless, it appears safe to say that FEM approaches hold great promise as the analytical tool of choice for polymeric reinforced retaining walls of the type constructed for the prediction exercise. FEM are particularly attractive if predictions of wall performance at a non-failure state are the goal. From the practitioners perspective, however, there are strong arguments for the continued use of limit equilibrium methods. Certainly, if avoidance of catastrophic failure is the only concern then the errors associated with this class of solutions may be justified. Such a justification is warranted provided it is not "excessive" (that term varying from case to case). It is felt that this set of predictions has focused on the magnitude of errors. Limiting equilibrium methods cannot be a rational design approach if "factors of safety" are simply compounded. These approaches must be used with constraint and knowledge. It is hoped that further analysis of the wealth of information found in these proceedings may shed additional light in this direction.

Finally, the authors would like to direct the readers to other papers by symposium participants that employ limit equilibrium-based approaches to solve the general problem. These include Jewell (1987a,b), Bonaparte and Schmertmann (1987), Gourc et al. (1987) and Delmas et al. (1986).

ACKNOWLEDGEMENTS

The authors would like to thank all those Workshop participants who gave predictions of the performance of the RMC trial walls. In addition, the authors would like to express their appreciation to all the participants for their contribution to the discussions that were held during the Workshop particularly with regard to interpretation of trial wall behaviour and results of analytical models. Finally the authors would like to thank J. DiPietrantonio who drafted the figures for this paper and Miss Martina Lahaie who typed the manuscript.

REFERENCES

1. Bathurst, R.J. and Jarrett, P.M. (1986). Class A Prediction Exercise for Reinforced Earth Walls, Bulletin No. 1 for NATO Advanced Research Workshop, Application of Polymeric Reinforcement in Soil Retaining Structures, Depts. of Civil Engineering, RMC and University of Strathclyde.

2. Bathurst, R.J., Wawrychuk, W.F. and Jarrett, P.M. (1987). Laboratory Investigation of Two Large-Scale Geogrid Reinforced Soil Walls, Application of Polymeric Reinforcement in Soil Retaining Structures,

NATO Advanced Research Workshop, Royal Military College of Canada, June 1987.

3. Bathurst, R.J. and Benjamin, D.J. (1987). Preliminary Assessment of Sidewall Friction on Large-Scale Wall Models in the RMC, Test Facility, Application of Polymeric Reinforcement in Soil Retaining Structures, NATO Advanced Research Workshop, Royal Military College of Canada, June 1987.

4. Bonaparte, R. and Schmertmann, G.R. (1987). Reinforcement Extensibility in Reinforced Soil Wall Design, Application of Polymeric Reinforcement in Soil Retaining Structures, NATO Advanced Research Workshop, Royal Military College of Canada, June 1987.

5. Delmas, P., Berche, J.C. and Gourc, J.P. (1986). Le Dimensionnement des Ouvrages Renforces par Geotextile: Programme Cartage", Bulletin de liaison des Laboratoires des Ponts et Chaussées, 142 mars, avril 1986.

6. Gourc, J.P., Ratel, A. and Gotteland, Ph. (1987). Design of Reinforced Soil Retaining Walls: Analysis and Comparison of Existing Design Methods and Proposal for a New Approach, Application of Polymeric Reinforcement in Soil Retaining Structures, NATO Advanced Research Workshop, Royal Military College of Canada, June 1987.

7. Jewell, R.A. (1987a). Reinforced Soil Wall Analysis and Behaviour, Application of Polymeric Reinforcement in Soil Retaining Structures, NATO Advanced Research Workshop, Royal Military College of Canada, June 1987.

8. Jewell, R.A. (1987b). Analysis and Predicted Behaviour for the Royal Military College Trial Wall, Application of Polymeric Reinforcement in Soil Retaining Structures, NATO Advanced Research Workshop, Royal Military College of Canada, June 1987.

AN ASSESSMENT OF THE RESISTANCE OF TENSAR SR2 TO PHYSICAL DAMAGE DURING THE CONSTRUCTION AND TESTING OF A REINFORCED SOIL WALL

D I BUSH - NETLON LIMITED, ENGLAND
D B G SWAN - UNIVERSITY OF STRATHCLYDE, SCOTLAND

1. INTRODUCTION

Damage during construction procedures is one of the factors affecting the physical properties of geotextiles and related products used to reinforce soil structures.

This paper makes an assessment of the damage to and change in physical properties of the 'Tensar' SR2 geogrids used as reinforcing elements in an incremental panel wall constucted at the Royal Military College, Kingston, Ontario.

Prior to shipment the "as produced" reinforcement was sampled. Constant rate of strain tensile and sustained load tests (creep) tensile tests were carried out on those samples. At the end of the experimental programme recovery of the reinforcing elements was supervised. A visual assessment of the damage to the "recovered" reinforçement was carried out and the tensile tests were repeated on representative samples.

2. CONSTRUCTION, SURCHARGING AND DISMANTLING OF THE REINFORCED SOIL WALL

'Tensar' SR2 reinforcing elements were installed and tensioned to remove the slack. An angular gravelly sand, with maximum particle size of 6mm, was placed by mechanical shovel. The fill was compacted into 125mm thick layers with six passes of a vibratory plate compactor in accordance with the Department of Transport Specification for Road and Bridge Works (1)

The reinforced soil wall was 3 metres high and surcharged with an overburden pressure of 50kPa for a period of 500 hours.

The overburden was removed from the wall with shovels and each level of reinforcement was carefully exposed for inspection, recovery and testing.

P. M. Jarrett and A. McGown (eds.), The Application of Polymeric Reinforcement in Soil Retaining Structures, 173–180.

3. TEST METHODS AND RESULTS

3.1. Sampling

Samples of the "as produced" geogrid were taken at random from the four rolls of Tensar SR2 prior to shipment. One roll was used to construct the "incremental" panel wall and the "recovered" specimens were all from this roll.

3.2. Constant rate of strain tensile tests

Samples of 'Tensar' SR2 geogrid approximately 480mm long x 400mm wide were held in specially designed jaws, described by McGown et al (2). They were subjected to a constant rate of strain of 2% per minute in a standard INSTRON 1107 tensile test machine. Load and extension between the jaws were recorded on an automatic chart recorder. Load/strain graphs are shown in Fig.1.

All tests were carried out at 20° + 2°C.

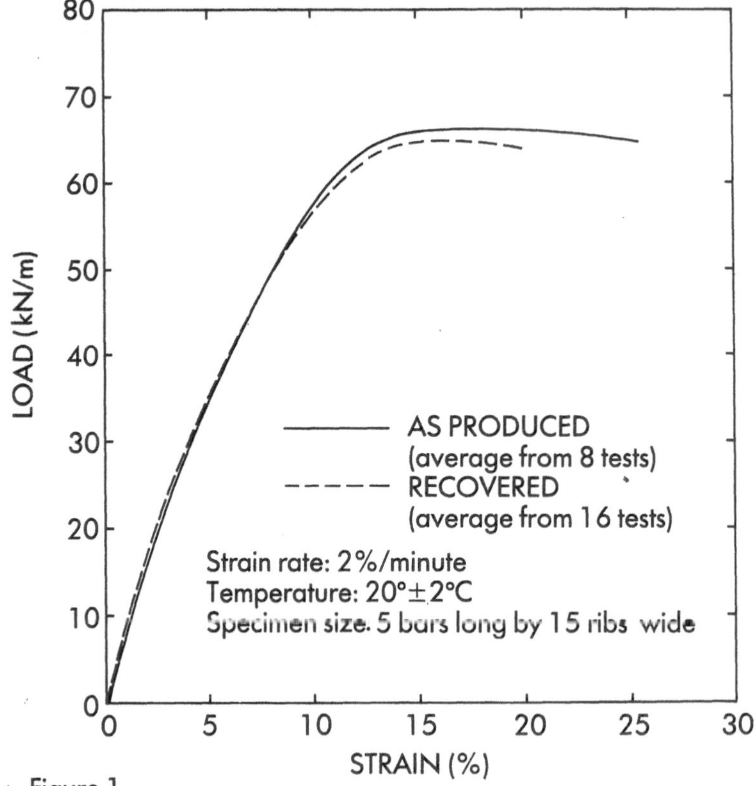

Figure 1

AVERAGE LOAD-STRAIN CURVES FROM CONSTANT RATE OF STRAIN TENSILE TESTS ON TENSAR SR2

The results for "as produced" and "recovered" reinforcement are shown in tables 1 and 2 respectively and summarised in table 3. Values of load and strain are recorded at peak load and rupture. Stiffness (i.e. secant value) is reported at 5% and 10% strain.

Specimen Number	Peak Load Data		Stiffness Values (kN/m) at Various Strains		Rupture Data	
	Load (kN/m)	Strain (%)	5%	10%	Load (kN/m)	Strain (%)
1	67.17	14.5	691	584	65.62	24.1
2	67.47	14.7	698	588	65.72	23.8
3	66.00	15.7	680	576	64.35	25.2
4	66.29	15.1	678	575	64.32	26.0
5	64.83	16.6	653	555	63.30	27.0
6	65.12	16.1	656	557	63.26	26.8
7	66.29	15.1	675	574	64.87	25.6
8	66.00	15.7	677	575	64.53	24.8

Table 1: Data from Constant Rate of Strain Testing of As Produced Tensar SR2 Geogrid used in 'Incremental Panel' Wall.

Specimen Number	Position in Wall	Peak Load Data		Stiffness Values at Various Strains (kN/m)		Rupture Data	
		Load (kN/m)	Strain (%)	5%	10%	Load (kN/m)	Strain (%)
1	P.1 (Right); B.17–21 R.12–28	65.85	16.85	710	567	65.12	19.10
2	P.1 (Right); B.22–26 R.12–28	65.85	16.45	698	567	64.24	20.55
3	P.1 (Left); B.3–7 R.1–17	65.56	15.95	698	564	63.95	19.20
4	P. 1 (Left); B.8–12 R.1–17	65.70	13.95	719	578	As Peak Load	As Peak Load
5	P.2 (Left); B.8–12 R.1–17	66.73	15.80	686	565	65.12	19.80
6	P.2 (Left); B. 3–7 R.1–17	55.73	10.05	692	557	As Peak Load	As Peak Load
7	P.2 (Right); B.24–28 R.12–28	66.59	17.35	692	563	65.41	20.90
8	P.2 (Right); B.19–23 R.12–28	65.41	16.20	686	575	63.65	20.90
9	P.3 (Right); B.14–18 R.1–17	65.18	14.65	698	569	As Peak Load	As Peak Load
10	P.3 (Right); B.19–23 R.12–28	65.41	15.90	704	565	As Peak Load	As Peak Load
11.	P.3 (Left); B.3–9 R.12–28	66.00	16.15	702	558	As Peak Load	As Peak Load
12	P.3 (Left); B.8–12 R.1–17	64.97	16.20	681	558	64.53	18.20
13	P.4 (Right); B.13–17 R.12–28	65.12	16.10	704	570	63.95	20.15
14	P.4 (Right); B.23–27 R.12–28	66.59	15.95	669	563	64.53	21.95
15	P.4 (Left); B.3–7 R.1–17	61.89	12.80	692	563	As Peak Load	As Peak Load
16	P.4 (Left); B.8–12 R.1–17	65.27	16.45	681	557	63.80	21.10

Table 2: Data from Constant Rate of Strain Testing of Damaged Tensar SR2 Geogrid used in 'Incremental Panel' Retaining Wall

Key: P. = Panel Number
B. = Bars
R. = Ribs

	Number of Specimens	Peak Load			Strain at Peak Load (%)	Rupture Load (kN/m)	Strain at Rupture (%)
		Mean, $\bar{x}$ (kN/m)	Standard Deviation, σ_n-1 (kN/m)	Coefficient of Variation $\frac{\sigma_n-1}{\bar{x}}$ (%)			
"As Produced"	8	66.15	0.90	1.4	16.2	64.50	25.4
"Recovered"	16	64.87	2.67	4.1	15.4	64.01	20.2

	5% Stiffness			10% Stiffness		
	Mean, $\bar{x}$ (kN/m)	Standard Deviation, σ_n-1 (kN/m)	Coefficient of Variation $\frac{\sigma_n-1}{\bar{x}}$ (%)	Mean, $\bar{x}$ (kN/m)	Standard Deviation σ_n-1 (kN/m)	Coefficient of Variation $\frac{\sigma_n-1}{\bar{x}}$ (%)
"As Produced"	676	15.4	2.3	573	11.6	2.0
"Recovered"	694	12.4	1.8	565	6.1	1.1

Table 3: Constant Rate of Strain Data from Tensar SR2 Geogrid used in 'Incremental Panel' Retaining Wall.

3.3. Sustained load (creep) tensile tests

To determine the load-strain-time properties of 'Tensar' SR2 geogrids, sustained load tensile tests were carried out (McGown et al (2)). The load was applied smoothly within 5 seconds to a specimen approximately 600mm long by 400mm wide and it remained constant for the duration of the test. Extension measurements were taken of the specimen at pre-defined time intervals varying from 0.2 seconds during the first ten seconds of the test to once per day after the first 24 hours. All tests were carried out at 20° + 1°C.

Four loaded levels were used to compare the "as produced" geogrid physical properties with "recovered" geogrids. The strain-time graphs are shown in Fig.2 for each load level. The variation of rate of strain with strain is shown in Fig.3. Isochronous curves are shown in Fig.4.

3.4. Visual assessment of construction damage

Standard terms are used to express the physical damage sustained during construction by each geogrid specimen. These are:

3.4.1. GENERAL ABRASION is a subjective description of the condition of the surface of the test specimen. The amount of abrasion is the extent of the abraded area expressed as a percentage of the whole specimen area.

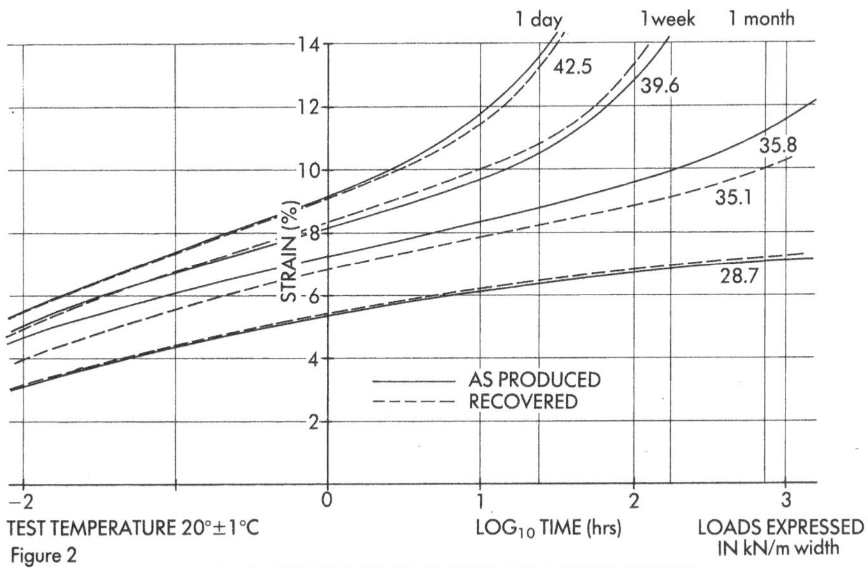

Figure 2
COMPARISON OF LOAD-STRAIN-TIME CURVES FOR 'TENSAR' SR2

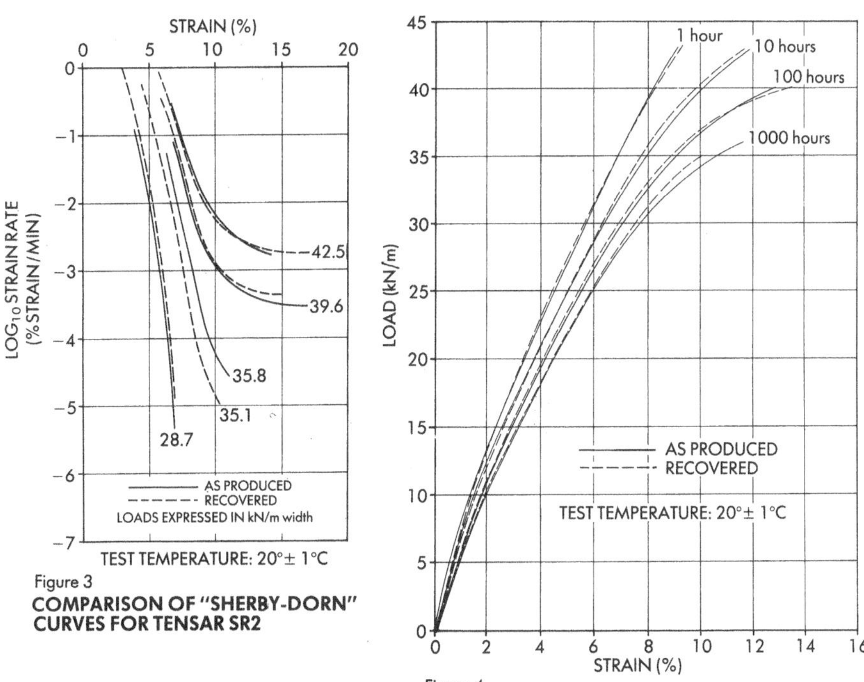

Figure 3
COMPARISON OF "SHERBY-DORN" CURVES FOR TENSAR SR2

Figure 4
COMPARISON OF ISOCHRONOUS LOAD-STRAIN CURVES FOR TENSAR SR2

3.4.2. a SPLIT is identified by the passage of light through the polymer structure when the specimen is held up against a white background. It can occur in the longitudinal ribs or the transverse bar.

3.4.3. a BRUISE is a splayed or flattened area of a bar or rib.

3.4.3. a CUT is the total severence of a rib.

3.4.4. EDGE FIBRILLATION is caused by a sharp object moving between or across the ribs, shaving away the rib edges.

The amount of damage sustained by each of the "recovered" test specimens and their position in the wall is recorded in Table 4.

Panel 1 is the lowest level of geogrid and Panel 4 the highest level above the wall base.

	Specimen Number	Position in Wall	Number of				General
			Splits	Bruises	Cuts	Edge Fibrillation	Abrasion (% of Total Area)
Constant Rate of Strain Specimens	1	P.1 (Right); B.17–21 R.12–28	0	0	0	0	20
	2	P.1 (Right); B.22–26 R.12–28	0	0	0	0	20
	3	P.1 (Left); B.3–7 R.1–17	0	0	0	0	20
	4	P.1 (Left); B.8–12 R.1–17	0	0	0	0	20
	5	P.2 (Left); B.8–12 R.1–17	0	0	0	0	10
	6	P.2 (Left); B.3–7 R.1–17	0	0	0	6	20
	7	P.2 (Right); B.24–28 R.12–28	0	0	0	0	10
	8	P.2 (Right); B.19–23 R.12–28	0	0	0	0	10
	9	P.3 (Right); B.14–18 R.1–17	0	1	0	0	10
	10	P.3 (Right); B.19–23 R.12–28	0	0	0	0	10
	11	P.3 (Left); B.3–9 R.12–28	1	0	0	0	20
	12	P.3 (Left); B.8–12 R.1–17	0	0	0	0	10
	13	P.4 (Right); B.13–17 R.12–28	0	0	0	0	10
	14	P.4 (Right); B.23–27 R.12–28	0	0	0	0	10
	15	P.4 (Left); B.3–7 R.1–17	0	0	0	6	10
	16	P.4 (Left); B.8–12 R.1–17	0	0	0	0	10
Sustained Load (Creep) Specimens	1	P.1 (Centre); B.24–29 R.1–17	0	0	0	0	10
	2	P.2 (Centre); B.23–28 R.28–44	0	0	0	0	10
	3	P.3 (Centre); B.24–29 R.28–44	0	0	0	0	10
	4	P.4 (Centre); B.21–29 R.1–17	0	0	0	0	10

Table 4: Amount of Visually Assessed Damage from Selected Tensar SR2 Test Specimens.

4. DISCUSSION

The size of the test specimen was chosen so that the behaviour of the nominal one metre wide 'Tensar' SR2 reinforcing element would be replicated and the influence of the construction damage would be properly represented. Extension measurements during creep tests were taken on the specimen to eliminate possible jaw effects.

The geogrids suffered very little damage due to construction, surcharging and recovery. There was only one split and one bruised rib in the 1,500 ribs inspected. The surface of the specimens suffered between 10% and 20% general abrasion which was to be expected for granular fills of 6mm maximum particle size. Greatest levels of abrasion occurred on the lowest levels of reinforcement. The edge fibrillation recorded on 6 ribs of each of two specimens is thought to have occurred during recovery of the reinforcement and to have been caused by a shovel.

In the constant rate of strain tests the "as produced" specimens show a slightly higher average peak load and average strain at peak load than the "recovered" specimens. Both "recovered" specimens with edge fibrillation damage show significantly reduced load and strain at peak. The "recovered" specimens with a bruise and a split show no significant difference, a result confirmed by those specimens suffering only general abrasion. Thus the reduction of the average peak load and corresponding strain of the "recovered" specimens is, therefore, due to "edge fibrillation". The values of stiffness at 5% and 10% strain show no significant difference between the "as produced" and all "recovered" specimens. The "recovered" specimens show a lower variation in stiffness and this may be due to them being selected from the same roll of 'Tensar' SR2 rather than taken at random from four rolls as the "as produced" specimens were.

In the sustained load (creep) tensile tests there is close agreement between the strain-time performance of the "as produced" and "recovered" specimens. Examination of the rate of strain - strain curves shows the instability limit strain of 'Tensar' SR2 to be approximately 18% for "as produced" and "recovered" specimens. The isochronous stiffness curves confirm the close agreement in the strain-time performance of both specimens.

In the analysis of reinforced soil structures the value long-term strength, for the appropriate design conditions, has a partial factor of safety applied to take account of variations occurring within the production process and the influence of construction the value of this partial factor of safety was recommended by Netlon Limited(3) as in the range of 1.25 to 1.4. This value was extrapolated from full-scale trials using well-graded fills of maximum particle size in the range 10mm to 125mm and appears to be conservative.

The properties of the "as produced" 'Tensar' SR2 geogrids were given to participants for predictions. These can be used to represent the in service geogrids.

5. CONCLUSIONS

This project offered the opportunity to examine 'Tensar' SR2 geogrids used as reinforcing elements of a soil retaining wall constructed under well defined conditions. Geogrids recovered from this wall suffered very little damage in the gravelly sand fill due to the construction, testing and recovery processes. There was no significant change in their physical properties derived from constant rate of strain tensile testing and from long-term sustained load tests. The properties of the "as produced" geogrids can, therefore, be used with confidence in the predictions of wall behaviour.

For design purposes, a partial factor of safety is proposed to take account of variations in production quality and damage due to construction processes. For retaining walls constructed from gravelly sand fills under well defined conditions the manufacturer's recommended value of this partial factor of safety appears to be very conservative.

ACKNOWLEDGEMENTS

This work was carried out as part of the interaction between the University of Strathclyde and Netlon Limited within the S.E.R.C. Teaching Company Scheme.

The authors wish to thank Professor A McGown and Dr K Z Andrawes of the University of Strathclyde for their helpful advice during this scheme. Thanks are also due to the staff of the Royal Military College, under Professor P M Jarrett and Dr R Bathurst, who assisted in recovery of the geogrids.

REFERENCES

1. Department of Transport: Specification for Road and Bridge Works, London, 1976.

2. McGown, A, Andrawes, KZ, Yeo, KC and DuBois, D. The load-strain-time behaviour of 'Tensar' geogrids: Polymer grid reinforcement, Thomas Telford Limited, London, 1985.

3. Test Methods and physical properties of 'Tensar' geogrids: Netlon Limited, 1986.

PRELIMINARY ASSESSMENT OF SIDEWALL FRICTION ON LARGE-SCALE WALL MODELS IN THE RMC TEST FACILITY

Richard J. Bathurst and **Daniel J. Benjamin**

Civil Engineering Department
Royal Military College of Canada

1.0 Introduction

1.1 General

The use of laboratory models to study the behaviour of soil retaining wall systems is common in the literature. Typically, experiments are performed in parallel-sided boxes of finite width with one of the perpendicular faces representing the model wall under study (e.g. Rowe, 1971). Indeed, the RMC test facility is a rather large example of this common laboratory approach. Unfortunetly, test facilities of this type can be expected to introduce some deviation in model response from the behaviour of the same system with infinite width due to *sidewall* friction. Strategies to reduce the influence of the apparatus edge effects include the use of friction reducing surfaces at the sidewalls and articulated model facings. Both approaches have been employed in the tests reported in the companion paper by Bathurst, Wawrychuk and Jarrett.

Bransby and Smith (1975) reported the results of small-scale retaining wall tests carried out using a sand backfill and rigid facings that were allowed to rotate outwards about the toe. The test results gave a reduction in the active earth pressure coefficient (K_a) of 14% for a wall height to width ratio $H/w = 2$ and a sidewall friction angle of $\phi_{sw} = 5.7°$.

As a result of discussions held during the NATO Workshop a preliminary investigation has recently been completed to study the contribution of sidewall friction to the performance of retaining wall models in the RMC Test Facility. The investigation comprised the following:

a) Direct shear tests to examine the shear-deformation behaviour of the test facility sand/polyethylene/plexiglas sidewall interface;
b) A large-scale *unreinforced* wall test to determine boundary forces acting on a model propped wall prior to and at soil failure;
c) A stability analysis of the entire block of unreinforced soil to estimate the contribution of sidewall friction to the vertical equilibrium of the soil mass after surcharging;
d) An analysis of the stability of the unreinforced test at limiting equilibrium to estimate the contribution of sidewall friction to wedge stability using direct shear test data.

2.0 Test Results

2.1 Direct Shear Tests

The results of direct shear tests carried out on samples of well-compacted dry sand are plotted on Figure 1. The results include direct shear tests performed at RMC (Lescoutre, 1986) using a small shear box apparatus ($36\,cm^2$ in plan area) and tests carried out at the University of Oxford using a larger direct shear apparatus ($393\,cm^2$ in plan area). The large apparatus test results are also reported by Jewell (1987b) in these proceedings. The combined sets of data give a peak friction angle for dense RMC

P. M. Jarrett and A. McGown (eds.), The Application of Polymeric Reinforcement in Soil Retaining Structures, 181–192.

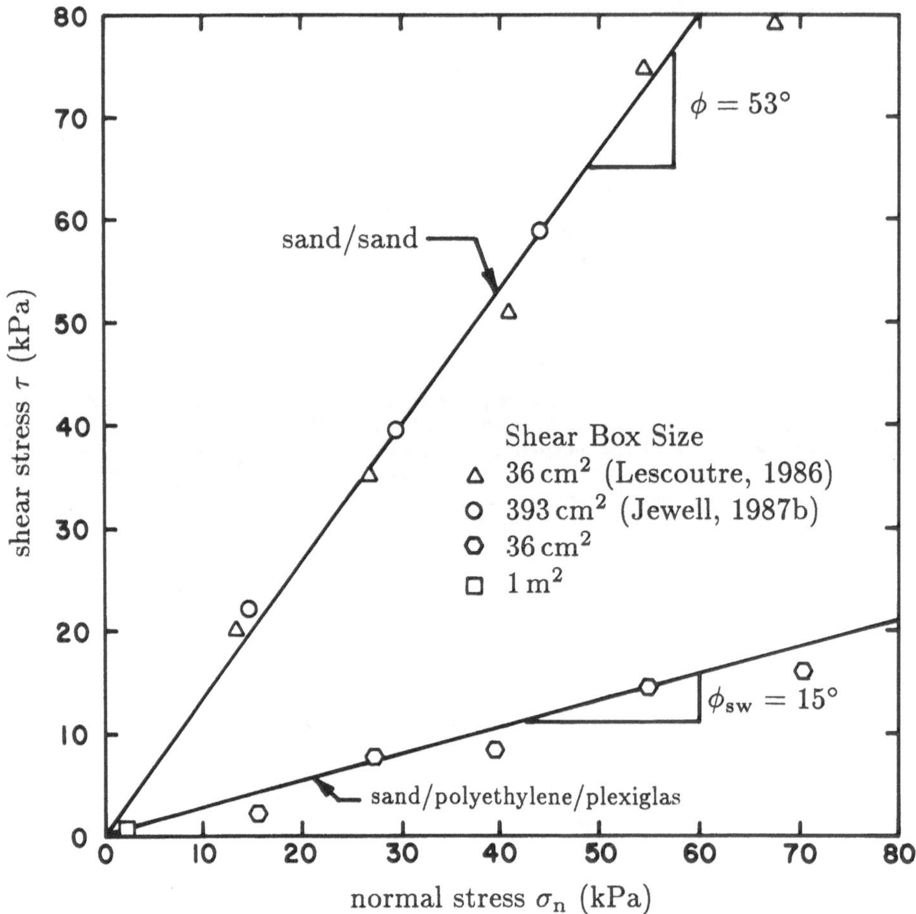

Figure 1 Results of Direct Shear Tests for Dense Sand Samples

sand of $\phi = 53°$. Also shown on the figure are the results of direct shear tests carried out to determine the shear-deformation response of the sidewall surface/soil interface. The shearing plane in these tests comprised two layers of 0.135 mm thick polyethylene sheeting over plexiglas with the sheeting left unrestrained. The data from small-scale tests gives a peak sidewall friction angle of $\phi_{sw} = 15°$. This peak friction angle is lower than the value $\phi_{sw} = 20°$ that was determined from the result of a large (1m x 1m) direct shear test reported in *Bulletin 1* by Bathurst and Jarrett (1986) (see Figure 2). This test was carried out under very low normal stress and was undertaken to give a rough estimate of the sidewall friction angle for prediction purposes and to confirm that the polyethylene/plexiglas construction did reduce sidewall friction in the test facility. The figure also shows that an attempt to further reduce sidewall friction by introducing silicon oil between composite layers was not successful at very low normal stress. Based on the comments made above it is reasonable to assume that the value $\phi_{sw} = 20°$ is an upper-bound estimate of sidewall friction angle for the trial walls reported by Bathurst et al. in these proceedings.

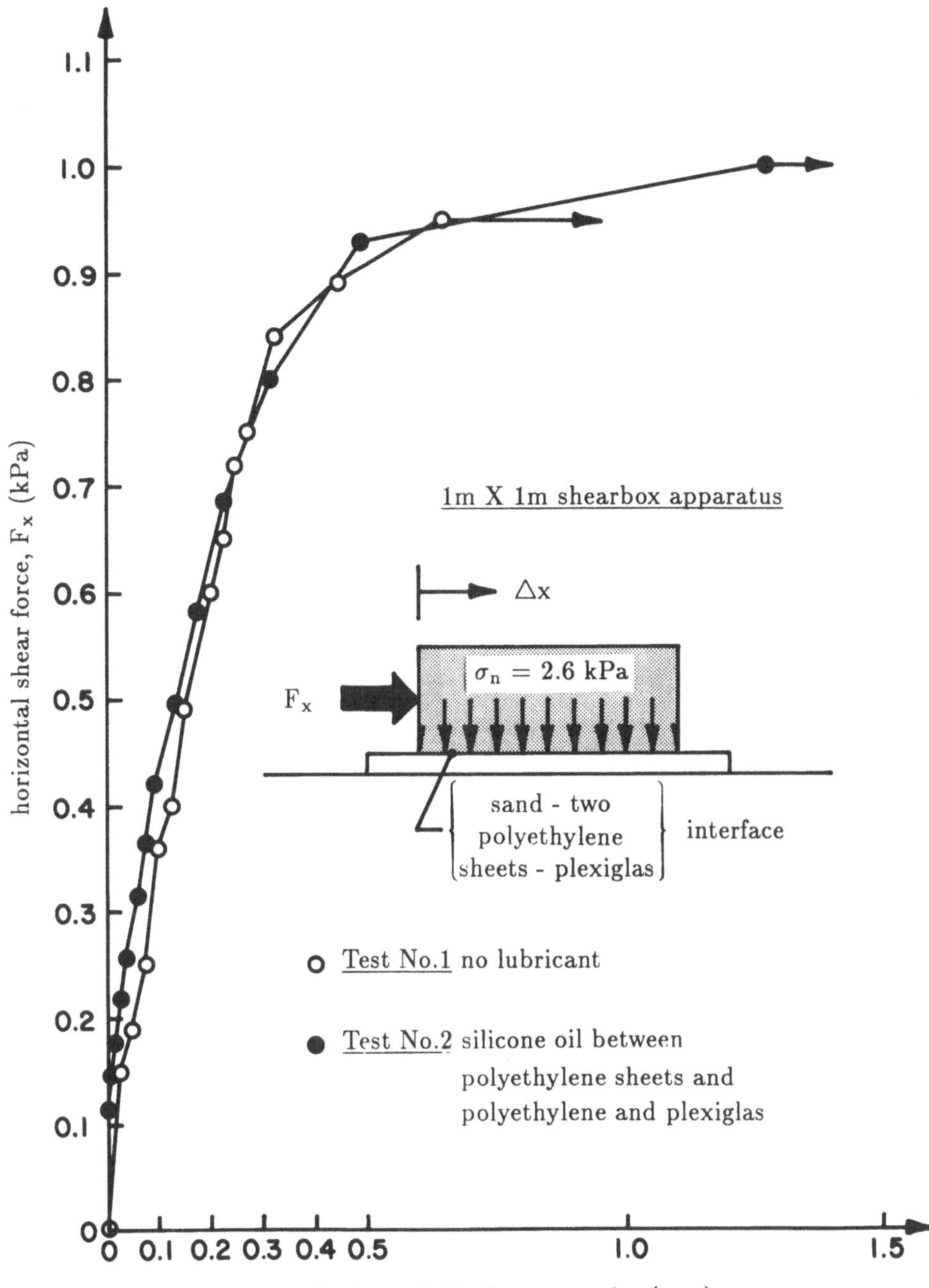

Figure 2 Large-Scale Direct Shear Tests

Figure 3 shows typical load-deformation behavior of direct shear tests carried out using dense sand samples under similar normal stress levels for sand alone and sand/polyethylene/plexiglas tests. The figure illustrates that full sidewall friction capacity is mobilized after relatively little deformation. The data from the large test on Figure 2 shows that only about 0.5 mm of shear displacement is required to achieve this condition. Furthermore, both the large-scale and small-scale tests did not show any systematic reduction in ϕ_{sw} at large deformations. The figure also shows that sand dilation in the sand/polyethylene/plexiglas tests was not recorded. This result is consistent with the visual observation that shearing in the direct shear apparatus occurred between the layers of polyethylene sheeting. The implication to the RMC facility is that full sidewall friction capacity will be mobilized during placement and compaction of the RMC sand and during surcharging of the retaining wall models.

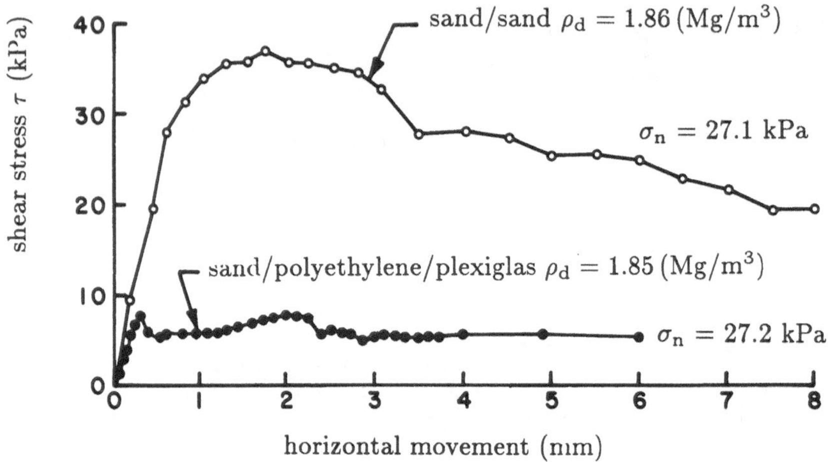

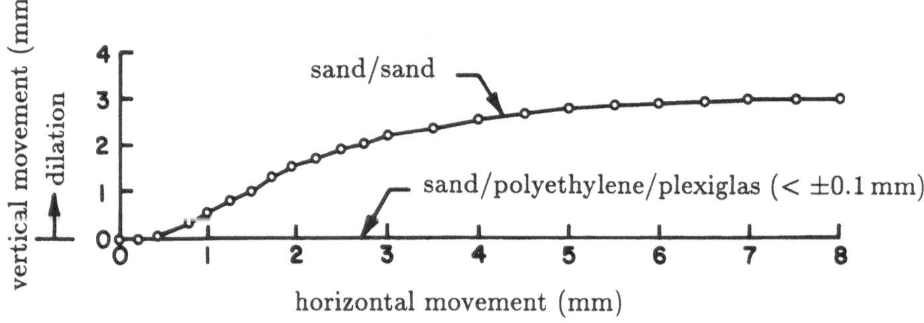

Figure 3 Load-Deformation from Direct Shear Tests

2.2 Large-Scale Unreinforced Test

A large-scale unreinforced test was carried out to measure retaining wall forces and vertical earth pressures within the soil mass. Measured values have been used to assess the influence of sidewall boundary conditions on retaining wall behaviour. The general test arrangement is given on Figure 4. The wall was supported during sand backfill placement as indicated on the figure. A computer-controlled actuator was used to provide a lateral prop support to the rigid wall facing at a point 2/3 from the base. The facing was constructed with the geogrid/wall connections in place in order to generate facing/soil interaction in a manner similar to that expected in reinforced systems. In addition, the wall was restrained vertically and horizontally at the toe at all stages in the test and load cells were used to record forces at the facing panel supports. Following construction, the soil behind the wall was surcharged to 22 kPa and the wall forces and vertical earth pressures recorded. Next, the full width of wall facing was allowed to rotate outwards about the toe at a controlled rate of 1.25×10^{-4}rad/min (i.e. 0.25 mm/min at the actuator) until an active earth pressure condition resulted (i.e. soil failure).

The soil behind the wall was determined to have failed after a toe rotation of about 0.01 radians or 20 mm movement at the actuator location. At failure, the total horizontal and vertical forces acting on the wall were measured to be about 10.4 kN/m and 6.7 kN/m width respectively. The magnitude of wall rotation required to generate the active condition is similar to values reported by Smith (1972) using a dense dry sand backfill and rigid retaining wall models.

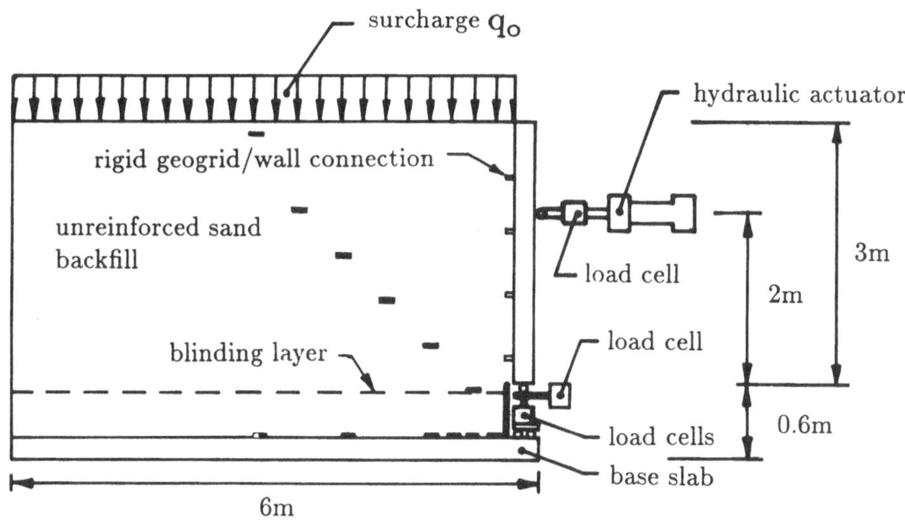

Figure 4 General Arrangement for Large-Scale Unreinforced Propped Wall Test

3.0 Results of Analysis

3.1 Vertical Stability

The total force contributions to vertical equilibrium of the retained soil within the test facility can be estimated from the results of the earth pressure cells located at the bottom of the test facility and vertical force measured at the bottom of the propped wall. The distribution of vertical earth pressures recorded below a portion of the retained soil mass after the soil was surcharged to 22 kPa can be related to Figure 5a. The figure shows the fraction (R) of surcharge pressure and soil self-weight recorded by the earth pressure cells where:

$$R = \frac{q_b}{\gamma h + q_o} \tag{1}$$

and q_b denotes base pressure. It is clear from the figure that losses occur. These losses are pronounced at the wall facing due to the relatively high wall/soil interaction developed at this interface (i.e. $\delta = 33°$) and are diminished further from the facing panels. A reasonable average R factor is about 0.88 or 0.9 indicating that about 10 to 12% of vertical earth pressure at locations well within the block of soil is lost to sidewall friction. It is possible that a portion of this under-registration is also due to cell/soil interaction. It should be noted that the surcharge pressure ($q_o = 22\,\text{kPa}$) in these calculations has been corrected for the influence of airbag coverage which does not extend completely to the sides of test facility boundaries when the airbags are inflated.

The reduction in total soil self-weight that is recorded at the base of the test facility can be denoted by X_{sw}. In addition, there will be a portion of the total surcharge force that can be expected to be carried by the sidewalls due to sidewall friction. This component of total vertical force is given the term X_q. Consider first the contribution of sidewall friction generated due to soil self-weight: The unit sidewall friction f_{sw} can be expressed as:

$$f_{sw} = K_{sw} q_z \tan \phi_{sw} \tag{2}$$

Here $q_z = \gamma z$ and is the vertical stress due to soil self-weight acting at depth z below the top of the wall and K_{sw} is the coefficient of *sidewall* earth pressure. Integration of equation (2) over both sidewalls gives a total sidewall force due to soil block self-weight according to:

$$X_{sw} = K_{sw} \gamma H^2 L \tan \phi_{sw} \tag{3}$$

where L is the length of the soil block and H is the soil block height. The additional sidewall friction generated due to surcharge loading can be accounted for by adopting an approach similar to that proposed by Jewell (1987b) in this proceedings. The attenuated vertical stress q_z acting at depth z can be described by the relationship:

$$q_z = q_o e^{-C_1 z} \tag{4}$$

where:

$$C_1 = \frac{2K_{sw} \tan \phi_{sw}}{w} \tag{5}$$

and w is the width of the soil block. The total surcharge force carried by the test facility sidewalls becomes:

$$X_q = q_o w L \left(1 - e^{-C_1 H}\right) \tag{6}$$

Assuming that the vertical force acting on the lagging board wall at the back of the test facility is equivalent to that measured on the front wall, it is possible to calculate the value of K_{sw} required to satisfy vertical equilibrium. The results of these calculations are shown on Figure 5b for $\phi_{sw} = 15°$ and $20°$. The figure shows that K_{sw} varies from about 0.23 to .52 for the range of R and ϕ_{sw} considered. This range includes the value of 0.4 that Jewell (1987b) estimates to be appropriate for the RMC sand under near plane-strain conditions.

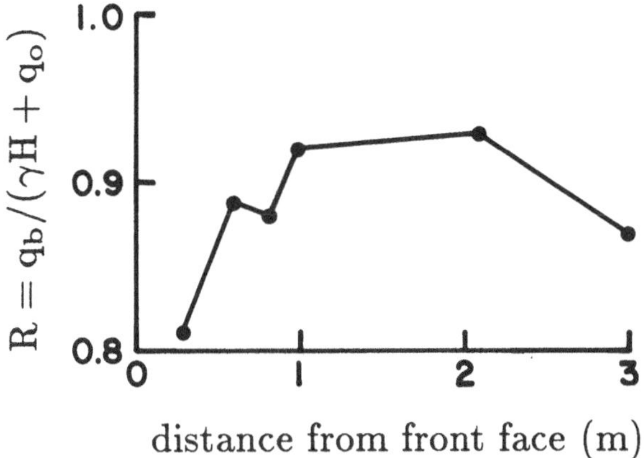

a) Distribution of Vertical Earth Pressures at Soil Base

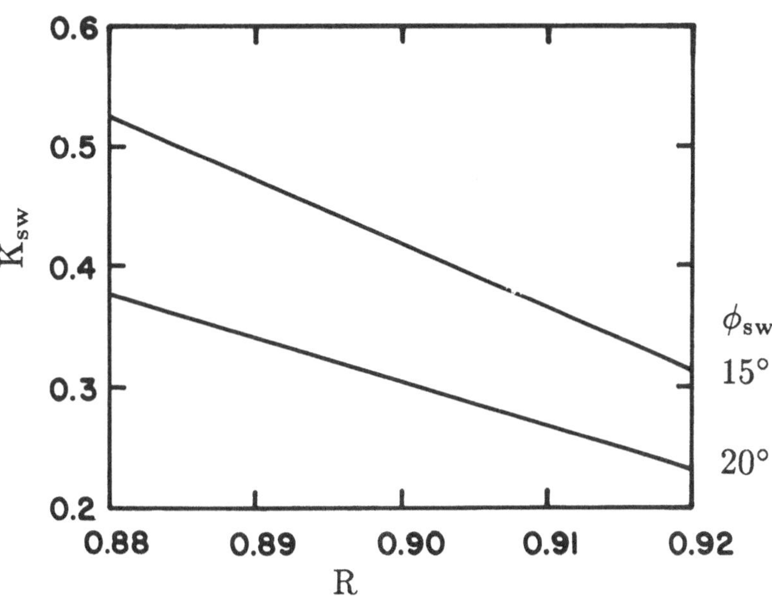

b) Coefficient K_{sw} versus R and ϕ_{sw}

Figure 5 Results of Vertical Stability Analysis

3.2 Wedge Stability at Limiting Equilibrium

A simple Coulomb wedge approach can be used to make an assessment of the contribution of sidewall friction to model wall stability at limiting equilibrium.

The horizontal and vertical equilibrium of a critical soil wedge subject to sidewall tractions can be related to the system of forces shown on the diagram in Figure 6. In this system the *total* sidewall resisting force due to wedge self-weight is again denoted as X_{sw} but is assumed to act parallel to the plane of soil failure. As in the previous analysis, sidewall friction can be expected to reduce the effect of any surcharge q_o acting at the surface of soil. This mechanism can be accounted for by introducing a vertical force vector X_q acting in the opposite direction to the wedge self-weight vector W. If the influence of sidewall friction is neglected the solution to this class of problem can be found in soil mechanics textbooks. The orientation of the critical failure plane in these problems is at (or close to) $\theta = (\pi/4 + \phi/2)$.

Consider first the contribution of soil self-weight to sidewall friction: Integration of equation (2) over both sidewalls gives a total sidewall force due to wedge self-weight according to:

$$X_{sw} = \frac{K_{sw}\gamma H^3}{3} \tan\phi_{sw} \tan(\frac{\pi}{2} - \theta) \tag{7}$$

The additional sidewall friction generated due to surcharge loading can be accounted for by adopting the approach proposed by Jewell (1987b) in this proceedings. The attenuated vertical stress q_z acting at depth z can be described by the relationship:

$$q_z = q_o e^{-C_2 z} \tag{8}$$

where:

$$C_2 = \frac{2K_{sw}}{w} \tan\phi_{sw} \sin(\frac{\pi}{2} - \theta) \tag{9}$$

Equations (2), (7) and (8) lead to an expression for the total surcharge force taken by the sidewalls:

$$X_q = C_2 w q_o \tan(\frac{\pi}{2} - \theta) \int_0^H (H - z) e^{-C_2 z}\, dz \tag{10}$$

The *total* active force P_A acting on the wall can now be expressed as:

$$P_A = \frac{W + q_o Bw - X_q - X_{sw}[\sin(\theta) + \frac{A}{D}\cos(\theta)]}{\sin(\delta) + \frac{A}{D}\cos(\delta)} \tag{11}$$

Here:

$$\begin{aligned} A &= \cos(\theta) + \sin(\theta)\tan(\phi) \\ B &= H\tan(\frac{\pi}{2} - \theta) \\ D &= \sin(\theta) - \cos(\theta)\tan(\phi) \\ W &= \gamma HBw/2 \end{aligned} \tag{12}$$

The critical solution can be determined numerically by varying θ to find $dP_A/d\theta = 0$.

With the exception of the coefficient of sidewall earth pressure K_{sw}, all variables in equation (11) are known or can be estimated with some confidence from tests carried out at RMC. Various solutions to equation (11) in terms of horizontal and vertical components of P_A are given in Figure 7. Values of $P_{Ah} = P_A\cos(\delta)$ and $P_{Av} = P_A\sin(\delta)$ are plotted against assumed values of K_{sw} using a range of soil friction angles ϕ and sidewall friction angles $\phi_{sw} = 15$ and 20°. The wall facing/soil friction angle δ was calculated directly from the measured wall forces at failure. Superimposed on the figure are the measured ranges of wall forces acting at limiting equilibrium in the large-scale unreinforced wall test. The $K_{sw} = 0$ condition represents the ideal situation of no sidewall friction. As may be expected, the measured facing loads fall below these values indicating that sidewall friction does occur. The lateral confinement of the sand soil in the RMC test facility suggests that the fully-mobilized soil friction angle is somewhat higher than the value of 53° measured in direct shear tests (e.g. Rowe, 1969). Jewell (1987a,b) suggests that a value of 55 or 56° may be appropriate. If this is the case, the percentage reduction in P_A due to sidewall friction is of the order of 12 to 18%. For $\phi = 55°$ and sidewall friction angles of $\phi_{sw} = 15$ to 20° the corresponding range for K_{sw} is 0.22 to 0.5. This range of values is in agreement with the results of the vertical stability analysis described earlier and the value of 0.4 proposed by Jewell (1987b).

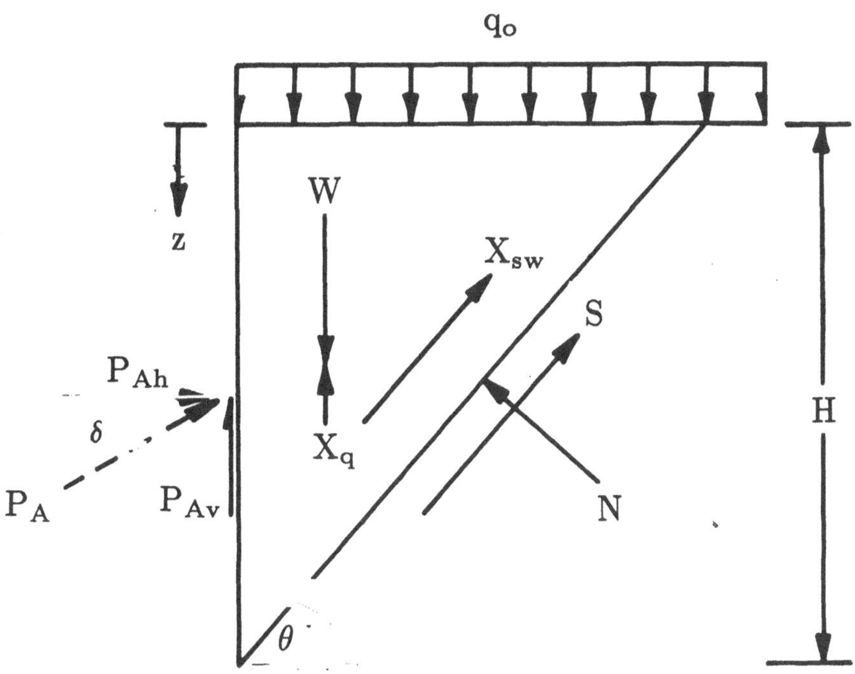

Figure 6 Coulomb Wedge Analysis with Contribution of Sidewall Friction Forces

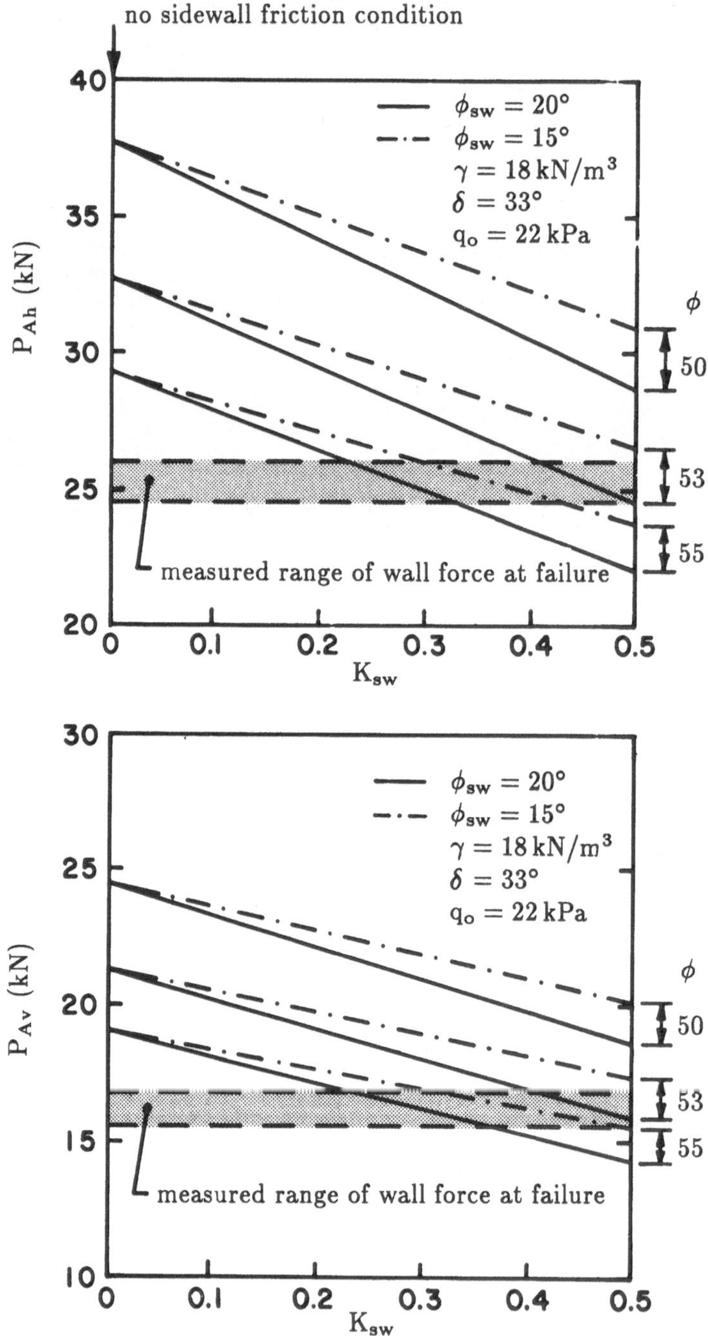

Figure 7 Results of Stability Analysis and Measured Wall Forces

4.0 Summary of Results and Implications to RMC Trial Walls

The results of the laboratory tests and analyses carried out to investigate sidewall friction contribution to test walls constructed in the current RMC test facility suggest that sidewall friction is fully-mobilized at all stages in construction and surcharging. The fully-mobilized friction angle is $\phi_{sw} = 15°$ and can be assumed to operate at large deformations including soil failure. The results of stability calculations assuming $\phi_{sw} = 15°$ show that a reasonable value for the coefficient of *sidewall* earth pressure K_{sw} is 0.4 which is in agreement with the value proposed by Jewell (1987b) based on the results of laboratory plane-strain tests. Measurements taken during surcharging of the unreinforced wall to $q_o = 22$ kPa showed that the sidewall friction effect is responsible for a reduction of about 10 to 12% in vertical earth pressure at the base of the test facility.

The limiting equilibrium wedge analysis was shown to give a correct prediction of wall forces for reasonable values of system parameters. Stability analyses can be performed for a hypothetical *unreinforced* test carried out in the RMC Test Facility with and without sidewall friction and subject to a surcharge pressure of $q_o = 50$ kPa. Assuming the following parameter values: $H = 3$m, $\gamma = 18\,\text{kN/m}^3$, $\delta = 33°$, $K_{sw} = 0.4$ and $\phi = 55°$ the analyses show that the active earth pressure force P_A is reduced below the ideal no-friction condition by 14% for $\phi_{sw} = 15°$. If it is assumed that the sidewall force contributions calculated for this unreinforced case are applicable to the same soil mass in a reinforced condition then, the 14% reduction in P_A can be used as a preliminary estimate to account for the additional capacity of the RMC trial walls due to sidewall friction.

Acknowledgements

The authors would like to acknowledge the contribution of J. Bell (Research Assistant) who helped with many aspects of the test program reported here. The authors would also like to thank J. DiPietrantonio who drafted the figures and Dr. P.M. Jarrett for his input during this investigation. Funding for this investigation was provided through the ARP program and the Chief of Construction and Properties, Dept. of National Defence (Canada).

References

1. BATHURST, R.J. and JARRETT, P.M. (1986)
Class A Prediction Exercise for Reinforced Earth Walls,
Bulletin No.1 for NATO Advanced Research Workshop, Application of Polymeric Reinforcement in Soil Retaining Structures
Departments of Civil Engineering RMC and the University of Strathclyde

2. BATHURST, R.J., WAWRYCHUK, W.F., and JARRETT, P.M. (1987)
Laboratory Investigation of Two Large-Scale Geogrid Reinforced Soil Walls
Application of Polymeric Reinforcement in Soil Retaining Structures
NATO Advanced Research Workshop, Royal Military College of Canada
June 1987

3. BRANSBY, P.L. and SMITH, I.A.A. (1975)
Side Friction in Model Retaining-Wall Experiments
Journal of the Geotechnical Engineering Division, ASCE,GT7

4. JEWELL, R.A. (1987a)
Reinforced Soil Wall Analysis and Behaviour
Application of Polymeric Reinforcement in Soil Retaining Structures
NATO Advanced Research Workshop, Royal Military College of Canada
June 1987

5. JEWELL, R.A. (1987b)
Analysis and Predicted Behaviour for the Royal Military College Trial Wall
Application of Polymeric Reinforcement in Soil Retaining Structures
NATO Advanced Research Workshop, Royal Military College of Canada
June 1987

6. LESCOUTRE, S.R. (1986)
The Development of a Large-Scale Test Facility for Reinforced Soil Retaining Walls
M.Eng. thesis, Royal Military College of Canada, Kingston

7. ROWE, P.W. (1969)
The Relation Between the Shear Strength of Sands in Triaxial Compression, Plane Strain and Direct Shear
Géotechnique 19, No.1, pp 75-86.

8. ROWE, P.W. (1971)
Large Scale Laboratory Model Retaining Wall Apparatus
Proceedings of the Roscoe Memorial Symposium, Cambridge University, March 1971

9. SMITH, I.M. (1972)
Discussion, 5'th European Conference on Soil Mechanics and and Foundation Engineering, Madrid, 1972

ANALYSIS AND PREDICTED BEHAVIOUR FOR THE ROYAL MILITARY COLLEGE TRIAL WALL

R.A. JEWELL
University of Oxford

Abstract

Predictions are made for the incremental and propped reinforced soil walls built at the Royal Military College, Kingston, and these illustrate the theory and analysis presented in a companion paper. New ideas are applied to the interpretation of shear box data to derive parameters for the sand, and supplementary test data are reported. The predictions are based on two "bounding" equilibrium states for reinforced soil walls, and the influence on these of the parameters and boundary conditions in the trial walls are investigated. The side wall friction appears to have a major influence: a closed form analysis is presented for side wall friction so that it can be taken into account for both the self weight and surcharge loading. The predictions are then presented and compared with the measured data. As all aspects of the wall behaviour under different loading conditions were predicted this provides a comprehensive test of the analysis. The finding is that the analysis captures almost all the main features of the observed behaviour, particularly the patterns of behaviour, but also the magnitudes. Further, it provides a natural link between the results of the two different trial walls.

INTRODUCTION

1.1 Introduction

This is a companion paper to "Reinforced soil wall analysis and behaviour" by R.A. Jewell published in the same proceedings, and hereafter referred to as the *companion paper*. Although all the theoretical concepts and the analysis are described in the *companion paper* additional details about the theory are brought out by the prediction for the RMC trial wall.

The data for the trial wall were presented in *Bulletin 1* by Bathurst and Jarrett (1986).

P. M. Jarrett and A. McGown (eds.), The Application of Polymeric Reinforcement in Soil Retaining Structures, 193–235.

1.2 Organisation of the predictions

The predictions for the incremental trial wall, built at the Royal Military College (RMC) in July 1987, are set out in the following manner.

First the **properties of the materials** used for the construction are examined and parameters evaluated for the subsequent analysis. Additional test data for the sand are presented from tests carried out independently at Oxford University.

The geometry of the trial wall and the theoretical distribution of the soil stresses due to self weight and surcharge loading are then discussed. The allocation of the required soil stresses to the individual layers of reinforcement is described.

A complete set of predictions for the central loading case (surcharge 12 kPa, duration 100 hours) is then made assuming **ideal boundary conditions** - that means no benefit is allowed from the boundaries of the wall, importantly the rough rigid base, the interaction between the facing and the base and the side wall friction in the test wall.

The **additional outward movement** induced by the construction of the incremental wall is then calculated for purely translating and purely rotating facing panels. This allows the overall face movements for the wall to be estimated, to complete the set of predictions.

The likely influence of the actual **boundary conditions** in the trial wall are then discussed. The rough, strong base and the connection between the face and the base are both likely to reduce the reinforcement force magnitude and distribution in the lower reinforcement layers. The choice is made to ignore these two beneficial factors because of the lack of an analysis or basic data for their evaluation.

The influence of the measured **side wall friction** is anticipated to be significant, particularly with surcharge loading. A simple closed form analysis is derived to allow side wall friction to be taken into account.

The predictions for the wall are then repeated allowing for side wall friction. For this analysis the (predicted) most likely shearing resistance for the sand is used, and the calculations carried out for the central loading case and for the **end of the test** (surcharge 50 kPa, duration 1000 hours).

Finally the **strain compatibility** in the trial wall is examined using the stress-strain properties for the sand deduced from the direct shear tests. The evaluation of this important aspect of reinforced soil behaviour is dealt with at the end of the paper because there are no direct data for the stress-strain characteristics of the sand. The interpretation for direct shear test data proposed in the *companion paper* is illustrated and used with an elastic plastic model for the sand.

1.3 Measured behaviour

In preparing the final manuscript for publication, after the trial wall has been built, it has been possible to include a section at the end of the paper commenting on the measured behaviour compared with the predicted behaviour. A general report on the predictions is published elsewhere in the *proceedings*.

SOIL MATERIAL PROPERTIES

2.1 Soil shearing resistance

The two important values of frictional shearing resistance are the **peak** angle of friction and the **critical state** angle of friction. The mobilised angle of friction in the reinforced soil wall will probably be in this range. The magnitude of tensile strain in sand below the critical state stress ratio on the stress paths relevant for a reinforced soil wall is small relative to the working tensile strain in polymeric reinforcement.

The peak shearing resistance depends on the combination of mean stress (**pressure**) and specific volume (**density**) in the soil. The pressure and density combinations for the sand fill in the trial wall will be estimated to allow the appropriate peak shearing resistance to be evaluated.

The shearing resistance also depends on the loading conditions and stress path in the soil. For the **incremental wall** there is a **loading** stress path (increasing mean stress and stress ratio) under **plane strain** conditions. The direct shear test data which are available for the sand are relevant to these conditions.

For the **propped wall** there is an **unloading** stress path (decreasing mean stress and increasing stress ratio) under **plane strain** conditions. The standard direct shear test does not follow such a stress path. A biaxial test with a constant vertical applied stress and reducing lateral stress would provide more relevant data.

2.2 Material mineralogy, grading and density limits

The data are provided in *bulletin 1* except where stated (Bathurst and Jarrett, 1986). The sand is a mix of feldspar and quartz, with relatively angular particles between medium sand and fine gravel size. The measured maximum and minimum dry densities were 19.2 kN/m3 and 15.9 kN/m3.

The (dry) density for the compacted sand in the trial wall is expected to be 17.6 kN/m3, a relative density of 52%. The bulk density of the slightly moist sand with the anticipated 3% moisture content has been taken as 18 kN/m3 for the prediction.

Because the soil is relatively coarse grained the low moisture content will probably not significantly influence the effective stresses in the soil. The properties for the dry sand were used for the prediction.

2.3 Critical state angle of friction

One measurement for the plane strain critical state angle of friction comes from direct shear tests at large shear displacement, although the measurement is imprecise at large displacement. Five of the direct shear tests reported in *bulletin 1* were sheared to sufficiently large displacement. The average critical state angle of friction from these tests was ϕ_{cv} = 38°. The results are included in **Table 1.**

Four additional direct shear tests on the RMC sand were carried out independently at Oxford University, as described in the next section. Three tests in a larger shear apparatus gave an average measured critical state angle of friction ϕ_{cv} = 38.5°, and the one test in a standard direct shear apparatus gave ϕ_{cv} = 38°.

The link between soil mineralogy and the critical state angle of friction has been highlighted recently in Bolton's (1986) correlation of data for sand. Bolton indicates that the presence of feldspar in an otherwise quartz sand gives $\phi_{cv} \approx$ 36° to 37°. For pure felspathic sand the critical state angle of friction is higher still $\phi_{cv} \approx$ 39° to 40°, Koerner (1968). Hence a critical state angle of friction between 36° and 40° is consistent with the mineralogy of the RMC sand.

Cornforth (1973) suggested a simple approximate measurement for the critical state angle of friction. He proposed that the angle of repose of a loosely tipped slope of dry sand subject to excavation at the toe would approximately equal ϕ_{cv}. This test was carried out at Oxford with the dry RMC sand, in a plane strain glass tank 150 mm wide, and gave an angle of repose 38° ± 1°.

Critical state strength for the RMC sand

The conclusion is that the feldspar content of the RMC sand results in a higher critical state angle of friction than the typical values for quartz sand. The plane strain critical state angle of friction for the RMC sand is likely to be in the range ϕ_{cv} = 37° to 39°, and the mean value was used in the predictions.

Table 1. Analysis of the direct shear test results given in *bulletin 1*.

Dry density :kN/m3	16.6	17.5	18.3	18.3
Relative density :%	21	52	73	73
Vertical stress :kPa	$(\phi_{ds})_p$	$(\phi_{ds})_p$	$(\phi_{ds})_p$	$(\phi_{ds})_{cv}$
13.5	46.5	53	56.5	38.5
27.1	42	50	54	35.5
40.6	46	47	53.5	41
54.2	43.5	49	54	38
67.7	43	45	50	38.5

2.4 Peak angle of friction

Standard (small) direct shear tests

Fifteen standard (small) direct shear tests on the RMC sand were reported in *bulletin 1.* These comprise five tests under different stresses, each repeated at three relative densities. The data have been analysed to give the peak (secant) angle of friction for the sand as a function of the relative density and the applied vertical stress, **Table 1.**

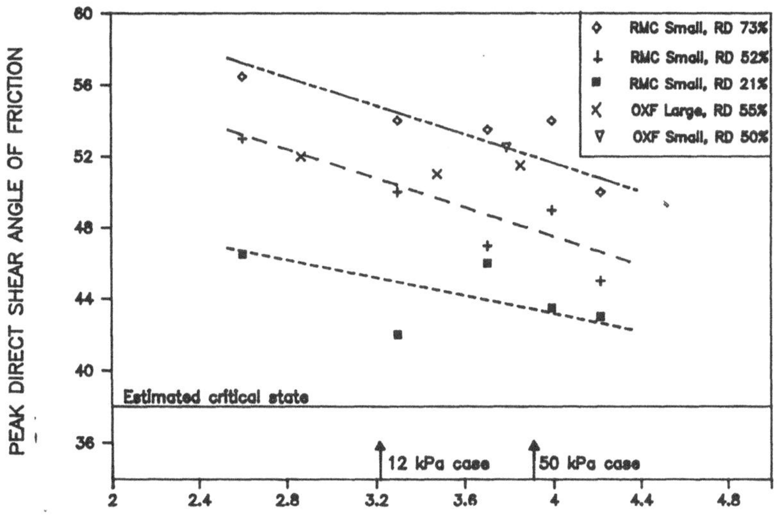

Fig. 1 Variation of the peak direct shear friction angle with pressure and density, for the RMC sand.

Frictional shearing resistance above the critical state angle of friction is derived from the sand dilatancy, which is related to the density of the soil and the mean stress level. To illustrate the relationship, the peak, direct shear angles of friction in **Table 1** are shown plotted in Fig. 1 as a function of the relative density of the sand and the logarithm of the applied vertical stress.

The peak, direct shear angle of friction can be selected from Fig. 1 for any desired combination of relative density and mean pressure in the soil within the range of the test data.

Direct shear tests at Oxford

The ratio of the length of the test specimen to the mean particle size is about 45 for the RMC sand in a standard direct shear apparatus. To check for possible scale effects, three direct shear tests were carried out at Oxford University in a 254 mm long, 152 mm wide and 152 mm deep direct shear apparatus. (Jewell and Wroth (1987) give details of the apparatus).

The conventional test arrangement was adopted (to be comparable with the RMC tests) with a rough, rigid top platen not secured to the top half of the apparatus. The tests were on dry sand with samples prepared by tamping in five layers with a 1.5 Kg tamper.

The test results are given on Fig. 2, and the measured peak and critical state angles of friction, and the peak angles of dilation are reported in **Table 2.**

One standard (60 mm by 60 mm) direct shear test was also carried out at Oxford University for comparison purposes with the three larger tests, and the fifteen RMC tests. The result is also shown in Fig. 2, and the measured parameters included in **Table 2.**

The measured peak, direct shear angles of friction in the Oxford tests are shown in Fig. 1 and compare well with the previously reported results. This provides confidence in the direct shear strength data for the RMC sand.

Table 2. Results from the direct shear tests at Oxford.

Relative density :%	σ_v :kPa	$(\phi_{ds})_p$	$(\phi_{ds})_{cv}$	$(dy/dx)_{max}$	Apparatus :mmxmm
55	17.5	52	38	.33	254x152
55	32.5	51	39.5	.35	254x152
55	47	51.5	38	.30	254x152
50	44	52.5	38	.40	60x60

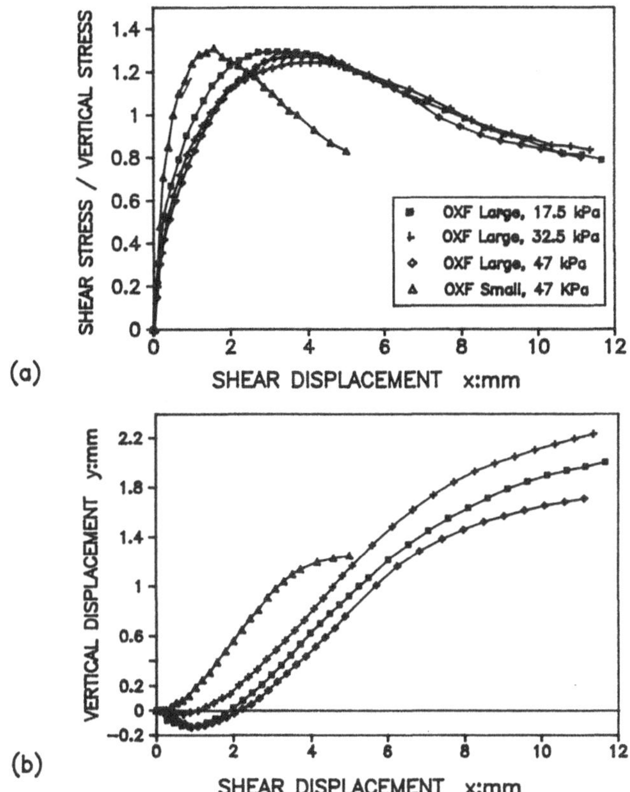

Fig. 2. Direct shear tests on the RMC sand completed at Oxford University.

Peak, direct shear angle of friction

The expected relative density for the compacted sand in the trial wall is 52% (section 2.2). The vertical stress at half the depth in the wall with the 12 kPa and 50 kPa surcharge loads will be approximately $(\sigma_v)_{12} = 40\ kPa$ and $(\sigma_v)_{50} = 80\ kPa$. Assuming a (conservative) value of the active earth pressure $K_a = 0.25$, the estimated mean stress in the sand for the two loading cases is $s_{12} = 25\ kPa$ and $s_{50} = 50\ kPa$.

The peak, direct shear angle of friction values for these combinations of relative density and mean stress level are, from Fig. 1., $(\phi_{ds})_{12} = 51°$ and $(\phi_{ds})_{50} = 49°$.

Plane strain angle of friction

The **plane strain** angle of friction, which is the relevant value for the prediction, can never be smaller than the **direct shear** angle of friction deduced from the standard interpretation for a standard direct shear test (see the *companion paper*). How much larger is it likely to be?.

One approach is to select the boundary conditions for a direct shear test to constrain the sand to deform in as uniform a manner as possible across the centre of the apparatus. For dense sand this may be achieved by fixing the top loading platen to the top half of the shear box, forcing the top half of the apparatus to move as a unit, symmetrically with the bottom half. The relative displacement measurements in the test should then more closely represent the angle of dilation in the sand. This is discussed by Jewell and Wroth (1987).

The measured angle of dilation and the critical state angle of friction for the sand can then be combined using a flow rule to estimate the plane strain angle of friction. It will be proposed in a future publication that this test arrangement and accompanying analysis will usually be conservative, only at best giving the plane strain angle of friction.

One test with these boundary conditions was performed at Oxford University on the RMC sand in the 250 mm by 152 mm shear apparatus. The sample density was 18.2 kN/m3, a relative density 70%, and it was tested with a vertical stress 30 kPa. The test results are given in Fig. 3. The measured variation of the direct shear angle of friction throughout the test is shown together with the plane strain angle of friction calculated from Bolton's (1986) simple flow rule

$$\phi_{ps} = \phi_{cv} + 0.8\psi \qquad (1)$$

with the measured angle of dilation and the expected critical state angle of friction for the RMC sand (section 2.3).

These results indicate a peak, **plane strain** angle of friction for the RMC sand of the order $\phi_{ps} \approx 55°$ to $56°$. In other words, the plane strain angle of friction is likely to be at least $5°$ higher than the measured direct shear angle of friction.

Selected friction angles

The discussion in the *companion paper* suggests that the mobilised stress ratio in the sand fill in the trial wall is likely to be high and close to peak.

A range of mobilised friction angles are used for the first set of predictions with **ideal boundary conditions** to indicate the significance of the angle of friction. The range used is $\phi = 45°$, $50°$ & $55°$.

Lower mobilised angles of friction are adopted for the calculation of the construction induced movements, because these occur with lower strain in the soil. The values $\phi = 30°$, $35°$ & $40°$ were selected.

For the actual prediction of the wall behaviour allowing for the expected boundary conditions a mobilised **plane strain** angle of friction $\phi = 50°$ was selected. This is below the peak, plane strain angle of friction for the sand.

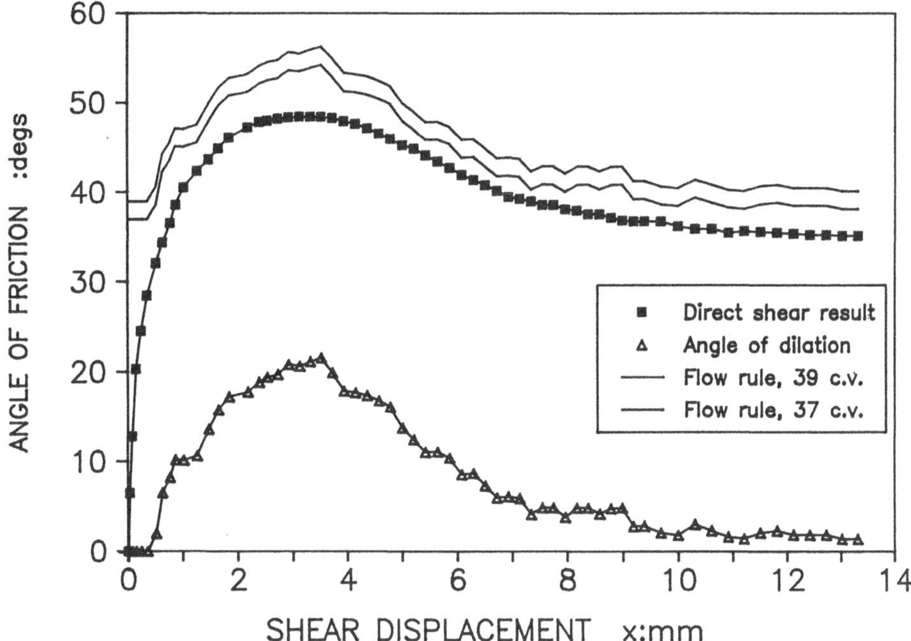

Fig. 3. Direct shear test on the RMC sand with enforced symmetrical deformation. Interpretation for the plane strain angle of friction using a flow rule.

The same friction angle was adopted for both the 12 kPa and the 50 kPa surcharge load cases. This was because the additional soil strain under the higher surcharge is anticipated to cause additional soil shearing resistance to be mobilised in the sand counterbalancing (to well within the accuracy of the prediction) the influence of the increased mean stress level in the soil.

The angle of dilation corresponding with the mobilised angle of friction is $\psi = 15°$, from Bolton's flow rule and the estimated critical state angle of friction.

2.5 Estimated soil shear modulus for compatibility

A simple elastic plastic model for the sand is proposed for making estimates of strain compatibility. An interpretation for the direct shear test has been presented to quantify the relationship between the mobilised soil frictional resistance and the principal tensile strain. The relevant equations were derived in the *companion paper*.

The shear modulus for use with the compatibility curve is estimated from the stage of the direct shear test between the critical state stress ratio, marking the end of the substantial rotation of principal axes at the beginning of the test, and the selected mobilised plane strain angle of friction.

The shear displacement developed between the critical state stress ratio and the anticipated **plane strain** mobilised stress ratio ($\phi = 50°$) is approximately 1mm, Fig. 3. The sample is 150 mm thick.

The change in stress ratio is

$$\Delta(t/s) = \sin\phi - \sin\phi_{cv} = \sin 50° - \sin 38° = 0.15 \tag{2}$$

The estimated shear strain is

$$\Delta(\gamma) = \frac{\delta x}{H} = \frac{1}{150} = 7 \quad 10^{-3} \tag{3}$$

The estimated shear modulus for a mobilised angle of friction 50° is

$$(G/s)_{50} = \frac{\Delta(t/s)}{\Delta(\gamma)} = 23 \tag{4}$$

The initial shear modulus is approximately twice as large, Fig. 3, giving a range for use with the compatibility curve $(G/s) \approx 20$ *to* 50.

The magnitude of principal strain required to reach the mobilised angle of friction $\phi = 50°$, with a shear modulus $(G/s) = 23$, is

$$(\epsilon_3)_{50} = \frac{\Delta(t/s)}{2(G/s)} = 3 \quad 10^{-3} \tag{5}$$

The analysis indicates about 0.3% tensile strain required to mobilise the working shearing resistance in the sand.

Selected shear modulus

The selected shear modulus for the sand to allow for an analysis of **strain ccmpatibility** is in the range $(G/s) \approx 20$ to 50.

This indicates an anticipated principal tensile strain required in the sand to mobilise an angle of friction $\phi = 50°$ of the order $\epsilon_3 \approx 0.2\%$ to 0.4%.

REINFORCEMENT MATERIAL PROPERTIES

3.1 Reinforcement stiffness

Preliminary estimates suggest that the wall will achieve equilibrium with a reinforcement force in each layer of the order 5 kN/m or less. No assessment is made, therefore, for the reinforcement strength characteristics as loss of stability by reinforcement rupture is not in question.

The isochronous load extension curves at 20°C for the "ex-works" Tensar SR2 reinforcement given in *bulletin 1* (*Fig. 7c*) are the correct data from which to estimate the reinforcement **stiffness** for the appropriate durations of sustained load. The expected **design temperature** 20° to 22°C and **mechanical damage** to the reinforcement material during construction may reduce the stiffness somewhat, but this should be negligible in the trial.

The isochronous load extension data has few measurements at lower loads, having been derived with higher reinforcement forces in mind than those in the RMC trial walls. The lowest test load reported for the Tensar SR2 is 13.2 kN/m, and the lowest measured strain is at this load after 1 hour, and is just less than 2%. The isochronous stiffness data which are also given are for 10% extension.

The 100 hour isochronous curve was used to derive the reinforcement stiffness for the central loading case (12 kPa for 100 hours) even though reinforcement lower in the wall would have been loaded for longer. Similarly the 1000 hour curve was used for the final loading case (50 kPa after 1000 hours) as this was the longest time for which load extension data is given. No attempt was made to extrapolate to the stiffness at the expected cumulative time 2300 hours for all the loading steps through the trial to the end of the test.

Stiffness values were determined from the extensions at a load 5 kN/m after 100 hours and 1000 hours. The magnitude of extension was 0.8% and 0.9% respectively (*Bulletin 1, Fig. 7c*), which gives reinforcement stiffness values $K_{100} = 625\,kN/m$ and $K_{1000} = 550\,kN/m$.

Selected reinforcement stiffness

The selected values of reinforcement stiffness for the two loading cases are $K_{100} = 625\,kN/m$ and $K_{1000} = 550\,kN/m$.

The K_{100} stiffness was also used for estimating the movements induced by incremental construction, although a shorter loading period would have been more appropriate. The approximate nature of the calculation for incremental movement, however, does not justify the greater refinement.

GEOMETRY OF THE TRIAL WALL AND LOAD DISTRIBUTION TO THE REINFORCEMENT

4.1 Local equilibrium

The concept of reinforcement layers maintaining local equilibrium in the reinforced soil has well as overall equilibrium was discussed in the *companion paper*. The central notion is that each reinforcement layer provides the required horizontal stresses in the soil for half the spacing to the next reinforcement layer above and below.

The geometry for the incremental trial wall is shown in Fig. 4a. For convenience the surcharge is assumed to be applied exactly on the top of the wall at 3.0 m height. There are four 0.75 m facing panels and the reinforcement is attached at the lower third point of each, 0.25 m above the panel base. The vertical depth of soil locally supported by each reinforcement layer in metres is 0.625, 0.75, 0.75 and 0.875 for layers 1 to 4 respectively, Fig. 4a.

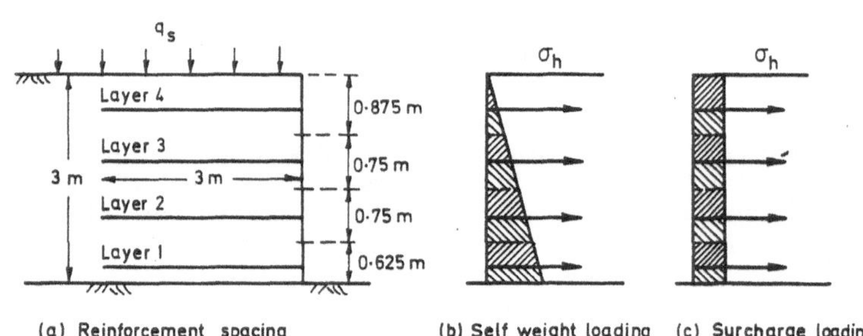

Fig. 4. Local equilibrium for the reinforcement layers in the RMC trial wall.

It is interesting to look at the percentage of the gross required force that each layer must support with **ideal boundary conditions**. The horizontal required soil stress distribution is triangular for the self weight loading, and uniform for the surcharge loading, Figs. 4b and c. The percentage of the load supported by each reinforcement layer is given in **Table 3**. This shows that the self weight loading is unevenly distributed to the lower layers (as expected with uniform spacing) but that the surcharge loading is quite evenly distributed to the four reinforcement layers.

Influence of the boundaries

The pattern of loading will be shown later to change quite significantly when side-wall friction is included.

Meanwhile, there is a estimate for the influence of the base boundary in the trial. If the rough, strong base boundary to the wall acts like a strong reinforcement layer it would reduce the load taken in the lowest reinforcement **layer 1**. In the incremental wall the base would hold the 0.125 m of soil above it in equilibrium. The required soil stresses in this zone amount to 8% of the total for self weight loading and 4% of the total for surcharge loading. The resulting percentage reduction in the force in the lowest reinforcement **layer 1** would be 22% and 19% for the self weight and surcharge loading cases respectively. This is a significant change.

Table 3. Percentage of the gross required force carried by each reinforcement layer: ideal boundary conditions.

Loading by:	Self weight	Surcharge
Layer 4	9 %	29 %
Layer 3	21 %	25 %
Layer 2	33 %	25 %
Layer 1	37 %	21 %
Total	100 %	100 %

PREDICTION WITH IDEAL BOUNDARY CONDITIONS

Central loading case: 12 kPa surcharge applied for 100 hours

5.1 Introduction

The aim is to predict the reinforcement forces and force distribution, estimate the corresponding reinforcement strains and hence the overall elongation in each reinforcement layer at the end of the loading period. This will be completed for three mobilised soil strengths covering the likely range for the trial wall. This will indicate the sensitivity of the predicted behaviour to the soil strength properties.

The additional outward movements caused by the incremental construction are estimated in section 6.4, and these allow the overall outward movement of the incremental wall face to be derived.

Two idealised equilibrium states will be analysed, the **ideal length** and **truncated length** cases described in the *companion paper*.

5.2 Force in the reinforcement

The maximum force in each reinforcement layer is uniform in the zone between the most critical mechanism and the wall face and is the same magnitude for both equilibrium cases. The required horizontal soil stress for equilibrium in this zone is given by the standard Rankine active earth pressure coefficient for a smooth wall.

Table 4. Maximum force in the reinforcement in kN/m: 12 kPa surcharge, $\phi = 50°$, ideal boundary conditions.

Loading by:	Self weight	Surcharge	Total
Layer 4	0.95	1.36	2.31
Layer 3	2.21	1.17	3.38
Layer 2	3.47	1.17	4.64
Layer 1	3.90	0.98	4.88
Total	10.53	4.68	15.21

The calculated forces in each reinforcement layer developed by the self weight loading and the surcharge loading for $\phi = 50°$ ($K_a = 0.13$) are reported in **Table 4.** The distribution of the gross required force between the four reinforcement layers in the wall is illustrated in Fig. 5.

The earth pressure coefficients for the other two soil strengths in the range are $K_a = 0.17$ and 0.10 for $\phi = 45°$ and $55°$ respectively. The magnitude of the calculated reinforcement forces are also illustrated in Fig. 5.

The range for the anticipated mobilised soil strength changes the predicted reinforcement forces by about 50%.

5.3 Strain in the reinforcement

The reinforcement stiffness for the central loading case is $K_{100} = 625\,kN/m$ (section 3.1). The maximum reinforcement strain is linearly related to the maximum reinforcement force by the reinforcement stiffness, so that the set of predicted maximum strains can be shown with the addition of a new scale at the top of Fig. 5.

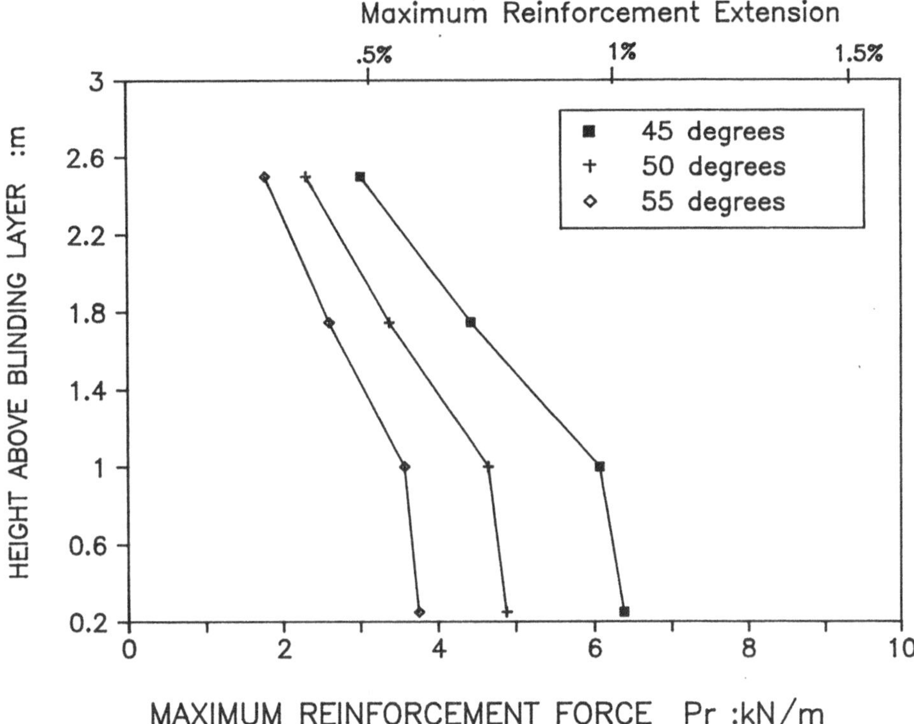

Fig. 5. Maximum force in the reinforcement layers: 12 kPa surcharge, ideal boundary conditions.

The range of predicted maximum reinforcement strain is 0.3% to 1.0%, well below the 10% limit strain for the Tensar SR2 reinforcement.

5.4 Stressed reinforcement length

Two idealised states of equilibrium in a reinforced soil wall are examined. Neither is likely to represent the actual case but they are likely to <u>bound</u> it. The ideas are described in the *companion paper* where the equations and design charts are given.

The two equilibrium states considered are the **ideal length** and the **truncated length** cases. The zone in the soil where reinforcement force is required for these two cases, and the position of the **most critical mechanism** through the toe, have been calculated for the three soil strengths and are shown in Fig. 6.

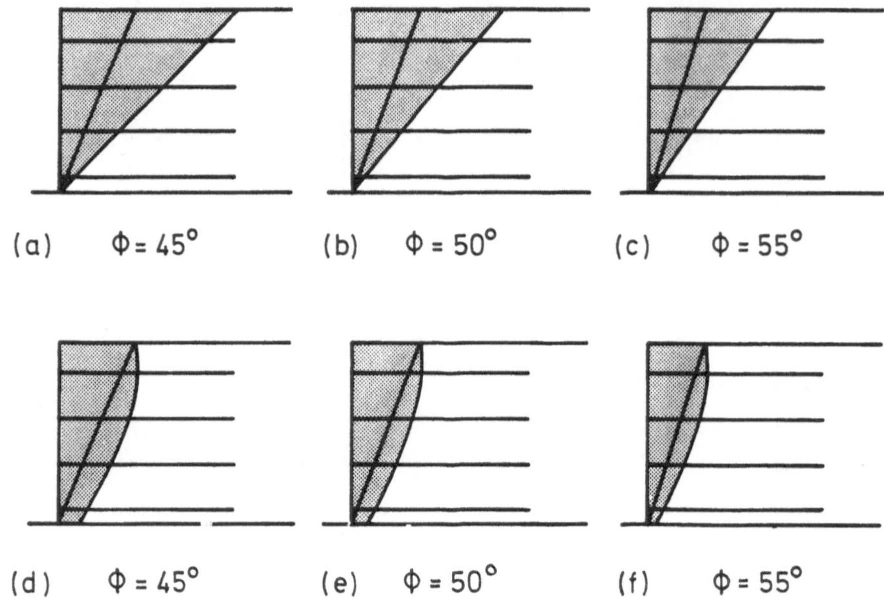

Fig. 6. Predicted zones of reinforcement force for (a to c) the ideal length and (d to f) the truncated length cases.

One point from Fig. 6 is that the location of the required reinforcement force all lies well within the available reinforcement length in the trial wall. This fact, combined with the relatively small reinforcement force magnitudes in the wall, suggests that **bond** between the reinforcement and the soil is not a problem, and relative slippage between the soil and the reinforcement in the trial wall is unlikely.

There is another implication from the above, which is that there is unlikely to be movement in the wall due to deformation in the soil behind the reinforced zone. As there is a rigid foundation beneath the trial wall, this implies that the movement at the face of the wall will be caused only by the elongation in the reinforcement layers and the additional outward movement resulting from the incremental construction.

5.5 Reinforcement force distribution

The equations giving the distribution of force along the reinforcement in the **ideal length** case are given in the *companion paper*. The force is uniform towards the face and decreases gradually between the **most critical mechanism** and the **locus of zero required force**. The magnitude of the maximum force in each layer was calculated in section 5.2.

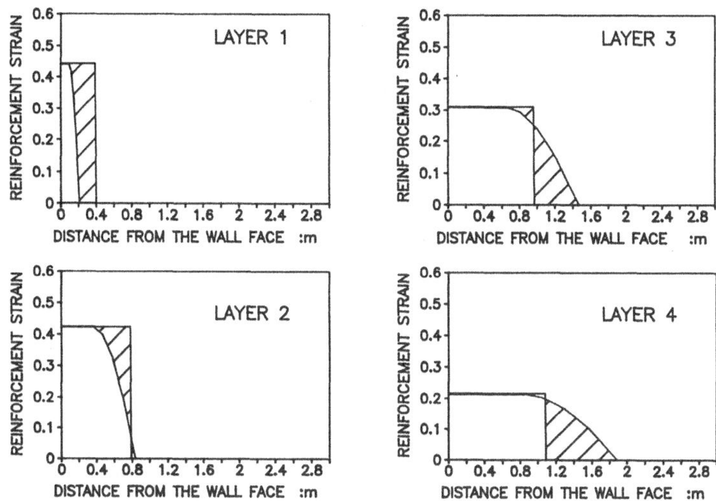

Fig. 7. Predicted reinforcement force distributions: 12 kPa surcharge, ideal boundary conditions.

The **truncated length** case involves an idealised reinforcement force distribution with a uniform force (equal to the maximum force) along the whole length of the reinforcement. Clearly a bond length would be required on the end of the reinforcement layers to generate the reinforcement force if this limiting state of equilibrium were to be achieved.

The two "bounding" reinforcement force profiles are shown plotted together in Fig. 7a to d for the four reinforcement layers in the trial wall and for the case $\phi = 50°$. The range between the two force profiles has been shaded in the figure. The actual distribution of the reinforcement force would be expected to lie within the shaded zone.

There are two points which should be brought out:

Firstly, the "bounds" only apply to the **distribution** of the reinforcement force behind the most critical mechanism, the **maximum force** in each layer is the same for the two equilibrium states.

Secondly, the shape of the predicted force profile changes with the elevation of the reinforcement layer in the wall, Fig. 7. The stressed reinforcement length is greater towards the top of the wall. The predicted "rate of change" in the reinforcement force (ie the force transfer with the soil) is much steeper lower in the wall.

5.6 Reinforcement elongation

The effect of a bond length on the overall elongation in the reinforcement is small for the **truncated length** case for wide width polymer reinforcement materials. These reinforcements typically have a high bond capacity compared with the working force magnitude. The bond length is ignored in the calculation of the reinforcement elongations in the truncated length case. This is consistent because the **truncated length** case aims to represent the minimum likely reinforcement length, and hence the minimum likely reinforcement elongation.

Table 5. Calculation of reinforcement elongations: 12 kPa surcharge for 100 hours duration.

Reinforcement	Elevation :m	$\frac{z}{H}$	$\frac{\delta K}{HP}$	P :kN	δ :mm
Layer 4	2.50	0.17	0.36	2.30	4.0
Layer 3	1.75	0.42	0.33	3.38	5.2
Layer 2	1.00	0.67	0.26	4.64	5.8
Layer 1	0.25	0.92	0.13	4.88	3.1

Notes: $K = 625\, kN/m$; $H = 3000\, mm$.

The distribution of elongation in the reinforcement layers can be derived directly from the non-dimensional charts given in the *companion paper*. The elongation is simply scaled from the chart at the elevation of the reinforcement layer in the wall. The magnitude of the strain is calculated by using the maximum reinforcement force and the reinforcement stiffness for the particular wall.

The procedure is illustrated in Fig. 8a for the **truncated length** case and a mobilised soil strength $\phi = 50°$. The calculated reinforcement elongations are shown in Fig. 8b as the implied horizontal movement at the face, measured in mm. The numbers are entered in **Table 5.** The procedure may be repeated for the **ideal length** case, and for the other mobilised soil strengths, using the appropriate charts in the *companion paper.*

The results for **both equilibrium cases** are shown plotted together for the **three soil strengths** in Fig. 9. The actual distribution of reinforcement elongation is expected to lie between the two predictions, and this range has been shaded.

One point to notice is the relatively small difference between the predicted reinforcement elongation for the two equilibrium states at about half the wall height. The main difference between the two equilibrium states occurs at the top and bottom of the wall. The difference between the two "bounding" predictions also indicates the **order of accuracy** which might be anticipated from the prediction.

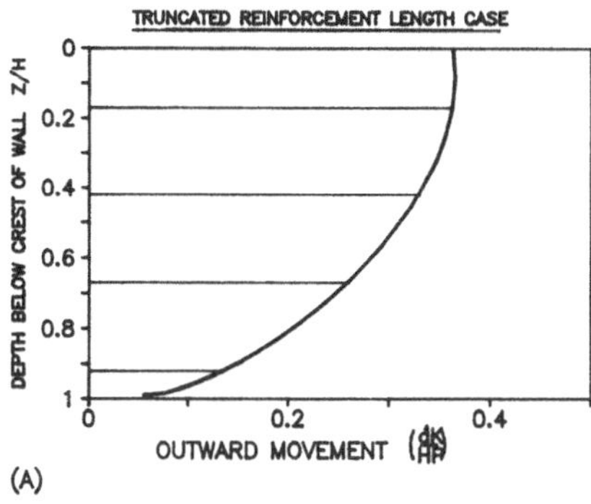

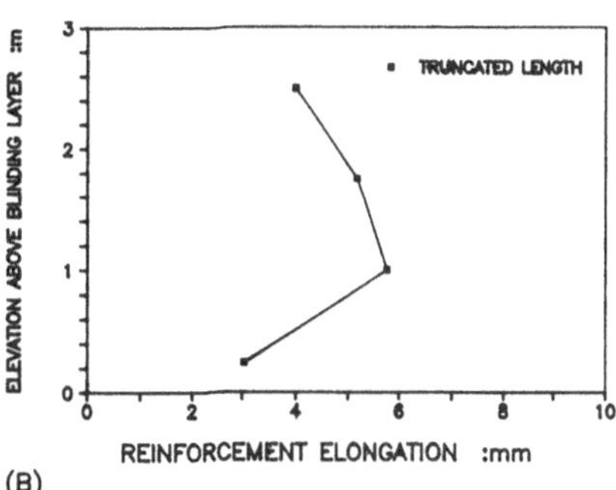

Fig. 8. (a) Use of a non dimensional chart (b) to calculate reinforcement elongations.

Another important point to remember concerning these results is that the reinforcement elongations do not represent the total outward movement at the face. The influence of the "moving datum" due to incremental construction has to be added.

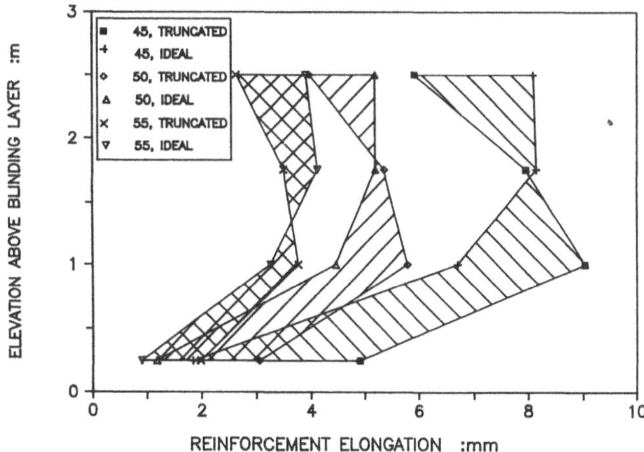

Fig. 9. Range of reinforcement elongations: 12 kPa surcharge and ideal boundary conditions.

ADDITIONAL MOVEMENT FROM INCREMENTAL CONSTRUCTION

6.1 Introduction

The additional movement from incremental construction represents a "moving datum" for the reinforcement layers because they are attached to panels which are aligned over the previous layer of panels which have already moved outwards.

At any elevation where there is reinforcement in the wall, the net surveyed outward movement of the face from the vertical line through the original position of the lowest panels is made up from (1) the elongation in the reinforcement layer and (2) the initial position of the unstressed (or lightly prestressed) reinforcement layer from the vertical line.

6.2 Panel geometry and movement

The trial incremental wall is built from four 0.75 m high panels each with one reinforcement layer attached, which means that the incremental displacement will be the same for each construction lift. The reinforcement is attached to the panel at the lower third point, and the layers of panels are not connected top to bottom. This gives each panel the maximum freedom of movement.

The actual panel movement in the trial wall is unlikely to be pure **translation** because the net soil stress on the panel will act higher than the lower third position of the equilibrating reinforcement force. The panel movement is also unlikely to be pure **rotation** which would occur if the bottom of the panel were fixed to the top of the panels in the layer below. Both these extreme cases are calculated, so that the actual net movement would be expected to lie somewhere in the range.

6.3 Soil strength and compaction load

Incremental construction movement occurs during initial filling and compaction. The deformations in the soil and the force in the reinforcement are relatively small. Therefore the appropriate mobilised angle of friction for the soil should approximately represent "at rest" conditions or a little higher. For the RMC sand this would be equivalent to $\phi \approx 35°$. A range of values $\phi = 30°, 35° \& 40°$ are used below for comparison.

The other parameter which must be evaluated is the uniform surcharge equivalent to the compaction and construction loading. For the trial wall the dead weight of the vibrating plate compactor is 1 kN/m2. The static equivalent to the dynamic load depends on the frequency and amplitude of vibration. Construction loading also comes from people and equipment moving the fill, and any temporary overfilling. The choice of equivalent load to represent all of these for the prediction must be somewhat speculative. A value 4 kN/m2 was selected (equivalent to 0.2 m of temporary overfilling).

6.4 Incremental construction movement

The values of the parameters for the trial wall ($H_{inc} = 0.75m$, $q_s = 4kN/m2, n = 1, \gamma = 18kN/m3, K = 625kN/m$) can be substituted directly into the equations in the *companion paper, section 5.2,* to determine the incremental displacement in each layer of construction.

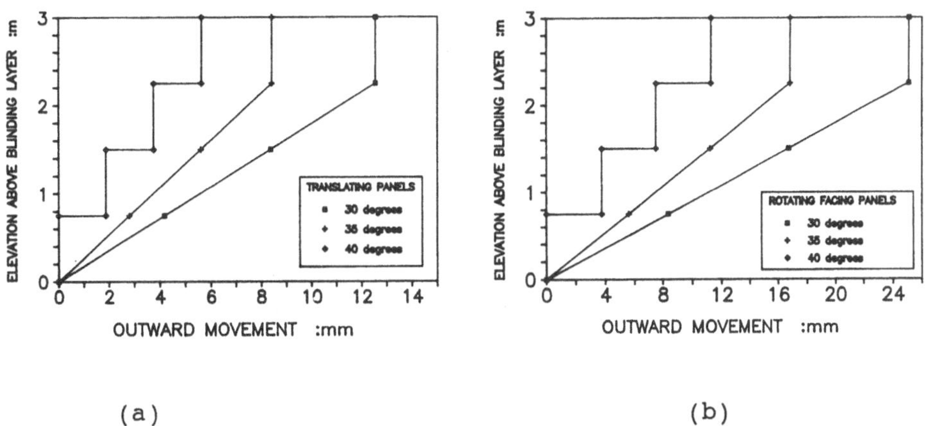

Fig. 10. Incremental construction induced movement: (a) translating and (b) rotating panels.

The results are plotted in Fig. 10a and b as the cumulative movement due to incremental construction at each panel level. Both the pure **translation** and the pure **rotation** results are given, the actual panel movement is likely to be between these two extremes. The data points in the figures show the initial position for each panel with respect to the vertical line through the initial position of the toe of the wall. The calculated outward movement of the top of each panel is about 0.2% to 0.5% of the panel height.

6.5 Overall movement of the wall face

The overall outward movement at the wall face can now be determined. In the trial wall this movement is due to the elongation in the reinforcement (calculated in section 5.6) and the moving datum caused by incremental construction (calculated above). There is no movement due to foundation settlement because of the rigid base, and negligible movement from the unreinforced fill behind the reinforced zone because of the long reinforcement length.

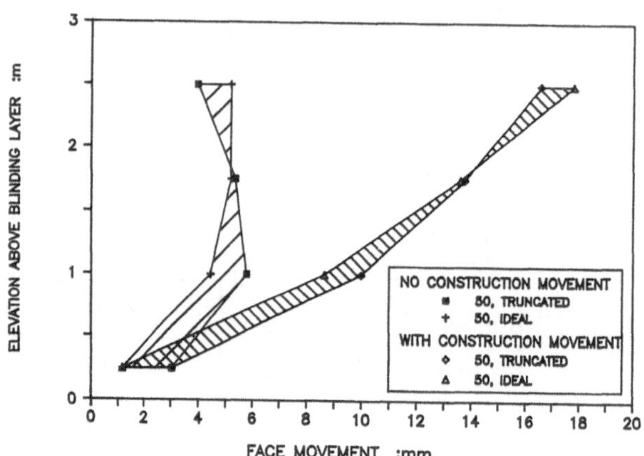

Fig. 11. Total outward movement at the face due to reinforcement elongation and incremental construction: 12 kPa surcharge and $\phi = 50°$.

The predicted total outward movement at the wall face with ideal boundary conditions is shown in Fig. 11 for both postulated equilibrium states, and for the case of the final mobilised soil strength $\phi = 50°$ and with the <u>average</u> translational and rotational incremental panel displacement calculated with a mobilised soil strength $\phi = 35°$. These two components of movement were recorded in Figs. 9 and 10. Results for other combinations can be derived directly from these.

SIDE WALL FRICTION IN THE TRIAL WALL

7.1 Boundary conditions in the trial wall

The influence of boundary conditions were reviewed in the *companion paper, section 9.* The calculations for the incremental wall have been based on assumed **ideal boundary conditions.** That is the face is assumed to have no stiffness and to ideally provide the required horizontal stress on the boundary of the soil exactly balanced by the connecting reinforcement force. There is assumed to be no net vertical and horizontal force in the face transmitted through the connection between the face and the rigid base. Finally, the rough strong base to the wall is assumed not to influence the force in the lower reinforcement layers.

In the incremental wall the facing panels do act freely from one another which is probably as close to the **ideal** assumptions as possible. Thus the interaction of the facing and the base should only influence the force in the lowest reinforcement layer. The maximum possible reduction in the reinforcement force in the lower layer due to the rough base was estimated to be about 20% (section 4.1).

7.2 Side wall friction

The problem always encountered with laboratory and field trials is at the side boundaries to the structure. Side wall friction can provide significant stabilising forces in the soil. Attention was paid to the possible influence of the side walls for the RMC trial and a measured value of side wall friction $\phi_{sw} = 20°$ is reported from a large scale shear box test on the prepared boundary (*bulletin 1*).

Bransby and Smith (1975) made detailed experimental observations and presented results from a **numerical analysis** for the influence of side wall friction on model retaining wall tests.

An analysis is derived below to evaluate the influence of side wall friction in the RMC trial wall. A simple closed form analysis is presented which slightly underpredicts the results of the numerical solution by Bransby and Smith (1975).

7.3 Side wall friction: analysis for self weight loading

If there is side wall friction then the shear stresses generated between the side wall and the deforming soil will act to resist the deformation in the soil. Because there are additional resisting forces, the side wall forces reduce the wall pressure required to support the soil, or reduce the required reinforcement forces in a reinforced soil wall.

Bransby and Smith (1975) show that the side wall shear stresses change both the failure mechanisms and the stresses in the soil. For a simple analysis an assumption can be made that the critical failure mechanism in the soil is not changed, and that the side wall friction only acts to alter the overall force equilibrium. The calculation below is for the triangular wedge of soil between the most critical plane and the wall face, the equilibrium of which determines the maximum reinforcement force.

The **most critical plane** and the equilibrium force polygon for the soil wedge are shown in Fig. 12a and b. The required horizontal force for equilibrium P_a is provided by the reinforcement layers. A side wall force P_s acting in the opposite direction to the movement of the soil wedge changes the force equilibrium as shown in Fig. 12c and d. For a simple analysis the soil movement can be assumed to be parallel to the most critical mechanism.

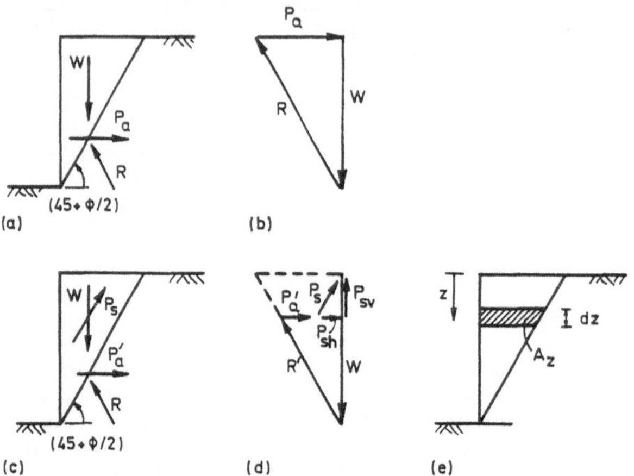

Fig. 12. The influence of side wall forces on self weight equilibrium in a retaining wall.

The side wall force is derived from the normal stress and the frictional shearing resistance between the sand and the side wall. The force is found from summing the shearing stress on the side wall over the cross sectional area of the critical wedge, as shown in Fig. 12e.

The intermediate stress σ_2 has been observed to remain a constant function of the mean stress s for **plane strain** loading,

$$\sigma_2 / s = K_2 \tag{6}$$

and Stroud (1971), for example, measured $K_2 = 0.74$ for sand in **plane strain**.

The mean stress s_z at any depth in the critical wedge is a function of the vertical stress in the soil and the active earth pressure coefficient

$$s_z = \frac{(1 + K_a)}{2} \gamma z \tag{7}$$

The gross side wall force P_{sw} on a wall of height H can be found from integration over the cross section of the critical wedge to give

$$P_{sw} = \frac{K_2(1 + K_a)}{2} \tan \phi_{sw} \frac{WH}{3} \tag{8}$$

where ϕ_{sw} is the mobilised angle of friction on the side wall, and W is the weight of a unit thickness of the critical wedge.

The net side wall force per unit thickness of soil P_s is a function of the height to width ratio for the wall H/w. From eqn. (8) and remembering there are two sides,

$$\frac{P_s}{W} = \frac{K_2(1+K_a)}{2}\tan\phi_{sw}\frac{2H}{3w} \tag{9}$$

The reduced active earth pressure coefficient due to side wall friction K_a' is a direct function of the net side wall force. The wedge equilibrium forces are shown in Figs. 12b and d. The active earth pressure coefficient is

$$\frac{P_a}{W} = \sqrt{K_a} \tag{10}$$

and the apparent earth pressure coefficient is defined in terms of the total soil weight so that

$$\frac{P_a'}{W} = \sqrt{K_a'} \tag{11}$$

Resolving the net side wall force into horizontal and vertical components, and by symmetry (Fig. 12d and eqn. (10))

$$\frac{(P_a' + P_{sh})}{(W - P_{sv})} = \sqrt{K_a} \tag{12}$$

which gives a direct expression for the reduced active earth pressure coefficient, from equations (11) and (12),

$$\sqrt{K_a'} = \sqrt{K_a}\left(1 - \frac{P_{sv}}{W}\right) - \frac{P_{sh}}{W} \tag{13}$$

The components of the net side wall force are, from Fig. 12,

$$P_{sh} = P_s\cos(45 + \phi/2)$$

$$P_{sv} = P_s\sin(45 + \phi/2)$$

Comparison with Bransby and Smith (1975)

Two numerical results for the influence of side wall friction on smooth retaining walls, the case above, are given by Bransby and Smith. They give the percentage reduction in the active earth pressure coefficient for soil with an angle of friction $\phi = 35°$ and $50°$, for a wall geometry $H/w = 2$ and with side wall friction $\phi_{sw} = 5.71°$ (ie. $\mu = 0.1$). Their calculation is for an intermediate stress ratio $K_2 = 0.37$. The results from the Bransby and Smith numerical analysis are compared with the simple closed form analysis in **Table 6.**

Table 6. Percentage reduction in the active earth pressure coefficient.

Angle of friction:	35°	50°
Bransby and Smith	-11 %	-13 %
Eqns: (9) & (13)	-11 %	-10 %

Note: Case with ϕ_{sw}= 5.71°; H/w= 2.

The two analyses compare well although the simple closed form analysis giving a slightly lower reduction in the active earth pressure coefficient. One obvious conservatism in the simple analysis for self weight loading is that it ignores the slight reduction in the vertical stress in the soil due to "arching" between the side walls.

Application to the RMC trial wall

The RMC trial wall has a height to width ratio H/w= 1.25, and Bransby and Smith's results indicate a reduction in the active earth pressure coefficient of approximately 7% for "full friction". The simple analysis gives a similar result, a 6.5% reduction for ϕ= 50°.

The numerical results presented by Bransby and Smith (1975) are not directly applicable to the RMC trial wall, however, because of the two following points.

(1) "Full friction" in Bransby and Smith is an angle of side wall friction ϕ_{sw}= 5.71° for sand against glass. When the measured value of side wall friction for the RMC boundaries is allowed ϕ_{sw}= 20° the percentage reduction in the earth pressure coefficient changes from 6.5% to 22.5%.

(2) Bransby and Smith appear to have estimated the intermediate principal stress with a coefficient K_2= 0.37. The data for sand in plane strain generally indicates a higher value $K_2 \approx$ 0.65 to 0.75.

The relationship between the parameter **b**

$$b = \frac{\sigma_2 - \sigma_3}{\sigma_1 - \sigma_3}$$

and K_2 for sand with a mobilised angle of friction ϕ can be deduced from a Mohr circle

$$b = 0.5 + \frac{K_2 - 1}{2\sin\phi}$$

Values of K_2= 0.65 to 0.75 for sand with ϕ= 40° to 50° correspond with values b= 0.25 to 0.35 which is the commonly observed range for plane strain tests on sand.

Increasing the intermediate stress to $K_2 = 0.7$ in the closed form analysis increases the predicted influence of the side wall friction further from 22.5% ((1) above) to 40%.

Apparent active earth pressure coefficient

The result is that the influence of side wall friction on the self weight loading of the sand in the RMC trial wall is likely to be significant. If the measured side wall friction were mobilised in the trial wall, the apparent active earth pressure coefficient would be 40% less than the expected earth pressure coefficient without side wall friction,

$$K_a' = 0.6K_a$$

7.4 Side wall friction: analysis for surcharge loading

Without side wall friction, a uniform surcharge loading increases the vertical effective stress in the soil equally at every depth behind the wall. Side wall friction reduces the net vertical stress in the soil due to the surcharge. This is in addition to reducing the active earth pressure coefficient, as described above.

To allow for this effect in the RMC trial, an analysis is required for the reduction in the net vertical surcharge stress in the soil at any depth due to the side wall friction. Once again, a straightforward closed form analysis is presented.

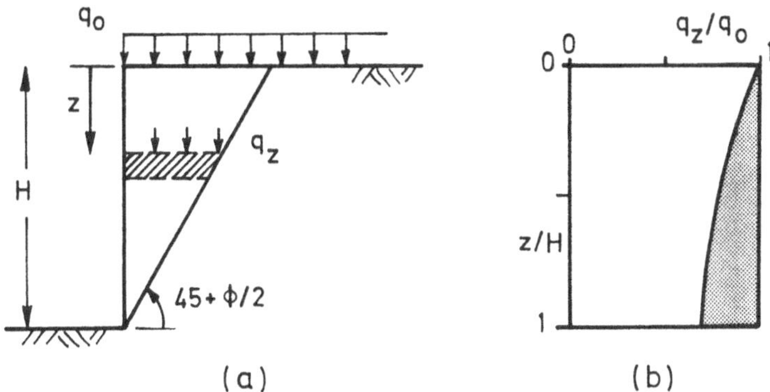

Fig. 13. (a) Illustration of the soil stresses due to surcharge loading and (b) the reduction due to side wall friction.

A vertical wall supporting a uniform surface surcharge q_o is shown in Fig. 13a. At a depth z below the wall crest the vertical stress due to the surcharge loading is q_z. Following the same analysis as before, the intermediate stress due to the surcharge load is

$$\left(\sigma_2\right)_z = K_2 \frac{\left(1+K_a\right)}{2} q_z \tag{14}$$

The vertical component of the side wall shear stress acts cumulatively to reduce the vertical stress in the soil. Remembering there are two sides to the trial wall of width w, the rate of change of the net vertical loading can be expressed as

$$\left(\frac{dq}{dz}\right)_z = -\left(\frac{K_2\left(1+K_a\right)}{2} q_z \tan\phi_{sw}\right)\frac{2}{w}\cos(45+\phi/2) \tag{15}$$

which has the form

$$\left(\frac{dq}{dz}\right)_z = -Cq_z$$

The solution to equation (15) is

$$q_z = q_o e^{-Cz} \tag{16}$$

where the constant C is defined above.

Results for surcharge loading

The net vertical stress due to the surcharge loading can be represented by the ratio q_z/q_o. The RMC trial wall is 2.4 m wide. With an angle of friction in the soil $\phi = 50°$ and using the measured angle of friction for the side wall $\phi_{sw} = 20°$ the constant $C = 0.11$, assuming a value $K_2 = 0.7$ for the intermediate stress. The resulting variation in the net vertical stress due to surcharge loading with depth is shown in Fig. 13b.

These results indicate that the vertical surcharge loading would be reduced by approximately 30% at the bottom of the RMC trial wall due to the side wall friction.

7.5 Implications for the RMC trial wall

The analysis presented above indicates that side wall friction is likely to have an important influence on the RMC trial wall.

For self weight loading in the wall the side wall friction effectively provides about 40% of the required horizontal stresses for equilibrium, or, in other words, results in an apparent active earth pressure coefficient $K_a' = 0.6K_a$ (section 7.3). This reduction in the required horizontal stresses for equilibrium is shown in Fig. 14a.

The side friction also leads to "arching" for any vertical surcharge applied to the RMC trial wall. At a depth 3 m below the crest of the wall the side walls support about 30% of the applied surface surcharge load, (section 7.4). This was shown in Fig. 13.

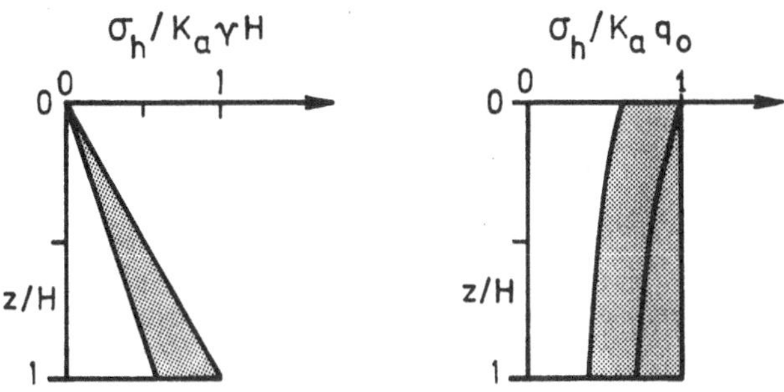

(a) Self weight loading (b) Surcharge loading

Fig. 14. Illustration of the reduction in the required soil stresses due to side wall friction in the RMC trial wall.

The reduced vertical stress also requires proportionally less horizontal stress to maintain equilibrium in the soil, because of the side walls. The apparent active earth pressure coefficient is the same as that calculated for the self weight loading, 40% less than the normal value, giving the required horizontal stress for equilibrium shown in Fig. 14b.

PREDICTION ALLOWING FOR SIDE WALL FRICTION

12 kPa applied for 100 hours
50 kPa surcharge applied for 1000 hours

8.1 Introduction

A prediction for the RMC trial wall allowing for the influence of the side wall friction can now be completed. The prediction is for the expected **plane strain** mobilised angle of friction in the soil $\phi = 50°$ (section 2.4), and for the measured angle of side wall friction $\phi_{sw} = 20°$. The prediction is for the central loading case (12 kPa/100 hours) and for the end of the test (50 kPa/1000 hours).

For both loading cases the difference between the prediction with and without the side wall friction is indicated.

An implicit assumption in the analysis for side wall friction is that it only reduces the magnitude of the required reinforcement forces for equilibrium and does not change the location in the soil where these forces are required. This assumption is likely to be conservative. The side wall friction is likely to reduce the overall elongation in the reinforcement layers more than predicted.

8.2 Force in the reinforcement

Self weight loading

The distribution of the gross required horizontal force for equilibrium to the individual reinforcement layers was described in section 4.1. Side wall friction reduces the gross required force but not the distribution of the force.

The percentage of the **reduced** gross required force due to self weight carried by each of the four reinforcement layers is given in **Table 7**, together with the force magnitudes. The force in each layer can be compared with the results in **Table 4**.

Table 7. Maximum reinforcement force: 12 kPa surcharge and side wall friction.

Loading	Self weight	Surcharge	Self weight :kN	Surcharge :kN	Total :kN
Layer 4	9 %	33 %	0.57	0.78	1.35
Layer 3	21 %	26 %	1.33	0.61	1.94
Layer 2	33 %	24 %	2.08	0.56	2.65
Layer 1	37 %	18 %	2.34	0.43	2.77
Total	100 %	100 %	6.32	2.38	8.70

Surcharge loading

The same form of presentation can be used for the surcharge loading. In this case the gross required force is reduced by more than the 40% due to the apparent earth pressure coefficient. The reduction of stress in the soil due to the "arching" over the side walls (section 7.4) gives an additional overall reduction in the gross required force of 15%. The resulting reduction in the gross required horizontal force to support the surcharge load is 49%.

The "arching" of the surcharge loading alters the distribution of the gross required force between the reinforcement layers. The distribution **between the reinforcement layers** of the reduced gross required force due to surcharge loading is given in **Table 7.** The magnitude of the resulting forces in the reinforcement layers for 12 kPa surcharge loading are also given. The values can be compared with the results in **Table 4.**

Total reinforcement force

The resulting maximum force in the four reinforcement layers for the central loading case and at the end of the test are shown plotted in Fig. 15a and b, where they are compared with the values without side wall friction.

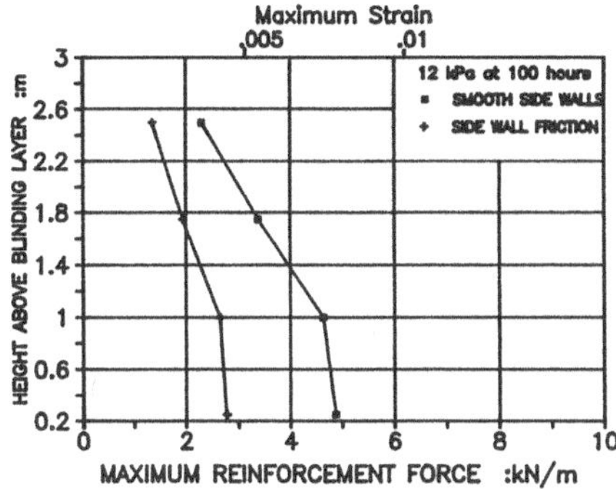

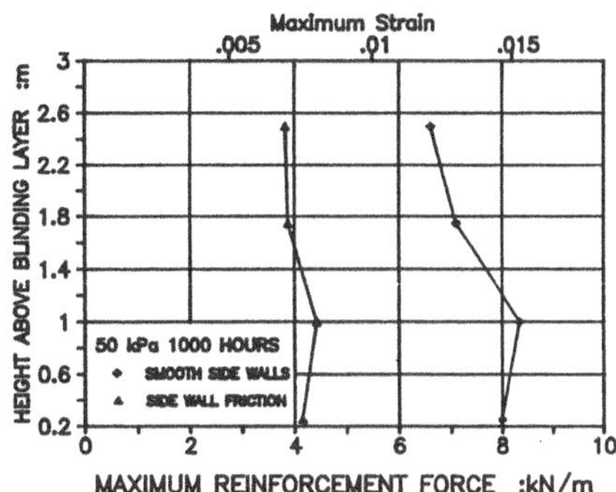

Fig. 15. Maximum reinforcement force with and without side wall friction (a) 12 kPa surcharge and (b) 50 kPa surcharge.

8.3 Strain in the reinforcement

The maximum strain in the reinforcement depends on the maximum force and the reinforcement stiffness (section 3.1). Using the appropriate value of stiffness for the two loading cases ($K_{100} = 625kN/m$ and $K_{1000} = 550kN/m$) gives the set of maximum reinforcement strains shown on the top scale in Fig. 15.

Table 8. Calculation of reinforcement elongations; 12 kPa for 100 hours and side wall friction.

Reinforcement	Elevation :m	$\frac{z}{H}$	$\frac{\delta K}{HP}$	P :kN	δ :mm
Layer 4	2.50	0.17	0.36	1.35	2.3
Layer 3	1.75	0.42	0.33	1.94	3.0
Layer 2	1.00	0.67	0.26	2.65	3.3
Layer 1	0.25	0.92	0.13	2.77	1.7

Notes: $K = 625\,kN/m$; $H = 3000\,mm$.

8.4 Reinforcement elongation

The procedure for evaluating the total elongation in each reinforcement layer is the same as before (section 5.6). In the case of side wall friction the reinforcement force is lower, which reduces the overall elongation.

The calculation for the central loading case and the **truncated length** equilibrium case, and allowing for side wall friction, is set out in **Table 8**. First the depth of each reinforcement layer beneath the wall crest is determined and the non dimensional outward movement read from the relevant chart (Fig. 8a for this case, but see the *companion paper* for the other charts). The maximum force in each layer is recorded in **Table 7**, and the reinforcement stiffness and the wall height are the same for all the layers. The magnitude of the outward movement can be directly evaluated, **Table 8**.

The predicted set of reinforcement elongations for the two equilibrium states of **ideal length** and **truncated length** are shown in Figs. 16 and 17. The difference due to the side wall friction is indicated for the two loading cases.

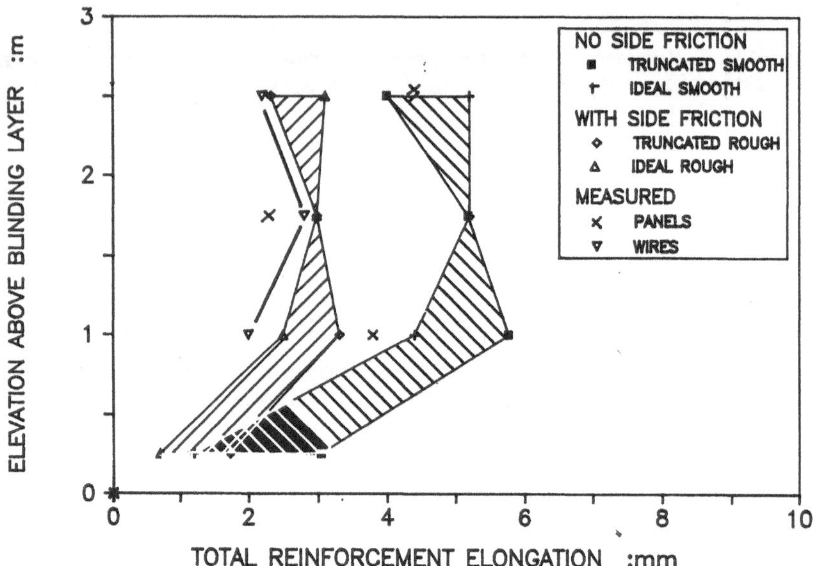

Fig. 16. Predicted reinforcement elongations for the 12 kPa surcharge loading, compared with the measured values.

8.5 Incremental construction movement

The height to width ratio for each incremental lift is high enough to allow the influence of the side wall friction to be ignored. The previously calculated movements caused by incremental construction can be used (section 6.4).

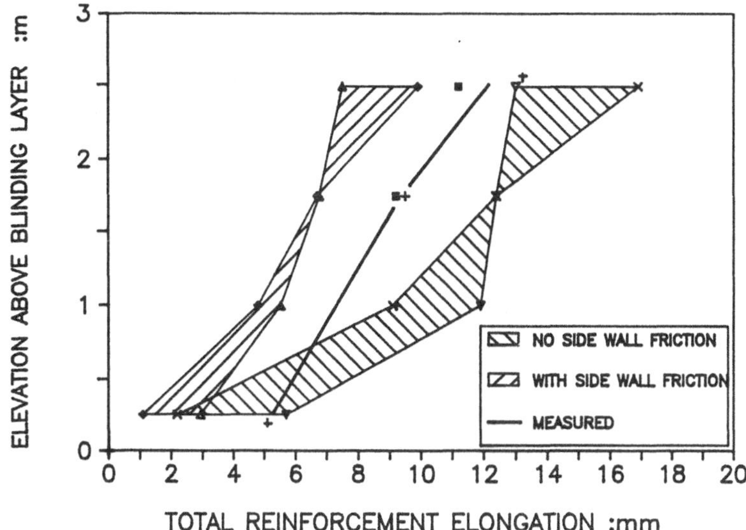

Fig. 17. Predicted reinforcement elongations for the 50 kPa surcharge loading, compared with the measured values.

PREDICTED BEHAVIOUR FOR THE INCREMENTAL TRIAL WALL

9.1 Introduction

The analysis for walls involves two idealised equilibrium states which are expected to bound the actual equilibrium in the reinforced soil. The data from the measured behaviour of the trial walls have been added to the figures in this and the previous section.

9.2 Deformations

The total deformation at the wall face is made up from the reinforcement elongations and the "moving datum" caused by incremental construction.

Reinforcement elongation

The predicted reinforcement elongations for the **central loading case** are shown in Fig. 16. The results for the ideal length case are also shown, as are the maximum likely elongations that were predicted assuming no side wall friction.

The predicted reinforcement elongations at the **end of the test** are shown in Fig. 17. Again, the companion **ideal length** case is also given, as are the maximum likely elongations calculated ignoring the side wall friction.

Incremental construction movement

Because the facing panels can either translate or rotate there is a range of possible movement due to incremental construction. The predicted range is shown in Fig. 18, based on a mobilised angle of friction $\phi = 35°$ consistent with the initial filling and compaction of the sand (section 6.3).

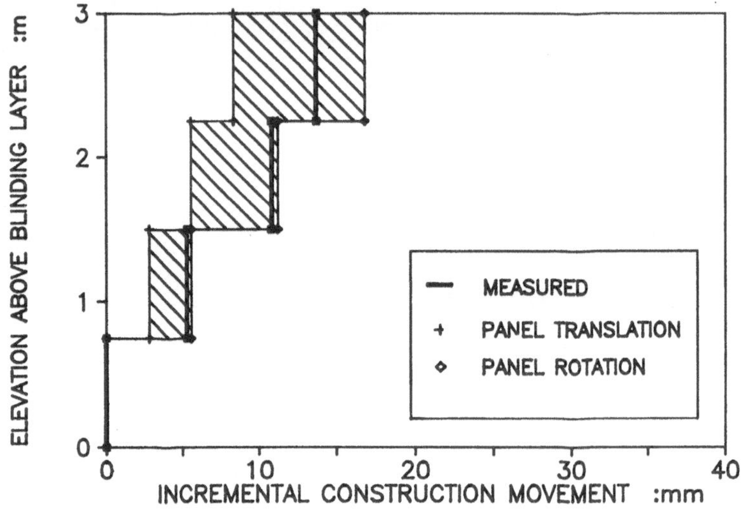

Fig. 18. Predicted incremental construction movement, compared with the measured values.

9.3 Reinforcement force and strain distributions

In the zone between the most critical plane and the wall face the predicted reinforcement force is uniform, and the magnitude is the same for both the **ideal length** and the **truncated length** equilibrium states. The variation of the reinforcement force behind the most critical plane does depend on the the equilibrium case, and the variation for both cases is shown.

The predictions are presented in terms of the tensile reinforcement strain as this is what is actually measured. The corresponding force is governed by the appropriate reinforcement stiffness (section 3.1). As usual, the prediction is for the case with side wall friction.

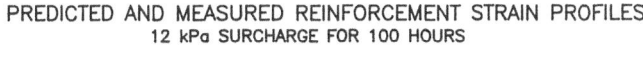

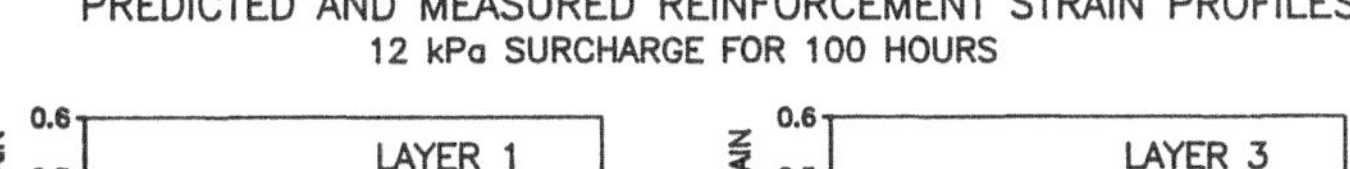

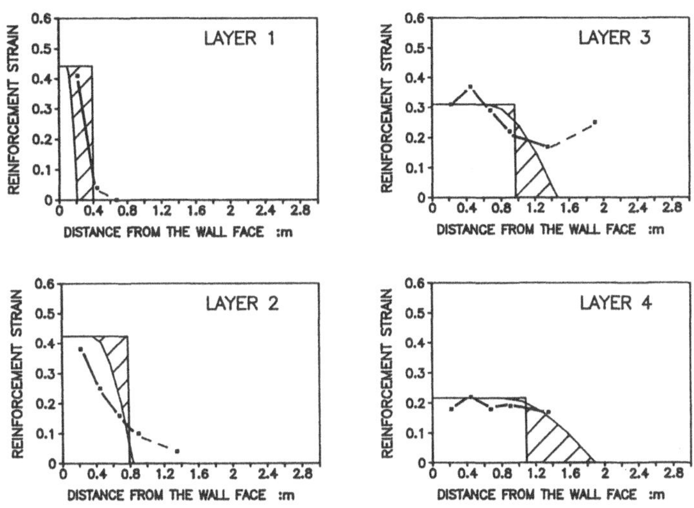

Note. Reinforcement strain is in units of percent (%).

Fig. 19. Predicted reinforcement strain profiles for the 12 kPa surcharge loading, compared with the measured values.

The predicted reinforcement strain in the uniform zone at the front of the wall, and the predicted variation of strain behind this zone for the **ideal length** and the **truncated length** cases, are shown in Fig. 19. These predictions are for the central loading case and for the four reinforcement layers. The reinforcement stiffness $K_{100} = 625\ kN/m$.

The corresponding predictions at the end of the test are shown in Fig. 20. The reinforcement stiffness $K_{1000} = 550\ kN/m$.

9.4 Stress on the base

Because of the side wall friction, the predicted vertical stress on the base of the wall is less than the sum of the overburden and the surcharge stress. The equilibrium in the reinforced soil has a uniform vertical stress on the lower boundary.

In the simple analysis for side wall friction the reduction in the self weight vertical stresses was ignored, "arching" was only calculated for the surcharge loading. The reduced vertical stress at the base of the wall due to surcharge loading by 30%, (section 7.5). Thus the total predicted uniform vertical stress for the two loading cases is 62 kPa and 89 kPa.

PREDICTED AND MEASURED REINFORCEMENT STRAIN PROFILES
50 kPa SURCHARGE FOR 1000 HOURS

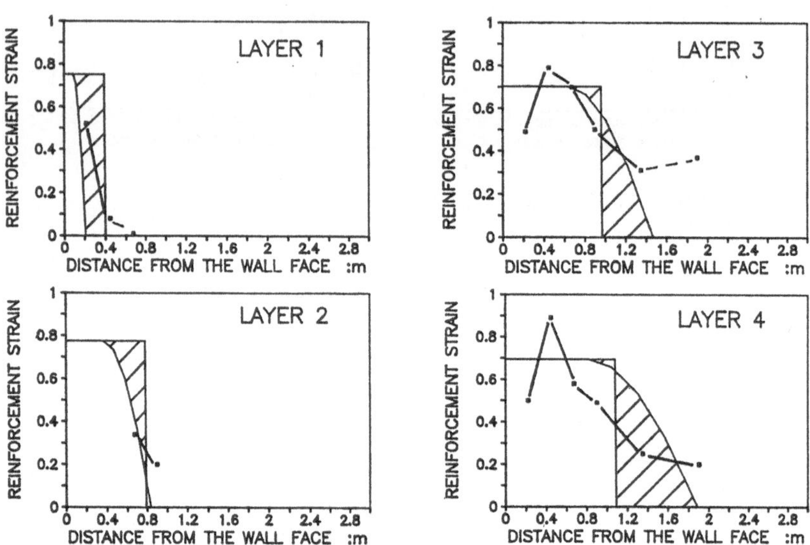

Note. Reinforcement strain is in units of percent (%).

Fig. 20. Predicted reinforcement strain profiles for the 50 kPa surcharge loading, compared with the measured values.

10.1 Magnitude of face movement

Attention in this paper has been given to the incremental wall, the most usual form of reinforced soil wall construction. The analysis and prediction for the incremental wall has provided much material, and space does not permit a similar analysis for the propped wall.

However the RMC trial walls do provide an outstanding example of the contrast between an **incremental** and a **propped wall.** In an incrementally constructed wall there is a "moving datum" which results in deformation at the face additional to the deformation caused by the reinforcement elongation. This additional movement does not occur in the propped wall because the propping supports the face so that the initially "unstressed" reinforcement layers all align on a vertical line through the toe (or at the angle of the propped face).

In the RMC trial the predicted incremental movement due to construction is about a factor of four greater than the predicted movement due to the reinforcement elongation (Figs. 16 and 18). This means that the movement in the propped wall should be only about 20% of the movement in the incremental wall - a very significant difference.

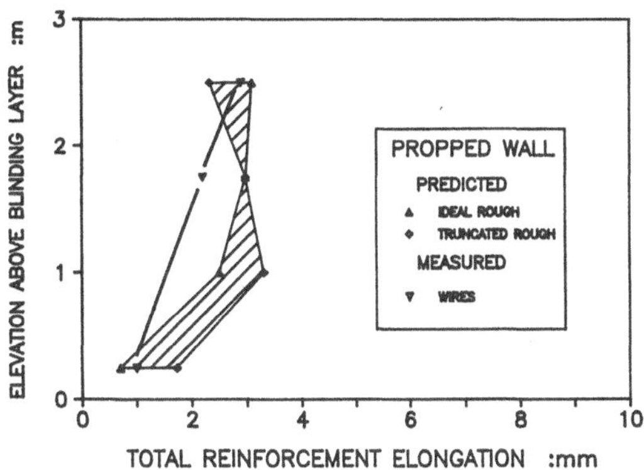

Fig. 21. Predicted face movement for the propped wall under 12 kPa surcharge, compared with the measured movement.

The predicted face movements for the propped wall at the end of the 12 kPa surcharge loading, and at the end of the test, are shown in Figs. 21 and 22. The predicted movement is due only to the reinforcement elongation and is exactly equal to the equivalent prediction for the incremental wall. The prediction allows for the measured side wall friction.

10.2 Influence of the stiff face boundary

As discussed in the *companion paper,* one possible detrimental consequence of propped wall construction with a continuous and stiff face is that the face deformation is constrained to remain approximately linear. The analysis that has been presented indicates that this type of face movement only occurs freely for the **ideal length** equilibrium state combined with **ideally spaced** reinforcement. When more uniform reinforcement spacing is used the face movement towards the top of the wall is reduced. The results in Fig. 21 are a typical example of this.

The consequence is that close to the top of a propped wall the stiff face might tend to "pull the reinforcement outwards" to meet the required boundary displacement, thereby causing higher than expected reinforcement force close to the face at the top of the wall.

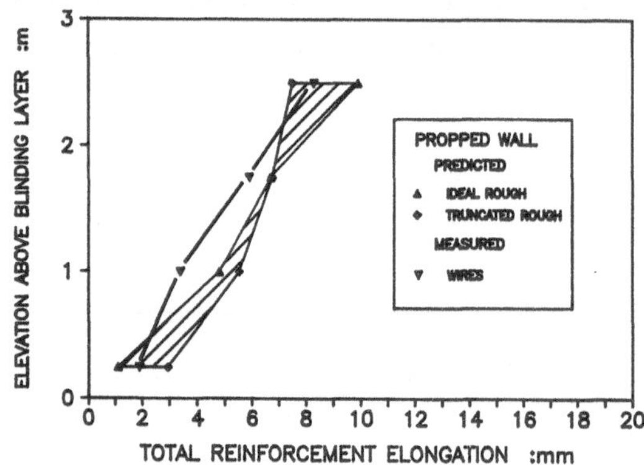

Fig. 22. Predicted face movement for the propped wall under 50 kPa surcharge, compared with the measured movement.

By exact analogy, towards the mid-height of the wall the outward movement would be reduced by the face boundary condition thereby locally reducing the reinforcement force close to the face. The discrepancy between the predicted movement with an **ideal face boundary** and the linear movement with a stiff continuous facing is illustrated in Fig. 21 for the RMC propped wall.

The conclusion to be drawn is that this phenomenon should be considered in the design of propped walls to ensure that local over-stressing at the reinforcement face connection towards the top of the wall does not occur.

STRAIN COMPATIBILITY

11.1 Introduction

The aim of this section is to relate the measured soil and reinforcement properties on a compatibility curve for the incremental wall (*companion paper, section 6.2*). Only the overall compatibility for the whole wall will be considered - a more detailed analysis would consider equilibrium in each layer. The influence of the measured side wall friction is included in the analysis.

11.2 Material properties

The soil is initially at a low stress ratio on filling the wall, corresponding to approximately K_0 conditions. The initial mobilised angle of friction $\phi = 35°$ was chosen (section 6.3). This is somewhat less than the critical state angle of friction.

The simple model for sand involves elastic shear to the (peak) mobilised stress ratio, followed by perfectly plastic shear with a constant angle of dilation. The predictions were based on a mobilised angle of friction $\phi = 50°$ which has a corresponding angle of dilation $\psi = 15°$ (section 2.4).

The shear modulus for the sand deforming from the initial to the final stress ratio was estimated as $(G/s) \approx 23$. The initial stiffness was about twice as high giving the range $(G/s) \approx 20$ to 50 (section 2.5).

11.3 Reinforcement properties

The load extension properties are assumed to be linear and the reinforcement stiffness varies with time under load. The reinforcement stiffness values used for the prediction are $K_{100} = 625\ kN/m$ and $K_{1000} = 550\ kN/m$ (section 3.1).

11.4 Compatibility curve

The compatibility curve depicts the relationship between the **required force** for equilibrium in the structure and the **available force** from the reinforcement. The link is the tensile strain in the soil in the direction of the reinforcement. The overall compatibility curve considers equilibrium along the **most critical plane**. Because the principal stresses are vertical and horizontal on this plane, the relevant tensile strain is the minor principal strain for vertical walls.

The simple compatibility curve is drawn ignoring the construction loading sequence *(companion paper)*. Drawn this way it gives a much more clearly defined equilibrium point.

The self weight and surcharge loading are used to determine the **required force**. The required force depends directly on the shearing resistance in the soil. The link between the shearing resistance and the tensile strain in the soil for the trial wall is predicted from the simple soil model and the analysis of the direct shear test results.

The **available force** depends on the magnitude of the tensile strain, the stiffness of the reinforcement and the number of reinforcement layers.

Central loading case

The compatibility curve for the **central loading case** (12 kPa and 100 hours) is shown in Fig. 23. The variation of the required force is (almost) bilinear reflecting the elastic plastic model for the sand. The measured range of the soil shear modulus is also shown.

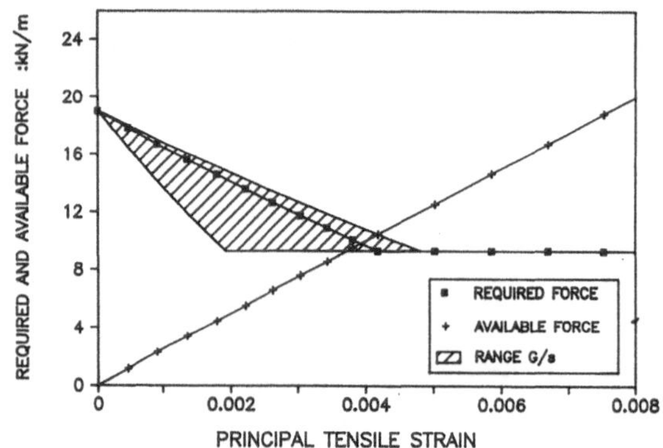

Fig. 23. Compatibility curve showing the predicted overall equilibrium point under 12 kPa surcharge loading.

The compatibility curve indicates equilibrium in the wall, where the gross **required** and **available forces** are equal, almost exactly at the point when the soil reaches the plateau of continued deformation at a constant angle of friction $\phi = 50°$. The predicted tensile strain at the equilibrium point is 0.4%, which is consistent with the predicted reinforcement strains in Fig. 19.

The compatibility curve at the **end of the test** (50 kPa and 1000 hours) is shown in Fig. 24. The equilibrium point occurs at a larger tensile strain 0.8%, which is more than sufficient for the reinforcement to mobilise the plateau stress ratio. Again, the magnitude of the equilibrium tensile strain is consistent with the earlier predicted reinforcement strains, Fig. 20.

The measurements made on the trial walls have been plotted with the predictions in Figs. 16 to 22 for ease of comparison.

12.1 Implications from "back analysis"

Leroueil and Tavenas (1981), in a review of the developments that can be achieved in geotechnical engineering from field observation, have highlighted the possible pitfalls of drawing conclusions from back analyses. One major pitfall is from

> "..the common practice to obtain a good fit only for one or two parameters of the field behaviour: ..evaluating settlements but not lateral displacements ..construction pore pressures but not deformations ..strength parameters but not stress strain response up to failure."

> "..the most serious shortcoming *(of the above methodology)* is that basic soil mechanics principles are frequently not satisfied by the results: .. many cases are reported of 'correct' settlements computed together with erroneous pore pressures and effective stresses."

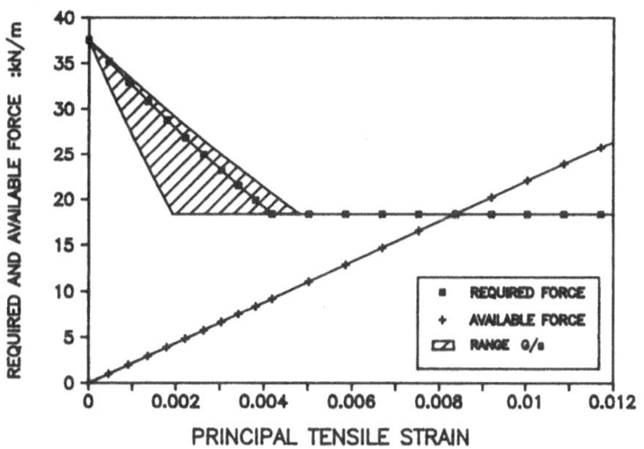

Fig. 24. Compatibility curve showing the predicted overall equilibrium point under 50 kPa surcharge loading.

In their guidelines for the proper use of back analysis, Leroueil and Tavenas emphasise that "soil mechanics principles must apply to the results of back analyses ...**all interrelated parameters must be back analysed with equal success**..", adding, "..a back analysis successful on only 50% of the parameters is nothing but 100% wrong." They go on to say that "..time is a systematic parameter in all geotechnical problems and the validity of back analyses should be checked at various stages of the process under consideration".

Against these criteria, the required predictions and the measurements recorded for the RMC trial walls present a fairly rigourous test of any proposed analysis. Except for vertical settlements, every aspect of the reinforced soil wall behaviour was requested to be predicted at different times through the loading history of the wall.

12.2 Observation on the predictions

The agreement between the predictions from the analysis which has been presented and the measured behaviour plotted in Figs. 16 to 24 is satisfactory in two respects. (1) The predicted pattern of behaviour as well as the magnitude is similar to the measurements. (2) The above observation applies equally to the two loading cases. Also important is that the analysis allows for the measured material properties and boundary conditions, and provides a link between the measurements in the two separate trial walls with a different face and construction boundary condition.

12.3 Observation on the measurements

An excellent feature of the RMC trial walls was that several of the measurements were made by more than one independent means: for example, the reinforcement elongation was measured directly with steel wires, estimated by integrating the measured strain profile along the reinforcement and interpreted from the movement of the facing panels. This provides confidence in the magnitude of the measured parameter but also indicates the order of accuracy that can be attributed to any measurement.

The trial walls clearly show that there is "scatter" in the behaviour of a reinforced soil wall introduced by the variable influences of construction. Therefore an "exact" representation of the behaviour by a single analysis is perhaps an unrealistic goal. If this is correct, then an analysis that "bounds" the likely behaviour (such as proposed in this paper) might indeed turn out to be the most satisfactory approach.

12.4 Class A Prediction

The author's class A submission to the NATO prediction exercise varied in some minor details from the results presented in this paper but was based on the truncated length equilibrium state allowing for side friction. In that analysis a mobilised plane strain angle of friction for the sand of $\phi = 50°$ and the measured side wall friction of $\phi_{SW} = 20°$ was used. A maximum likely limit was also predicted with the side wall friction set to zero.

Acknowledgements

The author would like to thank Dr. R. Kaniraj of the Indian Institute of Technology, who carried out four of the direct shear tests reported in the paper while at Oxford University on a British Technical Cooperation Award.

The foresight and hard work of Richard Bathurst, Peter Jarrett and Alan McGown, and their co-workers, was essential in that they made the prediction symposium possible, designed the trials to provide the detailed data of wall performance, and provided the forum for the exchange of knowledge that seems certain to advance the state of the art.

References

Bathurst, R.J. & Jarrett, P.M. (1986). Class A prediction exercise for reinforced earth walls. *Bulletin No. 1 for NATO Advanced Research Workshop, Application of Polymeric Reinforcement in Soil Retaining Structures,* Royal Military College, Kingston.

Bransby, P.L. & Smith, I.A.A. (1975). Side friction in model retaining wall experiments. *Journal of Geotechnical Engineering,* **ASCE GT7**, July, 615-632.

Bolton, M.D. (1986). The strength and dilatancy of sands. *Geotechnique* **36,** No. 1, 65-78.

Cornforth, D.H. (1973). Prediction of drained strength of sands from relative density measurements. ***ASTM*** *Spec. Tech. Publ.* 523, 281-303.

Jewell, R.A. (1987). Reinforced soil wall analysis and behaviour. *Proc. NATO Advanced Research Workshop, Application of Polymeric Reinforcement in Soil Retaining Structures,* Martinus Nijhoff.

Jewell, R.A. & Wroth, C.P. (1987) Direct shear tests on reinforced sand. *Geotechnique* **37**, No. 1., 53-68.

Koerner, R.M. (1968). *The behaviour of cohesionless soils formed from various materials.* PhD Thesis, Duke University. Soil Mechanics Series No. 16.

Leroueil, S. & Tavenas, F. (1981). Pitfalls of back analyses. *Proc. 10 Int. Conf. Soil Mech. & Fndn. Engng., Stockholm,* **Vol 1,** 185-190.

Stroud, M.A. (1971). *The behaviour of sand at low stress level in the simple shear apparatus.* PhD Thesis, University of Cambridge.

Case Histories and Full Scale Trials

POLYMERICALLY REINFORCED RETAINING WALLS AND SLOPES IN NORTH AMERICA

M.A. Yako, Student, Purdue University, Lafayette, Indiana, USA
B.R. Christopher, Principal Engineer, STS Consultants, Ltd. Northbrook, Illinois, USA

1. INTRODUCTION

This paper presents a summary of reported polymerically reinforced soil wall and slope projects constructed in North America. The projects included walls and slopes where the reinforcement was used to resist lateral earth pressures and prevent internal failure through the face or toe of the structure. Projects where reinforcement has been used to provide increased stability against deep seated failure of embankments were not included. Specific emphasis was placed on projects that have been instrumented and monitored such that performance assessments could be made.

The projects were obtained through a review of published case histories and through interviews with public agencies, geosynthetic manufacturers, and prominent design engineers. The projects were reviewed with respect to four categories:

- General Information
- Design Information
- Reinforcement Properties and Design Methodology
- Instrumentation

2. PROJECT SUMMARIES

The project summaries for each specific review category are contained in the attached Tables 1 through 4. Table 1 includes general project information and provides references where additional information may be obtained. Table 2 provides the design requirements for the specific projects including height, surcharge loading, and foundation and backfill characteristics. Table 3 presents information pertaining to the actual properties of the reinforcement, and the design methodology used to determine the reinforcement requirements. Finally, Table 4 presents project instrumentation and monitoring information. Unfortunately, only a few cases were reviewed which contained this extremely valuable and needed information.

In all, summaries of 54 projects are provided in the tables. The summaries include 13 projects that were at least

P. M. Jarrett and A. McGown (eds.), The Application of Polymeric Reinforcement in Soil Retaining Structures, 239–283.

TABLE 1. GENERAL INFORMATION

CASE No.	PROJECT NAME	LOCATION	YEAR CONSTRUCTED	STRUCTURE	HEIGHT (m)	LENGTH (m)
1	Illinois River Road	Siskiyou national forest Oregon	1974	Roadway widening	3.0	20
2	Olympic Natl. Forest Road	Shelton Washington	1975	Timber haul road wall	0.9 to 6.1	50
"	"	"	"	"	"	"
3	Theory & princ. of reinforced earth	WES Vicksburg Mississippi	1974-1976	Test wall	Design:3.7 Failed:3.0	4.9
4	Cortright Lava Timber Sale	Gifford Pinchot Natl. forest WA	1979	Roadway widening	6.7	30
5	New York Route 22	Columbia County NY	1980	slide correction in hill 2 walls	4.9	33 and 46
6	Mt. Baker Natl. Forest	Mt. Baker Natl. Forest WA	1980	roadway widening across slide area	1.8	
7	Camp Hill Road	Willamette Natl. Forest OR	1981	roadway widening across slide area	8.5	88
8	State route 203	Mamoth lakes CA	1981	Road widening 2 walls	0.9 to 3.7	230
9	Interstate I-70 test wall	Glenwood Canyon CO	1982	Highway embankment test wall	4.6	91
	"	"	"	"	"	"
	"	"	"	"	"	"
	"	"	"	"	"	"
10	CP Rail	Waterdown Ontario	1982	Slope failure repair	4.6 and 5.5	91
11	Highway US 69	Beaumont TX	1982	Slope failure repair	6.7	--
12	Wauwatosa	Wauwatosa WI	1982	Retaining wall	8.2	64
13	Devil's Punch Bowl Slide Correction	Devil's Punch Bowl State Park OR	1983	Landslide stabilization	9.1	Bottom 21 Top 52
14	Hillcrest Road	San Pablo CA	1983	Landslide stabilization for roadway	11.9	150

FACING	TYPE	REINFORCEMENT BRAND	POLYMER	COMMENTS	REFERENCES
Wrapped,1:8 slope gunite	Nonwoven needlepunched	Fibertex 400	Polypropylene	1st fabric wall in U.S.	1,2
Wrapped,vert. CSSI asphalt emulsion coat	Nonwoven needlepunched	Fibertex 400 and 600	Polypropylene	Durability treat/untreated fabric	1,3,4,8
"	Nonwoven needlepunched	Bidim C-28 & C-38	Polyester	"	"
Alcoa T11 high strength panels,vertical	4 ply strip	Nylon heavy duty	Neoprene coated nylon fabric	Failure maybe large deform of ties	5,6
Wrapped asphalt emulsion coating	Nonwoven needlepunched	Fibertex 200	Polypropylene		8
Wrapped,1:3 slope reinforce gunite	Nonwoven needlepunched	Bidim C-34	Polyester	Crushed stone below lift to decrease sliding	7,8
Wrapped asphalt emulsion coating	Woven slit film	Supac 5W	Polypropylene	Sawdust backfill	8
Asphalt emulsion,latex paint	Woven slit film	Supac 5W	Polypropylene	Sawdust backfill for slope stability	8
Railroad ties	Tire sidewall w/ tie back anchor bar			withstood earthquake of 5.8	9
Wrapped 1:5 slope reinforce gunite	Nonwoven needlepunched	Fibertex 200 and 400	Polypropylene	Wall incorp four nonwoven geotextiles	8,10,11
"	Nonwoven needlepunched	Supac 4NP & 6NP	Polypropylene	Each in 2 weights in 10 test segments	"
"	Nonwoven needlepunched	Trevira S1115 & S1127	Polyester	"	"
"	Nonwoven heatbonded	Typar 3401 & 3601	Polypropylene	"	"
Wrapped 1.5:1 slope	Geogrid	Tensar SR2	Polyethylene HDPE	Hi ground water pressure caused initial failure	12,13,14
Slope 3:1 Tensar SS2 on slope face	Geogrid	Tensar SS2	Polypropylene	On site silty clay excavated & recompacted	15
Wrapped asphalt emulsion coating	Nonwoven needlepunched	Supac 12NP	Polypropylene	--	Western builder 1983
Wrapped 1:6 slope UV stable grid	Geogrid	Tensar SR2	Polyethylene HDPE	First geogrid retaining wall in U.S.	8,16,17
Wrapped 1:4 slope	Geogrid	Tensar SR2 and SS2	Polyethylene HDPE Polypropylene	First use of geogrids to repair landslide	18

TABLE 1. GENERAL INFORMATION (cont'd)

CASE No.	PROJECT NAME	LOCATION	YEAR CONSTRUCTED	STRUCTURE	HEIGHT (m)	LENGTH (m)
15	U.S. 26	Jewell Jct. OR	1983	Temporary wall for stage construction	4.9	114
16	Danapoint	Danapoint CA	1983	Retained fill for parking lot	3.0	91
17	Route 84	LaHonda CA	1984	Slope failure repair	14.0	70
18	Tarque Verde Road	Tuscon AZ	1984-1985	Grade sep. for major intchnge 46 walls	0.9 to 6.1	1600
19	Jasper Asphalt Plant Operation	Lithonia GA	1985	Material handling platform	6.1	32
20	Gaspe ° Peninsula Seawall	Gaspe ° Peninsula Quebec	1985	coastal protection	5.3	99
21	Stump Springs Road	Sierra Natl. Forest	1985	retained fill	5.5	24
22	Rush Canyon	Angeles Natl. Forest	1985	retained fill	6.1	15
23	SR516	King County WA	1985	temporary wall	2.4	30
24	SR522	King County Wa	1985	temporary walls (5)	2.1 to 3.4	1372
25	Fish Creek Road	Umpqua Natl. Forest OR	1985	earth reinforcement		
26	Cascasde Dam Hydroelectric Facility	Cascade MI	1986	crane operating platform	3.0	
27	Interstate I-70	Glenwood Canyon CO	1986	retaining wall for roadway	4.6	61
28	TH 35E	St. Paul MN	1986	retaining wall for roadway	1.5 to 4.3	300+
29	Coalbank Slough Bridge	Coos County OR	1986	temporary walls (2)	1.8 to 3.0 1.8 to 5.5	30 33
30	Coquille River Bridge Detour	Coos County OR	1986	temporary wall for detour	3.0	9.1
31	Lemoyne	Lemoyne PA	1986	retained fill for parking lot	4.0	82
32	SR 4	Pacific County WA	1986	temporary walls (2)	3.0	120

FACING	TYPE	REINFORCEMENT BRAND	POLYMER	COMMENTS	REFERENCES
Wrapped 1:4 slope no facing	Woven	Mirafi 600x	Polypropylene	Sandy clay silt backfill	8
Wrapped gunite	Nonwoven needlepunched	Supac 12NP	Polypropylene	--	--
Wrapped 1.5:1 to 1:1 slope compacted straw	Geogrid	Tensar SR2	Polyethylene HDPE	Slip out caused by intense storms	19
full height pre-cast concre panels	geogrid	Tensar SR2	polyethylene HDPE	first full height concrete tilt up face in U.S.	20,21,22
cruciform pre-cast concre panesls	geogrid	Tensar SR2	polyethylene HDPE	use of geogrids w/incr wall panels	20,23
concrete wave deflector panels	geogrid	Tensar SR2	polyethylene HDPE	1st N. Amer/ sea wall w/polymer based soil reinforce	24
slope 1:1 CE111 Tensar on slope	geogrid	Tensar SS1 & SS2	polypropylene		
slope 1.5:1	geogrid	Tensar SS2	polypropylene		
wrapped no facing	uv stabilized woven slit film	Permea-tex 2300	polypropylene	granular backfill	
wrapped no facing	uv stabilized woven slit film	Permea-tex 2300	polypropylene	granular backfill	
	geogrid	Tensar			
wrapped wood facing attached after construct	geogrid	Signode TNX250	polyester	construce 4 days w/ inexper crew	25
wrapped gunite	non woven heat bonded	Typar 3601	polypropylene	based on results of glenwood canyon test walls	
treated wood	uv stabilized woven	Nicolon Geolon 500	polypropylene		
	non woven needlepunched	Supac 8NP	polypropylene	1 foot layer spacing	
				1 to 2 foot layer spacing	
wrapped gunite	woven	Mirafi 1200HP	polypropylene		
				granular backfill	

TABLE 1. GENERAL INFORMATION (cont'd)

CASE No.	PROJECT NAME	LOCATION	YEAR CONSTRUCTED	STRUCTURE	HEIGHT (m)	LENGTH (m)
33	SR 2	Snohomish & King Counties WA	1986	walls temp (8) permanent (2)	1.5 to 6.7	490
34	36th Avenue Regrade	Anchorage Alaska	Started 1986	retained fill for bike path	1.2 to 1.8	130
35	Devon Geogrid Test Fill	Devon Alberta	Started 1986	test slopes	11.9	72
36	I 210 over Southern Pacific RR	Lake Charles LA	under construction	retaining wall	4.6	253
37	I 20 over Chicago mill box culvert	Talullah LA	under construction	reinforced slope	3.4 to 6.1	1400
38	NATO Advanced research workshop	Royal Military College Kingsto Ontario	testing in progress	test walls (2)	2.7	2.4
39	North Halawa Valley access road	Oahu HA	to be constructed	retaining walls for approach roadway	3.0 to 7.3	854
40	FHWA test walls	Algoquin IL	under construction	retaining walls	6.1	11
	"	"	"	"	3 at 6.1	11
	"	"	"	"	6.1	11
	"	"	"	"	6.1	11
	"	"	"	"	2 at 7.6	15
	"	"	"	"	2 at 7.6	15
41	Conrail	Philadelphia PA	1987	rail line	6.7	90
42	Maryland Avenue	Christina DL	1986	parking lot	1.2-3.7	110

FACING		REINFORCEMENT		COMMENTS	REFERENCES
	TYPE	BRAND	POLYMER		
wrapped permanent walls gunite	uv stabilized woven slit film	Permea-tex 2300	polypropylene	granular backfill	
wrapped wood facing attached after construct	geogrid	Mirafi Paragrid 100/25S	polyethylene	for support of bike path	
1:1 slope	geogrid	Mirafi Paragrid 50S/ 50S & 5T	polyester	attempt to determine soil fabric	26
"	geogrid	Signode TNX 5001 & TNX 250	polyester	interaction mechanism for cohesive soils	"
"	geogrid	Tensar SR1 & SR2 SS1	polyethylene HDPE polypropylene	"	"
"	glasgrid	Bay Mills Glasgrid	modified asphaltic glass	"	"
wrapped gunite	geogrid	Tensar SR2	polyethylene HDPE	experimental project	
slope 4:1 no facing	geogrid	Tensar SS2	polypropylene	experimental project	
incremental propped vertica timber panels	geogrid	Tensar SR2	polyethylene HDPE	experimental project	27
wrapped 1:5 slope reinforce gunite	non woven		polypropylene	most extensive use of fabric reinforcing	
articulating precast concrete panels	steel strips	reinforced earth	steel	evaluate design metholology	28
"	bar mats	VSL	steel	"	"
"	geogrid	Tensar SR2		"	"
"	non woven geotextile	Quline	polyester	"	"
1H:1V slope	woven geotextile geogrid	Amoco 2006 Signode	polypropylene	"	"
1H:2V slope	"	"	"	"	"
wrap around w/ precast concret bk false face	geogrid	Tensar SR2	HDPE	--	36
gabion	geogrid	Tensar SR2	HDPE	--	36

TABLE 1. GENERAL INFORMATION (cont'd)

CASE No.	PROJECT NAME	LOCATION	YEAR CONSTRUCTED	STRUCTURE	HEIGHT (m)	LENGTH (m)
43	Rotterdam	Rottc.-dam NY	1986-87	stream diversion & parking lot	5.5	180
44	Waterford Harbor	Houston TX	1986	marina bulkhead	2.4	3350
45	Wendy°s	Mobile AL	1986	driveway & parking lot	1.8-3.7	74
46	Residence Inn	Tyler TX	1985	land development	6.1	104
47A	Elliott Plaza	Westbank British Columbia	1985	prototype retaining wall	2.4	33
47B	"	"	"	"	2.4	33
48	Smith	Kelowna British Columbia	1985	retaining wall	2.4	30
49A	Naramatta Road	Pentkton British Columbia	1985	"	1.8	29
49B	"	"	"	"	2.4	32
50	Gorman Kiln	Westbank British Columbia	1986	retaining wall	1.8-4.9	44
51	Gorman Shop	Westbank British Columbia	1986	"	3.7-4.9	32
52	Borges Rehabilitation	Vernon British Columbia	1986	"	3.1-4.9	15
53	CNR Overpass	Vernon British Columbia	1986	"	2.4	162
54	Torhielm	Kelonna British Columbia	1987	"	1.2-6.1	99

FACING	TYPE	REINFORCEMENT BRAND	POLYMER	COMMENTS	REFERENCES
full height concrete	geogrid	Tensar SR2	HDPE	--	36
full height concrete	geogrid	Tensar SR2	HDPE	--	36
masonry blocks	geogrid	Tensar SR2	HDPE	--	36
full height concrete	geogrid	Tensar SR2	HDPE	--	36
wafflecrete	geogrid	Tensar SR2	polyethylene HDPE		
"	"	"	"		
"	"	"	"		
"	"	"	"		
"	"	"	"		
wafflecrete	geogrid	Tensar SR2	polyethylene HDPE	reinforce rear wall of dry kiln to support wall	
"	"	"	"		
cast in place concrete	"	"	"	rehabilitation of failing cast in place con. wall	
wafflecrete	"	"	"		
"	"	Tensar SR1 SR2	"		

TABLE 2. DESIGN INFORMATION

CASE No.	HEIGHT (m)	REINFORCEMENT SPACING (m)	REINFORCEMENT LENGTH (m)	SURCHARGE LOAD	FOUNDATION				
					SOIL TYPE	STRATIFICATION	WATER TABLE	c kPa	0
1	3.0	0.23 - 0.30	3.0	1 meter of fill dual tandem axel loads	rock				
2	0.9 - 6.1	0.23 - 0.30	4.0	930 kN on 4 axels	weathered rock				
3	design 3.7 failed 3.0	0.6 vertical 1.2 horiz.	3.0		lean clay				
4	6.7	0.23	4.0	311 kN dual tandem wheel load	rock				
5	4.9	0.15 - 0.23	3.7		stiff to hard silty clay w/ sand & gravel	uniform	high water table		
7	8.5	0.3	4.0 at base 11.9 at top	311 kN dual tandem wheel load		old landslide		5	38
9	4.6	0.23 - 0.40	4.0		highly compress compressible silts and clays	uniform, thick deposits			
10	4.6 & 5.5	1.2	3.7 - 5.5	30 kN/m	clayey silt till		high water table	6.9	31
11	6.7	0.61	0.9		silty Beaumont clay		high water table	20	15
12	8.2	0.2 - 0.4	3.4		dolomitic limestone bedrock				
13	9.1	0.3 - 0.9	4.9		soft gray shale	shale-fractured and weathered			
14	11.9	0.4 - 0.6							
16	3.0	0.20	2.4	13.6 kPa & 160 kN dual tandem wheel	silty sand	uniform		0	30
17	14.0	0.61	6.1						
18	0.9 - 6.1	0.3 - 0.9	3.7	12.0 kPa	collapsible soils				
19	6.1	0.3 - 0.6	3.7	12.9 kPa	soft clay w/ crushed stone on top				24
20	5.3	0.3	3.5	21.0 kPa	dense sand and gravel		at surface	0	30

		REINFORCED FILL					BACKFILL				
d kN/m³	w (%)	SOIL TYPE	c° kPa	0°	d kN/m³	w (%)	SOIL TYPE	c° kPa	0°	d kN/m³	w (%)
		silty sand, gravel size rock fragments	0	38	18.5		silty sand, gravel size rock fragments	0	38	18.5	
		uniform crushed basalt	0	31	19.6		uniform crushed basalt	0	31	19.6	
		clean uniform concrete sand	0	36	15.3	4.5	lean clay				
		75mm minus crushed rock	0	35	17.3		75mm minus crushed rock	0	35	17.3	
		open graded crushed stone	0	33			open graded crushed stone	0	33		
		wood chips	0	32	5.8		wood chips	0	32	5.8	
		rounded,well graded clean sand & gravel		35	20.4						
		granular	0	35	19.9						
		silty Beaumont clay	20	15			silty Beaumont clay	20	15		
		25mm crushed limestone	0	40	18.1		silty clay	144	10	22.0	16
							native soil	9.6	30		
		crushed basalt	0	40	22.0						
		on-site silty clay	24	20							
		25mm crushed gravel	0	35	21.2		25mm crushed gravel	0	35	21.2	
		granular	2.4	32							
		granular	0	34	19.6		granular	0	34	19.6	
		granular	0	40	21.2		granular	0	40	21.2	
		well graded sandy gravel	0	34	21.0		well graded sandy gravel	0	30	21.0	

TABLE 2. DESIGN INFORMATION (cont'd)

CASE No.	HEIGHT (m)	REINFORCEMENT SPACING (m)	REINFORCEMENT LENGTH (m)	SURCHARGE LOAD	FOUNDATION				
					SOIL TYPE	STRATIFICATION	WATER TABLE	c kPa	0
21	5.5	1.8 1.8	3.0 0.9	H-20 loading	SM-GM	landslide deposit		0	34
22	6.1	0.76	4.0	H-20 loading	SM-SC	uniform		24	20
26	3.0	0.3	2.4 at base 4.6 at top	840 kN crane, 180 kN lifting capacity	medium dense fine sand w/ some clay			0	30
28	1.5 - 4.3	0.3 - 0.6			clay	uniform bedrock at 3 to 4.6 m		33 - 57	20
31	4.0	0.15	3	15.9 kPa & 160 N dual tandem axle	shale rock fragments & silt	uniform	--	0	20
34	1.2 - 1.8	0.61	2.9	--	--	stratified	--		
35	11.9	primary 2.0 secondary 1 tertiary 0.3	primary 13 secndry 3-5 tertiary 1.5	--	soft silty clay	stratified	at minus 6.1 m	20	27
36	4.6	0.23 - 0.61	3.8	--	clay	stratified	at minus 0.3 m	50	0
37	3.4 - 6.1	0.61	varies	--	clay	stratified	at minus 2.1 m	100	0
38	2.7	0.76	3.0	0.76 m of sand & inflatable air bags	sand blinding layer	0.24 m thick	--	0	43
39	3.0 - 7.3	0.15 - 0.30	2.3 - 3.4	--	weathered rock	--	--	--	--
40	6.1	0.38 & 0.76	4.3	3m sand	gravely sand	uniform	at minus 0.5 m	0	39.0
"	7.6	0.76	4.3	--	gravely sand	uniform	at minus 0.5 m	0	39.5
41	6.2	--	--	railroad	--	--	--	--	--
42	1.2-3.7	0.91	--	parking lot	--	--	--	--	--
43	5.5	0.3-0.6-1.2	4.4	12 kPa	GW	GW 1st 1m clay & silt	--	0	45
44	2.4	--	3.2	4.8 kPa	CH	piling below	--	6.2	16
46	6.1	0.2-1.2	4.6	0 kPa	SM-ML	--	--	0	30

		REINFORCED FILL					BACKFILL				
d kN/m³	w (%)	SOIL TYPE	c° kPa	0°	d kN/m³	w (%)	SOIL TYPE	c° kPa	0°	d kN/m³	w (%)
18.8		SM-GM	0	34	18.8						
18.1		SM-SC	24	20	18.1						
		fine to coarse sand & gravel	0	37	19.6		medium dense fine sand w/ some clay	0	30		
		granular	0	30- 35	18.1		granular	0	30- 35	18.1	
21.2	--	sand & gravel, 10% fines	0	35	21.2	--	shale rock fragments & silt	0	20	21.2	
--	--	75mm minus sand & gravel	--	--	--	--	--				
--	35	inorganic clay of lo to medium plasticity	13	22	15.7	21.5 OMC	inorganic clay of low to med plasticity	13	22	15.7	21.5 OMC
15.7	25	sand	0	30	--	--	inorganic clay of low to med plasticity				
13.3	37	clay	5 ----- 0	0 ----- 12	13.3	--	--				
17.0	3	washed sand w/ some gravel	0	41	16.0	3	washed sand w/ some gravel	0	41	16.0	3
--	--	granular	--	--	--	--	weathered rock				
20.2		gravel sand silt	0 0 13	42 39.0 35.5	18.0 20.2 17.0	 2 19	gravely sand	0	39	20.2	2
20.2		gravely samd silt	0 13	39.0 35.5	20.2 17.0	2 19	gravely sand silt	0 13	39 39.5	20.2 17.0	2 19
--	--	--	--	--	--	--	--	--	--	--	--
--	--	--	--	--	--	--	--	--	--	--	--
--	--	SM	0	35	18.1	--	SM	0	30	18.1	--
--	--	CH	6.2	16	19.2	--	CH	6.2	16	19.2	--
18.8	--	SM-ML	0	30	18.8	--	SM-ML	0	30	18.8	--

TABLE 2. DESIGN INFORMATION (cont'd)

CASE No.	HEIGHT (m)	REINFORCEMENT SPACING (m)	REINFORCEMENT LENGTH (m)	SURCHARGE LOAD	FOUNDATION				
					SOIL TYPE	STRATIFICATION	WATER TABLE	c kPa	0
47A	2.4	0.6 & 1.2	1.4	--	GP	variable	--	0	33.5
47B	2.4	0.6 & 1.2	2.8	2.4m wall above	GP	variable	--	0	33.5
48	2.4	0.6 & 1.2	1.8	--	SP	fine river sand	-1m	0	--
49A	1.8	0.6	1.1	--	GP	--	--	--	--
49B	2.4	0.6 & 1.2	1.3	--	GP	--	--	--	--
50	1.8-4.9	0.6 & 1.2	2.9	4.8kPa of backfill	SP	--	--	0	35
51	3.7-4.9	0.6 & 1.2	2.9	B.C.F.S. L-45	GP	--	--	0	35
52	3.0-4.9	0.6 & 1.2	2.9	4.8kPa	--	--	--	--	--
53	2.4	0.6	3.0	12kPa backfill 33.7 degree slope	SP	variable	--	0	35
54	1.2-6.1	0.6	3.2	2.4kPa	SP	varved	--	0	30

		REINFORCED FILL					BACKFILL				
d kN/m³	w (%)	SOIL TYPE	c° kPa	0°	d kN/m³	w (%)	SOIL TYPE	c° kPa	0°	d kN/m³	w (%)
20.4	--	GP	0	33.5	20.4	--	GP	0	33.5	20.4	--
20.4	--	GP	0	33.5	20.4	--	GP	0	33.5	20.4	--
--	--	SP	0	33.5	19.6	--	SP	0	33.5	19.6	--
--	--	GP	0	32	17.3	--	GP	0	32	17.3	--
--	--	GP	0	32	17.3	--	GP	0	32	17.3	--
20.4	--	SP	0	35	20.4	--	SP	0	35	20.4	--
20.4	--	GP	0	35	20.4	--	GP	0	35	20.4	--
--	--	--	--	--	--	--	--	--	--	--	--
19.6	--	GP	0	35	19.6	--	GP	0	35	19.6	--
18.8	--	SP	0	30	18.8	--	SP	0	30	18.8	--

TABLE 3. REINFORCEMENT PROPERTIES AND DESIGN METHODOLOGY

CASE No.	REINFORCEMENT PROPERTIES: BRAND	METHOD OF STRENGTH EVALUATION	TENSILE STRENGTH kN/m	MAXIMUM ELONGATION (%)	DESIGN PROPERTIES: STRENGTH T(kN/m)	εe (%)	REDUCTION FACTORS: DURABILITY	CREEP	DESIGN METHOLOGY: EXTERNAL STABILITY	INTERNAL STABILITY: MODEL	VERT. STRESS	H.STRESS DISTR	PULLOUT
1	Fibertex 400	25mm strip test	12.3	200	7.5	35			sliding overturning	tieback wedge	overburden	Ko	2/3 0
2	Fibertex 400	25mm strip test	11.4	166					sliding overturning	tieback wedge	overburden	Ko	2/3 0
	Fibertex 600	"	17.1						"	"	"	"	"
	Bidim C-28	"	10.7	60					"	"	"	"	"
	Bidim C-38	"	18.9						"	"	"	"	"
3	heavy duty nylon strips	webbing capstan grip tst & mod grab/50mm wide	190	14.5						tieback wedge		Ko and Ka	
4	Fibertex 200	25mm strip test 100mm strip tst grab test	5.2 8.4 (605 N)		5.2 8.4 (605 N)				sliding overturning				
5	Bidim C-34	grab test	(1000 N)		13.1					tieback wedge			2/3 0
7	Supac 5W	grab test strip test	(670 N) 23		12.3				sliding overturning landslide	tieback wedge	overburden	Ko	2/3 0
9	Fibertex 200 Fibertex 400	wide width tst w=200mm l=100mm	5.8 10.0	140 145	3.2 5.4					tieback wedge		Ko	
	Supac 4NP Supac 6NP	t=10%	12.6 24.3	65 60	5.1 9.8					"		"	
	Trevira S1115 Trevira S1127	"	6.7 16.8	80 75	4.4 10.9					"		"	
	Typar 3401 Typar 3601	"	7.7 12.4	60 55	3.2 4.2					"		"	
10	Tensar SR2	QC single rib W=1,L=3 ribs t=50mm/min	79	12	16.1				global	"			0.8 tan 0
12	Supac 12NP	manufacturer 25mm grab test	(1470 N)		12.8				slide/overturn short/long term slope stability	"	overburden	Ko	pullout factor
13	Tensar SR2	QC single rib W=1,L=3 ribs t=50mm/min	79	12	15.8			40%		"			0
16	Supac 12NP	manufacturer 25mm grab test	(1470 N)		12.8				sliding bearing capacity	"	overburden	Ko	pullout factor
17	Tensaar SR2				6.7			15%	global				pullout test
18	Tensar SR2	index test W=15,L=5 ribs t=2%/min	70	16-25	29	10			sliding bearing overturning global stability	tieback wedge	trapezoidal	K-stress ratio	0.9 tan 0
19	Tensar SR2	index test W=15,L=5 ribs t=2%/min	70	16-25	29	10			sliding bearing overturning global stability	tieback wedge	trapezoidal	K-stress ratio	0.9 tan 0

TABLE 3. REINFORCEMENT PROPERTIES AND DESIGN METHODOLOGY (cont'd)

CASE No.	BRAND	REINFORCEMENT PROPERTIES: METHOD OF STRENGTH EVALUATION	REINFORCEMENT PROPERTIES: TENSILE STRENGTH kN/m	REINFORCEMENT PROPERTIES: MAXIMUM ELONGATION (%)	DESIGN PROPERTIES: STRENGTH T(kN/m)	DESIGN PROPERTIES: θe (%)	DESIGN PROPERTIES: REDUCTION FACTORS DURABILITY	DESIGN PROPERTIES: REDUCTION FACTORS CREEP	DESIGN METHOLOGY: EXTERNAL STABILITY	DESIGN METHOLOGY: INTERNAL STABILITY MODEL	DESIGN METHOLOGY: INTERNAL STABILITY VERT. STRESS	DESIGN METHOLOGY: INTERNAL STABILITY H.STRESS DISTR	DESIGN METHOLOGY: INTERNAL STABILITY PULLOUT
20	Tensar SR2	10000 hr. creep tst extap. to 70 yr. design	29		21.5				sliding bearing overturning global stability	tie-back wedge			0.9 tan 0
21	Tensar SS1 amd SS2	manufacturers literature			12.8			incorp. / design strength	sliding bearing overturning	pseudo tie-back wedge	Meyerhof	Ka	
22	Tensar SS2	manufacturers literature			12.8			incorp. / design strength	sliding bearing overturning	pdeudo tie-back wedge	Myerhof	Ka	
26	Signode TNX250	wide width ASTM D-4595	48	9.3	14	5	1/3	1/3	sliding bearing overturning deep seat slope	tie-back wedge	Myerhof and Trapezoidal	Ka and K-stress ratio	0 pullout test
27	Typar 3601	wide width W=200mm,L=100mm t=10%	12.4	55	5.1					tie-back wedge			
28	Nicolon Geolon 500	wide width grab test	29 35	10	26.2	10			sliding bearing capoacity				
31	Mirafi 1200 HP	Manufacturer 1"x1" jaw test	79		17.5				sliding bearing capacity	tie-back wedge	overburden	Ko	- pullout factor
43	Tensar SR2	10000 hr., 1000000 hr. extrapolated	79	10	29	--	1.0	1.0	sliding-1.5 bearing-2.0 overturning-2.0	tieback wedge	Meyerhof	Ka	test
44	Tensar SR2	"	79	10	29	--	1.0	1.0	sliding-1.5 bearing-1.5 overturning-1.5	tieback wedge	Trapezoidal	Ka	test
46	Tensar SR2	"	79	10	29	--	1.0	1.0	sliding-1.5 bearing-2.0 overturning-2.0	tieback wedge	Trapezoidal	Ka	test
47 thru 53	Tensar SR2	Mfg. based on 1% or less creep in 120 yrs	--	--	16.1	--	--	Incorp design strength	sliding bearing cap. overturn slip failure	tieback wedge	Meyerhof	Ka	--
54	Tensar SR1 SR2	Mfg. based on 1% or less creep in 120 yrs	--	--	8.1 16.1	--	--	incorp design strength	sliding bearing cap. overturn slip failure	tieback wedge	Meyerhof	Ka	--

TABLE 4. INSTRUMENTATION

CASE No.	LATERAL MOVEMENT			OVERALL SETTLEMENT		
	STRAIN IN REINFORCEMENT	FACE	IN AND BEHIND STRUCTURE	INCREMENTAL SETTLEMENT OF WALL	SURFICIAL SETTLEMENT OF WALL	FOUNDATION SETTLEMENT (differential)
2		transit	26 horizontal inclinometer tubes w/slip rings	26 horizontal inclinometers	level	level
		6mm to 52 mm	0 to 52mm	15 to 30mm	0	0
3	indirectly calculated	observation				
		bottom 57 mm top 152 mm				
5			slip tube			slope indicator & settlement devices
			0 to 6.3mm			6.3mm to 33.5mm
9		survey posts and laser targets	30 horizontal inclinometer-extensometer	5 manometers	observational	5 vertical inclin-ometers sondex 5 manometers
					0.3m over 90m length after first 3 months	large amount of consolidation 0.6m at W. end wall
17		survey reference points	extensometers	inclinometers	survey reference points	inclinometers
18	resistance strain gauges & induct.coils	observational				
	line of max strain 18 to 25 from vertical	65 mm				
19	resistance strain gauges & induct.coils	observational				
	data inconclusive	within anticipated range				

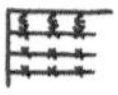

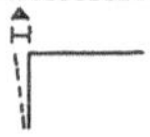

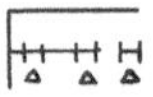

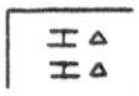

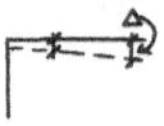

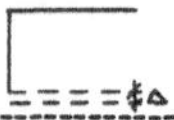

TOTAL STRESSES: BASE OF WALL	TOTAL STRESSES: FACE OF WALL	INTERNAL STRESSES (soil)	LOAD DISTRIBUTION IN REINFORCEMEN	CHANGES OF LOAD DISTRIBUTION WITH TIME	PORE PRESSURE RESPONSE BELOW WALL	TEMPERATURE (if instrumentation affected)	COMMENTS
	6 WES pressure cells-center line 1ft.from face	6 WES pressure cells					excessive deformation of the reinforcing strips beyond that necessary to create Rankine state
	measured pressure less than predicted except at failure						
		load cells			piesometers		
		loads larger near back of wall			excess pore press. did not build at loading		
							results of the instrumentation program were not available
load cells	load cells	load cells	strain gauges	strain gauges		18 SINCO resistance thermom probes	monitoring program still in progress
approached predict press. at toe of wall	approached predict press. at toe of wall		Rankine wedge defined	Rankine wedge defined			
load cells	load cells	load cells	strain gauges	strain gauges			monitoring program still in progress
approached predict press. at toe of wall	approached predict press. at toe of wall		data inconclusive	data inconclusive			

TABLE 4. INSTRUMENTATION (cont'd)

CASE No.	LATERAL MOVEMENT			OVERALL SETTLEMENT		
	STRAIN IN REINFORCEMENT	FACE	IN AND BEHIND STRUCTURE	INCREMENTAL SETTLEMENT OF WALL	SURFICIAL SETTLEMENT OF WALL	FOUNDATION SETTLEMENT (differential)
20	strain gauges glued to the reinforcement	observational	muti position mech. extensom & observation	observational		
	0.75%	47mm	37mm	17mm to 36mm		
26	micromesurement & bison induct strain gauges	observational	wire extensometers		observational	
	1% upper level 0.85% lower level of wall					
35	elect resistanc & bison strain gauges	horizontal extensometer	horiz. extensom and vertical inclinometer	vert. extensom and horiz. inclinometer	vert. extensom and horiz. inclinometer	vert. magnetic extensom & inclinometer
36	strain gauges		inclinometers			
37	strain gauges		inclinometers			
38		observational			observational	

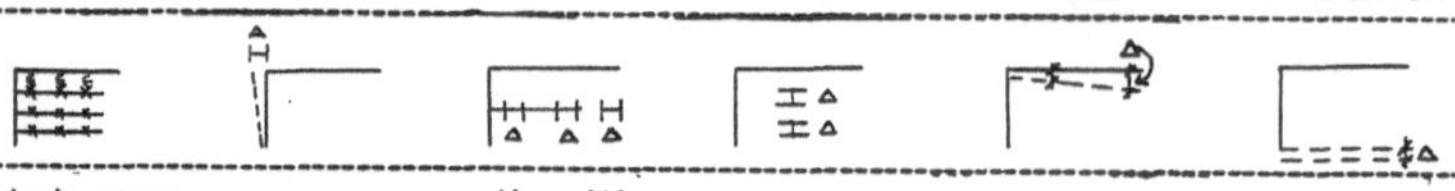

TOTAL STRESSES: BASE OF WALL	TOTAL STRESSES: FACE OF WALL	INTERNAL STRESSES (soil)	LOAD DISTRIBUTION IN REINFORCEMEN	CHANGES OF LOAD DISTRIBUTION WITH TIME	PORE PRESSURE RESPONSE BELOW WALL	TEMPERATURE (if instrumentation affected)	COMMENTS
σ_V	σ_H				μ	T°	
	peter stress cells		stain gauges	strain gauges	six peter pneumatic piezometers	thermocouples	
	erratic results		rankine failure surface defined	no conclusion			
	load cell at tie rod location		strain gauges	strain gauges			wall performed satisfactory during crane loading
			no conclusion	no conclusion			
			strain gauges	strain gauges	20 SINCO pneumatic piezometers	frost heave gauges with thermocouples	construction not complete
					piezometers		
load cells	@ panel geogrid locations		strain gauges	strain gauges		thermocouples	

TABLE 5 TENSAR GEOGRID REINFORCED SOIL RETAINING WALLS
NORTH AMERICAN SUMMARY

FACING:	WRAP-AROUND	GABION
APPROXIMATE NUMBER OF WALLS CONSTRUCTED (+ or -30%):	40	10
STRUCTURES:	Temporary Const. Land Development Blast Protection Railroad support Walls over high settlement areas Landslide repair	Land Development Erosion Control Walls over high settlement areas Landscaping
TYPICAL HEIGHT RANGE:	8' to 30'	8' to 30'
REINFORCEMENT - GEOGRID		
BRAND:	a. Tensar SR1, SR2, or SR3 b. Tensar SS2	Tensar SR1, SR2, or SR3
POLYMER:	a. HDPE b. Polypropylene	HDPE
TYPICAL REINFORCEMENT SPACINGS:	8" to 24"	18" or 36"
FOUNDATION SOILS:	Any soil type	Any soil type

<u>NOTES</u>:

Embedment Lengths: L=0.6 to 1.0 x Height; L=f (soil strength, surcharge loads, etc.); Typically L=0.7 x Height

Reinforced Soils: Any Soil Type, Excluding Organics

Retained Backfill Soils: Any Soil Type, Excluding Organics

Design Tensile Load: Approx. 40% x Ultimate Strength for SR products and approx. 25% x Ultimate Strength for SS products prior to factoring

Reduction Factors: For Durability 1.0, Creep 1.0 as it is accounted for in determining Design Tensile Load, Site Damage typically 1.0, Range from 1.0 to 1.35 Used

TIMBER	MASONRY	FULL-HT. PRECAST CONCRETE	ARTICULATED CONCRETE OR MASONRY UNITS
10	15	30	16
Landscaping Land Development	Land Development Highways	Highway walls Land Development Bridge wing walls Harbor walls	Highway walls Land Development Walls over high settlement Seawalls
5' to 18'	4' to 24'	3' to 20'	12' to 24'
a. Tensar SR1, SR2, or SR3 b. Tensar SS2 a. HDPE b. Polypropylene	Tensar SR1, SR2, and SR3 HDPE	Tensar SR2 and SR3 HDPE	Tensar SR2 and SR3 HDPE
7", 14", or 21"	8", 16", 24", or 32"	8" to 4'0"	12" to 24"
Any soil type excluding organics.			

External Stability: Typical Factors of Safety - Sliding: 1.5
Bearing: 1.5 to 2.0
Overturning: 2.0
Global: 1.3 to 1.5

Internal Stability: Mode - Tie-back wedge

Vertical Stress: Trapazoidal distribution approximately through 1985, Meyerhof type distribution since then

Horizontal Stress Distribution: K_a

partially instrumented, of which four were research walls and nine were production walls. Major research projects that are currently underway (Projects No. 35, 38 and 40), but have not been completed, were also included in the summaries, as they will undoubtedly complement the currently availabile information.

It is believed that the projects surveyed are only a portion of the total number of walls and slopes reinforced with polymeric materials in North America. For example the attached Table 5 provided by the Tensar Corporation presents a summary of structures where their products have been used and includes an estimation of 121 projects. Likely there are numerous other unreported reinforced soil projects where other geosynthetics have been used. Our best estimate of polymerical reinforced soil walls and slopes that have been constructed in North America is on the order of 200.

The following section provides an overview of projects that were either unique or where significant technical information was available.

2.1 Overview of Technically Significant and Unique Projects

The Illinois River Road Wall project, constructed in 1974 in the Siskiyou National Forest, Oregon, was the first reported full-size geosynthetic retained soil wall built in North America. The United States Forest Service [1] designed and built the wall to restore the width of Illinois River Road which had been reduced by erosion. The 2.7 m high and 10 m long wall was designed using geotextile reinforcement. A conversation with the designers indicates that the wall is still in service and has performed satisfactorily since construction in 1974.

Between 1974 and 1976, the United States Army Engineers Waterways Experiment Station in Vicksburg, Mississippi investigated the effect of horizontal reinforcement on the stability of earth masses [5, 6]. Two instrumented retaining walls were constructed. The walls were to be 3.7 m high, 4.9 m long and 3.0 m deep. One wall was reinforced with neoprene coated nylon fabric strips and the other with galvanized steel strips. The fabric strip reinforced wall failed when the wall height reached 3.0 m. The exact cause of the failure was unknown, but is believed to be the result of significant elongations in the reinforcement beyond that which was needed to create active failure in the soil. There was also some indication that the elongation could have been concentrated at the face connections. Lateral pressures were measured at the face of the wall. The measurements indicated lateral pressures slightly less than those predicted by Rankine theory prior to failure and equal to or greater than the predicted lateral pressures at failure.

In 1981, the California Department of Transportation built a tire-anchored timber wall on State Route 203 in Mammoth Lakes [9]. The tire-anchored timber wall incorporated two waste materials - used automobile tire side walls and used railroad ties. Two walls varying in height from 1 to 3.7 m and totaling more than 230 m in length were constructed. The walls performed far better than expected, including withstanding an earthquake of 5.8 Richter magnitude in 1981 with no visible evidence of damage. Several other walls were constructed by Caltrans using the same system as the Mammoth Lakes wall. The walls were constructed in the following locations in California: Route 101 in Marin County (1983), Route 1 in Santa Cruz County (1983) and Big Basin Park in Santa Cruz County (1984). No other information was available concerning the projects.

The first geogrid reinforced wall constructed in the United States served to stabilize a landslide near the entrance to Devil's Punch Bowl State Park in Oregon in 1983. The project, a FHWA Experimental Features Project, was to assess the construction problems of near-vertical walls with geogrids. The geogrid wall was chosen over the other alternatives for two reasons [16, 17]:

1. It had the lowest estimated cost $\$260/m^2$, and

2. The open face structure of the geogrid allowed establishment of vegetation, which provided a natural appearance compatible with the surrounding state park.

The geogrid wall was less expensive than the geotextile wall, since the geotextile wall required a protective facing.

In 1982, the Colorado Division of Highways authorized the construction of a 4.6 m high, 91 m long geotextile reinforced soil wall test section in Glenwood Canyon, Colorado [9, 10]. The 91 m long wall was divided into ten, 9.1 m sections, each of which incorporated the use of a different style or weight of non-woven geotextile. The purpose of the test wall was to determine the validity of the design methodology in evaluating the properties of the reinforcement, including creep. All sections of the wall were fully instrumented as indicated in Table 4. Two sections of the test wall were intentionally underdesigned. These two sections were then surcharged with 5.2 m of fill in an attempt to produce failure; however, under the full height of the surcharge, insignificant internal and external deformations were observed. As of 1975, creep related movements have not been observed in any of the wall sections, even though the soft silt and clay foundation soils have been compressed approximately 0.6 m at one end of the wall [11]. Fabric samples have been taken on a yearly basis and tested for strength loss by Professor Bell of Oregon State University.

The results of the testing program were not available at the time of this paper. In 1986, the Colorado DOH constructed a geotextile reinforced wall in conjunction with the I-70 project through Glenwood Canyon. The wall, designed in accordance with the findings of the 1982 test wall, has performed satisfactory to date.

A series of high density polyethylene (HDPE) geogrid structures using full height concrete tilt up facing panels were constructed between 1984 and 1985 in Tuscon, Arizona [20. 21, 22]. A total of 46 walls ranging in height from 1 m to 6.1 m and totaling over 1600 m in length were constructed for a series of grade separations for a major thoroughfare. Two wall sections were heavily instrumented and only limited results have been reported to date. A preliminary assessment of lateral and vertical pressures indicated that the lateral pressures at the face of the wall were less than those predicted by either the Rankine, Meyerhof or trapezoidal pressure distribution theories. However, vertical pressures at a distance of approximately 0.3 m behind the wall were found to exceed those predicted by either the trapezoidal, Meyerhof or overburden pressure theories. Local strain measurements on the reinforcement indicated a wedge type distribution of peak strains over the height of the wall that agreed with a Rankine type failure surface.

The first reported polymer-based reinforced soil sea wall in North America was constructed along the coast of the Gaspe' Peninsula in the Canadian province of Quebec in 1985 [24]. The 5.2 m high and 100 m long sea wall protects Highway 132 and the adjacent land from storm waves which can reach heights of 3 m. The sea wall was constructed using sectional articulating type precast concrete facing panels and high density polyethylene geogrids. An extensive monitoring program was established to aid in evaluating the performance of the sea wall. The sea wall was reportedly performing as anticipated after 1.5 years of service.

To allow access of equipment for reconstruction of the Cascade Dam hydro-electric facility in Cascade in Cascade, Michigan, the owner, STS Consultants, Ltd. designed and constructed a reinforced soil wall using polyester geogrids to increase the width of the crest of the dam [25]. The soil wall was required to support a 840kN crane with a 100kN lifting capacity. The wall constructed in 1986 was fully instrumented to monitor internal and external lateral movement as well as strain levels in the geogrid elements. The wall was monitored during construction and over the five month period following construction up until and including the operation of the high capcity crane. Preliminary data indicated no apparent strain over the five month period following construction. The strain level during external loading by the crane was found to be close to that predicted by theory at the base of the wall, but exceeding that predicted by theory in the upper levels of the reinforcement. Monitoring is being continued, so that long-term performance, including stress relaxation after removal of the crane, can be evaluated.

Several future projects of interest were also reviewed. More than 39 kilometers of approach roadways will be constructed in conjunction with the Interstate Route H-3 project on the Hawaiian Island of Oahu [28]. The interstate will connect the northeast coast of the island with the rest of the island. The west access road alone, referred to as the North Halawa Valley access road, will require 6.3 kilometers of retaining walls ranging in heights up to 8 m. The North Halawa Valley access road, when constructed, will be the most extensive use of geotextile fabric retaining walls to date. Environmental concerns have currently delayed construction.

Presently under construction is a 12 m high, 42 m wide and 72 m long geogrid reinforced test slope in Devon, Alberta [26]. The slope is constructed at 1 horizontal to 1 vertical with native cohesive soils. Three test sections were established to compare the performance of three different geogrid materials. The test sections were desgined with the minimum number of geogrid layers and minimum length of reinforcement, so that each reinforcing layer acts independently. The primary objective of the test fill is to evaluate the mechanisms whereby geogrid reinforcement can control lateral strains in a cohesive soil mass.

Another series of test walls is being constructed under the sponsorshiop of the FHWA in Algonquin, IL [20]. The test sections will consist of six, 6 m high by 11 m wide walls and four, 7.6 m high by 15.2 m wide slopes. The walls will be constructed using four types of reinforcement (metallic strips, metal bar mats, a polymeric geogrid and a needle punched non-woven geotextile). The soil will principally be free draining granular materials with soil variation evaluated for the bar mat reinforcement. A woven geotextile and a geogrid will be used to construct 1 horizontal to 2 vertical slopes with gravelly sand and 1 horizontal to 1 vertical slopes using silt. The structures will be fully instrumented to monitor the stress distribution during construction and subsequent external loading. The project is to be constructed in the summer of 1987 with monitoring completed by the summer of 1988. The project will conclude a three year study to develop reinforced soil design guidelines.

A test wall is also being constructed in conjunction with the NATO workshop [38] for which this paper was prepared. The results of this test should be reported elsewhere in these proceedings.

3. CURRENT TRENDS IN RETAINING STRUCTURE DESIGN

This next section describes current trends in the design of geosynthetic retaining structures. The following factors will be discussed:

1. Reinforcing Materials
2. Facing Systems
3. Backfill Materials and Interaction Mechanisms
4. Design Methodology and Factors of Safety
5. Project Costs

3.1 Reinforcing Materials

Of the projects surveyed, nine utilized woven geotextiles, twelve utilized non-woven geotextiles and thirty-two utilized geogrids. Other polymeric systems that were used included heavy duty nylon reinforcing strips and tire sidewalls with anchor bars. From the introduction of fabric reinforced retaining structures in North American in 1974 (Project No. 1), non-woven geotextiles were primarily used until the introduction of geogrids in 1982 (Project Nos. 11 and 15). Woven geotextiles were found to have had limited use as wall reinforcement. Their primary application has been in temporary wall construction.

The current domination of geogrids is probably due to several factors. Firstly, the Tensar Corporation, the principal geogrid manufacturer, has provided a strong technical approach in its promotional efforts, an approach which had previously been absent from the geotextile manufacturers. Secondly, the apparent high strengths of geogrids may have appeared attractive to designers, because they could reduce the number of layers of required reinforcement.

The ability to use thicker lifts when geogrids are utilized perhaps appeared attractive to the contractor from the stand point of reducing construction time. However, as noted in the Glenwood Canyon project [9, 10, 16, 17] the forming systems which had previously proven adequate for geotextiles were not stiff enough to handle the large lift thicknesses. Therefore, new forming systems were required. These new forming systems were much more complicated and required substantially more time to reset from one lift to the next [16, 17]. Thus, time saved because of the reduced number of lifts was absorbed by the increased forming time.

Other factors influencing the choice of reinforcement include the unit costs and the need for protection of the face material. From the projects reviewed, geotextiles tend to have a lower unit cost than geogrids as summarized in Section 3.5. Geogrids, apparently due to their thickness, have been found to be less susceptible to UV deterioration than most geotextiles [29].

It is expected that newer higher strength geogrids and geotextiles will permit the construction of even higher walls than previously constructed using geotextiles. Of the reported projects the tallest geogrid reinforced wall was 9.1 m, the tallest geotextile reinforced wall was 8.5 m and the tallest geogrid reinforced slope was 14 m.

3.2 Facing Systems

All geosynthetic materials experience a loss in strength and ductility when exposed to ultra-violet radiation. Therefore, it is necessary to use either UV-stabilized synthetics for temporary applications or some other type of permanent facing system for long-term applications. The facing systems used in the projects evaluated included: gunite, emulsified asphalt coatings, latex paint, timber, aluminum panels, natural vegetation, and sectional articulating type and full-height concrete panels. The most common system used for geotextile reinforced walls was gunite. Gunite appears to be a very satisfactory facing material. The gunite facing system utilized for the Glenwood Canyon test wall remained intact even after the wall experienced differential settlements in excess of 2 ft. Gunite facing systems also protect the geosynthetics from vandalism. The emulsified asphalt coatings do not always adhere to the geotextile [8]. For example, the Camp Hill project utilized a woven slit film geotextile which was to be coated with an emulsified asphalt; however, when the asphalt failed to adhere to the fabric, the fabric face was then painted with latex paint.

Several geogrid walls and slopes were constructed using vegetation for cover to reduce UV exposure. As indicated in Project No. 18, full height concrete panels have also been used. One innovative approach (Project No. 26) which can be applied to both geotextile and geogrid walls is the post construction attachment of facing panels.

3.3 Backfill Materials and Interaction Mechanisms

A majority of the 54 projects surveyed utilized granular backfills. Granular backfill was used for all vertical wall projects. Six of the slope projects utilized cohesive backfills and two slope projects used wood chips. Information was not available for 11 projects.

Although six of the projects surveyed utilized cohesive backfill materials, unfortunately none of the projects appeared to be instrumented and long-term performance data was not available. Hopefully, the findings of the Devon Test Fill project will lead to a better understanding of the soil-reinforcement interaction for such structures.

In determining the pullout resistance for the projects surveyed, several different approaches were utilized in determining the soil-fabric friction angle. The approaches included using two-thirds of the angle of internal friction of the soil (Ø), (0.8 to 0.9 tan(Ø)), and the full soil-reinforcement interaction angle obtained directly from pullout tests.

3.4 Design Methodology and Factors of Safety

The tieback wedge method of analysis using a Rankine failure surface was the only reported method used in analyzing the wall structures. Intenral vertical stress distributions were determined using either the trapezoidal, overburden or Meyerhof stress distributions. A stress ratio determined from an assumed bilinear failure surface after Jewell (1984) was used in several geogrid slope cases. Both active and at-rest earth pressures were used in the design of retaining structures with flexible facing materials.

Table No. 6 presents the factors of safety applied when evaluating external and internal stability for the various projects. The range and average of the factors of safety used in the design of the various permanent walls and the slopes are included in the table:

TABLE NO. 6: SUMMARY OF REPORTED FACTORS OF SAFETY

	Stability Factor	Factor of Safety
1.	Sliding:	Range 1.2 to 2.0, Average 1.54
2.	Overturning:	Range 1.2 to 2.8, Average 1.66
3.	Bearing Capacity:	Range 1.5 to 2.0, Average 1.89
4.	Global Stability:	Range 1.3 to 1.5, Average 1.40
5.	Reinforcement Rupture:	Range 1.5 to 4.5, Average 2.34
6.	Pullout Resistance:	Range 1.5 to 2.0, Average 1.68

Although in several cases the factors of safety are substantially less than those normally recommended for external stability consideration, the adequacy of the design cannot be determined without studying the actual design procedure since other conservative assumptions may have been made.

3.5 Project Costs

Table No. 7 presents the reported cost data for a number of the projects. The total costs include fabric, backfill and equipment and labor costs per square meter of wall face. The costs range from $39/m^2$ to $450/m^2$. The average cost of the permanent walls was $158/m^2$, not including project numbers 4, 5 and 6 for the reasons mentioned below and project number 8 since tire reinforcement was utilized. The average cost of the temporary walls was $83/m^2$.

The lower bound cost is for the SR 522 project in Washington [No. 24]. Five temporary walls were constructed using UV-stabilized geotextiles, which resulted in a cost savings since facing materials were not required. The New York Route 22 project [No. 5] was constructed at a cost of $450/m^2$. The high cost resulted from the use of expensive crushed stone backfill ($24/cy). Chassie (1984) noted that the high costs of Projects No. 4 and 6 may have resulted from the walls being changed ordered into the project.

TABLE NO. 7: PROJECT COSTS

Project Number	Total Project Cost ($/m^2$ of wall face)
1	129
2	127
4	320
5	450
6	312
7	203
8	270
9	134
13	255
15	92
23	46
24	39
27	129
29	96
30	160
32	65
33	109

Figure number 1 compares the cost of various retaining wall systems, including the average costs for the projects surveyed. The cost data for projects 1, 7, 9, 13 and 27 was superimposed on the cost comparison chart [33]. For walls less than 30 ft in height, geosynthetic reinforced walls would appear to be more economical than metallic reinforced soil walls.

4. TECHNICAL ASSESSMENT OF MONITORED PROJECTS

Of the projects reviewed, thirteen were instrumented to varying degrees beyond simply taking surface survey measurements. As previously indicated, Table 4 summarizes the instrumentation program for those projects. The table was set up to include the full compliment of instrumentation that might be considered on a project to assess the principal design assumptions, the principal design considerations include:

1. The location of the potential failure surface corresponding to the location of maximum stresses within the reinforcement.

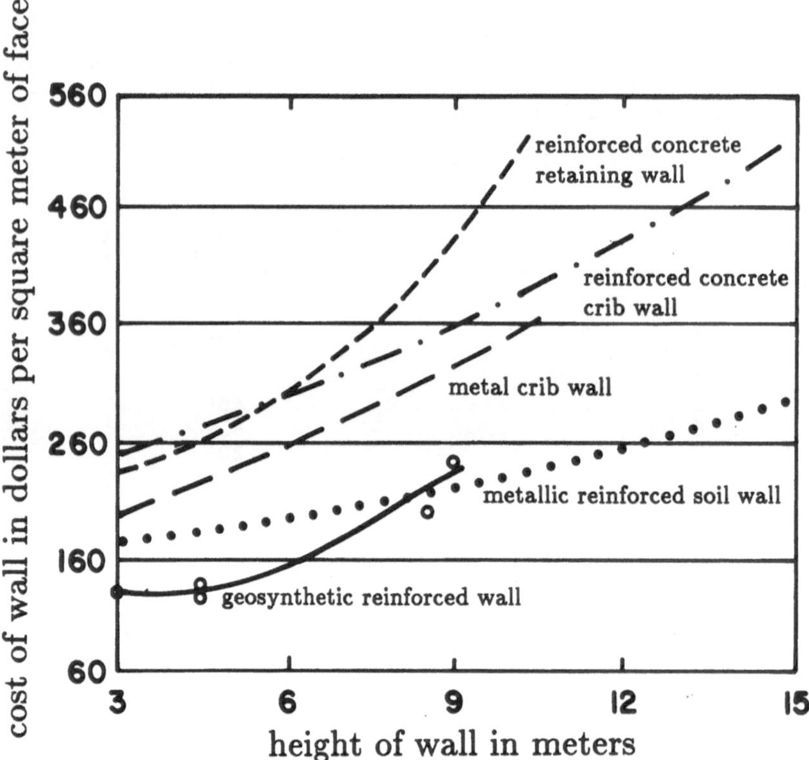

FIGURE 1 Derived from a comparison prepared by the California Department of Transportation in cooperation with the Federal Highway Administration, this chart compares the materials and backfill costs of competing systems. (adapted from vsl corp., 1981)

2. The stress level resisted by the reinforcement as determined from lateral earth pressure within the reinforced mass based on the vertical pressure at critical levels.

3. Both short term and long-term horizontal displacements.

4. External structural stability including the magnitude of vertical displacement and the foundation support stability.

As indicated in Table 4, most of the project instrumentation programs concentrated on lateral movement and overall settlement. Vertical movement was typically measured by external surveys and internal measurements using varying combinations of settlement plates and anchors, vertically spaced inductance coils, and horizontal inclinometers. Lateral deformation was usually measured with some form of extensometer and/or inclinometers. In many cases the results of such measurements were found inconclusive as the instruments were not sensitive enough to measure movements that were usually smaller than anticipated.

A full compliment of instrumentation, including the measurement of strains in the reinforcement and soil pressures in the reinforced soil mass was noted for eight of the projects (Numbers 3, 9, 18, 19, 20, 26 and 38). Project No. 35 was also fully instrumented; however, results were not available at the time of this writing. Due to significant variations in the structures instrumented, including the type of backfill, reinforcement, facing panels and foundation conditions, it is very difficult, if not impossbile, to develop definitive conclusions from the instrumentation results of the projects surveyed. Therefore, the following includes a synopsis of the results rather than a comparison of the results.

In the Waterways Experiment Station Project [3], a failure of the polymer strip type reinforcing was observed. Observations indicated the failure may have been due to excessive deformation of the reinforcement near the face of the wall and possibly at facing connections. Pressure cells embedded in the soil near the face of the wall indicated an increase in pressure with increased overburden as shown in Figure 2. However, the measured lateral pressures were less than those predicted by Rankine theory. The pressures increased significantly as the wall started to fail. The pressure measurements indicated that the reinforcement became more effective with increased lateral movement. As the reinforcement started to yield, the lateral stress was apparently then transferred to the wall face.

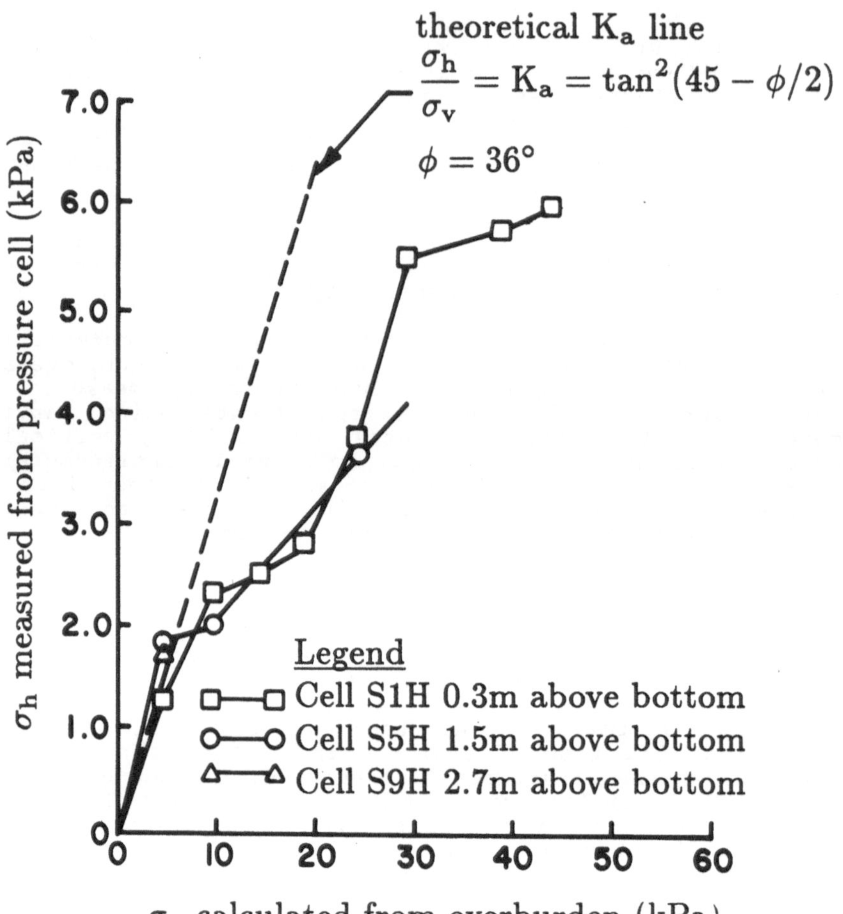

FIGURE 2 Comparison between the measured horizontal stress and the calculated vertical stress for wall reinforced with rubber-coated strips. (Case No. 3)

For the Glenwood Canyon Walls [9], very small strains were measured in high elongation needle punched non-woven fabrics, even though the design methodology would indicated factors of safety near 1. The results were attributed to the apparent improvement in stress-strain properties of the geotextiles under the confined soil condition. This project emphasizes the need to use confined stress-strain values in predicting performance of structures constructed with non-woven geotextiles.

As previously indicated, in the Tarque Verde Project [18], full-height tilt up concrete facing panels were used. Strain measurements on the geogrid reinforcement indicated that the stress levels in the reinforcement were less than anticipated. The location of maximum reinforcement stress was found to be 18 to 19 degrees from the vertical, as measured using bonded resistant strain gages and 24 to 25 degrees from the vertical (or approaching a Rankine surface) as measured with inductance coils. The researchers felt that the bonded resistant strain gages provided the most reliable data. The location of maximum tensile strains at a steeper surface than predicted by Rankine would tend to agree with model studies of full height facing panels (Christopher, 1987). As shown in Figure 3, lateral stress measurements using soil embedment cells near the face of the wall found low stress levels, again indicating that the reinforcing was adequately resisting lateral stress within the reinforced soil mass. Higher lateral stress levels were observed near the base of the wall which may have been due to embedment of the face at the base and connection of the facing panel with an embedded footing. As the face was not free to move at the base, the reinforcing could not sufficiently deform to carry significant stress. Vertical soil stresses measured with embedded pressure cells found greater pressures at the base of the wall than predicted by trapezoidal distribution, overburden stress calculations or Meyerhof distribution (Figure 4). However, the stress levels were surprising lower near the face of the wall. This was attributed to wall friction and to soil arching over the load cells.

The Lithonia Project [19] was similar to the Tarque Verde Project, except that articulating facing panels were utilized. A good summary of the two projects was provided by Berg, et al (1986). The results were less conclusive than the Tarque Verde Project, however, again lower stress levels were measured near the face of the wall (Figure 5) again indicating resistance to lateral stresses by the reinforcement. The vertical pressures measured below the base of the wall (Figure 6) appeared to be less than those measured at the Tarque Verde project, however, data was limited to only two locations.

Probably the most extensively instrumented wall was the Gaspe Peninsula Reinforced Soil Seawall Project [20]. Strain measurements indicated that the line of maximum tension in

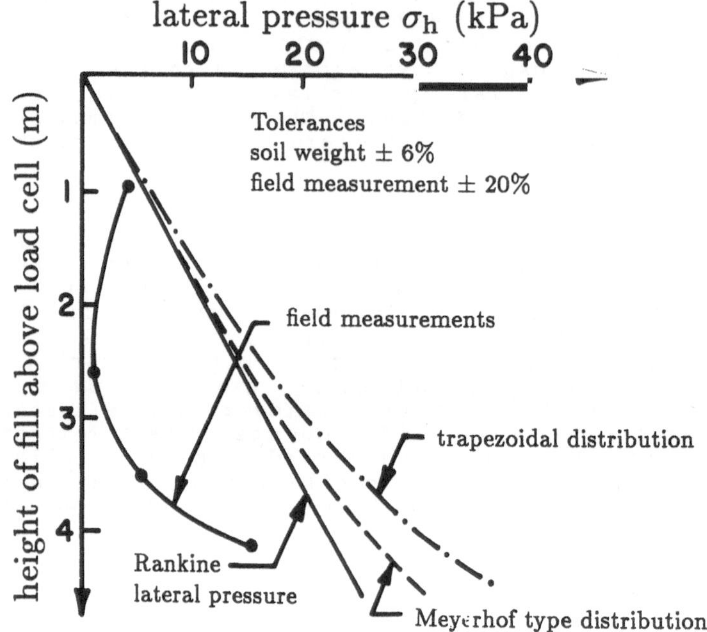

FIGURE 3 Comparison of Measured Horizontal Soil Pressures to Predicted Values - Tucson Wall (Case No. 18)

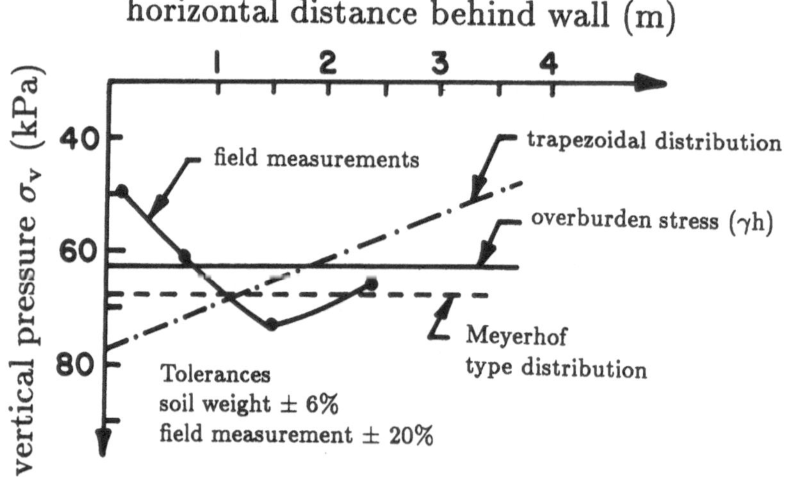

FIGURE 4 Comparison of Measured Vertical Soil Pressures to Predicted Values - Tucson Wall (Case No. 18)

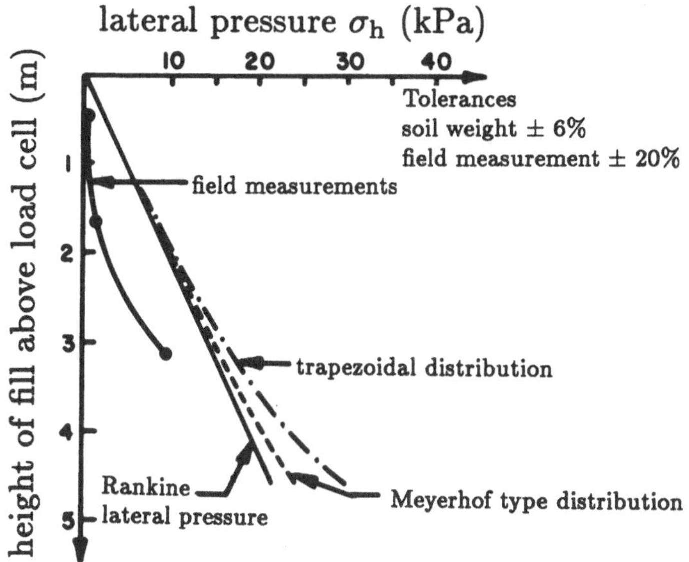

FIGURE 5 Comparison of Measured Horizontal Soil Pressures to Predicted Values - Lithonia Wall (Case No. 19)

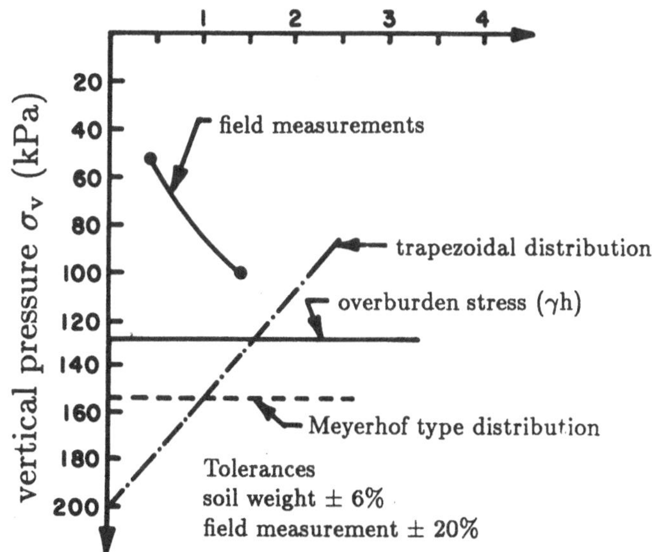

FIGURE 6 Comparison of Measured Vertical Soil Pressure to Predicted Values - Lithonia Wall (Case No. 19)

the reinforcement was very close to a Rankine surface (Figures 7 and 8). The measurements indicated that the tensile loads in the geogrid reinforcement was below tolerable levels. Future measurements at this project will be of significant interest as the polymer reinforcement is subjected to cyclic loading.

The Cascade Project [26] is significant in that it was exposed to very high temporary surcharge loads. In all of the previous projects, it was reported that the stress in the reinforcement was less than that predicted by the design model. For this project, the instrumentation results would indicate that stress levels were very close to those predicted and, during maximum surcharge loading, were slightly higher than anticipated. The locus of maximum stresses obtained from strain gage measurements (Figure 9) found the highest stress levels at the top of the wall towards the back of the surcharge loads, similar to that observed by Wichter et al, 1986.

It is very apparent that more controlled research test walls are needed to evaluate design methodologies and develop predicted capabilities. Even so, for simple structures (i.e. walls less than 30 ft height with low surcharge loading, granular backfill, and facing elements that are free to move) current design methods using unconfined stress-strain data appear conservative and are thus adequate. However, for continued development and better predictions of performance, an extensive amount of research is still needed.

5. CONCLUSIONS

Polymeric reinforcement is gaining wide spread use and is helping to solve many problems that could not formerly be solved using rigid retaining structures. It is estimated that over 200 structures have been constructed in North America. Applications of polymeric reinforcement range from support of a bicycle path over peat deposits in Anchorage, Alaska to protecting a sea coast in Quebec, Canada. Innovative uses of these materials can help solve construction problems, such as the Cascade Michigan Dam project where a geogrid wall was used to widen a dam and supported a 840 kN crane with a 100 kN lifting capacity for a total surcharge of 200 kN/m^2 during lifting.

Polymerically reinforced retaining structures have also been shown to be very cost effective alternatives to conventional retaining structures. Temporary walls have been constructed at costs as low as \$39/m^2 of wall face. Costs for permanent walls range from \$130 to \$260/m^2 of wall face.

To the author's knowledge, all of the projects surveyed have performed satisfactorily in serving their intended purpose. Unfortunately, problems associated with construction or

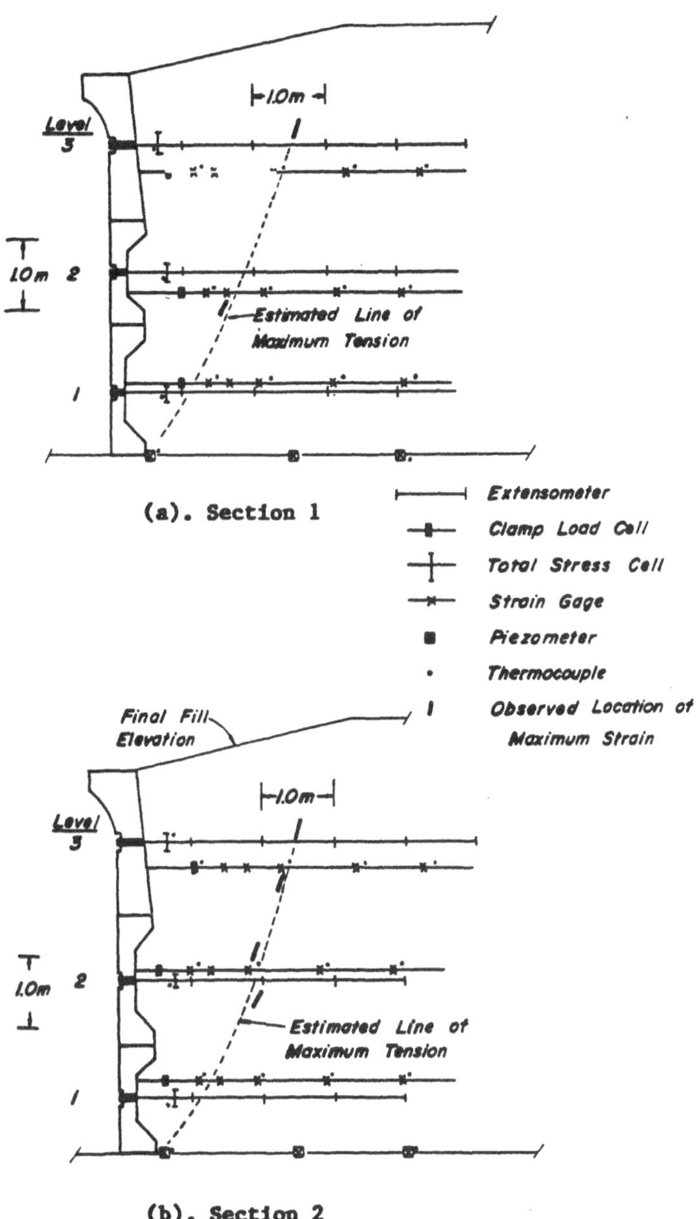

FIGURE 7 Instrumentation Layouts (Case No. 20)

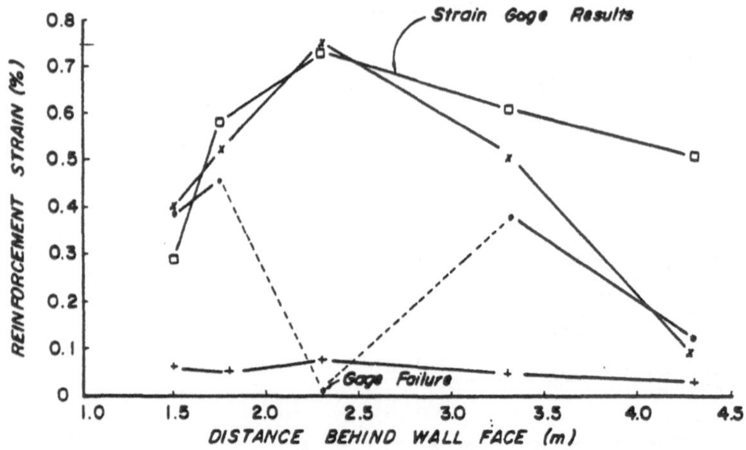

(a). - Geogrid Strain, Section 2, Level 3

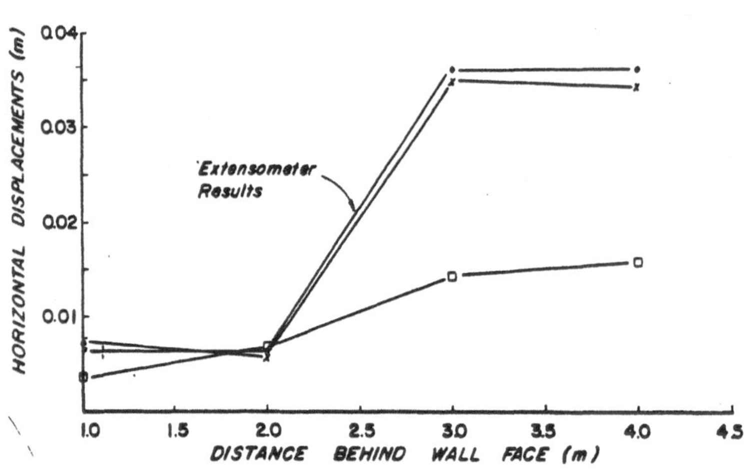

(b). - Wall Fill Strain, Section 2, Level 3

Date of Reading	
+———+ 85/08/01	Fill Close to Bottom of Top Panel (El. 2.8)
□———□ 85/08/06	Fill at Top of Wall (El. 4.3)
x———x 85/10/31	Fill to Road Elevation (El. 5.1)
•———• 86/06/11	Reading After First Winter

FIGURE 8 Strain Measurements (Case No. 20)

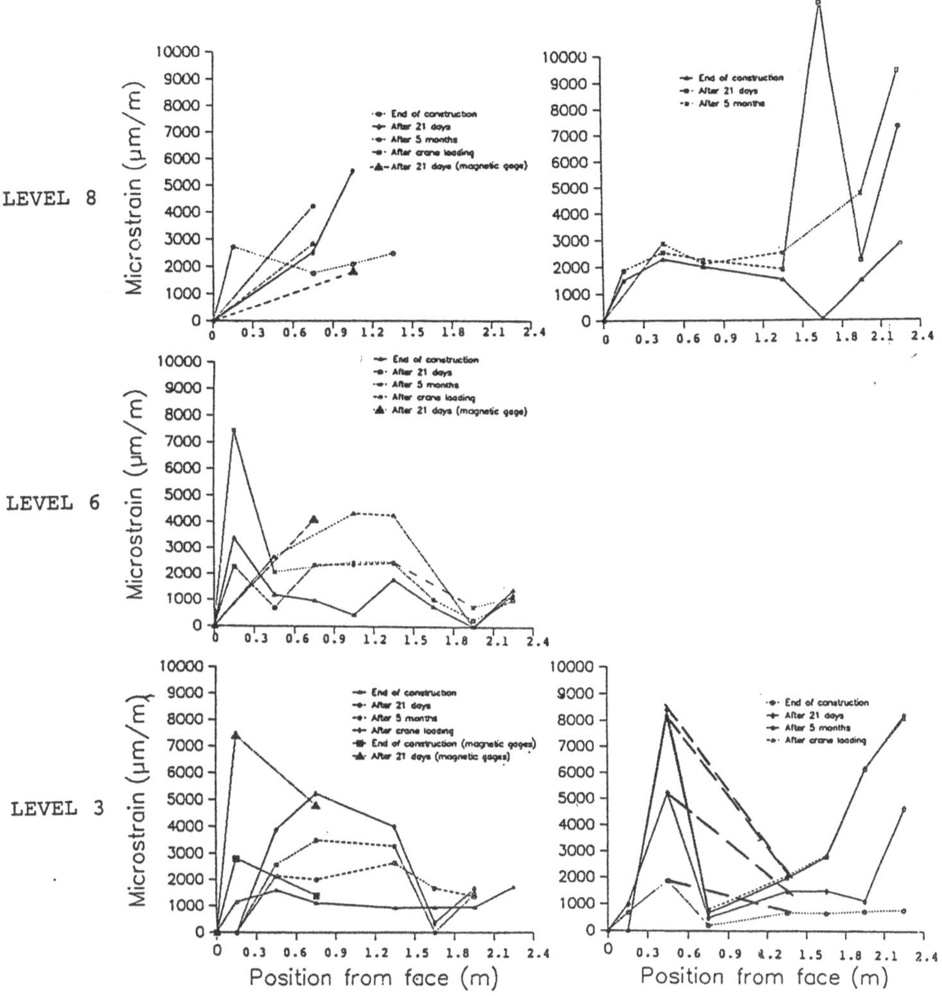

FIGURE 9 Strain Distribution Along Reinforcement During Monitoring Events (Case No. 26)

performance were not reported for the projects reviewed. Information concerning problems would have provided valuable information for future users.

The authors will continue to update the summary tables as information on new projects and additional information on existing projects become available. Any help from the readers in this effort would be appreciated.

ACKNOWLEDGEMENTS

The authors wish to thank all of the individuals who provided project information and assistance in preparation of the summary tables, Terry Harrison for typing the manuscript, Maria Flessas for reviewing the manuscript and Horward Faye for computerization of the summary tables.

REFERENCES

1. Bell, J.R., and Steward, J.E., "Construction and Observations of Fabric Retained Soil Walls," Proceedings, International Conference on the Use of Fabrics in Geotechnics, Paris, France, Vol. 1, April 1977, pp. 123-128.

2. Bell, J.R., A.N. Stilley and B. Vandre, "Fabric Retained Earth Walls," Proceedings, 13th Annual Geology and Soils Engineering Symposium, Moscow, Idaho, 1975, pp. 271-287.

3. Steward, J.E. and J. Mohney, "Trial Use, Results and Experience Using Geotextiles for Low-Volume Forest Roads," Proceedings, 2nd International Conference on Geotextiles, Las Vegas, Nevada, Vol. 2, August 1982, pp.335-340.

4. Mohney, J., "Fabric Retaining Wall, Olympic National Forest," Highway Focus, U.S. DOT, Federal Highway Administration, Vol. 9, No. 1, May 1977, pp. 88-103.

5. Al-Hussaini, M.M., "Field Experiment of Fabric Reinforced Earth Wall," Proceedings, International Conference on the Use of Fabrics in Geotechnics, Paris, France, Vol. 1, April 1977, pp. 119-121.

6. Al-Hussaini, M.M. and E.B. Perry, "Effect of Horizontal Reinforcement on Stability of Earth Masses, "Technical Report No. S-76-11, 1976, U.S. Army Engineer Waterways Experiment Station, Vicksburg, Mississippi.

7. Douglas, G.E., "Design and Construction of Fabric Reinforced Retaining Walls by New York State," Transportation Research Board, Transportation Research Record 872, Washington, D.C., 1982, pp. 32-37.

8. Chassie, R.G., "Geotextile Retaining Walls, Some Case History Examples," prepared for presentation at the 1984 NW Roads and Streets Conference, Corvallis, Oregon, February.

9. Walkinshaw, J.L. and K.A. Jackura, "New Developments in Retaining Wall Design 38th California Transportation and Public Works Conference," Oakland, California, May 1986.

10. Bell, J.R., R.K. Barrett and A.C. Ruckman, "Geotextile Earth-Reinforced Retaining Wall Tests: Glenwood Canyon, Coloardo," Transportation Research Board, Transportation Research Record 916, Washington D.C., 1983, pp. 59-69.

11. Barrett, R.K., "Geotextiles in Earth Reinforcement," Geotechnical Fabrics Report, March/April 1985, pp. 15-19.

12. Busbridge, J.R., "Stabilization of CP Rail Slip at Waterdown, Ontario, Using Tensar Grid," Proceedings of the Symposium on Polymer Grid Reinforcement in Civil Engineering, London, U.K., 1984, Paper No. 2.3.

13. Fluet, J.E., "Geogrids Enhance Track Stability," Railway Track and Structures, June 1984.

14. Tensar Corporation, "Case Study: CP Rail Slope Repair," Morrow, Georgia.

15. Tensar Corporation, "Case Study: U.S. 69 Slope Repair," Morrow, Georgia.

16. Bell, J.R., T. Szymoniak and G.R. Thommen, Construction of a Steep Geogrid Retaining Wall for an Oregon Costal Highway, Proceedings of the Sympsium on Polymer Grid Reinforcement in Civil Engineering, London, U.K., 1984, Paper No. 6.3.

17. Szymoniak, T., et al, "A Geogrid-Reinforced Soil Wall for Landslide Correction on the Oregon Coast," Transportation Research Board, Transportation Research Board 965, Washington, D.C. 1984, pp. 47-55.

18. Bonaparte, R. and E. Margason, "Repair of Landslides in San Francisco Bay Area, "Proceedings of the Symposium on Polymer Grid Reinforcement in Civil Engineering, London, U.K., 1984, Paper No. 2.4.

19. Forsyth, R.A. and D.A. Bieber, "La Honda Slope Repair with Geogrid Reinforcement," Proceedings of the Symposium on Grid Reinforcement in Civil Engineering, London, U.K., 1984, Paper No. 2.2.

20. Berg, R.R., R. Bonaparte, R.P. Anderson and V.E. Chourey, "Design, Construction and Performance of Two Geogrid Reinforced Soil Retaining Walls," Proceedings, Third International Conference on Geotextiles, Vienna, Austria, 1986, pp. 401-406.

21. Anderson, R.P., "Soil Reinforcement Objective: Polymer Geogrid Replaces Galvanized Metal Strips in Concrete-Faced Retaining Walls," Geotechnical Fabrics Report, January/February, 1986.

22. "Geogrids Used to Reinforce Retaining Walls," Rocky Mountain Construction, May 1985.

23. "Jasper Construction Builds Retaining Wall Utilizing Tensar Geogrid as Reinforcement," Dixie Contractor, October 1985.

24. Berg, R.R., P. LaRochelle, R. Bonaparte and L. Tanguay, "Gaspe' Peninsula Reinforced Soil Seawall Case History," Proceedings, ASCE Symposium on Soil Improvement, Atlantic City, New Jersey, April 1987.

25. Christopher, B.R., "Geogrid Reinforced Soil Retaining Wall to Widen an Earth Dam and Support High Live Loads," Proceedings, Geosynthetics '87 Conference, New Orelans, Louisiana, February 1987.

26. Scott, J.D., et al, "Design of the Devon Geogrid Test Fill," Proceedings, Geosynthetics '87 Confernece, New Orleans, Louisiana, February 1987.

27. "Class A Prediction Exercise for Reinforced Earth Walls," NATO Advanced Research Workshop to be held June, 1987, Royal Military College of Canada, Bulletin No. 1.

28. Christopher, B.R., Unpublished STS Consultants, Ltd. Report on Task C Field Test Wall Construction - Prepared for FHWA, Washington, D.C., Contract No. DTFH61-84-C-00073, "Behavior of Reinforced Soil", 1987.

29. Castelli, R. and G. Munfakh, "Geotextile Walls in Mountainous Terrain," Proceedings, Third International Conference on Geotextiels, Vienna, Austria, 1986, pp.459-463.

30. Christopher, B.R. and R.D. Holtz, "Geotextiles Engineering Manual," Prepared for FHWA, National Highway Institute, Washington, D.C., Contract No. DTFH61-80-C-00094, 1985.

31. Koerner, R.M., "Designing with Geosynthetics," Prentice Hall, Englewood, Cliffs, New Jersey, 1986.

32. Bell, J.R., "Design Criteria for Selected Geotextile Installations," Proceedings, First Canadian Symposium on Geotextiles, Calgary, Alberta, 1980.

33. Holtz, R.D. and B.B. Broms, "Walls Reinforced by Fabrics - Results of Model Test," Proceedings, International Conference on the Use of Fabrics in Geotechnis, Paris, France, Vol. 1, April 1977, pp. 113-117.

34. VSL Corporation, "VSL Retained Earth, Technical Data," Rock and Soil Stabilization Systems, Los Gratos, California, 1981.

35. Wichter, L. Risseeuw, P. and Gay, G., Grossversuch zum Tragverhalten einer Steilwand aus Gewebe und Mergel. Proceedings of the Third International Conference on Geotextiles, Vienna, Austria, 1986.

36. Personal Communication, Ryan R. Berg, The Tensar Corp., 1987.

GEOTEXTILE-REINFORCED RETAINING STRUCTURES : A FEW INSTRUMENTED EXAMPLES

Ph. DELMAS
Laboratoire Central des Ponts et Chaussées, France
J.C. BLIVET
Laboratoire Régional des Ponts et Chaussées de Rouen, France
Y. MATICHARD
Laboratoire Régional des Ponts et Chaussées de Nancy, France

1 - INTRODUCTION

The principle of retaining structures consisting of a reinforced soil mass has led to routine applications primarily through the work of la Terre Armée (TA). From this viewpoint, geotextile-reinforced structures may be compared to TA reinforced-earth structures.

These two processes do have a number of points in common : the retaining function is performed by a gravity structure of reinforced soil; reinforcement is by elements in which tensile stresses predominate, placed horizontally when the fill is laid down ; and the facing is generally vertical.

But the similarities between the two types of structure end here, and an analysis of their internal behaviour reveals differences.

In TA reinforced-earth structures, the shape of the reinforcing strips serves to mobilize the dilatancy of the soil at the level of local soil-reinforcement interactions (in the pull-out mode). The resulting additional local vertical stress at the reinforcement makes it possible to mobilize substantial friction in spite of the narrowness of the reinforcements, especially in the upper part of the wall where the vertical stresses are smallest. But this beneficial working mode calls for the use of soils that are dilatant, and so suitably prepared (Schlosser and Guilloux, 1981).

P. M. Jarrett and A. McGown (eds.), The Application of Polymeric Reinforcement in Soil Retaining Structures, 285–311.

Geotextiles are in sheet form, and because of their two-dimensional character do not generally make use of the dilatancy of the soil. It should, howerer, be noted that some geogrids can bring out the dilatant behaviour of the soil. But, generally speaking, because of the large contact area between the soil and the geotextile (figure 1), the additional normal stress that serves to mobilize dilatancy on strip reinforcements is not needed to mobilize the friction necessary for stability. With geotextiles, then, it will be possible to use non-dilitant soils, and in particular any material likely to be used as fill.

As for the tensile strength of the reinforcements, table 1 gives comparative strength values for one layer of reinforcements and a width of one metre. It can be seen that the two processes yield comparable values.

The stiffness of the reinforcements, again calculated for one layer and a width of one metre, is between 25 and 100 times smaller in the case of geotextiles. This results in differences in behaviour, characterized by larger strains within the geotextile-reinforced structure for similar forces.

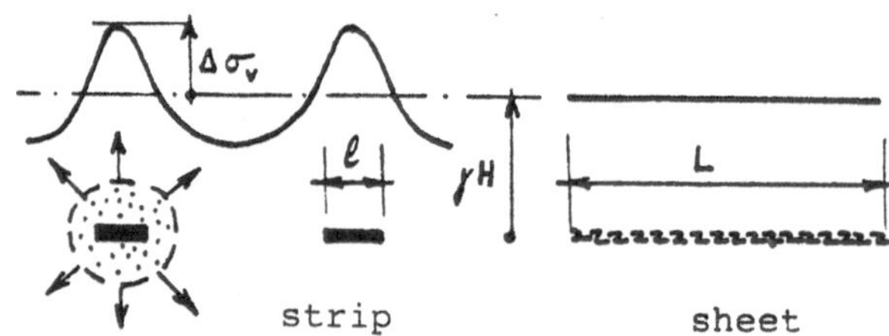

Figure 1 - Comparison of the friction behaviour of "Terre Armée" reinforcement and geotextiles

These findings explain why, in observations of geotextile-reinforced structures, an effort has been made to measure the strains of the reinforcements and of the soil. In the case of real structures, it is important to be able to make sure of a level of strain that is acceptable for the structure, whether locally, in the soil (to remain below the peak shear strength values), in the geotextile or at the contact surface, or more generally with respect to the superstructures or the facing.

Table 1 - Comparison of stiffness and strength values of TA reinforced-earth reinforcements and geotextiles (one layer one metre wide)

	"Terre armée" (b = 40 mm, spacing 1,20 m)	Geotextile
Maximum tensile strengh	57,8 kN/m	50 ~ 300 kN/m
Allowable tensile strengh (permanent use)	26,9 kN/m	5 ~ 60 kN/m
Stiffness	28 000 kN/m	500 ~ 2000 kN/m

Based on a few examples of instrumented real structures, we shall attempt in what follows to determine the influence on the behaviour of the structures of :

- the method of placement of the formwork and of the facing ;
- the use of clayey soil ;
- the long-term behaviour of the geotextiles.

For practical reasons, this analysis will be limited to the investigation of structures with flexible facings.

2 - INFLUENCE OF PLACEMENT METHOD ON THE BEHAVIOUR OF GEOTEXTILE-REINFORCED STRUCTURES

In practice, the stability of the structures with respect to the equilibrium of forces is ensured by the tensioning of the reinforcement sheets inside the soil mass. However, the overall behaviour of the structure will depend primarily on how the sheets of geotextile are tensioned and how the strains are induced. Indeed, for a given structure, overall static equilibrium can be ensured for different stress conditions in the geotextiles depending on how the strains occur in the soil mass.

Accordingly, placement will undeniably be a key stage in the generation of the forces in the reinforcement sheets. The few instrumented structures described here show, in effect, that in certain cases the final forces can be reached as soon as the construction of the structure is completed, independently of the of the subsequent loading conditions.

This is explained by the stiffness of the geotextiles, which means that rather large strains of the soil mass are required to generate the tensile forces that ensure stability.

It follows that correct placement of the sheet is critical: no pleats, if possible manual tensioning before compaction of the fill, etc. However, the type of compaction and the way in which the fill is cased will in general be the two factors that determine the final strains in the soil mass. In addition, the facing, and its relative stiffness, will also be a factor to be taken into account in understanding the limit-state strains of the structure.

As an example, we may mention the first geotextile-reinforced structure built. It is in many respects a reference structure that sheds some light on the behaviour of geotextiles used for reinforcement.

2-1 Embankment of A15 motorway

This structure, built on an experimental basis in 1971 (Puig and Blivet, 1973), is one of the first geotextile-reinforced retaining structures built with a clayey soil. At the site of the structure, the embankment of the A15 motorway is in a compressible valley with, under the surface layer of old fill, a 3-m layer of peat having the following mean properties : w = 300 % ; C = 3.2 ; e = 6.8 . The fill material is a clayey sand-gravel consisting of a mixture of weathered chalk, clay, and flint. The water content of the fraction under 20 mm ranged from 16 to 21 % at placement, corresponding to a saturation S_r of 94 to 98 % (at the normal Proctor optimum, w = 13 %, $_d$ = 19.0 kN/m3). This explains, in particular, the rubber cushion effect observed when the embankment was built up.

Figure 2 shows the standard cross-section of the structure, reinforced with layers of needle-punched polyester nonwoven (Bidim U34) every 50cm

The form used was a temporary bank having a batter of 3 in 2 (figure 2).

Seven month after the facing was uncovered, the structure was loaded with an embankment 4 m high having a batter of 3 in 2.

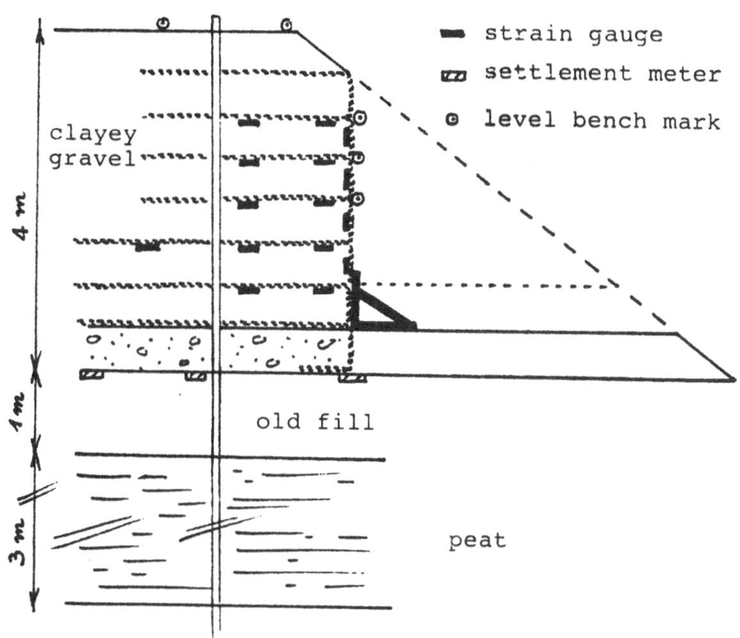

Figure 2 - A15 motorway : detail of construction with temporary bank and arrangement of measuring instruments

This experimental embankment, the total length of wich was 20 m, was instrumented with :

- settlement meters placed in the base of the embankment ;
- strain gauges placed on the reinforcements (these gauges were made of resistive constantan wire mounted on a plastic support ; they had a useful length of 6 cm and were attached to the geotextile by clamping their ends between two steel plates) ;
- an inclinometer tube placed in the body of the embankment
- level bench marks and plates on the facing.

Photo 1 - A15 motorway : overall view of completed structure

For all observations, the reference measurement was made on 23 March 1971 just after the wall was uncovered (photo 1).

Immediately after the removal of the "formwork" fill, measurements of the horizontal displacement of the facing revealed a displacement of 15 to 25 mm at the top of the structure and a displacement of 6 mm at the toe, so indicating tilting of the top.

After the top of the embankment was loaded, the measurements revealed an overall displacement of about 20 mm. These displacement values are also confirmed by the inclinometer tube in the soil mass.

It should be noted that the interpretation of these measurements is complicated by the presence of substantial settlements of the foundation soil : the total settlement, one month after loading, reached 1.10 m, with a maximum differential settlement of 0.25 m in 3 m

Even though the strain gauges on the geotextiles did not survive long (of the 35 placed, only 6 were still functioning three months after the removal of the "formwork" fill), they gave some idea of the strains and forces in the geotextile sheets. The following may be noted :

- in general, the strain of the geotextile increased as the distance from the facing decreased ;

- the maximum values measured (before surcharging) reached 1.7 to 1.8 % in the geotextiles at the toe of the structure.

In interpreting these findings, one must first of all bear in mind the very special mode of placement. Using an adjacent bank as a form prevents strain of the geotextiles before the removal of the form ; again, especially near the facing, the loss of compaction caused by the process interferes with the proper confinement of the end of the layer of material (photo 1). Moreover, the removal procedure used causes the wall to work as a trench retaining structure, by stressing first the uppermost layers of reinforcements, then the underlying layers, as evidenced by the tilting of the top when the form was removed.

It is reasonable to suppose that the tensile forces engendered in the reinforcements depend to a large extent on this particular stressing mode. Nor can it be ruled out that the lowest layers were stressed at the back of the soil mass to ensure the overall stability of the structure when the forms were removed.

So far, there are few instrumented works built using global forms, in all likelyhood because of the large force required for the temporary stability of the formwork. But the example of the "La Houpette" structure on National Highway 4, built in France in 1986 using this process, should be mentioned.

2.2 - La Houpette embankment

To carry out the widening of a road on an embankment 4m high and 300 m long, it was decided to build a geotextile-reinforced structure consisting of a vertical part 1.4 m high reinforced by two layers of UCO 44614 topped by a bank having a batter of 3 in 2 (figure 3). For practical reasons, the contractor chose to build the lower part in forms placed all at once.

The fill is anggregate having a continuous 0/250-mm grading and the following geotechnical properties : γ_d = 22 kN/m^3 c' = 0 kPa ; φ' 35° to 40°.

Because of the size and rather sharp angles of the grains of the fill material, a test of compaction on geotextile was carried out. Figure 4 gives the tensile curves of the geotextile before and after this compaction. It can be seen that, in this particular case, the compaction had relatively little effect on the stress/strain curve of the product.

However, in the course of specific tests carried out on various products placed between two layers of flint-bearing clay dumped from a height of 1.5 m and compacted to 95 % of the normal Proctor optimum in layers 30 cm thick, Perrier (1986) found that the compaction could decrease the maximum tensile strength of the geotextile by as much as 30 %, and alter its stiffness, either increasing or decreasing it (table 2).

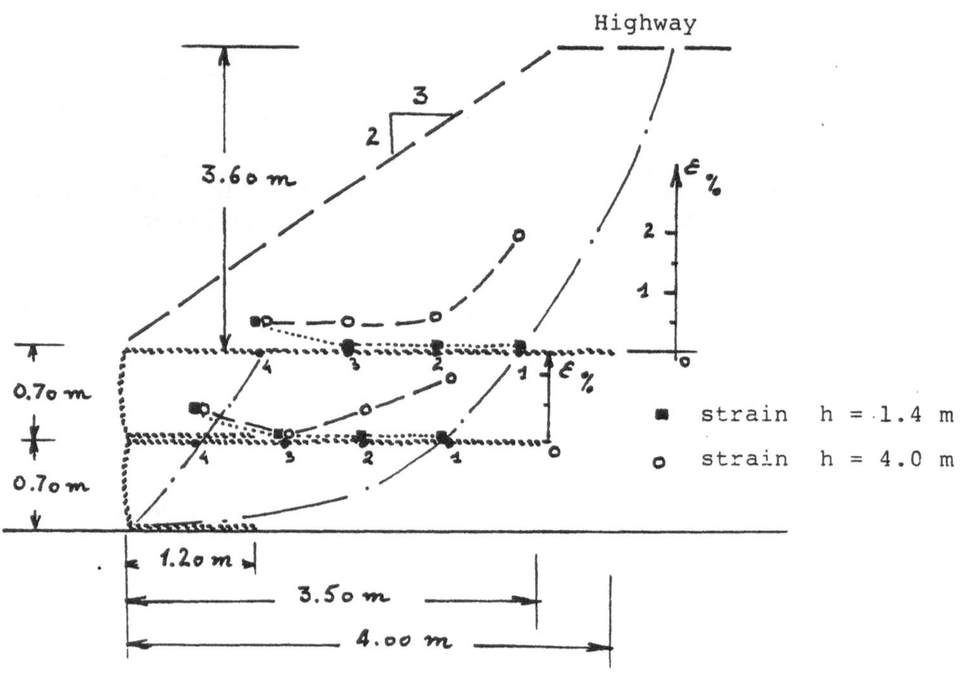

Figure 3 - National Highway 4 : standard cross section and arrangement of measuring instruments

This embankment was instrumented with cable-type strain measuring devices. Each measuring point consists of two cables attached to the geotextile 20 cm apart (in the direction of the stress), protected from the friction of the fill material by a sheath, that extend out through the front of the soil mass. The relative displacement of the two cables is measured with respect to a fixed reference about 1 m from the facing, with the same tension applied to both. This procedure has the advantage of being easy to implement and relatively inexpensive. It makes it possible to estimate strains that exceed 0.5 % and also, provided that the measurement bench mark is levelled, makes it possible to determine the absolute displacement of the points of the soil mass.

Table 2 - Influence of compaction on the tensile properties of the geotextile (after Perrier, 1986)

	Before compaction		Loose after compaction	
	$\bar{\alpha}_o$ (kN/m)	$\bar{\varepsilon}_o$ (%)	$\frac{\bar{\alpha}_o - \bar{\alpha}}{\bar{\alpha}_o}$ (%)	$\frac{\bar{\varepsilon}_o - \bar{\varepsilon}}{\bar{\varepsilon}_o}$ (%)
. NTA PES	41,8	51,2	8,1	21,9
. t PES ST	158,6	7,6	35,2	18,4
. t PES AJ	115,3	7,3	30,2	15,1
. t PP X	40,9	15,2	15,4	5,9
. t PP N	96,4	8,7	8,7	0,0
. t PET	55,0	22,5	21,5	19,1
. G PET	19,1	7,3	0,0	26,0

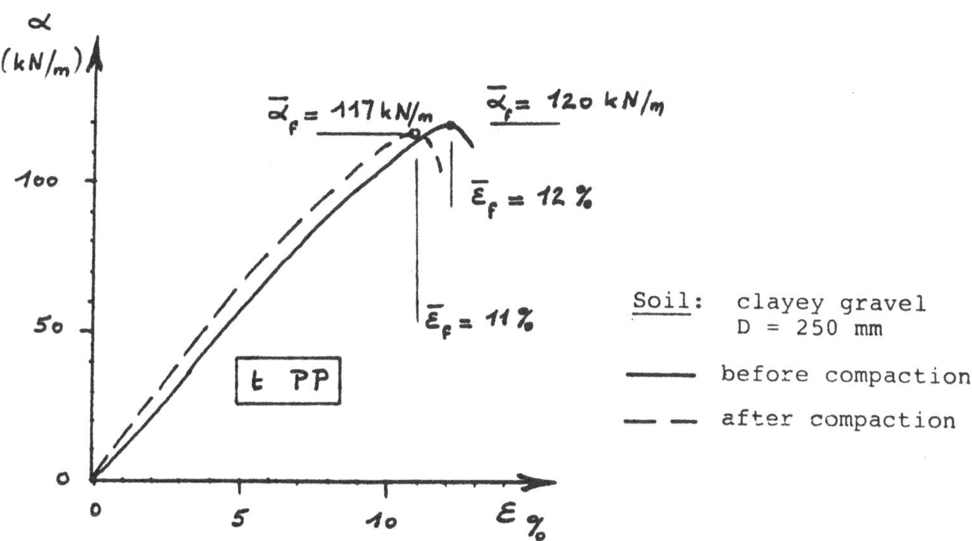

Figure 4 - National Highway 4 : comparison of tensile stress/ strain curves of the woven polypropylene geotextile before and after compaction

The deformations measured immediately after the removal of the forms from the two layers, before the upper embankment was built, and at the end of its construction, are given in table 3

Table 3 - National Highway 4 : deformations measured in geotextiles at measurement profile no. 2

Point of measurement	1		2		3		4		Facing
Height of embankment (m)	1.4	4.0	1.4	4.0	1.4	4.0	1.4	4.0	
Geotextile 1	0	1.0	0	0,5	0	0	0.5	0.5	
Geotextile 2	0	2.0	0	0,5	0	0.5	0.5	0.5	

Otherwise, the absolute displacements recorded revealed, after the removal of the forms, an overall displacement of the sheets of geotextile 7 mm for the bottom sheet and 5 mm for the top sheet. This displacement was simultaneous with the strain of points no. 4 of both sheets, near the facing.

It will be noted that, as regards the strains of the sheets, the form removal stage corresponds with a mobilisation along a surface of maximum tension located near the facing, while the stage of embankment construction mobilizes the reinforcements farther back in the soil mass, so justifying the anchorage lengths determined in the preliminary design stage.

From these findings, we may conclude that the rather rudimentary fill placement method used at this site (photo 2) is doubtless sufficient to explain the overall displacements observed when the forms were removed. It would seem that no special care was taken with the compaction in the vicinity of the facing.

It would also seem that the principle of a single form for the whole facing leads, when the form is removed, to a mobilization of strains in the sheets in accordance with those anticipated in the preliminary design calculations. Having said this, we should point out that the final distribution of strains in the geotextiles in fact corresponds to the sum of the strains resulting from each of the two stages of constructions.

Photo 2 - National Highway 4 : placement of fill

Generally speaking, geotextile-reinforced structures are built with partial forms for each layer. In what follows, we shall not analyze structures built using forma supported only by the next lower layer, even though in historical fact this was one of the first processes used. This is because the use of a form supported by the next lower layer assumes that there is no deformation of this layer when its form is removed, or at least that any deformation is uniform, which is practically never the case.

Obtaining a vertical facing calls for a series of adjustements, all the more delicate in that a deviation from the profile caused by an outward displacement of the form cannot be corrected, as shown by the example of the CER's experimental structure at Rouen, the defects of which were aggravated by the use of very moist silty materials (photo 3). In short, this process is not a very reliable way of building structures with vertical facings.

We shall accordingly restrict ourselves to analyzing the deformations engendered by the process in which the form for each layer is placed directly above a fixed point and supported by a fixed reference, derived from the patent held by the Laboratoires des Ponts et Chaussées (1985) and worked by the MUR EBAL company.

In this connection, we may mention the Langres structure.

Photo 3 - Experimental structure built by CER of Rouen : the difficulty of obtaining a vertical facing using a form resting on the next lower layer

2.3 - Langres structure

Where it goes round the ramparts of the town of Langres, national highway 19 is on the uphill side of a retaining wall 4 m high and includes a sharp bend with its convex side outward. To correct the disorders of the existing wall, it was decided to build, under the cover of this structure, a geotextile-reinforced structure having its facing set back about 20 cm from the uphill facing of the existing structure, designed to take out thrust (Delmas et al., 1984).

The structure is reinforced by nine layers of geotextile. Because of its experimental character, three different geotextiles were used : Propex 6066, Tri X 13221, and UCO 44615. We give below only the results obtained on the first section, which is the most fully instrumented.

The structure was built using one form per layer, supported by the existing structure. Given the permanent character of this structure and the state of knowledge when it was built, the fill material chosen was a crushed limestone aggregate having a continuous 0/31.5-mm grading, placed at 95 % of the normal Proctor optimum, or w = 5 % ; γ_d = 20 kN/m^3. The shear strength was estimated to be : φ' = 45° ; c' = 0 kPa.

Because of the angular character of the fill material, a compaction test was carried out. The main results, given in table 4, show that there is little effect on the strength, and that the modulus at failure is reduced by an average of 9 % in the direction of stressing of the product.

Friction tests conducted using a large shear box (25 x 45 cm) yielded the following limits : tan Ø = 0.78 for a soil/ geotextile relative displacement "u_p" = 2 cm.

The instrumentation implanted at the site included glued strain gauges 10 mm long capable of measuring strains up to 10 %. The gauges were bonded to the geotextile on a rubber cement that ensured a suitable surface condition. A laboratory calibration was carried out because we had little experience with gauges of this type and, in particular, this way of bonding them. This revealed, notably, that while in the short term the measured strain values were reliable up to 2 %, in the longer term the creep of the cement made adequate precision impossible.

In addition, the deformations of the facing were measured using inclinometer tubes set in PVC tubes placed on the outside of the facing and attached by straps anchored in the structure.

Figure 5 shows the strains measured in the sheets when the road was reopened. It can be seen that the measured strains do not exhibit the distribution normally expected in reinforcing structures, and in particular exhibit no maximum. Moreover, the measurements made immediately after the removal of the form from the layer corresponding to the sheet measured show that, at this stage, from 70 to 95 % of the final strains have been reached, with the balance appearing when the the next layer is placed (table 5).

Table 4 - Langres : tensile properties of thegeotextiles before and after compaction

	Before compaction		Loose after compaction	
	$\bar{\alpha}_o$ (kN/m)	$\bar{\varepsilon}_o$ (%)	$\frac{\bar{\alpha}_o - \bar{\alpha}}{\bar{\alpha}_o}$ (%)	$\frac{\bar{\varepsilon}_o - \bar{\varepsilon}}{\bar{\varepsilon}_o}$ (%)
t PP	77,3	18,3	0	- 10,0

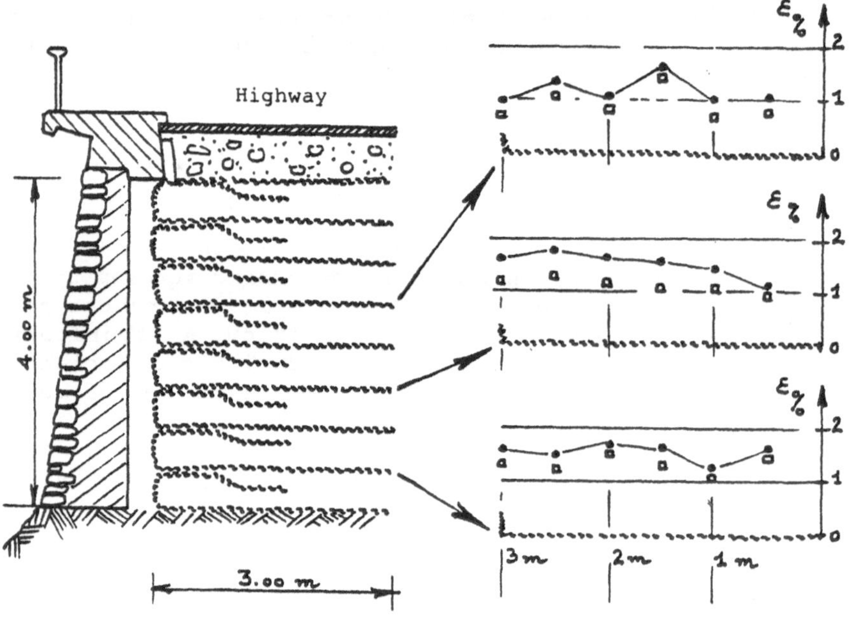

▫ strain after construction of layer
• strain at end of construction

Figure 5 - National Highway 19 (Langres) : strains measured in geotextiles immediately after the construction of the pavement and reopening of the road

Table 5 - Langres : evolution of strain mobilization in the layers of geotextile as the embankment was built up

Point of measurement	1	2	3	4	5	6
Geotextile 2						
ε_i (%)	1.52	1.10	1.36	1.56	1.28	1.44
ε_f (%)	1.63	1.34	1.61	1.65	1.45	1.63
$\varepsilon_i/\varepsilon_f$	0.93	0.82	0.84	0.94	0.88	0.88
Geotextile 4						
ε_i (%)	0.82	1.07	1.07	1.22	1.37	1.37
ε_f (%)	1.05	1.40	1.51	1.54	1.67	1.59
$\varepsilon_i/\varepsilon_f$	0.78	0.76	0.70	0.79	0.82	0.86
Geotextile 6						
ε_i (%)	0.73	0.71	1.40	0.78	1.04	0.76
ε_f (%)	1.07	1.03	1.60	1.04	1.34	0.94
$\varepsilon_i/\varepsilon_f$	0.69	0.68	0.87	0.75	0.77	0.80

ε_i : strain of geotextile at the end of layer construction

ε_f : strain of geotextile at the end of wall construction

It would seem, in this particular case, that the placement of the fill soil, and especially its compaction, account for a large share of the final strains measured in the structure.

As it happened, the absolute necessity of ensuring the verticality of the structure, to join up with the cantilever footway, together with the problems encountered when the form was removed from the first layer (because of slightly insufficient compaction), led the contractor to take special care in the compaction of the fill, especially near the facing. In addition to the desired density, this compaction resulted in a lateral deformation of the structure towards the wall in the layers from which the forms had already been removed. This deformation resulted in tensioning of the geotextile, which thereby confined the fill. This may explain the relatively uniform strain along the sheets of reinforcement.

It will also be noted that the phenomenon seems to have been enhanced by the fact he final strain valuees were reached as soon as the next layer was placed. The situation might be substantially different in the case of a very high embankment (at least in the bottom of the structure), or if the top of the structure carried a large surcharge.

From these few examples, it is manifest that the placement method has a large impact on the strains measured in the reinforcing sheets. However, other factors may further modify the strain distribution field in the reinforcements, such as particular external stresses applied to the structure or displacements of the facing.

To illustrate the former possibility, we may cite the example of the Lixing structure (Delmas and Matichard, 1986).

2.4 - Lixing structure

To repair a shear slide that occurred shortly after the excavation of a cutting for a new lane, a solution consisting of the building of two super-imposed geotextile-reinforced walls was proposed.

We shall deal here in particular with the upper structure, the function of which is to retain the slide mass that could not be excavated in the course of the work. To accomplish

this, the structure, which has a total height of 4.5 m and is 5 m wide, is embedded 1.5 m in the sound soil (figure 6). With respect to the slide, it acts as a stabilizing abutment.

However, the presence of horizontal sheets of geotextile along which the shear strength is substantially less than in the fill makes it possible for the rupture to propagate at the geotextile. This to some extent complicates the design calculations for the structure by making it necessary to estimate the forces induced in the structure by the slide mass and the magnitude of their role in the internal behaviour of the structure.

In the case of conventional abutments built with frictional materials, the forces induced on the repair by the slide mass may be estimated by a stability analysis along an imaginary slip surface prolonging the existing surface inside the abutment.

In the case of geotextile-reinforced abutments, this seems to be the most realistic approach, since the structure acts on the slide not as a rigid retaining structure but as an abutment capable of taking out a large shear stress.

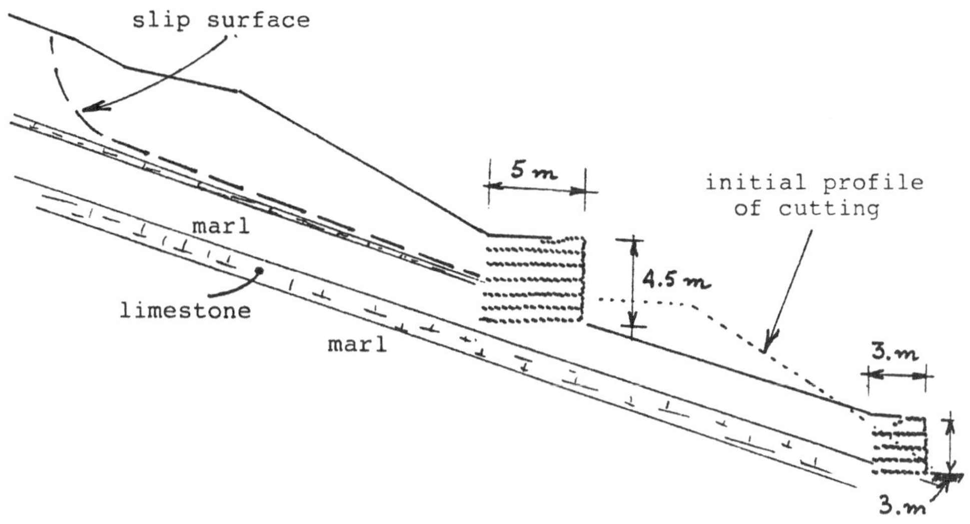

Figure 6 - Lixing : cross-section of proposed repair and slip surface

The geotextile used is woven strips of UCO 44614 polypropylene and the fill a sand (Vosgian sandstone) having the following mean mechanical properties : $c' = 0$ kPa, $\varphi' = 35°$ at 95 % of the normal Proctor optimum, or $\gamma_d = 17$ kN/m^3.

Compaction tests on the geotextile revealed no significant changes in the properties of the product.

The instrumentation of the structure consisted of glued strain gauges to measure the strains of the geotextiles and an inclinometer placed in the reinforced structure to measure its displacements (figure 7).

The conclusion to drawn from these measurements is that on the whole the strains in the geotextiles increased up to about ten days after the construction of the abutment. The inclinometers, on the other hand, indicated deferred strains : practically no displacement at the end of construction, so showing that the strains measured by the gauges were related primarily to the internal stability of the structure ; but, in the year following construction, up to July 1986, an overall displacement by 2.5 mm of the part located above the geotextile extending the old slip surface. This finding confirms the validity of designing the structure as an abutment, and also reveals the gradual mobilization of friction along the sheet of geotextile.

Although the strain gauges were no longer operational when the displacements were measured on the inclinometer tube, it is highly likely that these movements engendered in the sheets tensions distributed differently from those observed in the first stage. Finally, it will be noted that no significant movement of this structure has been observed since July 1986.

Otherwise, the deformations found at the facing, if large enough, may induce large modifications of the tensions in the reinforcements. In what follows, we shall consider only structures with flexible facings, the most likely to be deformed.

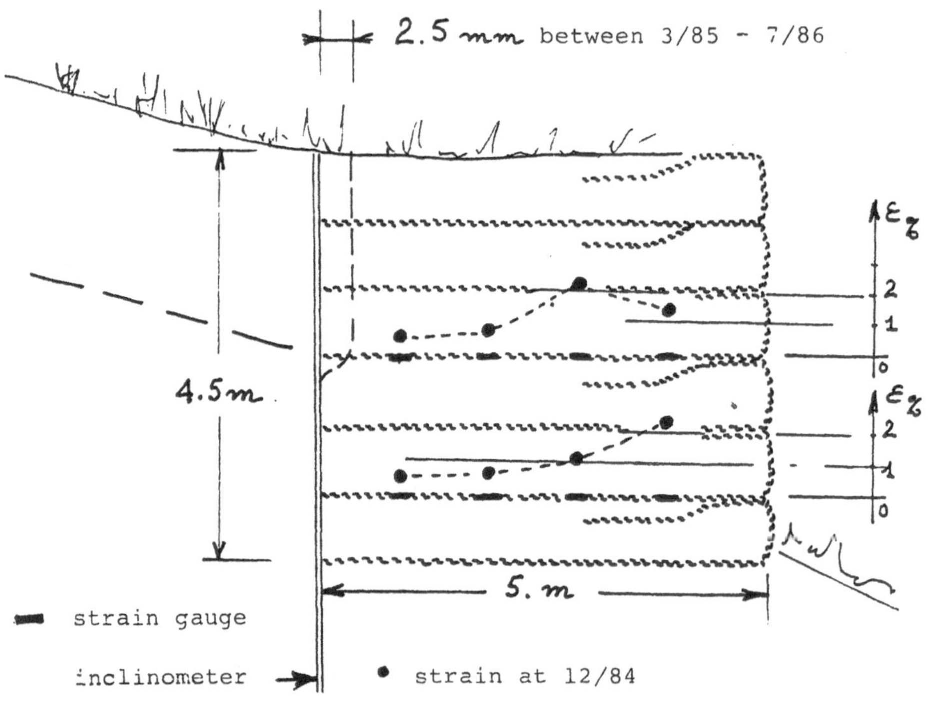

Figure 7 - Lixing : arrangement of measuring instruments and valuesof displacements and strains measured at the end of construction

Combined measurements of the displacements of the facing and of the reinforcing sheets have been made on only a few structures. Among them we may mention the Allevard structure, the Grenoble University Hospital Centre structure, both instrumented and monitored by the University of Grenoble, and a structure on the A7 motorway at La Galaure, monitored by the Lyon Laboratory. Although the facing measuring equipment used on the first two structures is not without interest, we shall consider here only the processing of measurements of the last structure, in which the observed movements have been found to affect the overall behaviour of the structure.

2.5 - Structure on A7 motorway

For the widening of the A7 motorway between Saint Rambert and La Galaure, an approach combining geotextile reinforcement with a prefabricated wall was chosen for the embankment portions. The structure consists of a soil mass reinforced by six layers of woven polypropylene geotextile (UCO 44615) together with a polypropylene-ethylene grid placed on the facing to retain the topsoil on the embankment slope, 50° from the horizontal. This structure, which has a mean height of 2.7 m, is topped by a prefabricated wall 1.5 m high to allow piping to pass (photo 4). The fill is a pea greavel having good mechanical properties : $c' = 0$ kPa, $\varphi' = 41°$, for a placement density $\gamma_d = 22.5$ kN/m^3.

The properties of the geotextiles used are summarized in table 6

The instrumentation placed on the structure (Marchal,1987) includes measurements of reinforcing sheet strains by inductive sensors, vertical strain measurements by horizontal inclinometers, measurements of the relative displacements of the reinforcing geotextile and the grid, and measurements of the rotation of the underlying retaining wall. The inductive sensors used here are attached to the geotextile by distribution plates 30 cm wide attached to the geotextile 50 cm apart. The long measurement baseline substantially decreases the uncertainty arising from the fact that the geotextile is not flat, but the main advantage of these sensors is good long-term reliability.

Figure 8 shows the strains measured on the sheets and the settlements of the sheets at the end of construction and six months later. It will be noted that the lack of ties between the geotextiles and the facing grid, together with the low coefficient of friction of the two products, results in a relative slippage of about 2 cm. The local decompaction at the facing leads to local settlements, together with a small local increase in the tensile forces near the end of the reinforcing sheet and a slight rotation of the upper retaining structure.

Photo 4 - A7 motorway : view of structure during construction

Table 6 - A7 motorway at La Galaure : mechanical properties of the geotextiles used

Tensile caracteristics (NF 38014)	$\overline{\sigma}$ (kN/m)	$\overline{\varepsilon}$ (%)
Geotextile t PP	72.1	13.0
Geogrid G PP PET	9.1	8.0

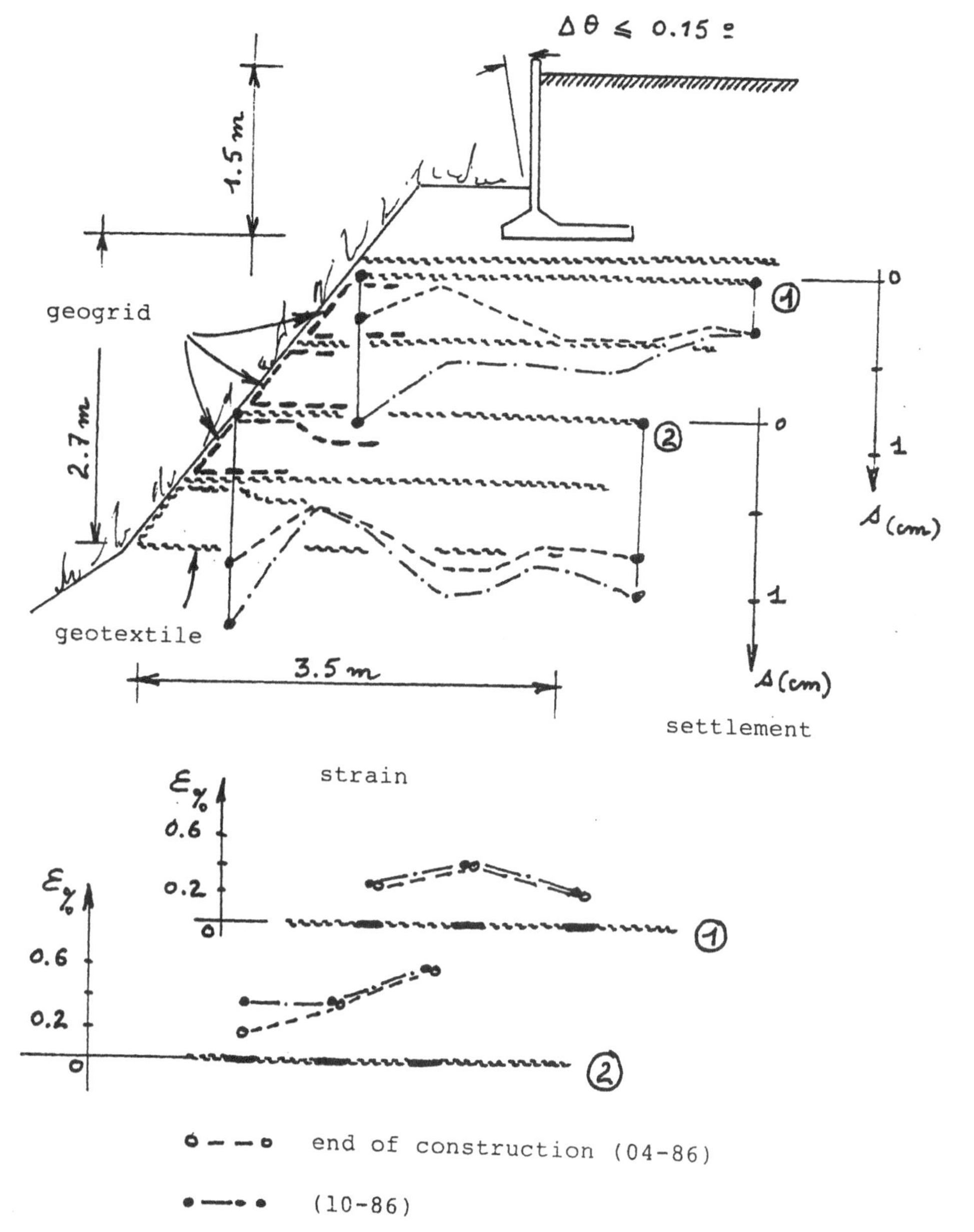

Figure 8 - A7 motorway : arrangement of measuring instruments and measurements made at the end of construction and six months later

It may be concluded from this particular example that proper compaction of the structure, in particular of the facing, is the key to the proper behaviour of the structure, and that the "reinforcing" and "facing" geotextiles must be tied together to confine the soil and pretension the sheet.

The conclusion to be drawn from the few examples given here is that while the tensions in the reinforcing sheets depend to a large extent on the internal stability of the structure, they are significantly influenced by the placement method and by the nature and timing of the stresses to which the structure is subjected. For this reason, the placement method must be chosen not only to facilitate the building of the structure, but also to ensure optimum tensioning of the sheets. The method mentioned above may be used (Delmas et al., 1986). Since each layer is placed behind a form supported by a fixed reference independent of the structure, it ensures a good final geometry and effective compaction of the fill near the facing. In addition, the absence of forms on the underlying layers allows deformations of the structure by the compaction and contributes to the pretensioning of the sheets. Along the same lines, the process proposed by Mc Gown et al. (1987) promotes tensioning of the sheets at the time of placement and serves to reduce deferred deformations of the structure.

3 - THE USE OF CLAYEY SOILS IN GEOTEXTILE-REINFORCED STRUCTURES

In so far as it is not necessary to use dilatant soils in geotextile-reinforced structures, soils having a large fraction of fine materials can be used. If the soil used meets the usual specifications for fills, its use in a reinforced structure does not, a priori, pose any special problems if its mechanical properties are properly taken into account in the design and if the water content of the soil at placement is not likely to result in pore overpressures during subsequent loading.

The example of the A15 motorway, mentioned above, is a case in point. In spite of placement in a moist condition (S_r = 94 to 98 %), which resulted in a rubber-cushion effect during compaction, no disorders were found at the time of placement. This may be explained by the use of a draining geotextile - here, a needle-bonded nonwoven - that served to prevent the creation of large pore overpressures and facilitated the consolidation of the fill material. The traces of deposited calcite on the samples taken in 1986 seem to confirm the draining action of the geotextile.

The example of the experimental embankment at Rouen provides some additional information about the actual behaviour of reinforced fine soils having a high water content (Blivet et al., 1986). This experimental structure, 5.6 m high, was built with a silt having a water content of w_{npo} + 5 % and was reinforced with various types of geotextiles. Here we shall consider only two of them, on which pore pressure measurements were made : an Enka woven polyester and a needle-bonded nonwoven polyester in conjuction with a Rhône Poulenc grid.

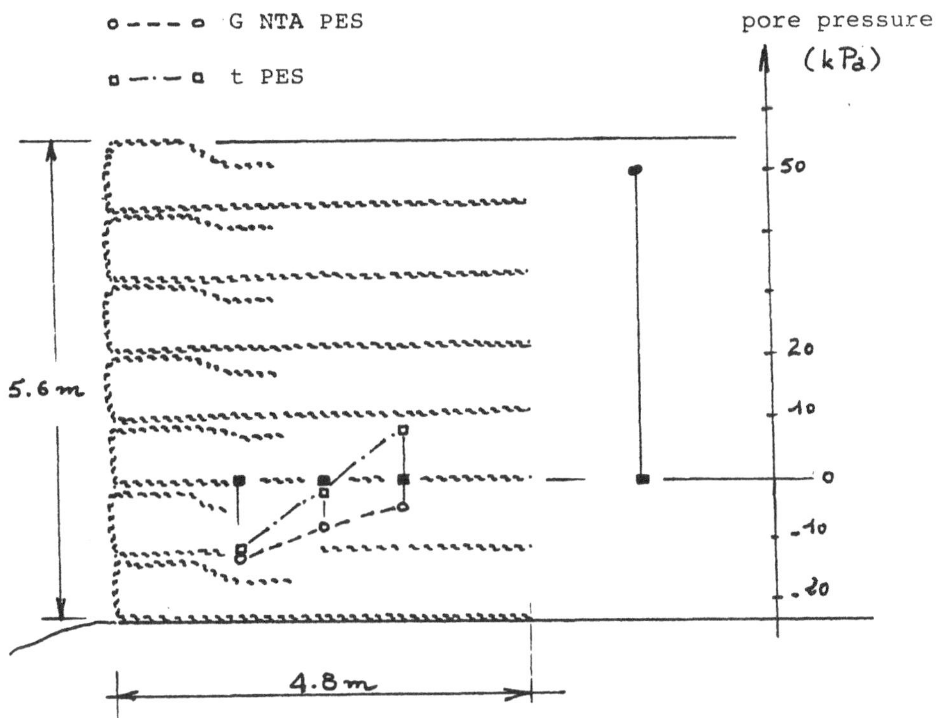

Figure 9 - Rouen experimental structure : arrangement of measuring instruments and measurement values 120 days after construction

Because of the type of soil used as fill, its poor mechanical properties, and the type of forms used, the placement strains were large.

Considering mainly the measurements made from the end of construction, figure 9 shows the difference between the behaviour of the products. The pressure sensor inside the embankment, outside the structures, reveals placement overpressures of as much as 50 kPa.

On the woven sheet, the pore pressures are positive at the back of the structure, then disappear and finally become negative near the facing.

On the composite geotextile, on the other hand, the pressures are negative over the whole length of the reinforcement.

This difference in local behaviour can lead to large changes in overall stability : on a nearby test section reinforced with a woven polyester with its surface treated to be non-wetting, the soil mass turned over because of anchorage failure. An after-the-fact calculation revealed an effective angle of friction of 5°, as against a soil-geotextile angle of friction of 21° in a drained condition.

We may conclude from these two examples that, while the use of soils having a large percentage of fines poses no special problems, it is necessary to make sure that the placement and compaction are not likely to engender pore pressures, or that the geotextile can dissipate them so as to ensure optimum local soil-geotextile friction.

4 - INFLUENCE OF THE LONG-TERM BEHAVIOUR OF THE GEOTEXTILES ON THE REINFORCED STRUCTURES

Without attempting an exhaustive analysis of the problems of the long-term behaviour of the geotextiles, one may attempt to learn something from a few examples.

With respect to creep behaviour, it will be noted that most authors agree that creep in geotextiles depends to a large extent on the load factor. In particular, some recognize that, below 10%, creep in polymers at ambient temperature is negligible. For guidance, table 7 gives the load factors of the geotextiles at the sites mentioned earlier.It can be seen that they are less than 10 % in all cases. Without attempting to predict the future, it would therefore seem that in these few structures the risk of creep is very small.

Moreover,the A15 motorway structure provides information about the ageing of geotextiles. This structure, stressed for eight months in 1971, has since been protected by a bank of fill. The sampling done in 1986, or 15 years after placement, provided an opportunity to measure the mechanical properties of the product. The loss of strength found, 21 %, agrees with the loss of strength routinely observed during placement (up to 30 % with angular materials). And the gain in stiffness found can be explained by the presence of calcite deposits in the geotextile. Chemical analyses of the fibres revealed no change of the crystalline structure (percentage, orientation), confirming the absence of chemical deterioration of this type of product in a soil of pH 10.

Table 7 - Maximum work factor measured on the geotextiles at a few instrumented sites

Location	Date	Geotextile	Soil	α/α_f	ε
A 15	1971	NTA PES	Clay (Ph 10)	0,06	1,8 %
Langres	1983	t PP	Limestone	0,08	1,6 %
Allevard	1983	t PP	Gravel clay	0,9	2 %
Lixing	1984	t PP	Sand clay	0,12	2,7 %
A 7	1986	t PP	Gravel clay	0,05	0,6 %

This finding confirms the results of earlier studies based on samples taken from actual sites (Sotton et al., 1983).

5 - CONCLUSION

The conclusion to be drawn from these few instrumented examples is that the tensile forces, and in particular their distribution in the structure, depend to a large extent on the conditions of placement, but also on the way the structure is stressed.

This must be taken into account in designing these structures, and for this reason preference should be given to design methods in which the actual conditions of placement can be simulated and taken into account. In this connection, we may note the interesting approach made possible by the "displacements" method (Gourc et al., 1986).

In addition, an attempt should be made to optimize the mobilization of the forces by chosing a type of form that allows pretensioning of the geotextile and holds deferred deformations of the structure to a minimum.

REFERENCES

1) BLIVET J.C., JOUVE P., MAILLOT R.(1986) Numerical modelization of earth reinforcement by geotextile : hydraulic function. C.R. IIIe Cong. Int. Geotextiles, Vienne, avril, IV, pp 1061-1066.

2) DELMAS P., FAVRE J.M.,MATICHARD Y., LEHMANN M., PRUDON R., REBUT P. (1984) Renforcement par géotextile d'un mur de soutènement sur la RN 19 à Langres. Revue Générale des Routes et Aérodromes, 609, juin, pp 61-66.

3) DELMAS P., MATICHARD Y.(1986) Landslides confortation with geotextile reinforced earth work. C.R. IIIe Cong. Int. Geotextiles, Vienne, avril, IV, pp 1091-1096.

4) DELMAS P., PUIG J., SCHAEFFNER M. (1986) Mise en oeuvre et parement des massifs de soutènement renforcés par des nappes.

5) GOURC J.P.,RATEL A., DELMAS P.,(1986) Design of fabric retaining walls : the "displacement method". C.R. IIIe Cong. Int. Geotextiles, Vienne, avril, IV, pp 1067-1072.

6) MARCHAL J.(1986) Autoroute A7 : Elargissement entre St Rambert d'Albon et la Galaure. Compte rendu de mesures. Rapport Labor. P. et Ch., 30 pp.

7) Mc GOWN A., ANDRAWES K.Z., MURRAY R.T. (1987) The influence of lateral boundary yielding on the stresses exerted by back-fills. C.R. coll. Interactions Sols-Structures, ENPC, mai, pp 585-592.

8) PERRIER H., LOZACH D. (1986) Essai de poinçonnement sur géotextile - Influence des sollicitations de compactage. Compte rendu de mesures. Rapp. Labor. P.et Ch., 45 pp.

9) PUIG J., BLIVET J.C. (1973) Remblai à talus vertical armé avec un textile synthétique. Bull. de Liaison P. et Ch., 64, mars-avril, pp 13-18.

10) SCHLOSSER F., GUILLOUX A. (1981) Le frottement dans le renforcement des sols. Revue franç. de Géotech., 16, pp 65-77.

11) SOTTON M., LECLERCQ B., PAUTE J.L., FAYOUX D. (1982) Quelques éléments de réponse au problème de durabilité des géotextiles. C.R. IIe Cong.Int. Géotextiles, Vol. 2, Las Vegas, août, pp 553-558.

GEOTEXTILE REINFORCED RETAINING WALLS: DISCUSSION OF INSTRUMENTED LARGE SCALE TEST WITH RESPECT TO THE VERIFICATION OF DESIGN CONCEPTS

B. GRAF, J.A. STUDER
GSS GLAUSER STUDER STUESSI, CONSULTING ENGINEERS INC., ZURICH, SWITZERLAND

1. INTRODUCTION

To conserve the landscape in the construction of new arterial roads it is becoming more and more necessary to employ deep cuts with a long service life and high reliability using steep artificial slopes for stabilization. Such slope stabilization can be built with relatively low expenditure, if the excavated material is used as fill material and the fill material is strengthened by geotextile reinforcement. High standards are required for the safety and durability of such permanent slopes. Besides the fact of relatively low costs for geotextile reinforced retaining walls (compared to the expenses for conventional retaining walls), the total construction expenditure can be high (long and deep cuts). The above mentioned requirements are arguments for the need of a consistent, verified design concept for geotextile reinforced slopes. To the best of our knowledge, such a concept does yet not exist.

On the other hand, geotextile reinforced structures have been built successfully for a long time. However, with respect to safety and reliability, this construction method has been mostly used and accepted for less important structures (for example low and/or temporary embankments).

Against the application of geotextile reinforcement for high and steep slopes, however, with a service life of 50-100 years, the following arguments are used:

- Insufficient permanence of geotextiles against UV-radiation,
- Susceptible in the case of acts of sabotage,
- Insufficient knowledge about ageing and related changes of material strength for geotextiles,
- Aesthetics,
- Maintenance.

Mostly it is suggested, to compensate for these almost incontrollable influences on the stability and durability of such structures, by appropriate reductions of the material strengths in the static analysis, appropriate construction methods (lining) and with the help of a long term monitoring of such structures. However, appropriate reduction of material strengths presumes realistic results for the forces and stresses from the static ananlysis.

P. M. Jarrett and A. McGown (eds.), The Application of Polymeric Reinforcement in Soil Retaining Structures, 313–337.

The limit analysis design of geotextile reinforced retaining walls is usually similar to that of conventional retaining walls (cf.(8)), whereas the application of such conventional analysis on reinforced walls has not yet been verified.

Because of the high costs of large scale tests, only a few fully instrumented and evaluated cases have been reported. For none of the published experiments limit states could be reached (cf.(13),(15)). The design, however, is mainly based on limit analysis.

From the foregoing it is clear that there is a need for additional theoretical and experimental investigations of the behaviour of geotextile reinforced retaining walls.

The aim of this paper is the discussion of available limit analysis design concepts and their experimental verification. In section 2 for several reasons mainly the design concept proposed by Gudehus and Schwing (8) will be presented. The reasons are: Concept is based on model test results, the most important aspects of stability are incorporated, partial factors of safety have been calculated with a probabilistic method by these authors. Section 3 deals with the experimental verification of the presented design concept, whereas a procedure for test evaluation is discussed, followed by the application of this procedure to the evaluation of published data from large scale tests. Finally, in section 3 the proposals for future theoretical and experimental investigations are summarized. Section 4 deals with case studies.

2. DESIGN OF REINFORCED ARTIFICIAL SLOPES

Design steps are discussed with respect to the standard reinforced wall shown in figure 2.1: cohesionless soil, no surcharge loads. This section contains the theoretical background needed for evaluation of field data.

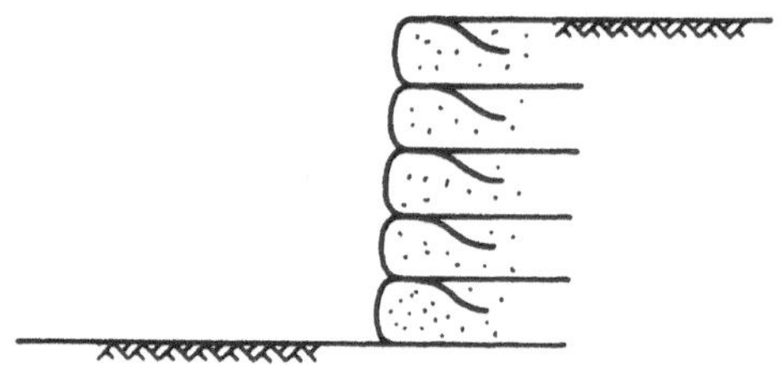

FIGURE 2.1. Standard geotextile reinforced retaining wall.

Today's design of reinforced artificial slopes is mostly based on limit-analysis models describing the failure of the structure. For displacement estimates empirical formulas are used. In the following the limit analysis design, determination of displacements and the statement of partial factors of safety are discussed.

2.1. Limit states

For convenience, limit states will be discussed separately for modes of so called internal-and external failure. External failure is characterized by failure along distinct slip surfaces, whereas all the other failure modes (for example zone failure) are called internal.

2.1.1. External failure. For external failure modes usually rigid-body failure mechanisms are assumed. Figure 2.2 shows some of these mechanisms. Figure 2.2a holds for failure along a cylindrical slip surface. Various design charts are based on this mechanism. These charts can be used for the determination of the design length and tensile strength of reinforcement (e.g. (5)).

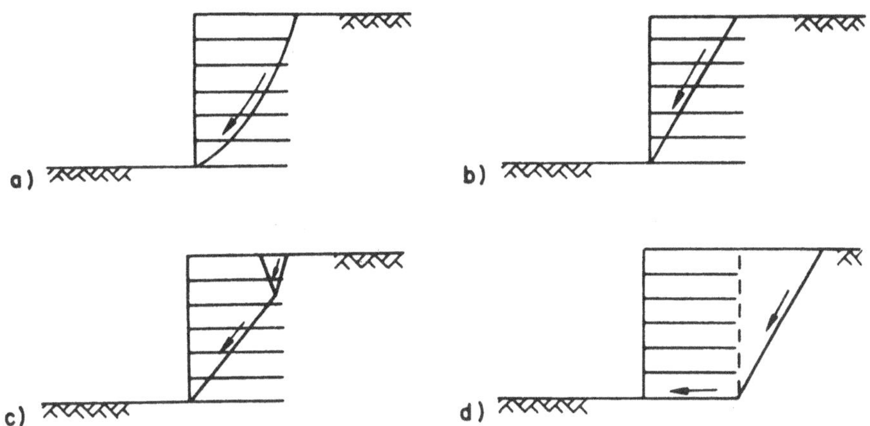

FIGURE 2.2. Failure for internal stability: a) Rotation, b) Translation, c) Translation, 2 Rigid bodies, d) Sliding.

English codes assume the Coulomb wedge from figure 2.2b. This kind of failure can be expected for embankments with local surcharge loads (10).

The so-called Dutch/Enka-Method (4) assumes a failure mode of 2 rigid-bodies under pure translation (figure 2.2c). This failure mechanism was originally observed in model and field tests concerning soil-nailing (7). Figure 2.2d shows the assumed failure mechanism for sliding. This kind of failure can be expected for insufficient design length of the reinforcement.

To the best of our knowledge, none of the discussed failure mechanism has been verified yet by experiments for geosynthetic reinforced retaining walls. For reasons of safety (in the sense of the so-called collapse theorems of the theory of plasticity), however, for kinematic solutions, the actual failure mechanism should be well approximated by the assumed one.

Gudehus und Schwing (8) published a paper with model test results for the failure mechanism. In these tests the stability of standard walls (cf. figure 2.1) was investigated as a plane strain problem giving regard to the geometrical and mechanical rules of similarity. The height of these walls was increased until the collapse of the structure occured. The observed failure mechanism (cf. figure 2.3) is different from the usually assumed ones shown in figure 2.2. For walls with no surcharge loads, cohesionless soil and sufficient internal stability, therefore, a failure mechanism of 2 rigidbodies under pure translation, can be assumed. More specific features of the observed mechanism will be discussed later.

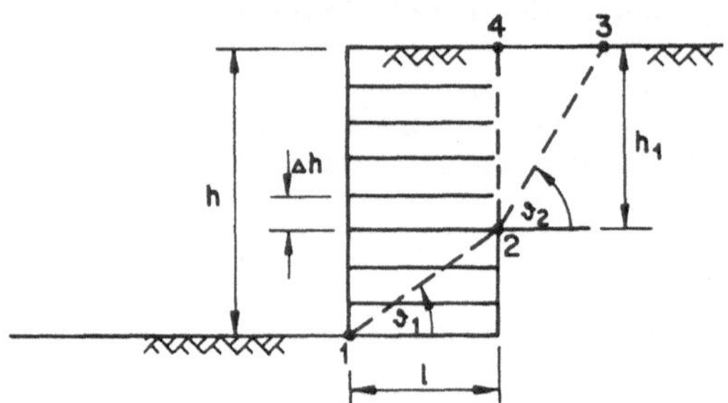

FIGURE 2.3. Observed failure mechanism (Gudehus and Schwing, (8)).

Rotational mechanisms or the Coulomb-wedge can be expected for restricted surcharge loads. For sail-nailing under these conditions (restricted surcharge loads) a mechanism of 2 rigid-bodies under rotation was observed (8). A large scale test from (15) supports the Coulomb-wedge assumption. Experimental verification is planned for geotextile reinforced retaining walls (8).

From model test observations and numerical studies the following knowledge has been obtained for the failure mechanism shown in figure 2.3 (cf.(8)):

- The slip surface $\overline{24}$ coincides with the ends of the reinforcement. This observation simplifies the formulation and evaluation of the limit state equation.

- For the angle ϑ_2 (cf. figure 2.3) the Coulomb solution can be assumed ($\vartheta_2 = 45° + \phi/2$; ϕ: friction angle). This result was derived by numerical studies and was experimentally verified.

- The failure mechanism from figure 2.3 contains the following failure modes as particular cases: overturning, sliding and bearing (vertical translation). For example, it can be shown with the help of the limit state equation, that for small $\Delta h/h \longleftrightarrow \vartheta_1 \rightarrow 0$ holds, i.e. in this case sliding is the relevant failure mode for the structure.

- Because of local base failure, the vertical stresses at the base of the wall are essentially smaller than the overburden-pressure (cf.(8), (3)). As a consequence of this experimental finding, Gudehus and Schwing (8) neglect in stability calculations the pull-out resistance of the reinforcement near the toe of the wall (up to 10% wall-height). In all other layers the pull-out resistance is:

$$T = 2\mu \sigma_i l_i, \qquad (2.1)$$

  where

  σ_i : overburden-pressure for reinforcement layer i,
  l_i : length of reinforcement layer i behind slip surface,
  μ : friction coefficient (soil/geosynthetic).

With the help of the abovementioned findings, the limit state equation can be easily formulated and evaluated. In this equation the slip-surface inclination ϑ_1 (cf. figure 2.3) occurs as the only unknown quantity. This inclination can be derived with the assumption, that the sum of the tensile forces should be maximum along the slip surface $\overline{12}$. This assumption has been experimentally verified (8). With the help of the abovementioned limit state equation, a design chart was made available by (8). This design chart is further discussed in section 2.3.

Other stability investigations, based on one of the failure modes shown in figure 2.2, will not be further discussed here, because they have not been experimentally verified. As already mentioned, the investigations of (8) are restricted to the standard structure from figure 2.1. The incorporation of surcharge loads is planned in future work by the last mentioned authors.

2.1.2. Internal stability. The following, so called internal failure modes will be discussed:

- Failure of lining due to earth pressure,
- Failure of reinforcement.

For design purposes the earth pressure and the maximum tensile force (related to internal stability) for the reinforcement must be known.

Field data for the earth pressure derived by (10) imply, that the use of active earth pressure loads leads to conservative design (cf. figure 2.4). For the upper part of the wall the measured pressures are only a little greater than the theoretical active pressures; for the lower part of the wall, however, the measured pressures are up to 50% lower than the active ones.

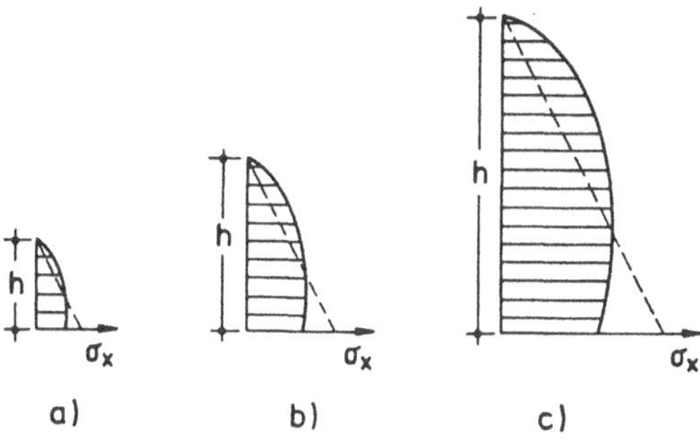

FIGURE 2.4. Earth-pressure distribution: a) h = 2 m, b) h = 4 m, c) h = 7 m. Dashed line: Active earth pressure (cf. (8)).

John (10) calculated the earth pressures, shown in figure 2.4, from the measured connector-forces (the reinforcement is fixed to the lining with the help of connectors). Such force-measurements are much more reliable than measurements with the help of earth pressure cells.

With increasing distance from the lining, the forces in the reinforcement increase (cf. (16) and figure 2.5), i.e. the maximum tensile forces are not equal to the earth pressure forces.

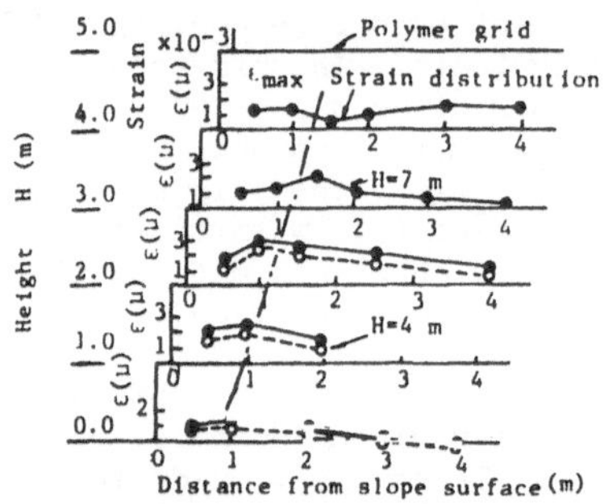

FIGURE 2.5. Measured strains for the reinforcement (Yamanouchi et al., (16)).

Firstly, the design earth pressure will be discussed. Most standards recommend a linear earth pressure distribution (cf. (10)). Gudehus and Schwing (8) assume a linear distribution down to 2/3 of the height of the wall and afterwards a constant value down to the toe (cf. figure 2.6). The inclination of the linear part of the distribution should be assumed like in the active case. The distribution from figure 2.6 was chosen as an approximation of the field data derived by (10). A more sophisticated approximation of the field data can be derived with the help of the so-called silo theory (8). By the assumption of an linear distribution, the pressures are clearly overestimated (cf. figure 2.4). The proposed distribution of (8) is more realistic, easy to use and, therefore, will be further discussed in the context of the evaluation of field data (cf. chapter 3).

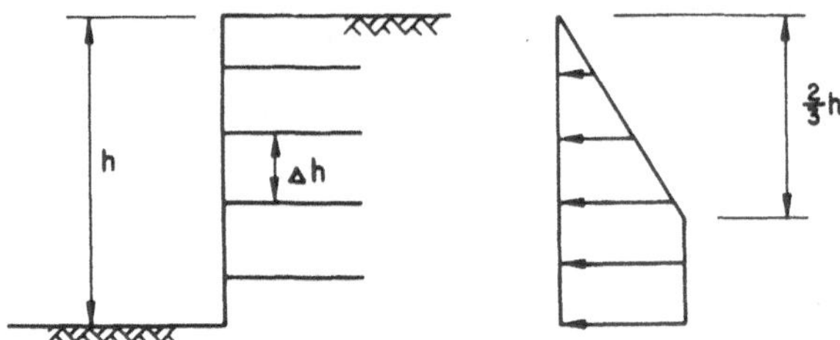

FIGURE 2.6. Design earth-pressure distribution (Gudehus and Schwing, (8)).

As mentioned before, the tensile forces in the reinforcement can increase essentially with growing distance from the wall face (cf. figure 2.5). There is as yet no theoretical prediction discribing this increase. Measurements from (15) indicate that the maximum tensile force can be up to 4 times higher than the resultand earth pressure force acting on the wall face. Gudehus and Schwing (8) assume the following correlation between the earth pressure and the design tensile force:

$$\max Z = 0.9\gamma \cdot h \cdot \Delta h \cdot K_{ah}; \tag{2.2}$$

$$K_{ah} = tg^2(45° - \phi/2); \tag{2.3}$$

i.e. the maximum tensile force is assumed to be correlated to the earth pressure for the depth of 0.9h (cf. figure 2.2). This correlation has not yet been verified by experiments.

2.1.3.Factors of safety. From the literature very little direct information can be gained about the factors of safety to apply for the design of geosynthetic reinforced walls.

For the design of reinforced retaining walls, the partial factors of safety for the material properties of soil, the friction coefficient between soil and geotextile, the material properties of the reinforcement and the surcharge loads are needed.

For the case of external stability, (8) calculated the partial factors of safety for the standard wall shown in figure 2.1. They assumed the friction coefficient and angle of friction, occurring in the limit equilibrium equation, to be statistically independent, random quantities (cf. (8)) for the assumed probability distributions) and derived the partial safety factors with the help of the second-moment reliability code. With respect to the mean values of friction angle and friction coefficient, the following results were derived:

> For the friction angle (of the soil) the partial safety factors are in the range of 1.1 - 1.2, for the friction coefficient in the range of 2-4. These ranges for the safety factors correspond to the range of 30°-40° for the mean values of friction angle of the soil.

Under the assumption that the tensile strength of the reinforcement exceeds the pull-out resistance, the tensile strength does not occur as a variable in the limit equilibrium equation. Gudehus and Schwing (8), therefore, assume a partial factor of safety for the calculated tensile force, within the range 3-4. The design tensile force has to be compared with the tensile strength of the reinforcement. In addition, to avoid creep of the reinforcement, the corresponding strain should be less than 5% (cf. also (17)).

The partial factors of safety proposed by (8) have to be verified by back-calculation of the reliability of well performing reinforced retaining walls. Besides this, surcharge loads should also be incorporated and the assumption of statistical data (distribution and corresponding parameters) should be verified with the help of experiments.

Besides the applied second-moment reliability code, other probabilistic calculations are possible (cf. (17)). The second-moment code, however, seems to be the most suitable method:

- Reliability of different types of constructional methods is comparable.

- Derivation of the widely accepted and easy to use partial factors of safety, on the basis of a consistent probability concept.

The above discussion holds for the partial factors of safety corresponding to external stability. For the case of internal stability and deformations, to the best of our knowledge, similar investigations do not exist.

By limiting the strains in the reinforcement, caused by the calculated tensile forces, to small values (5%), creep-deformations can successfully be prevented (cf. (17) and (16)). This holds for cohesionless fill materials and high strenght reinforcement, characterized by an initially steep inclination of the stress-strain curve. The creep problem, therefore, is mainly relevant for cohesive fill materials and/or continuous filement needle punched geotextile as reinforcement, which are characterized by a relatively low initial strength.

2.2 Displacements

2.2.1 Empirical estimates and possibilities of theoretical predictions.

The total displacements of a geosynthetic reinforced wall can be separated as follows:

- Settlement of the contact area,
- Deformations of the wall during construction,
- Deformations due to surcharge loads,
- Deformations caused by creep under gravitiy and surcharge loads.

The settlements of the contact area can be calculated conventionally and, therefore, will not be further discussed.

By experience it is well known that the largest deformations occur immediately (if creep is prevented). The amount of these deformations is largely influenced by the method of construction (cf. (8), (3)). The displacements of the head of a wall can reach up to 1-2 % of the height of the wall (8). A theoretical prediction of these displacements can hardly be done, also a prediction is not of great interest, because displacements of these amounts can be accepted for this construction method.

A good prediction of displacements caused by surcharge loads is still not possible. The application of numerical methods (Finite Elements) in the near future - with the goal of quantitatively good predictions - is prevented for several reasons (experience with constitutive equations, numerics for nonlinear problems). The prediction of such displacements, therefore, also must be based mostly on experience.

By restriction of the expected strains of the reinforcement (for the design tensile force) on small values (< 5 %), creep deformations can successfully be prevented (cf. (16)). This holds for cohesionless fill materials and strong reinforcements, characterized by an initially steep inclination of the stress-strain curve. The creep problem, therefore, is mainly relevant for cohesive fill materials and/or continuous filament needle punched geotextiles as reinforcement, which are characterized by a relatively low initial strength.

Displacements are mostly predicted with the help of the above mentioned empirical estimates. Yamanouchi et. al. (16) published the results of a FE-analysis predicting the displacements of a geotextile reinforced retaining wall under gravity and surcharge loads. For details of the modelling the reader should refer to Yamanouchi et. al. (16). For this retaining wall, the measured displacements were underestimated by a factor of 0.5 by the FE-analysis. Creep was not investigated. Qualitatively, however, some of the derived results show good agreement with the measurements:

- Distribution of tensile forces in a reinforcement layer,
- Distribution of tensile forces with respect to the height of the wall (cf. figure 2.5).

2.3. The design concept

In this chapter a short overview of the design concept, proposed by (8), will be given. This concept was derived for the standard retaining wall in figure 2.1: cohesionless soil, no concentrated surcharge loads. This concept will be used later on for the interpretation of field data.

The case of external stability will be dicussed first. In model tests the failure mechanism, shown in figure 2.3, was observed. With the help of the corresponding (to the failure mechanism) limit equilibrium equation, the design length l of the reinforcement layers and the tensile force Z can be calculated for the given geometry and soil data of a retaining wall. The design quantities l and Z (maximum of Z according to external or internal stability) are plotted in figure 2.7 versus the friction angle ϕ, with $\Delta h/h$ as contour lines. The design chart (figure 2.7) holds for friction coefficients (soil/geotextile):

$$\mu = 0.4 \, tg\,\phi; \tag{2.4}$$

i.e. the friction coefficient is already reduced by a partial safety factor (cf. chapter 2.1.3). Only the friction angle of the soil has to be reduced by a partial safety factor as mentioned in chapter 2.1.3, for the application of the design chart.

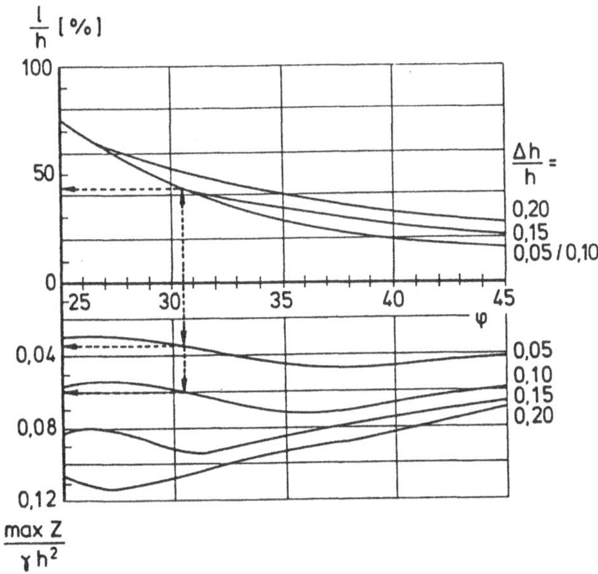

FIGURE 2.7. Design of a standard geotextile reinforced retaining wall (Gudehus and Schwing, (8)): Design chart (cf. figure 2.8 for situtation)

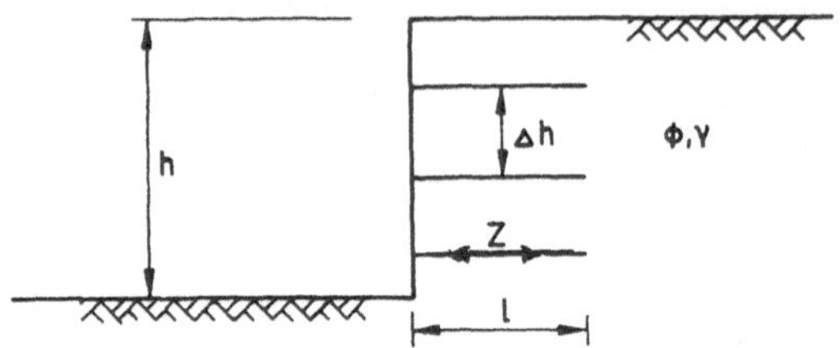

FIGURE 2.8. Situation corresponding to design chart

For reasons of internal stability, also the tensile force from equ. (2.2) has to be calculated. The design, finally, has to be based on the largest tensile force to the cases of internal or external stability. In figure 2.7 the maximum tensile forces are plotted versus the friction angle of the soil. This tensile force should be multiplied by a safety factor of 3-4 and these be compared to the tensile strength.

To avoid creep the strains caused by the tensile force should be at least less than 5%. The appropriate value for this strain limit can be gained from experiments.

According to experience, the displacements of the top of the wall can be expected to be in the range of 1%-2% of the height of the wall.

3 .EVALUATION OF INSTRUMENTED DATA

3.1.Principles of evaluation

Most of the input and output quantities of the design concept (for example tensile forces corresponding to limit states) cannot be measured directly in full scale tests. In addition, in most large scale tests the limit states which are assumed for design are not reached (cf. (13), (15)). In the following, therefore, firstly the possibilities of verification of the design concept are discussed.

3.1.1.Limit states

First of all, the case of external stability will be discussed. For verification of the design procedure, the following instrumented data and observations are needed:

- Geometry (height of retaining wall, length of reinforcement), loads and material properties corresponding to limit states,

- Position and inclination of slip surfaces,

- Tensile forces for the reinforcement layers intersected by slip surfaces.

The geometry and loads can be easily determined. As mentioned before, however, limit states can hardly be reached in large scale tests: Because of large displacements, the loading equipment and instrumentation fail before the limit state is reached (15). Large scale tests, therefore, are not very suitable for the determination of the critical geometry and load.

More suitable are model tests, which can be used for the systematic investigation of the behaviour of reinforced retaining walls in limit states (cf. (8)).

Now the determination of the failure mechanism will be discussed. Figure 3.1 shows the measured geotextile strains from a large scale test, presented as contour lines (cf. (15)). In this test, the ultimate surface load could not be reached. Therefore, the strains from figure 3.1 correspond to a pre-failure state. The expected failure plane ist also shown as a straight dashed line in figure 3.1. This line was drawn under the assumption, that a failure plane would intersect the reinforcement layers at the points of maximum strains. Even if the limit state was not reached in a large scale test, the strain measurements, therefore, can be used for deriving a first estimate of the position of potential slip surfaces. More conclusive results could be expected from data of slope indicator tests if there are any. Because of the large expenditure, however, large-scale test are not suitable for investigations of failure mechanismes. Model test are the better "tool" for such investigations, whereas large scale tests should be used for verification of model test results (adherence of similarity rules).

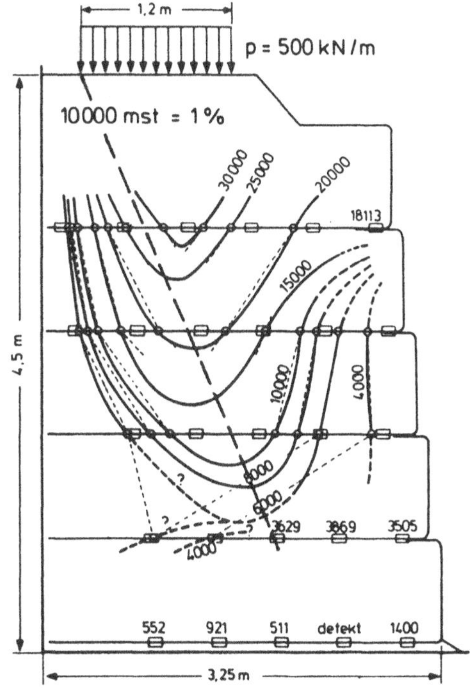

FIGURE 3.1. Contour lines for the geotextile strains (Wichter et al., (15)). Dashed line: expected slip surface.

Because of the large displacements, even for pre-failure states, it can be expected, that the tensile forces in the reinforcement can hardly be measured for ultimate states. The tensile forces resulting from calculations, therefore, can only be verified indirectely.

Large scale tests, therefore, are not well suited to the verification of the modelling of external stability. Large scale tests should be performed only in combination with model tests.

The design for internal stability depends on the following quantities:

- Earth pressure distribution,
- Maximum tensile force to be expected.

The earth pressure distribution, which is needed for the design of the lining and connectors (connection between lining and reinforcement), will be discussed first. With the help of force measurements in the connectors, the earth pressure distribution can be determined with high accuracy (cf. (10); (8)). Predicted earth pressure distributions, therefore, can easily be verified in large scale tests.

More difficult is the measurement of tensile forces in the reinforcement. Usually the strains are measured. Because of the complicated material behaviour of geotextiles (creep, temperature-effects, strain-rate dependency, influence of normal stresses on the material behaviour of needle-punched geotextiles) the back-calculation of forces is uncertain. Direct force measurements would lead to more reliable results. For the design of the reinforcement only the maximum tensile force is needed. The presented design concept assumes a correlation between this maximum tensile force and a resulting earth pressure force (cf. equ. (2.2)). This correlation has to be verified by experimental data. This data can be gained from large scale tests.

The pull-out resistance of reinforcement layers, incorporated in the limit state equation for external stability, depends strongly on the vertical stress assumed to act at a certain depth (cf. equ. (2.2). There are very little experimental data available for the distribution of vertical stresses with depth. Experimental data indicate that the vertical stress is considerably below the overburden pressure for points lying near the face of the wall. The assumption of a hydrostatic distribution for the vertical stresses with depth, therefore, would lead to an overestimation of the pull-out resistance. For this point to become more clear, measurements of the vertical stresses from large scale tests are needed.

3.1.2 Displacements. Up to now, the displacements cannot be predicted with the help of easy to use analytical or numerical methods. The displacements caused by gravity loads can be estimated with the help of empirical formulas. Creep is usually prevented by definition of a limit for the geotextile-strains to be expected under the calculated tensile forces. These formulas and limits will be verified later in connection with the evaluation of field data.

3.1.3 Safety requirements. Most of the papers on the subject of reinforced retaining walls deal with investigations of the statical aspects of external stability. There are only few papers, however, dealing with safety requirements and proposals for the use of partial factors of safety.

One of the open questions concerns the basic statistical data needed for the application of the second moment reliability code. Thus, very little is known about the probability distribution of the pull-out resistance. Also, back-calculations of the reliability of existing, well performing retaining walls have not yet been published. The question concerning partial safety factors, therefore, is closely related to experimental investigations. Future work on the subject of geotextile reinforced retaining walls, therefore, should also concentrate on the following topics:

- Determination of basic statistical data (type of probability distributions and their parameters),
- Documentation of well performing structures and back-calculation of reliability (cf.(2)).

3.2 Experimental results and predictions

From large scale tests or existing walls mostly only data for displacements are available. Only few data are available for verification of the force and pressure quantities resulting from the design concept for the case of internal and external stability (cf. table 3.1).

Authors	measured			observated failure mechanism	remark
	displacements	tensile forces	soil pressures		
Yamanouchi et al. (16)	x	x	-	-	-
Caroll and Richardson (3)	-	x	x	-	-
John (10)	-	-	x	-	only lateral pressures
Wichter et al.(15)	x	x	-	x	cohesive fill material

TABLE 3.1. List of published large scale tests.

For none of the large scale tests listed in table 3.1, data corresponding to the limit state are available. For some of these tests limit states should not be reached, the test reported by (15) had to be terminated for technical reasons before the limit state was reached. Consequently, the design concept concerning internal and external stability cannot be verified yet with the help of measurements from large scale tests. On the other hand, however, some of the available data indicate that the ultimate surface loads are strongly underestimated (cf. (15); (13)). This problem will be discussed in the next section.

The data from the first 3 tests, listed in table 3.1, correspond to service states. These data can be used for verification of the predictions for the earth pressure and maximum tensile force corresponding to internal stability (cf. section 2.1.2).

Finally, the predicted and measured displacements are compared.

3.2.1 Limit state. Several results from large scale tests indicate that the ultimate surface loads are considerably larger than the predicted ones (15); (13). The discussion of this contradiction between experiment and prediction will be based on the experimental and theoretical findings of Wichter et. al.(15).

Figure 3.2 shows the structural details, the position of the surface load, and the expected failure surface (cf. chapter 3.1.1) for the retaining wall. The test was carried out with cohesive fill material: $\phi = 21.5°$, $c = 40\ kN/m^2$, $\gamma = 20\ kN/m^3$. A Stabilenka 200-type reinforcement was used.

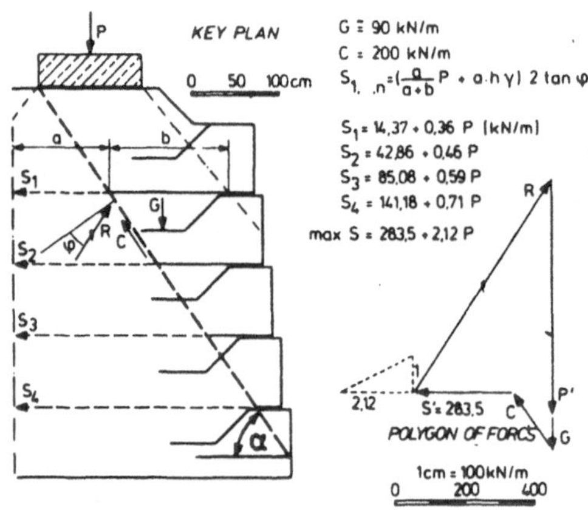

FIGURE 3.2. Prediction of ultimate load and tensile forces: Situation, slip surface, polygon of forces.

At the end of the test, a surface load of P = 500 kN/m was applied, but the ultimate state was not reached under this load.

Wichter et. al.(15) used the failure mechanism from figure 3.2 to predict the ultimate surface load. For the surface load a load distribution was assumed. Consequently, the total pull-out resistance (cf. S_1 $-S_4$ in figure 3.2) can be calculated as the sum of the following terms: A constant term due to gravity loads and a term which depends on the surface load. The formulas for the pull-out resistances are given in figure 3.2. Furthermore, in figure 3.2 G stands for the self-weight of the failure wedge, C for the cohesion-force on the failure plane and R for the resulting friction force on the failure plane. For the calculation of the ultimate surface load, the resulting pull-out resistance (max S = 283.5 + 2.12 P; cf. figure 3.2) was assumed. Formulation of the equilibrium equations and solving the equations for the unknown surface load P leads to the following result: Only for $P < 0$, i.e. tensile forces as surface loads, equilibrium can be reached. Wichter et al. (15) gives the following interpretation for this result: The reinforcement layers (for the assumed pull-out resistance) at the limit state cannot be pulled out all at once , i.e. there must be a so-called mechanism of intrinsic resistance.

On the other hand, each of the tensile forces S_1 to S_4 cannot exceed the tensile resistance of the geotextile (200 kN/m). Based on the information in figure 3.2 the pull-out resistance is greatest at position S_4. Wichter et al.(15) conclude from this that the tensile strength is first reached in position S_4 and then assume S_4 = constant = 200 kN/m. With this assumption a value of ultimate surface load of P = 83 kN/m can be calculated; i.e. it is very much smaller than the value observed experimentally (P = 500 kN/m). In addition, in the test the limit state was not reached. Thus there is a big contradiction between predicted and observed results.

This contradiction, however, is caused by a statical misinterpretation. Assuming max S (cf. figure 3.2) for the resultant pull-out resistance, the polygon of forces can be closed, if an additional force, the so-called "cause of failure-force" F_f (cf. (6)), is introduced (cf. figure 3.3). The polygon of forces, now, can be interpreted as follows: Failure occurs for a reduction of the pull-out resistance by the value of the force F_f, or, under the given gravity and surface loads the structure is stable, and an appropriately defined factor of safety, for the "pull-out case", will have a value greater than unity. This does not contradict the experimental findings.

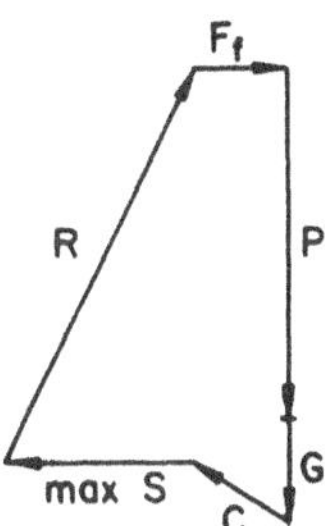

FIGURE 3.3. Polygon of forces, failure caused by decrease of pull-out resistance.

Also, failure caused by exceeding the tensile strength of the reinforcement has to be examined, which will lead to a different value for the factor of safety.

For design purposes, the following problem has to be solved:

1. Determine, for a given surface load P, the resultand tensile force of the reinforcement, which corresponds to the limit state.

2. Choose the material and geometrical data of the reinforcement in such a way that neither the tensile strength nor the pull-out resistance is exceeded by the predicted tensile force.

The solution of problem 1 leads to the following equation for the tensile force:

$$S_R = -C\cos\alpha + tg(\alpha - \phi)(P + G - C\sin\alpha) \tag{3.1}$$

α : shear-band inclination.

Using equ. (3.1) with the input-data: C = 200 kN/m, α = 55°, ϕ = 22.5°, P = 500 kN/m (largest applied load in test), G = 90 kN/m (self-weight of the failure wedge), S_R = 156 kN/m can be calculated for the resultand tensile force. This tensile force is even smaller than the tensile strength (200 kN/m) of a single reinforcement layer and very much smaller than the maximum available pull-out resistance of max S = 1343 kN/m. For the applied surface load of 500 kN/m, the limit state, therefore, could not be expected. With the help of the measured strains (cf. figure 3.1), back-calculation of the resultand tensile force leads to a value of approximately 100 kN/m. The measured (100 kN/m) and predicted (150 kN/m) tensile forces are at least not inconsistent.

Instability because of external failure is predicted for a surface load of 1510 kN/m (cf. equ. (3.1), if the tensile strength is assumed for the tensile forces in all reinforcement layers. This ultimate load for the retaining wall, however, can never be reached, because for P $\approx$ 500 kN/m base failure under this surface load occurs (cf. (15)).

For a suitable interpretation of the statics, therefore, the contradiction between experiment and prediction can be removed. Consequently, rigid body failure mechanisms are well suited for investigations of the stability of geotextile reinforced retaining walls. The last statement is also supported by the experimental and theoretical results reported by (8).

An open question, however, is the separation of the resultand tensile force into the single reinforcement layers. From measurements of the geotextile forces, corresponding to common service states, it is known that the tensile forces are increasing from the bottom to the top of a wall. Consequently, it can be expected that failure occurs first for the upper reinforcement layers (lowest pull-out resistance). This, however, should be incorporated in the formulation of safety requirements. This problem has not been treated yet in theoretical or experimental investigations.

3.2.2 Earth pressure distribution. The theoretical earth pressure distribution incorporated in the presented design concept was chosen with the help of the reported data from John (10) (cf. figure 2.6). More complete published data from experiments, which could be used for verification of the assumed distribution, are not yet available. The few additional data (cf. (1)), however, indicate the following:

- Near the top of a wall the active pressure can be assumed,
- Near the bottom the earth pressures are much smaller than in the active case.

The results of Berg (1), therefore, are consistent with the experimental findings of John (10) (cf. chapter 2.1.2). For verification of the assumed distribution more experimental data are needed.

3.2.3 Maximum tensile forces in the reinforcement. Very little experimental data are available for the distribution of tensile forces in the reinforcement (cf. table 3.1). In addition, the data reported by Wichter et. al. (15) were gained from a large scale test with cohesive fill. The presented design concept, however, is still restricted to cohesionless fill materials. For the discussion of tensile forces, therefore, only the experimental findings from (16) can be used (cf. table 3.1).

These data were gained from a retaining wall of 7 m height with a steep (1:0.2) face. The vertical distance of the polymer reinforcement layers is Δh = 1 m. Soil data: $\phi = 45°$; $\gamma = 14.4$ kN/m^3. The measured stresses and strains are due to gravity loads.

With the help of equ. (2.2) the following prediction for the maximum tensile force can be made:

$$Z = 0.9 \cdot 14.4 \cdot 7 \cdot 1 \cdot tg^2\,(45° - 45°/2) = 15.6 \text{ kN/m}.$$

The maximum measured strain was 0.3 %. The corresponding tensile force can be expected to be within the range 2.9 kN/m - 6.9 kN/m (cf. 16). A more accurate back-calculation of forces is for several reasons (small strains, inaccurate data for strain rates) not possible. The allowable tensile force (tensile force divided by a factor of safety) is 31.4 kN/m; i.e. the measured force is considerabley lower than the allowable tensile force.

With Z = 15.6 kN/m from equ. (2.2), the measured force is overestimated by a factor of 2.3 - 5.3. Reduction of the friction angle - as is common practice - by a partial factor of safety of 1.5 would lead to a overestimation by a factor 4.2 - 10.1. At least for cases without surface loads it can be expected that the predicted maximum forces are conservative.

For another large scale test (cf. (3)) very low strain values ($<$0.4 %) are also reported, whereas these values correspond to common service conditions. Under large surface loads (P = 500 kN/m) Wichter et. al. (15) measured a maximum strain of 1.5 %. For the last mentioned test cohesive fill material was used.

For all of these large scale tests the measured strains were markedly smaller than the ultimate design strains, assumed to prevent creep. The prediction of maximum tensile force, therefore, should be improved.

On the other hand, with the more sophisticated methods (for example FEM with accurate constitutive equations for soils and geotextiles, including nonlinear effects) needed for better prediction of stresses and strains, only little practical experience has been made and/or the application is expensive. Thus the application of more sophisticated methods cannot be expected in the near future. For this reason the common design practice should be improved with the help of experimental investigations. The aim of such investigations should be a reduction of the vertical spacing of the reinforcement layers. To attain this goal, reduction factors for the calculated maximum tensil forces should be defined with the help of experimental data.

3.2.4 Distribution of vertical stresses. Experimental results for the vertical pressures are reported by (3). In figure 3.4 the measured vertical pressures are plotted versus the horizontal distance from the wall face. For the location of the pressure cells an overburden pressure of 62 kN/m^2 is reported, i.e. the pressure cells were located at depths of 3-4 m.

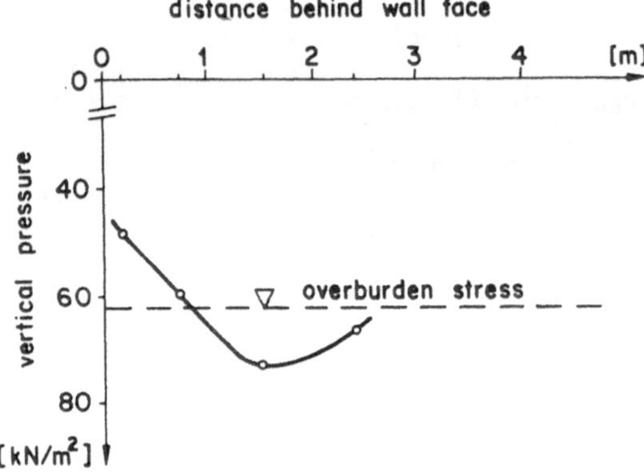

FIGURE 3.4. Measured vertical pressures (Caroll and Richardson, (3)).

As mentioned before in section 2.1.2, in the vicinity of the wall face the vertical pressures are much lower than the overburden pressure. Assuming the length of the reinforcement to be 40 % of the height of the wall (cf. (8)), for reasons of external stability, the reinforcement would end 1 - 1.2 m behind the wall face, and the vertical pressure over this area is substantially lower than the overburden pressure (cf. figure 3.4). This effect is especially pronounced in the vicinity of the bottom of the wall. Gudehus and Schwing (8), therefore, neglect the reinforcement layers, from the bottom up to 10 % of the height of the wall, in stability calculations. This foregoing, therefore, is consistent with experimental data.

3.2.5 Displacements. As mentioned before, the conventional design practice makes use of empirical displacement estimates:

- Horizontal displacements of the top of the wall in the order of magnitude of 1-2 % of the height of the wall after fade away of creep.

- By restriction of the design tensile force, as a certain fracture of the tensile strength, fade away of creep can be assumed to occur within a few months (cf. (16)). No prediction can yet be made for the elapse of displacements versus time.

For the discussed construction method the mentioned displacements of 1-2% are not critical at all. Failure because of creep has not been reported yet. On the other side, however, there is no experience for the behaviour of reinforced retaining walls with respect to full service life (50-100 years).

Gudehus and Schwing (8) simulated the case of an overstrained (8 %) reinforcement layer in model tests. For the displacements a linear increase with respect to time was observed. Failure can be expected to occur very soon. This model test result demonstrates the importance of adequate restrictions of creep deformations.

On the other hand, as already explained in detail in section 3.2.3, for conventionally designed reinforcement layers the strains can be expected to be less than 1.5 %; i.e. for such small strains creep is effectively prevented (this holds at least for the short term experience with this construction method).

To the best of our knowledge, long term displacement measurements (more than 5 years) are not available yet. Meanwhile, Wichter (14) started longterm displacement measurements for a retaining wall of 6 m height. Initially displacements are registrated every week, then once every 2 months, and after a 2 years period once every 2 years. Parallel to the displacement measurements, ageing of the reinforcement is investigated with the help of so called control strips, embedded into the structure. The described investigations are under work, results have not yet been published .

3.3 Conclusions and proposals concerning future theoretical and experimental work

The conclusions are presented in the sequence of the subsections of section 3.2.

The experimental investigation of limit states should primarily be based on model tests, as demonstrated by Gudehus and Schwing (8) for the standard retaining wall shown in figure 2.1. Additional model tests should be performed for investigation of failure due to surface loads. For verification (assumptions for the rules of similarity) of the model test results only few large scale tests are needed.

Only little experimental data are available for the earth pressure distribution. For verification of assumed distributions, therefore, additional data from large scale tests are needed. Force measurements in connectors are best suited for investigations of the earth pressure distribution.

For the maximum tensile force in a reinforcement layer, corresponding to the limit state of external stability, no experimental data are available yet. In particular with the help of testing results the problem could be solved, how the resulting tensile force (derived from a limit state equation) is distributed into the single reinforcement layers. Also consequences for the statement of safety requirements from such measurements can be expected (reinforcement pulled out collectively?; consequences resulting from failure of a single reinforcement layer?).

For the maximum tensile force, corresponding to internal stability, test results are available. The measured forces are overestimated by the prediction. Consequently, the number of necessary reinforcement layers is overestimated, leading to higher expenses. With the help of available experimental data and additional large scale tests correction factors for the predicted maximum tensile force could be defined.

For predicting the pull out resistance, good knowledge of the vertical stresses is essential. The few experimental data to this subject indicate that the vertical stresses, in particular in the vicinity of the bottom of the wall, are essentially lower than the overburden stress. For a better understanding of this problem, more experimental data from large scale tests are needed.

Displacement measurements in particular should be performed in the long term range with respect to the questions concerning ageing of geotextiles.

For the derivation of partial safety factors, more theoretical and experimental work has to be done:

- Basic statistic data from experiments (pull-out resistance),

- Back-calculation of the reliability of well performing retainig structures with the aim of calibration of the applied statistical methods,

- Partial safety factors for loads and material strengths.

4. CASE STUDIES

For the following case studies, 3 typical retaining walls, all located in Switzerland, have been selected.

For construction of a new communication road between Chiasso and St. Gotthard, a temporarily road had to be built on a steep hillside. As solution, a geotextile reinforced retaining wall was chosen (cf. figure 4.1). Design was based on internal stability (assumption of active earth pressure). Under the calculated tensile forces the design strains of the reinforcement were less than 5 %. Vertical displacements have been measured over a 2 years range (cf. (11)). The largest displacement was 18 mm.

For construction of a parking lot in St. Gallen, the retaining wall from figure 4.2 was chosen as construction method (12). The greatest height of the wall is C m.

The retaining wall from figure 4.3 was used for construction of a new railroad track (9). The geotextile reinforced wall was designed similar as an anchored wall. For more details concerning the fill material and the reinforcement see figure 4.3. The wall is not instrumented.

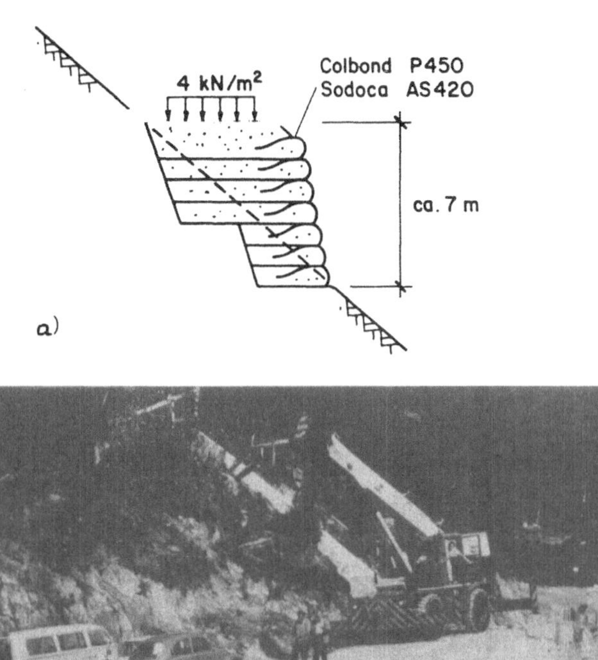

b)

FIGURE 4.1. Temporary road: a) Cross section, b) Road after construction

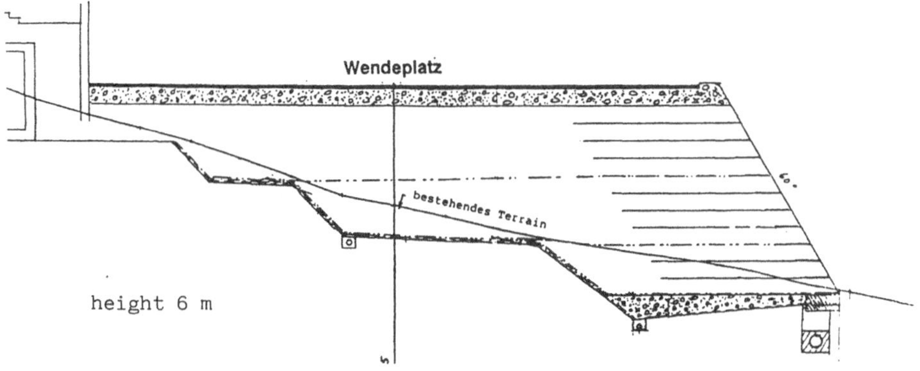

FIGURE 4.2. Parking lot

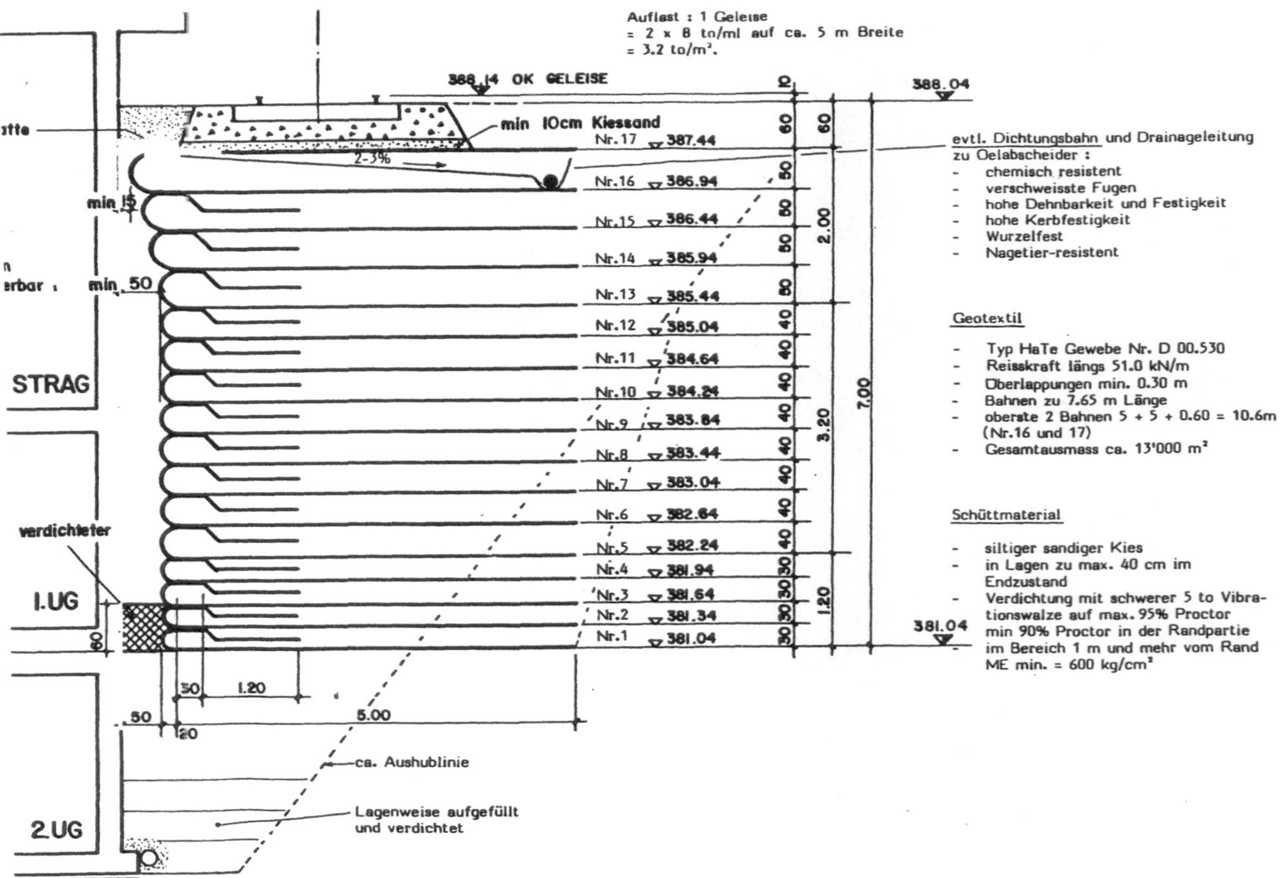

FIGURE 4.3. Cross section of geotextile reinforced retaining wall

5. SUMMARY

In this paper mainly the knowledge, which can be gained from large scale tests for the design of geotextile reinforced retaining walls, is discussed. This discussion is based on published design concepts and instrumented data. Proposals for future theoretical and experimental work are made.

The discussion of instrumented data is based on a design-proposal published by Gudehus and Schwing (8). The limit analysis design incorporated in this concept is verified by model tests. Partial factors of safety have been derived by the above mentioned authors by application of probability methods. This concept holds for cohesionless fill materials. Concentrated surface loads (strip loads) are not yet incorporated within this concept.

On the basis of the above briefly explained design concept, the evaluation and interpretation of instrumented data is discussed and applied on published data. The following results were derived:

- Ultimate loads, derived with the help of rigid body failure mechanisms, are consistent with observations from large scale tests. Reported inconsistents between predicted and measured loads (cf. (15)) can be shown to be a result of a static misinterpretation.

- For investigations of limit states model tests are better suited than large scale tests.

- The measured tensile forces (of the reinforcement) are overestimated by the prediction.

- As suggested by Gudehus and Schwing (8), the vertical stresses, in particular in the vicinity of the bottom of a wall, are remarkably smaller than the overburden pressure. This should be considered for estimates concerning the pull-out resistance.

The available instrumented data from large scale trials are insufficient for verification of the design concept. The following additional data are needed:

- Distribution of maximum tensile forces (with respect to depth) for the reinforcement layers in limit states (external stability),

- Earth pressure distribution,

- Maximum tensile forces corresponding to internal stability,

- Distribution of vertical stresses (with respect to depth),

- Long term displacement measurements (ageing of geotextiles),

- Basic statistic data, for example distribution and corresponding parameters for the pull-out resistance.

Future theoretical work should be concentrated on the formulation of safety requirements (partial factors of safety) needed for application of limit analysis design.

REFERENCES

1. Berg, R.R., Bonaparte, R., Anderson, R.P. and Chouery, V.E.: Design construction and performance of two geogrid reinforced soil retaining walls. Proceedings of the Third International Conference on Geotextiles, Vienna, Austria, 1986.

2. Breitschaft, G. and Hanisch, J.: Neues Sicherheitskonzept im Bauwesen aufgrund wahrscheinlichkeitstheoretischer Ueberlegungen - Folgerungen für den Grundbau unter Einbeziehung der Probennahme und der Versuchsauswertung. Proceedings of the Baugrundtagung, Berlin, 1980.

3. Caroll R.G. and Richardson, G.N..: Geosynthetic reinforced retaining walls. Proceedings of the Third International Conference on Geotextiles, Vienna, Austria, 1986.

4. Dutch/Enka Method. Calculation method to determine the internal stability of reinforced earth structures. Unpublished report of the Enka Industrial System, IEN 2.297 Ris/At., 1985.

5. Geotextilhandbuch. Published by Chemie Linz AG. Austria, 1986.

6. Goldscheider M.: Standsicherheitsnachweis mit zusammengesetzten Starrkörper-Bruchmechanismen. Geotechnik 2, 130-139, 1979

7. Gudehus, G. and Gässler, G.: Soil-Nailing - Some aspects of a new technique, ICSMFE X, Proc. Vol. 3, Sess 12, Stockholm, 665-670, 1981

8. Gudehus, G. and Schwing, E.: Standsicherheit kunststoffbewehrter Erdbauwerke an Geländesprüngen. Proceedings of the Baugrundtagung, Nürnberg, 1986.

9. Jaecklin, P.: Unpublished data - Jaecklin Consulting Eng., Ennetbaden, Switzerland, 1986.

10. John, N.W.M.: Geotextile reinforced soil walls in a tidal environment, Third International Conference on Geotextiles, Vienna, Austria, 1986.

11. v. Krannichfeldt: Unpublished data. Uffioio Strade Nazionali, Ticino, Switzerland, 1986.

12. Rüegger, R.: Unpublished data. Rüegger AG, Consulting Eng., St. Gallen, Switzerland, 1986.

13. Werner, G. and Resl, S.: Stabilitätsmechanismen in geotextilverstärkten Erdstützkonstruktionen. Third International Conference on Geotextiles, Vienna, Austria, 1986.

14. Wichter, L.: Geotextil-Erde-Stützwand als Dauerbauwerk. Bautechnik 62, 289-291, 1985.

15. Wichter, L. Risseeuw, P. and Gay, G.: Grossversuch zum Tragverhalten einer Steilwand aus Gewebe und Mergel. Proceedings of the Third International Conference on Geotextiles, Vienna, Austria, 1986.

16. Yamanouchi, T., Fukuda, N. and Ikegami, M.: Design and techniques of steep reinforced embankements without edge supportings. Third International Conference on Geotextiles, Vienna, Austria, 1986.

17. van Zanten, V.R.: Geotextiles and Geomembranes in Civil Engineering.

REVIEW OF SESSION

CASE HISTORIES AND FULL SCALE TRIALS

Data obtained from instrumented full scale polymer reinforced structures are of immense value in understanding the behaviour and verifying the design concepts for such structures. However, there are two major difficulties facing researchers, firstly shortage of funds for full instrumentation of trial structures and secondly lack of an agreed list of relevant case history data and a common format for their presentation. Consequently, as Christopher pointed out, of an estimated 200 slopes and walls constructed in the USA and Canada using polymeric materials, data on only 46 projects could be collected. Of these only 6 projects were fully instrumented. He presented these data in four comprehensive tables covering general information; design information; actual properties of the reinforcements and design methodology; and project instrumentation and monitoring information. This work represents a substantial step forward and workers in this field should be encouraged to extend such information so that a comprehensive data bank of case histories is constructed. It is important however, that when consulting such data the user is aware of all the parameters and assumptions used in the design.

Delmas has provided valuable data on full scale geotextile reinforced structures in an attempt to determine the influence on their behaviour of a) the method of placement of the formwork and the facing; b) the use of clayey soil and c) the long term behaviour of the geotextile. Five instrumented structures were described in detail. The results illustrated, perhaps not surprisingly, that the tensions in the reinforcing sheets were significantly influenced by the placement method and by the nature and timing of the stresses to which the structure is subjected. He recommended the use of placement methods which not only facilitate the building of the structure but also ensure the optimum tensioning of the geotextile. By monitoring the pore water pressures in two structures he concluded that the use of soils having a large percentage of fines posed no special problems provided that the placement and compaction do not cause build up of pressure and the geotextile is capable of dissipating any increase in the pore water pressure. This is important to ensure optimum soil-geotextile friction. It was interesting to learn that ageing of geotextiles, which was examined by testing samples 15 years after placement, caused loss in strength and in this particular case a gain in stiffness.

Studer emphasised the importance of instrumented large scale tests in selecting appropriate design procedures and safety factors, he highlighted the lack of reliable comprehensive data in the literature and presented a detailed list of the required data. He argued however, that large scale tests are not suitable for the determination of the critical geometry and

P. M. Jarrett and A. McGown (eds.), The Application of Polymeric Reinforcement in Soil Retaining Structures, 339–340.

load and that model testing should be used for the systematic investigation of the behaviour of reinforced walls in limit state and the determination of the failure mechanism. It is interesting to note that although the measured vertical stresses at the base in the vicinity of the wall were significantly smaller than the overburden pressure, it is accepted that the average value of the vertical stress at the base is equal to the overburden pressure.

Leflaive presented an interesting contribution regarding full size testing of Texsol walls. It might be helpful to point out that Texsol is a mixture of synthetic fibres and granular soil prepared in the field using special equipment. The walls were 3 m high and surcharged up to failure. The main and intriguing conclusions from these tests were that the failure was not characterised by a shear plane but occurred by the walls overturning and that the deformations were extremely small up to almost 90% of the failure load. It will be interesting to see the results of the planned centrifuge tests on this material. Jones pointed out certain geometric similarity between this technique and that used by the Victorians of placing stones on steep slopes. After 100 years or so, which is a considerable period of time, these stones needed replacement.

Leflaive drew attention to a very interesting observation regarding the determination of the factor of safety. It was found that when a yarn of the polymer was statically loaded with 60% of its failure load it fails in about two days. On the other hand, a Texsol sample loaded with 60% of its failure load for one year did not fail. This implies that the yarns in the Texsol sample are not loaded at 60% of their failure load but reach this value only when the load on the sample is close to the failure load. Although these findings seem consistent with the fact that strains in Texsol are very low prior to failure, they make the determination of the factor of safety quite difficult.

Berg gave a detailed review of the design, construction and performance of a 5.3 m high and 99 m long polymer-geogrid reinforced soil seawall. The wall is exposed to severe climatic conditions involving wave action and freeze-thaw cycles. Two sections of the wall were instrumented in order to monitor movements of the wall face and the reinforced soil mass. In addition attempts were made to measure the tensile loads in the geogrid layers. This is an extremely valuable example of a well documented case history and Mr. Berg's plan to continue monitoring the performance of the wall in future will certainly add to its value.

Scott reported on a large fully instrumented trial embankment to be constructed in Devon, Alberta in the summer of 1987. A series of sections are to be reinforced with various grids in an embankment 12 m high built from cohesive fill at a slope of 1:1.

In conclusion, this session has demonstrated the value of full scale trials and highlighted the difficulties encountered in monitoring their performance. It also emphasized the urgent need for developing standard guidelines on testing methodology and data collection. This information should be well documented in a comprehensive data bank of case histories.

Evaluation of Material Properties

THE PURPOSE OF MATERIALS EVALUATION AND RECOMMENDATIONS FOR MATERIALS TESTING

E. LEFLAIVE, Laboratoire Central des Ponts et Chaussées, France

A. MCGOWN, University of Strathclyde, Scotland

INTRODUCTION

Reinforced soil wall structures comprise a number of components and materials; the soil backfill, reinforcements, facing units, joints and connections, foundation for the facing units and the subsoil on which the structure is built. All are important to the efficient performance of the structure and should be properly evaluated by appropriate testing. However, it is the use of polymeric reinforcements which is the specific point of interest for this Workshop thus this paper deals only with the evaluation and testing of these. Further, detailed descriptions of test methods are not included as they are not considered to be the points at issue, rather it is the choice of which test methods that should be used that is critical.

The report is written in two parts: the first deals with the purpose of materials evaluation and the second provides a recommended approach to materials testing.

PART A: BACKGROUND - PURPOSE OF MATERIALS EVALUATION

For soil retaining structures reinforced by geotextiles the question of evaluation of the reinforcing materials has to be examined with reference to three different aspects: design concepts, construction practices and testing requirements for other end uses.

1. Design Concepts

The design of any reinforced soil retaining structure requires the study of both its external and internal stability.

For the purposes of external stability calculations, the retaining stucture is generally considered to be a rigid body, therefore only the properties of the surrounding soil are generally required to be known, apart that is from the mass (weight) of the structure. There is, however, a growing awareness that the overall deformability of the structure will influence such factors as base pressure distributions and lateral earth pressures and that these may in turn influence the external stability of the structure. In such cases the properties of the soil and reinforcement forming the stucture may have to be taken into account. This is not

P. M. Jarrett and A. McGown (eds.), The Application of Polymeric Reinforcement in Soil Retaining Structures, 343–355.

current practice and is only mentioned here to point out the possiblity of future developments. Thus external stability will not be discussed further in this report.

In contrast, a full knowledge of all the materials properties is certainly necessary for the calculation of the internal stability of reinforced soil retaining structures. The specific data required for the reinforcing of materials depends very much on the analytical approaches taken. These may be classified as follows:

1.1 Discrete Constituent Materials Approach

The principle of this approach is that an analytical model of the system is established within which discrete sets of test data are used for the in-isolation properties of the reinforcements and for their interface interaction characteristics with the soil.

This philosophy is quite common in engineering. In the field of geotextiles, it has been very clearly expressed by Studer (1982) on behalf of a Swiss group, at the Second International Conference on Geotextiles:

"The basic idea of this concept is as follows: if the properties of the soil around the geotextile are well known and if a model of the soil-geotextile interaction does exist it is always possible to investigate the behaviour of the soil-geotextile system analytically at a later stage. Thus an analogous concept like that used for composite structural materials, (e.g. reinforced concrete), is applied. However, in the field of geotextiles, analytical models are still at an elementary stage. Nonetheless, it is the committee´s belief that this is the only physically correct procedure from an engineering point of view".

It follows from this approach that all specifications for geotextiles used as reinforcements should:

"rely on standard tests on the geotextile only, in spite of the fact that a geotextile is never used as a pure construction material but always in conjunction with soil."

It may be noted that, strictly speaking, frictional characteristics have to be measured with the reinforcement in contact with another material, whether it be a soil or any other frictional surface; but as long as the friction test remains a standard test with standard components the general approach is not modified.

This paper from Switzerland was presented at a Session 2B of the Second Internation Conference on Geotextiles, Las Vegas, (1982), which was devoted to International Standards. Apart from one paper from Finland, dealing only with the use of non-woven geotextiles in road construction, and one Dutch paper on vertical drain specifications, six other papers gave general considerations to the specification and testing of geotextiles as discrete constituent materials and listed recommended testing procedures. These papers were from U.K., Germany (BRD), France, U.S.A., and from two international societies; PIARC and RILEM.

The German, French, PIARC and RILEM papers essentially supported the Swiss paper and presented testing methods only for geotextiles in-isolation. However, both the U.K. and U.S.A. papers mentioned that in some circumstances a choice may have to be made between data from in-soil and in-isolation testing. The paper from U.K. highlighted this to the greater extent.

1.2 Equivalent Composite Materials Approach

The principle of this approach is to replace the system formed by the soil and its reinforcement by an equivalent homogeneous material. In practice it can be applied only in situations where the distance between reinforcing elements is very small compared to the retaining structure dimensions, in other words, the scale at which the composite is formed has a different order of magnitude from that of the structure. In geotextile technology, "Texsol" and "Mesh Elements" systems have been proposed that satify this condition and are or could be used for retaining structures. Such systems have not been included in the testing program associated with the Workshop, nevertheless they are an application of polymeric reinforcement in soil retaining structures.

Soil-reinforcement interactions (friction, local stress effects, structural interactions, etc.) are automatically incorporated in the testing employed for the equivalent composite materials approach. For this reason, this approach, when it can be applied, is the simplest one. Indeed, it is nothing but a common application of mechanics of solids to real materials, albeit that the composite materials are neither continuous nor homogeneous.

1.3 System Simulation Approach

A third approach to the design is the system simulation approach. In some ways it can be considered as an intermediate solution to be used when both the discrete constituent and the equivalent composite materials methods are neither applicable nor available. The objective of this approach is to simulate the system behaviour, in part or as a whole, and to use it to predict the behaviour of a family of structures having in common a number of features with the simulated system. Again, it is a very common design approach when engineers have to give answers to complex problems where all the phenomena involved are not fully understood and cannot be easily measured separately.

In practice simulation testing for reinforced soil structures may be carried out in two ways. One is to test the reinforcments in the laboratory when incorporated into the soil and stress environments in which they will be used in the structure. This is done in order to obtain results that are more directly representative of their actual operational behaviour. The other is to test prototype structures, either at full size or at model scale. These model tests must not be confused with research work on models as it is intended that these will provide data which can be used directly in the design of an actual project.

The system simulation approach was strongly recommended by Murray and McGown (1982). In this paper, geotextile tests are divided into two groups: Index Tests and Design Data Tests. In-isolation tests and in-soil tests using standard soils, form the index tests group, which are considered to be valid only for quality control, (during and after manufacture), and to have little validity for design purposes. Design data tests it is suggested can be obtained from in-soil tests using the soil of the project and from prototype trials. However, it is recognised that developments and experience may make possible the use of index test results in design so that in the future, the differences between this simulation approach and the discrete constituent materials approach may greatly reduce.

2. Construction practices

The point here is that the analytical approaches used in the design process should be viewed only as the theoretical operation leading to an overall understanding of the stability of the structure and to its basic dimensioning. In contrast, practical engineering requires that what happens during construction must be taken into account as it may affect the resulting engineered structure in many ways, among which the material properties are an important aspect. Therefore materials evaluation must indicate to what extent reinforcing elements may be modified or damaged by planned or unexpected events during the building of the structure.

Thus, material testing must give the engineer information on possible material modifications due to construction procedures, and the tests required to do so may be quite different from those envisaged for calculation needs. Theoreticians may sometimes consider these tests to be of minor importance and include into the Factor of Safety such uncertainties on the actual material performance. However, experience shows that such material modifications must be fully assessed and independently allowed for. Failures in structures have often resulted from material damage due to poor field workmanship or unexpected building conditions.

Once in the ground the reinforcements may then suffer mechanical and chemical degradation. For geotextile reinforcements the most obvious risks are mechanical degradation due to accidental tensioning or tearing, puncturing and abrasion. In some cases a possible reduction of performance may also result from loss of permeability due to clogging where the reinforcement is also expected to have a draining effect. Problems related to joints and connections may also have to be considered from the construction point of view. It is also possible that degradation may result from changes imposed on the polymeric constituents of the reinforcement due to temperature changes, chemical and biological attacks, exposure to light, fire or even animal attacks. When buried in most soil structures the resistance to degradation from all of these agencies is considered to be very high but the need to develop test methods to justify this confidence must be one of the most urgent requirements in this field of study.

Another question to be discussed, relates to the behaviour of the reinforcements if they are not laid within geometrical tolerances. Avoiding waves or folds leading to a "Slack" effect may be viewed essentially as an inspection problem, but it may also be related to material properties and so special tests (e.g. flexibility). The possibility that a geotextile can conform to a very uneven surface can be measured by flexibility tests to a certain extent but more fundamentally by evaluating its ability to change its area during different types of tensile tests. This leads to an interesting question, viz., should the designer look for a fairly rigid reinforcing material to obtain a good laying geometry or for a geotextile that can easily conform to an uneven surface, to insure good contact and so interfacial friction conditions? This choice may need to be clarified by simulation testing. It is unlikely that standardised index testing will assist in this.

3. Testing Requirements of Other End Uses

Testing methods must be appropriate for the particular application of geotextiles under consideration, i.e. soil reinforcement. But geotextiles are also used for other purposes and it would be unreasonable and inefficient in practice to devise independently as many sets of testing methods as there are particular applications. Since properties that are useful for reinforcement design are also of interest in many other applications, although with different degrees of importance, it may be that a measure of compromise will have to be reached between the various field of application to limit the number of testing methods adopted. However, it is realised that this is not an easy goal to achieve because many aspects have to be taken into account. Most certainly, potential users of these materials must not be discouraged by an overcomplexity of test methods. Nonetheless, testing should be rigorous (i.e. time consuming or expensive), if its aim is to determine once and for all the technical characteristics of a given product. Whereas, it must not be time consuming or expensive if it is for identification purposes only. If tests have to be performed for each project then this is where a compromise must be found.

To avoid confusion between the different types of tests, definitions proposed by RILEM (1985) are restated here:

> "Test can be classified in 3 categories:
> - control tests in factories
> - reception or identification tests
> - suitability tests."

"A <u>control test in a factory</u> is a means of following the production of a geotextile. The manufacturer is responsible for the choice of his method.
The results of these controls are confidential.
External organisations can interact in two different ways with these procedures:

1) to recommend the control of certain parameters because they appear to be significant. (This would simply be in the form of a recommendation).

2) to establish agreement procedures or quality assessment certification to verify that a manufacturer possesses all the necessary manufacturing knowhow and the material testing facilites for quality control."

"A <u>reception or identification test</u> is devoted to the identification of a product by the end user and involves identification of:

- raw material;
- manufacturing procedure;
- physical properties (components and final product).

These characteristics are given by the Manufacturers and are their responsibility. They may be best presented on technical cards and include data which can be very simply verified by Contractors and other end users. Obviously there must be a good agreement between Manufacturers and Contracters regarding the tests procedures employed to obtain these data."

"A <u>suitability test</u> is devoted to the selection of a product for a given end use. The test to be chosen depends on:

- the function(s) to be employed
- the analytical procedure to be used.

It must be noted here that this classification is independent of the different analytical approaches mentioned previously. It does not tell whether a suitability test is a simulation test, an in-isolation test or a test on the composite. Its usefulness is to avoid misunderstandings due to possible confusion between factory control, commercial and technical identification and description, and suitability evaluation for the design of a project."

An additional factor is that data may require to be presented in different fashions for different applications, for example:

(i) Information may be presented quantitatively (e.g. mass per unit area = 250 g/m^2) or qualitatively (e.g. the geotextile is made of polyester). The first type involves a measurement; the second requires a test to identify a quality.

(ii) Information may also be given by a test which may be quite specific or has a wide range of applications; in that sense a test may have an extensive value or be specific, a point discussed previously.

(iii) In view of a given function or behaviour, information given by a test may be directly applicable for a classification or a calculation, may have an indirect significance (e.g. whether the geotextile is woven or non-woven may be useful for interpreting tensile and friction test results), or may have no direct significance; in that case it still may remain useful to validate other data by cross-referencing.

This last point is important for the reason that good design and good material evaluation must reduce to a minimum the probability of making a mistake in the choice of materials. In this sense, some data may be useful in spite of the fact that they do not directly fit into a calculation method; therefore these data must not be neglected and must be used as long as they are easily obtained.

4. Summary

It may be stated that data on materials may interact in many different ways within the design process and that material evaluation must give to the Engineer sufficient information to enable him to:

a) check that all the data available are consistent and, thus, reduce the probability of an error on material identification and properties.

b) refer to existing projects built with similar materials and, thus, make use of experience gained in the past.

c) dispose of all data required by the reinforcement dimensioning method(s) to be used.

d) assess possible material modification, degradation and/or misuse due to construction conditions, and evaluate potential construction problems.

PART B: RECOMMENDED APPROACH TO MATERIALS TESTING

1. Identification

Identifying a number of possibly suitable reinforcing materials is the first step in the process of selecting a product for a given project. The identification of these products should include a commerical reference, name and number, a description of the product and a list of basic physical characteristics. The purpose of this, as mentioned previously, is to enable the Engineer to check that the data available is consistent with previously published data and to allow reference to past experience. Identification testing is best when it is rapid and easy to control, while giving the experienced Engineer a good picture of the product and allowing an assessment of its overall performance with sufficient accuracy for first stage design.

Identification testing for geotextile reinforcements should provide data on properties such as those shown in Table 1.

TABLE 1 IDENTIFICATION DATA FOR GEOTEXTILE REINFORCEMENTS

Mass per unit area
Nominal Thickness (or other relevant dimensions)
Tensile Strength and Strain to Failure (for method see Section 2)
Puncture Resistance
Tear Resistance
Abrasion Resistance

2. Reinforcement dimensioning data

There are two main groups of characteristics:

- tensile properties (short and long term loading)
- friction (or soil-geotextile interaction) properties.

2.1 Tensile Properties

2.1.1 Short Term Loading

There is a general agreement that the unidirectional constant rate of strain tensile test is the most appropriate test methodology to determine the short term tensile behaviour of geotextiles. A number of specific testing methods have been devised to approach as closely as possible the performance of a geotextile strained unidirectionally, i.e. with $\varepsilon_2 = 0$. The different methods applicable to geotextiles above vary according to:

- the system for the application of stress and strain
- the size and shape of sample
- the way of expressing test results.

Three main systems have been proposed for the application of stress and strain:

- cylindrical deformation by air or fluid pressure
- biaxial testing on cross-shaped sample
- uni-axial testing.

The first two systems are usually considered to be too cumbersome for commerical testing purposes. They are, however, of great interest for research testing because they make possible testing conditions difficult or impossible to obtain otherwise. Also the stresses and strains can be established in both principal directions, which is of basic importance for theoretical interpretation. On the other hand, they have some technical drawbacks, which are independent of their mechanical complexity, as follows:

Firstly, when cylindrical deformations are obtained on a rectangular sample submitted to a pressure on one side or by inflating a sleeve then:

- there is a difference of pressure between the two faces of the geotextile which is related to the strength of the material, the curvature of the sample during the test, sample size and geotextile deformability. This gradient of σ_3 cannot be avoided. On the other hand this type of system can easily be adapted to apply an additional pressure in the direction of the thickness of the material.
- if the deformation of the sample is close to cylindrical, ε_2 is small, but remains usually positive, contrary to the uniaxial testing where ε_2 is negative.

Secondly, although plane biaxial testing allows control of stresses and strains in the planar condition, it does not normally reach failure. These tests are therefore more of interest for modulus determination than for ultimate strength evaluation.

Thirdly, although testing with stress and strain measurement in both directions is necessary to study the mechanical behaviour of planar structures, little is known about the biaxial behaviour of such materials, particularly the effect of manufacturing technique on the interrelation between stresses and strains in the different directions of the plane. For example woven products are often considered to be highly anisotropic but it seems that this is not so much so when biaxial stresses are applied. Theoretical models of such orthotropic structures with tensile resistance only are not available. Even if one considers only resistance to deformation in the plane, neglecting bending and torsion, theory would be useful to correlate different testing conditions, such as tensile tests with different values of ε_2, and to analyse the relationships between traction, shear and area change.

As mentioned previously, present practice for geotextile reinforcement testing is to consider only uni-axial deformation and to approach this situation experimentally. This is done most simply by traction on rectangular samples in conditions such that ε_2 is small. In order to reduce ε_2 in non-woven products to a minimum, one solution is to use transverse brackets with pins opposing restraint of the material during testing. This system was developed by the University of Strathclyde and first reported on by Sissons (1977) for 200 x 200 mm samples and is used in Switzerland with 100 (width) x 200 (length) mm samples. Another approach is to confine the geotextile in-soil, but the effect of soil confinement has influences other than opposing lateral restraint. This mode of testing is discussed in a later section.

The most common solution for reducing the influence of ε_2 is to adopt a wide width sample, i.e. with width larger than the length. This approach is favoured by many engineers and researchers. Discussions on this test concentrate on the following points:

- width/length ratio
- rate of strain
- use of measured ε_2 values for correcting ε_1 values

It is very often stated that a width/length ratio of 2 (200 x 100 mm samples) is sufficient for most geotextiles, including non-wovens. The main advantage of using this is that it limits the force that the equipment must develop for strong materials, (as a first approximation force is proportional to sample width). However, it is considered by others that it is necessary to adopt a larger of width to length ratio say of 5, associated with a correction taking into account the measured value of ε_2 to approach the ideal situation of $\varepsilon_2 = 0$.

Applying this correction is equivalent to determining the area change of the sample during the test. It is found that the maximum area change is larger with a width/length ratio of 5 than with a ratio of 2. Corrected values of ε_1 are smaller than measured ε_1 values. Their significance is that they correspond to the maximum area change possible with the material and that shear deformation resulting in lateral restraint is compensated for.

In tests with width/length ratio of 2, measured values of ε_1 at a given percentage of failure, tend to be larger than with a width/length ratio of 5 because a larger shear deformation occurs but, at the same time, tends to be smaller because area change is smaller. Thus, these values are not really meaningful for geotextiles where large ε_2 values develop during traction. The ε_2 correction is not normally applied to test results obtained on samples with a width/length ratio of 2. In general, width/length strength values found with a ratio of 5, are 5 to 10% larger than with a ratio of 2 for non woven geotextiles.

It has been shown that for needle punched non-wovens that by using test specimens with a width/length ratio of 5 and making corrections to the measured values of ε_1, the load-strain behaviour obtained is very similar to that from in-soil testing with a width/length ratio of 2, a confining stress of 100 kN/m^2 and sand as the confining medium. However, the effect of soil confinement is known to vary with the soil type and applied stress level, thus no unique relationship exists between these two test methods.

The in-soil test method, like the tests using width samples with a length ratio of 5, is most applicable to needle punched non-woven geotextiles. In-soil testing on melt-bonded non-wovens produces data which is only a little different from that in-isolation, (the higher the degree of bonding, the smaller the effect), and there is almost no difference for woven geotextiles and grids or nets whose structure is aligned along and across the line of traction,although sample shape has some effect on elongation at failure.

The constant rate of strain at which the short term tests are carried out does vary from Country to Country but it is becoming more standardised and is most often now carried out at 10 per cent per minute. However, a recent proposal is that the test should bring the test specimen to failure in 2 minutes ± 10%. These rates of strain are too high for manual recording of test data and so automatic data recording is generally required.

The effect of the rate of strain during testing on the measured load-strain behaviour of geotextiles varies with both their polymer content and their structure. The behaviour of wovens, girds and nets is predominantly controlled by their polymer content and the effect of the rate of strain is significant whereas needle punched (and to a lesser extent melt-bonded) products, being structure controlled are relatively less sensitive to rate of strain, particularly at low strain levels. For the same reasons the effect of the temperature of testing is significant for wovens, grids and nets and less important for non-wovens.

2.1.2. Long Term Loading

Once again there is general agreement that unidirectional tensile tests should be used to determine the long term tensile behaviour of geotextiles, however, only a limited amount of long term testing has been carried out. What testing has been done has involved sustained constant loading at constant temperature on a number of different test specimens. The same discussion on sample width/length ratio and in-isolation to in-soil testing, applies to long-term testing as much as to short term testing and as before the most commonly adopted width/length ratio is 2, although for woven geotextiles much narrower test specimens are quite often used. Very few long term in-soil tests have been undertaken. Nevertheless, such tests on soil-textile composites as have been performed, show that a very different creep behaviour of the composite is obtained compared to the reinforcing element in-isolation.

Just as important as the test methodology for these long term tests is the method of presenting the test data. Very often the data is plotted as strain versus the log of time and a creep coefficient derived from the slope of the linear part of the plotted data. This requires that the initial non-linear part of the plot is ignored, however, this may represent the majority of the strains developed in the geotextile at that load level. Thus a material which exhibits very large strains initially but little strain thereafter, may appear to be a more efficient soil reinforcement than a material which has only limited initial strains and modest time dependent strains. For this reason the total strain from the time of loading should be measured and included in data presentation. A test methodology and method of data presentation which accomplishes this has been developed in U.K. between the University of Strathclyde and TRRL,

Murray and McGown (1987). It is this method which has been used to present the data for use in the predictions of the test walls which form the background to this Workshop.

Only by carrying out such long term tests can the long term tensile behaviour be correctly established but this type of testing is very time consuming and expensive. Thus as discussed in Part A it is not expected that this will be undertaken for routine purposes. Consequently it is most important that the data obtained from such testing be correlated very closely with the short term test data which will be routinely obtained in order that checks on the quality of materials can be properly established. Recent work suggests that checks on the short term ultimate strength are not a sufficient measure of material behaviour and that it is the entire load-strain curve that must be considered. This should be the subject of future studies.

2.2. Soil-Reinforcement Interaction

This is usually achieved by the development at the interface between the soil and the geotextile, but in certain products such as grids and nets interlocking between the soil particles may also develop. Two conflicting approaches exist for the determination of these interaction properties; pull out tests and direct shear tests, the choice of which test method to be adopted is often dictated in design methods. Much more work requires to be undertaken to relate these two test methods and to standardise on the specific test apparatus and techniques. Further the influence of such factors as soil type, compaction levels and methods, and reinforcements stiffness needs to be more fully investigated. Lastly, investigation of the influence of redistribution of stresses at the soil/reinforcement interface with time as the soil strains or as the reinforcement creeps or degrades, needs to be assessed.

3. Other Properties

Other properties which may be of critical significance to the long term operational behaviour of reinforcements may be discussed in two groups as follows:

3.1. Durability

The long term degradation of geotextiles is, as stated previously, one of the most obvious risks to their satisfactory long term behaviour. Unfortunately there is almost no agreement on suitable test methods to establish this behaviour. This does not mean that geotextiles will not perform satisfactorily or that many of the available, but unsuitable, tests should be adopted until suitable tests are developed. Rather it means that there is an urgent requirement for this aspect of materials evaluation and that new appropriate tests should be developed. Also monitoring of full scale structures should be urgently undertaken. As a first step in this direction the damage occurring to geotextiles during the construction stage should be determined.

3.2. Joints and Connections

Less obvious than durability but just as important to the long term behaviour of geotextile reinforcements is the behaviour of their joints (the linkages between reinforcements) and connections (the linkages between reinforcements and other components). Many different types of joints and connections exist including sewing, stapling, bodkins and glueing. The short term behaviour of joints has been reported on by Murray et al (1986) but no long term studies are known to have been undertaken either at laboratory or full scale.

Often joints can be avoided by careful dimensioning of reinforcements but connections are generally unavoidable. The various types of connections employed should be identified and appropriate means of measuring their properties in both the short and long term developed.

4. Summary

The testing requirements for the geotextile materials used as soil reinforcements have been identified but it has been shown that there are many areas where no suitable test methods presently exist. It would be most useful if in time a list of the testing requirements for geotextile reinforcements made together with the identity of both the available suitable test methods and test methods that require to be developed.

REFERENCES

1. MURRAY, R.T. and McGOWN, A.: The Selection of Testing Procedures for the Specification of Geotextiles, Proc. II Int. Conf. on Geotextiles, Las Vegas, Vol. II, pp. 291-296, 1982.
2. MURRAY, R.T. and McGOWN, A.: Geotextile Test Procedures: Background and Sustained Load Testing. Application Guide 5, Transport and Road Research Laboratories, Department of Transport, U.K. 12 pp. 1987.
3. MURRAY, R.T., McGOWN, A., ANDRAWES, K.Z. and SWAN, D.: Testing Joints in Geotextiles and Geogrids, Proc. III Int. Conf. on Geotextiles, Vienna, Vol. III, pp. 731-736, 1986.
4. R.I.L.E.M. Committee SM-47 (Synthetic Membranes): Report on Meeting at Milano, Italy, pp. 10, 1985.
5. Session 2B: International Standards, Proc. II Int. Conf. on Geotextiles, Las Vegas, Vol. II, pp. 291-333, 1982.
6. STRUDER, J.: Basic Principles Underlying the Swiss Guidelines for the Use of Geotextiles, Proc. II Int. Conf. on Geotextiles, Las Vegas, Vol. II, pp. 301-305, 1982.

REVIEW OF SESSION

EVALUATION OF MATERIAL PROPERTIES

When dealing with reinforced soil structures, knowledge of the behaviour of the different components of the structure is essential. This is gained by testing the materials either separately or in combination, using standard test methods and specially developed testing techniques. Due to the importance of this subject, it is not surprising that most of the published work deals with various aspects of material testing. Accordingly, and perhaps ironically, designers are now faced with the problem of choosing suitable tests from which the relevant design parameters are to be determined. This session deals with the wider aspect of these problems by examining materials evaluation which of course includes testing.

Leflaive started the session and stated that there are four purposes for material evaluation. These are a) to check the consistency of information in order to avoid mistakes in material selection, b) to enable the engineers to refer to past experience, c) to provide data for calculations, and d) to enable the prediction of construction problems.

He concentrated on the problem of obtaining the relevant data for calculations (or dimensioning). The specific data sought depend to a large extent on the design concept chosen. For example, in the "discrete constituent material" approach it is the test data relating to the properties of the individual materials together with the interface properties which are needed and used in the analytical model. Another design concept is the "equivalent composite material approach" which uses properties determined from tests performed on the mixture of soil and reinforcements. Although the validity of this approach is doubtful for multi-layered systems, it may be applicable to situations where the polymer is randomly mixed with the soil such as in "Texsol" and the "mesh element" system. Design parameters may also be obtained from model and prototype testing of reinforced structures and used in what is known as "system simulation" approach.

Leflaive raised the questions of materials evaluation with respect to construction damage, ease of laying, mechanical and chemical degradation, permeability, and construction of joints and connections. He stressed the need to develop test methods capable of assisting the engineers in assessing these aspects. It was interesting to learn from Bush that this type of evaluation is underway for geogrids. His brief presentation of the results obtained so far emphasised the value of such work in both design and construction.

McGown drew attention to the importance of testing the sub-soil. A compressible sub-soil can greatly influence the internal strains in the structure thus complicating the problem. He also pointed out the change

P. M. Jarrett and A. McGown (eds.), The Application of Polymeric Reinforcement in Soil Retaining Structures, 357–358.

which occurred in the facing elements and connections from the early compressible aluminum facings to present day large concrete units. These developments, he urged, should be taken into consideration when identifying the test requirements.

With regard to soil testing, McGown pointed out the importance of lateral (tensile) strains in reinforced structures and argued that it is therefore essential to study the lateral strain behaviour of a range of typical granular soils. He explained that the 10% criterion for tensile strains in reinforced systems originated from the fact that at about 10% lateral strain the angle of internal friction reaches its ϕ_{cv} value. This aspect of soil testing, particularly at low stresses, has received little attention so far.

When dealing with soil reinforcement interaction there is a lack of agreement whether the required design data should be obtained from pull out tests or shear box tests. Scott expressed the opinion that both tests are needed and presented details of a fully instrumented pull out test and a shear box test on compacted clay and geogrid. Milligan provided some valuable data regarding the interaction characteristics. He showed that for sheet reinforcements the results obtained from direct shear test and pull out test agree well, basically because the same interaction mechanism operates in both tests. This is not the case for grid reinforcement where the mechanisms are different due to the presence of opening in these materials. However, it is accepted that there is a lack in the understanding of the effects of stress level, soil type, compaction and reinforcement stiffness on the interface characteristics.

It was interesting to observe that when McGown explained the various methods for determining the stress-strain relationships of the reinforcements, the discussion became lively. Basically, he explained in detail the importance of testing wide width samples, the relevance of in-isolation and in-soil tests and the effects of the temperature and the strain rate on the measured parameters. He strongly recommended the use of sustained load creep tests, both in-isolation and in-soil depending on the soil and the geotextile types, for the determination of design data. He highlighted the importance of the test methodology and of adopting a valid technique for interpreting and processing the measurements. These data are best presented to the engineers in the form of a load-strain curve together with a time correction curve.

Leflaive argued in favour of testing wide width samples in-isolation explaining that by adopting a width/length ratio of 5 and making corrections to the measured strains, the results are similar to those obtained using in-soil testing. McGown however, pointed out that although this may be true in certain situations for tests conducted under constant rate of strain, it is not the case for creep testing. In fact, from the discussion which followed, it is apparent that the creep behaviour of these materials is not clearly understood and that further work in this area is urgently needed particularly concerning the effect of temperature.

This session has certainly confirmed the importance of material evaluation and highlighted the difficulties involved. The most important outcome is the identification of the areas where further research is needed in order to achieve standardisation in approach and the areas which require long term field observations.

EVALUATION OF MATERIAL PROPERTIES - DISCUSSION

DR. G.W.E. MILLIGAN,

University of Oxford, England.

In the analysis or design of a reinforced soil wall, the properties of interaction between the soil and the reinforcement must be taken into account, in addition to the properties of the soil and reinforcement materials. Two simple limiting modes of interaction may be identified (See Figure 1): -

(i) direct sliding, in which a block of reinforced soil moves across a plane of reinforcement;

(ii) Pull out, in which a layer of reinforcement pulls out of stationary soil within an anchor zone.

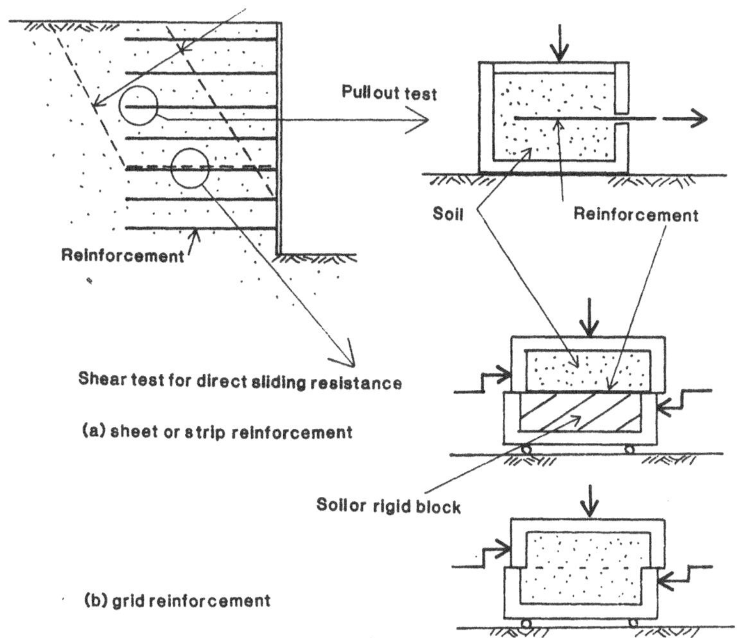

Fig. 1 Tests for soil-reinforcement interaction

P. M. Jarrett and A. McGown (eds.), The Application of Polymeric Reinforcement in Soil Retaining Structures, 359–361.

The tests normally performed to measure the interaction properties are either modifications of the direct shear test, or pull-out tests; it is important to be clear about the relevance of each.

The modified direct shear test will measure a coefficient of direct sliding for any type of reinforcement. With a geotextile the sliding is between soil and geotextile over the full plan area. With a geogrid the sliding resistance is generated mostly by soil sliding over soil in the apertures of the grid, and to a lesser extent by soil sliding over the material of the grid. The test is relatively simple to perform and interpret and is being adopted as a standard test in many countries.

The pull out test models directly the second interaction mode, but is more difficult to perform and interpret correctly. Special apparatus is required, and results may be greatly influenced by the boundary conditions imposed in the test. With geotextiles the interaction on either side of the reinforcement is very similar to that in the direct shear test; bond coefficients obtained from correctly performed and analysed pull out tests agree well with direct sliding coefficients from modified direct shear tests. There is therefore no need to perform pull-out tests for such materials; adequate data may be obtained from the much simpler direct shear tests.

With grids the mode of interaction during pull out is quite different from that in direct shear, as has been clearly established from tests using a photoelastic technique (Dyer 1985) to visualise the stress fields. In Figures 2 and 3 the light 'stripes' are aligned along the directions of major (compressive) stress, and their brightness gives a qualitative indication of the intensity of stress. A grid generates bond in pull out by concentrations of bearing stress against the transverse members of the grid, whereas direct sliding resistance is generated by soil/soil and soil/reinforcement shear. Bond resistance of grids can only be measured in pull out tests; with extensible reinforcement in particular the interpretation of such tests requires considerable care. For many design purposes it may be adequate to predict the bond characteristics using the approach of Jewell et all (1984).

References

Dyer, M.R. (1985). Observations of the stress distribution in crushed glass with applications to soil reinforcement. D. Phil. Thesis, University of Oxford.

Jewell, R.A., Milligan, G.W.E., Sarsby, R.W. and Dubois, D. (1984). Interaction between soils and geogrids. Symposium on Polymer Grid Reinforcement in Civil Engineering. I.C.E. London.

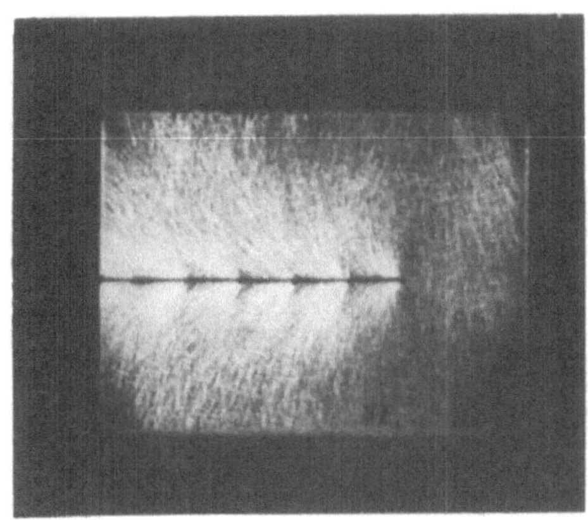

Fig. 2 Stresses in pull out test using a grid

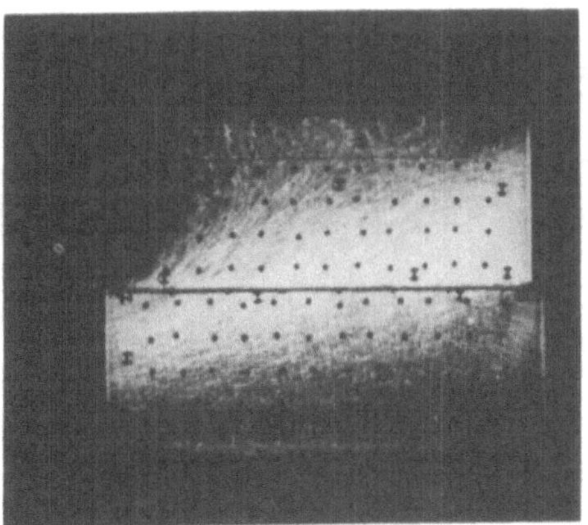

Fig. 3 Stresses in direct shear test using a grid

Analytical Techniques and Design Methods

REINFORCED SOIL WALL ANALYSIS AND BEHAVIOUR

R.A. Jewell
University of Oxford

Abstract

The paper considers the equilibrium in reinforced soil walls. Two solutions are presented which are expected to "bound" the likely state of equilibrium in a wall. The magnitude and distribution of the reinforcement forces and the soil stresses are derived analytically, allowing the movement of the wall face to be calculated. Two non dimensional charts are given for the calculation of outward movement. The resulting magnitude of movement predicted by the two equilibrium solutions is close, so that the likely range of movement in a wall is well defined by the analysis. Incremental wall construction causes a "moving datum" for the outward movement at the wall face due to reinforcement elongation. Thus, incremental construction results in additional movement at the wall face. This is the main distinction with propped walls which do not suffer from incremental construction movement. An analysis for the incremental movement is given. The selection of compatible values of mobilised soil shearing resistance and reinforcement force in a reinforced soil wall is discussed, and new ideas put forward for the way the soil is likely to behave in the reinforced zone. The application of the concepts and the analyses introduced in the paper are fully illustrated by the prediction for the Royal Military College trial wall which is presented in a companion paper.

INTRODUCTION

1.1 Design and prediction

The **detailed analysis** for reinforced soil is complex - like most soil mechanics - particularly when the soil is reinforced by relatively extensible polymeric materials. The derivation of **design methods** is more straight forward; simple conservative assumptions can be made to allow for uncertain material properties, boundary conditions, compatibility conditions, and additional generous safety margins can be applied.

P. M. Jarrett and A. McGown (eds.), The Application of Polymeric Reinforcement in Soil Retaining Structures, 365–408.

If a **prediction** of the equilibrium or serviceability state in a reinforced soil structure is to be made then there is no alternative but to try and evaluate the uncertainties. Conservative design simplifications applied to prediction would lead to overpredicted reinforcement forces and deformations. This paper examines the prediction of equilibrium in reinforced soil walls.

1.2 Organisation of the paper

In the paper the behaviour of reinforced soil walls is examined in several stages. The starting point is the concept of equilibrium in soil and in reinforced soil. This allows possible sets of forces required to maintain equilibrium in soil with a constant mobilised shearing resistance to be derived. These forces would be provided by the reinforcement.

Two possible equilibrium states for reinforced soil walls are derived. The magnitude and distribution of the reinforcement forces and the stresses in the soil are given. The two solutions represent two extreme cases for equilibrium in reinforced soil which, it is suggested, are likely to "bound" the actual equilibrium.

The next stage is to examine the magnitude of outward movement at the face of the wall which would be implied by the required reinforcement force distribution and the reinforcement stiffness. The outward movement can be determined analytically and the results for the two equilibrium solutions are presented in non dimensional charts. The deformation in the reinforced zone is assumed to be controlled by the reinforcement elongation; the soil shearing resistance is assumed constant and independent of the deformation.

The influence of incremental layered construction is examined next, particularly to see what datum for movement might be used (even for movement due only to reinforcement elongation) when the incremental facings are aligned over built facings which have already undergone movement.

To determine the likely equilibrium combination of the mobilised soil shearing resistance and reinforcement force in a reinforced soil wall, the relationship between the deformation and the shearing resistance in soil has to be examined. The link with the mobilised reinforcement force is through compatible strain. At best the reinforcement strain would equal the tensile strain occurring in the soil in the direction of the reinforcement. The overall equilibrium in a reinforced soil wall can be represented on a compatibility curve.

These ideas are fully illustrated by their application to the prediction of the behaviour of the incremental trial wall built at the Royal Military College, Kingston. The analysis is presented in a *companion paper* "Analysis and predicted behaviour for the Royal Military College trial wall".

2.1 Equilibrium in soil

Soil is a frictional material. An element of soil will remain in static equilibrium if the stress ratio remains below the available frictional shearing resistance. Frictional shearing resistance depends on the type of loading applied to the soil, and **plane strain** is the relevant loading condition for walls. Only frictional soils are considered in this paper. Frictional shearing resistance depends on **effective stresses** in the soil. For simplicity, the dry case is examined in this paper so that all the stresses are effective stresses.

The Mohr circle of stress represents the state of stress equilibrium in a soil element. The case of soil in equilibrium with a mobilised frictional resistance ϕ less than the plane strain frictional resistance ϕ_{ps} is shown in Fig. 1.

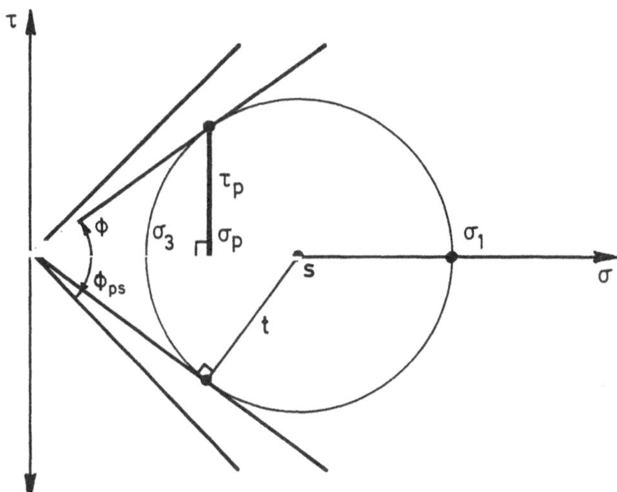

Fig. 1. Equilibrium stresses in frictional soil

The maximum stress ratio in the soil is often expressed as a ratio of the shear stress and normal stress acting on the plane of maximum stress obliquity, Fig. 1., giving

$$\frac{\tau_p}{\sigma_p} = \tan\phi \tag{1}$$

but the stress ratio in the soil can also be expressed more generally in terms of the principal stresses to give

$$\frac{t}{s} = \frac{\sigma_1 - \sigma_3}{\sigma_1 + \sigma_3} = \sin\phi \tag{2}$$

It is usual to define the factor of safety for the soil in terms of the ratio of the shearing stresses on the plane of maximum stress obliquity compared with the maximum available shearing resistance ϕ_{ps} , so that, for the plane strain case,

$$\text{Factor of safety} = \frac{\tan\phi_{ps}}{\tan\phi} \tag{3}$$

2.2 Equilibrium in reinforced soil

Reinforcement modifies the stresses in soil. A simple illustration is a direct shear test on an element of reinforced soil, as described by Jewell and Wroth (1987), Fig. 2. The reinforcement force P_r acting across the plane of maximum stress obliquity in the soil reduces the shear force that the soil must support and increases the normal force in the soil.

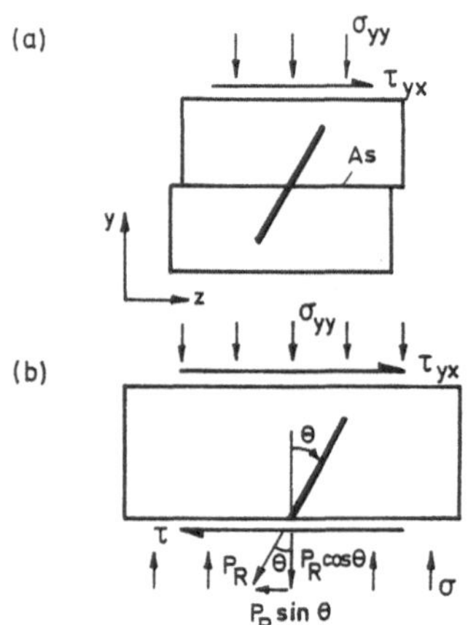

The effect of reinforcement is twofold:

(1) To reduce the stresses causing failure

$$\tau = \tau_{yx} - \frac{P_R \sin\theta}{As}$$

(2) To increase the stresses resisting failure

$$\sigma = \sigma_{yy} + \frac{P_R \cos\theta}{As}$$

Fig. 2. Reinforced soil in direct shear

The overall equilibrium in the soil should be expressed in terms of the forces acting across the plane in the soil being examined so as not to presume a uniform state of stress. Uniform stress is unlikely in reinforced soil.

However, it is a convenient concept to consider the influence of the reinforcement in terms of additional stresses in the soil. In the shear box, on a plan area of soil A_s, the additional resisting shear stress provided directly by the reinforcement is

$$\tau_R = \frac{P_r \sin\theta}{A_S} \tag{4}$$

and the additional normal stress provided by the reinforcement is

$$\sigma_R = \frac{P_r \cos\theta}{A_S} \tag{5}$$

These additional stress resultants acting in the soil enable the soil to remain in equilibrium under increased shear loading.

2.3 Equilibrium in a vertical reinforced soil wall

Previous concepts

The results of limit equilibrium analysis indicate three important zones of stress in the soil of a reinforced slope or wall, Fig. 3a.

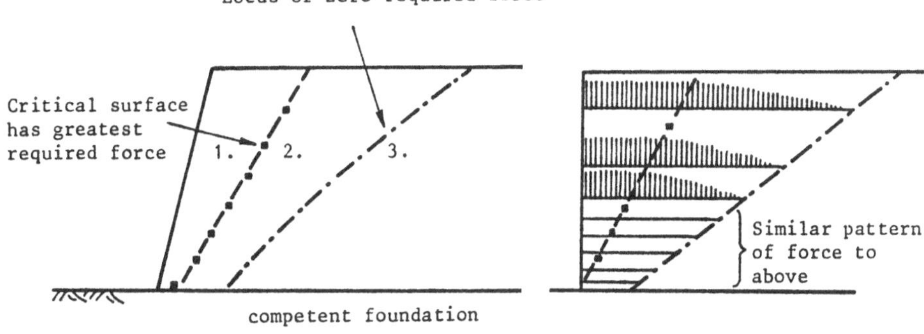

Fig. 3. (a) Zones of required force for equilibrium and (b) resulting reinforcement force distribution in a wall (Jewell,1985)

There are two surfaces through the soil that separate the three zones. Firstly, deep in the soil is a surface beyond which no additional stresses are required from the reinforcement to maintain equilibrium. This is a **locus of zero required force**. Secondly, through the toe of the slope there is a surface which requires the greatest total reinforcement force to maintain equilibrium. This is the **most**

critical surface through the toe, because the greatest reinforcement force is required to maintain equilibrium on this surface.

Reinforcement spacing; calculation of force in reinforcement layers

In fact, there are a whole family of geometrically **similar** surfaces parallel to the **most critical surface** intersecting the slope face above the toe.

For the vertical case the surfaces are plane and at an angle $\theta = (45+\phi/2)$ to the horizontal, Fig. 4a. The **similarity** of these surfaces through the soil implies that the reinforcement must be providing horizontal stress in the soil which increases linearly with depth, Fig. 4b.

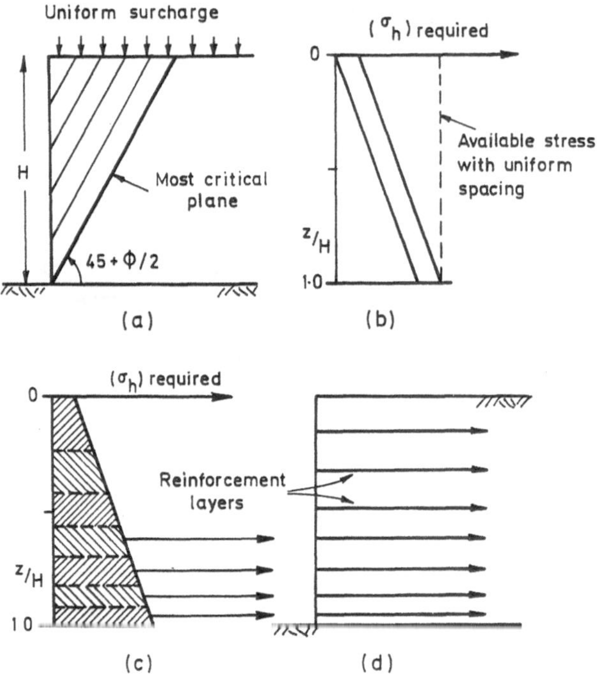

Fig. 4. (a) Most critical mechanism and similar surfaces through the face. (b) Triangular distribution of required stress from similarity. (c) Balanced distribution of reinforcement satisfying required stresses both locally and overall.

The stress required for equilibrium in the soil is provided by the force in the reinforcement layers. Jewell et al (1984) introduced the idea of a **balanced distribution** of reinforcement; the idea is that the reinforcement should be spaced to satisfy both **local** and **overall equilibrium.**

Locally each reinforcement layer can only provide the required soil stresses for one half the vertical spacing above and below the reinforcement layer. To make the most of reinforcement layers with a fixed allowable force the spacing must be reduced with depth so that the full allowable force can be mobilised in each layer, Fig. 4c. This would give an **ideal spacing** arrangement because each reinforcement layer is able to mobilise its full allowable force.

If **uniform spacing** was chosen for the reinforcement the required spacing would be set by the deepest reinforcement layer at the bottom of the wall. Half way up the wall the required reinforcement force would have dropped by half, Fig. 4b, and the reinforcement would have twice the required capacity. It is convenient to use a diagram of **required** soil stresses and **available** soil stress from the reinforcement, similar to Fig. 4b, to assess the chosen spacing arrangement. Conversly, this diagram can be used to determine the mobilised force in each layer in a chosen spacing arrangement.

In a balanced design the available stress provided by the reinforcement should always be greater than the required stress for local equilibrium at every depth in the wall, to avoid local overstressing of the reinforcement. Local rupture of a reinforcement can result in overall collapse, section 3.2.

In Fig. 3b the reinforcement extends back to the **locus of zero required force**. This reinforcement layout can be thought of as having **ideal length** because the reinforcement is provided everywhere in the soil that reinforcement force is required to locally maintain equilibrium. The case of equilibrium with reinforcement of ideal length is considered below.

2.4 Equilibrium with ideal length reinforcement

The starting point is soil with an assumed constant mobilised angle of friction and built to a vertical face. The aim is to derive a state of equilibrium for the soil everywhere and thereby to determine the required reinforcement layout.

The two bounding surfaces in the soil (see Fig. 3a) are both plane surfaces in this case. The **locus of zero required force** is a plane through the toe inclined to the horizontal at the angle of friction, plane **OE** in Fig. 5a. The **most critical surface** through the toe is the critical Coulomb wedge, a plane inclined at an angle $\theta = 45 + \phi/2$ to the horizontal, plane **OB** in Fig. 5a.

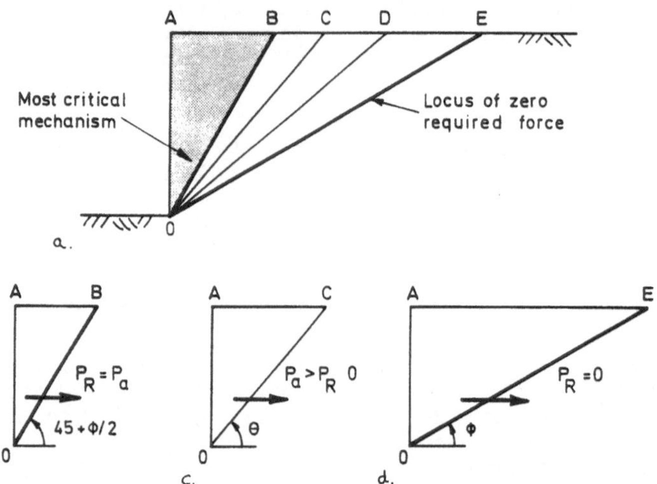

Fig. 5. Bounding surfaces for reinforcement of ideal length

The simplest way to consider the state of equilibrium is in terms of the **gross required** (reinforcement) **force** for equilibrium. A reinforcement force equal to the active force P_a is needed to hold the zone **OAB** in equilibrium on the plane **OB**, Fig. 5b. A smaller reinforcement force is needed to hold the zone **OAC** in equilibrium on the plane **OC**, Fig. 5c. Progressively smaller force is needed for equilibrium on flatter planes **OD**, until no reinforcement force at all is required for equilibrium of the block **OAE** on the plane **OE**, Fig. 5d.

The variation of the overall required reinforcement force depends on the angle θ to the horizontal of the plane surface in the soil through the toe, Fig. 6.

From a wedge analysis the overall required force for equilibrium is

$$\left(P_R\right)_\theta = W \tan(\theta - \phi) \tag{6}$$

which can be expressed as

$$\left(P_R\right)_\theta = \frac{\gamma H^2}{2}\frac{\tan(\theta - \phi)}{\tan\theta} \tag{7}$$

Equations (6) and (7) describe the variation of the required reinforcement force between the **most critical plane** through the toe and the **locus of zero required force**. The required reinforcement force is constant between the **most critical plane** and the **face**. The state of equilibrium, the reinforcement layout and the variation of force along the reinforcement is illustrated in Fig. 7 for reinforcement with **ideal length**.

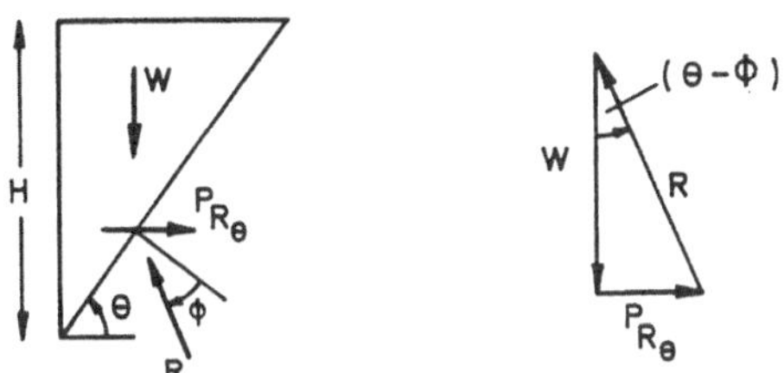

Fig. 6. Wedge analysis for the required reinforcement force

The **ideal length** state of equilibrium is the analytical counterpart to the pattern of reinforcement force in a reinforced soil wall deduced from limit equilibrium analysis and described by Jewell (1985), Fig. 3.

Equation (7) gives the variation of the total required reinforcement force with the angle θ through the toe. The expression for the force P_r in a single layer of reinforcement, which supports a force P at the most critical plane, is

$$P_{r\theta} = \frac{P}{K_a} \frac{\tan(\theta - \phi)}{\tan\theta} \tag{8}$$

which is valid in the range $(45 + \phi/2) \geq \theta \geq \phi$. Remember that $P_r = P$ when $\theta > 45 + \phi/2$.

Implied stress distribution for ideal length layout

The state of stress in the reinforced soil can also be identified. The Mohr circle for stress for soil elements along a line through the toe at an angle θ to the horizontal, which is a plane of maximum stress obliquity, is shown in Fig. 8a. From the pole for planes, it is possible to deduce the orientation of the second plane of maximum stress obliquity in the soil and determine the orientation of the principal stresses in the soil. The resulting stress distribution is illustrated in Fig. 8b by the pattern of planes of maximum stress obliquity and the principal stress directions.

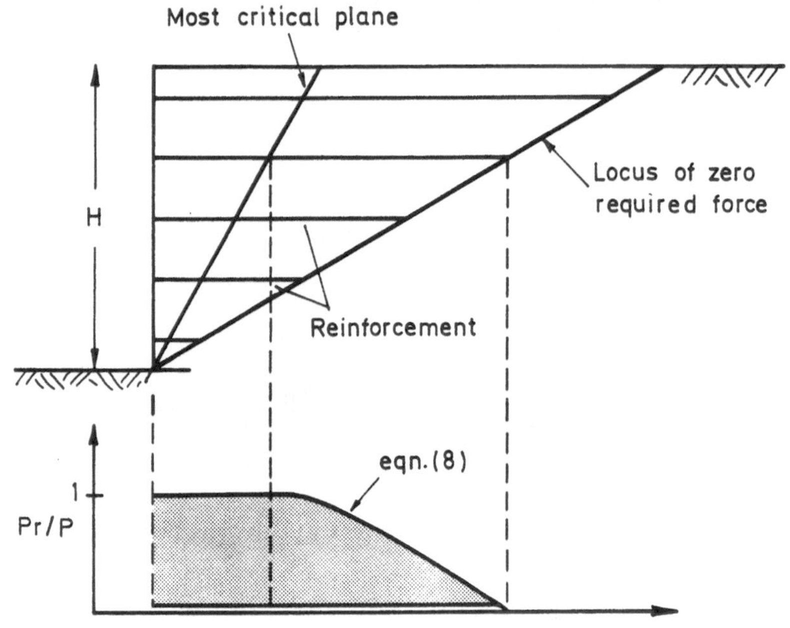

Fig. 7. Reinforcement layout and force distribution for the ideal length case.

2.5 Equilibrium with truncated length reinforcement

Consequences of truncation

Truncation of the reinforcement short of the **locus of zero required force** changes the pattern of equilibrium in the reinforced soil. For example, consider the reinforcement layout for ideal length illustrated in Fig. 7, but with the maximum reinforcement length limited to L_{max}. In the zone where the reinforcement has been omitted, the soil equilibrium will be equivalent to the active state behind a retaining wall, Fig. 9. The soil will exert active pressure on the back of the reinforced zone.

Fig. 9 illustrates the two **consequences of truncation** of the reinforcement length. Firstly, when reinforcement force is omitted higher up on a surface through the reinforced soil the lower reinforcement layers have to provide the additional force - the **required force is shed** to lower layers. Secondly, the active thrust from the unreinforced soil resting on the back of the reinforced zone changes the position of the locus of zero required force - it extends more deeply into the soil.

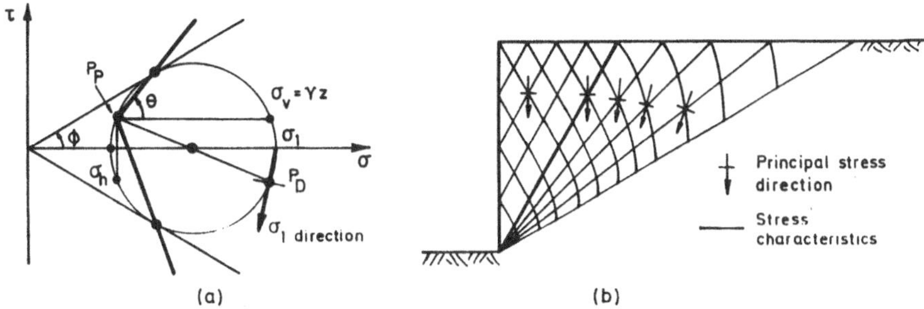

Fig. 8. Mohr circle for stress, and planes of maximum stress obliquity in the soil for the ideal length reinforcement case.

In spite of the imbalances which result from truncating the reinforcement length, it is likely that the truncated reinforcement layout will give a more efficient use of the reinforcement layers as explained below.

Truncation without increasing the required force

In the **ideal length** arrangement the **most critical mechanism** through the toe governs the gross required reinforcement force. Each reinforcement layer has an allowable force and spacing selected to hold the most critical mechanism in equilibrium. This magnitude of force is available along the whole length of the reinforcement where bond with the soil permits. Over much of the reinforcement length in the ideal length layout the mobilised reinforcement force is less than the available force.

Truncating the length of upper reinforcement layers sheds required force to lower reinforcement layers and requires additional force to be carried by the lower layers; but as long as the additional force does not increase the required force for any single layer above the allowable value then the reinforcement already has sufficient capacity to carry the additional load.

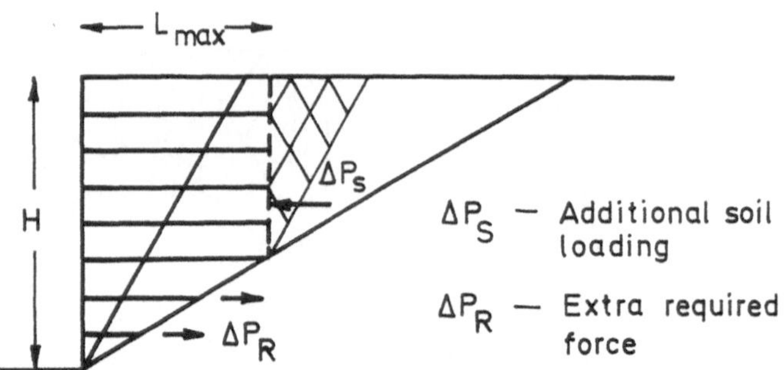

Fig. 9. Illustration of the consequence of truncating the reinforcement length.

In other words reinforcement length can be truncated higher in the wall without need for increase in the capacity of the reinforcement until the resulting load shedding brings the required force on lower reinforcement layers to the allowable force selected to support the most critical mechanism through the toe. Thus reinforcement length can be saved higher in the wall without requiring a change in the reinforcement properties or cross section.

It must be emphasised, however, that some additional reinforcement length is also required lower in the wall as a result of truncation.

Truncated length layout

To illustrate these ideas, consider the limiting case of reinforcement layers that can mobilise the allowable reinforcement force over a negligible bond length - so that the available reinforcement force is constant over the full length of the reinforement. Also, for simplicity, assume that the active state in the unreinforced soil behind the reinforced zone is equivalent to soil behind a smooth retaining wall (ie. with vertical and horizontal principal stresses).

The equilibrium in the zone between the front of the wall and the most critical mechanism is the same as before and the same reinforcement spacing and material properties would be chosen as for the layout for reinforcement of ideal length, zone **OAB** Fig. 10a.

With truncated reinforcement there will be two sections on any plane through the toe at an angle $\theta<(45+\phi/2)$ to the horizontal, Fig. 10b. On the portion **OC** the reinforcement will provide the full allowable force on the reinforcement layers intersected. On the portion **CC''** the soil is unreinforced and held in equilibrium by exerting active stresses on the back of the reinforced block, illustrated by the vertical plane **CB'**.

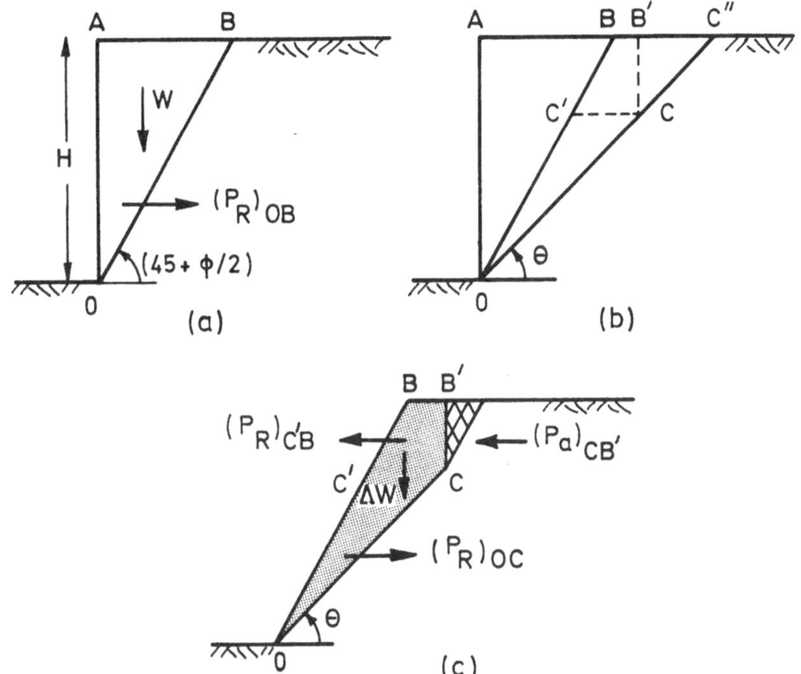

Fig. 10. Equilibrium with truncated reinforcement layers supporting the full active force and with negligible bond length

The position of **C** on a line at an angle θ can be found so that there is force equilibrium between the soil self weight forces, the active driving force of the unreinforced soil and the reinforcement layers acting with their full allowable force.

The equilibrium state for the soil segment **OBB'C** is illustrated in Fig. 10c. The segment has two net outward disturbing forces to support: firstly, the net reinforcement force which acts on **OB** but not on **OC**; secondly, the outward active earth force from the unreinforced soil acting on **CB'**.

The soil segment **OBB'C** can remain in equilibrium because the angle of the plane **OC** is less than **OB**, so that greater shearing resistance can be generated on **OC** by the same horizontal force. The greater net shearing resistance along **OC**, in combination with the additional self weight of the soil segment (ΔW), enables the additional outward thrusts on the soil to be held in equilibrium.

Repetition of the calculation on flatter soil segments, and with the simple assumption of a net active thrust for the smooth case from the unreinforced fill, provides a distribution of the reinforcement length with depth in the wall required to maintain force equilibrium. The results in Fig. 11a. are for $\phi = 30°$.

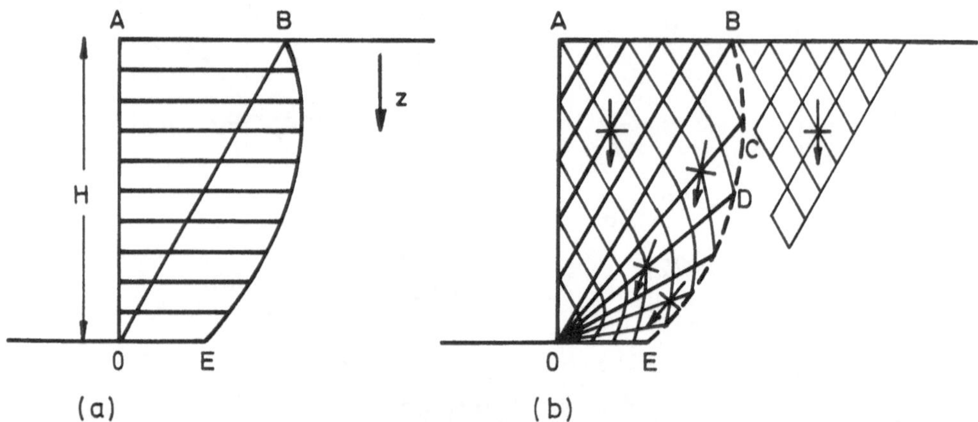

Fig. 11. (a) Required reinforcement length for truncated layers with negligible bond length. (b) Equivalent stress field in the reinforced soil.

The required reinforcement length to provide force equilibrium can be expressed analytically. The derivation is not repeated here. One way to express the result is with a pair of equations giving the required reinforcement length (L/H) at a depth below the crest of the wall (z/H) in terms of a plane through the roe at an angle θ to the horizontal. The expression for the required reinforcement length is

$$\frac{L}{H} = \frac{1}{\tan\theta}\left(1 - \frac{z}{H}\right) \tag{9}$$

which applies at a depth

$$\frac{z}{H} = \sqrt{\frac{\tan\theta\tan^2(45-\phi/2)-\tan(\theta-\phi)}{2\tan\theta\tan^2(45-\phi/2)-\tan(\theta-\phi)}} \tag{10}$$

Implied stress distribution for truncated length reinforcement

As explained for the **ideal length** case, each line segment through the toe is a plane of maximum stress obliquity and so the orientation of the principal axes of stress in the reinforced soil can be determined. The planes of maximum stress obliquity in the reinforced soil are drawn in Fig. 11b, where the directions of principal stress are indicated.

The simple assumption made about the unreinforced fill behind the reinforced zone results in vertical and horizontal principal stresses in the unreinforced soil. Because this was an **assumption** there is a lack of fit between the two zones of stress along the plane **BCDE**, Fig. 11b.

The implied stress distribution for the reinforced soil shows that the wall resists the outward thrust from the unreinforced soil by taking advantage of the additional shearing resistance available from the reinforcement forces acting on less steeply inclined planes through the wall toe.

The required reinforcement length distribution in Fig. 11a is shown for soil with a mobilised angle of friction 30°. The required reinforcement length in the **truncated length case** for soil friction angles in the range 20° to 50° are given in Fig. 12.

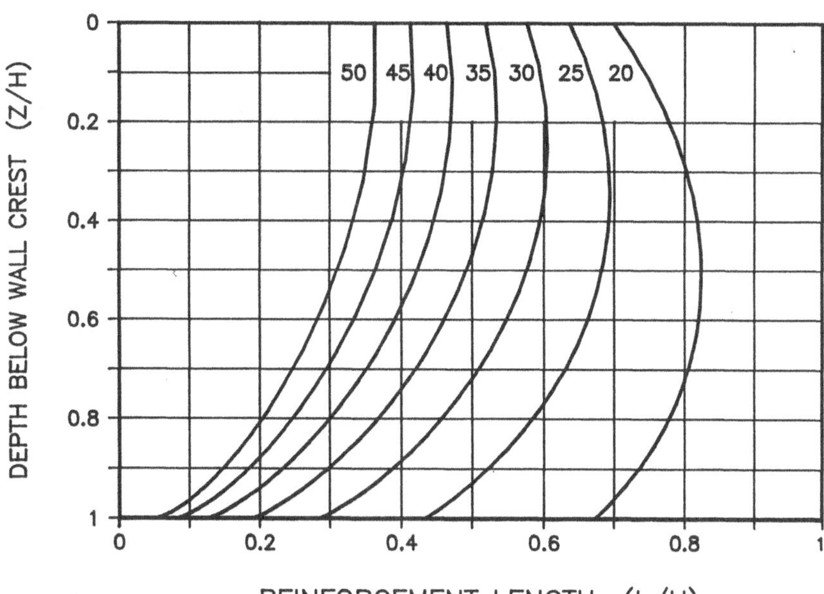

Fig. 12. Required minimum reinforcement length for force equilibrium (infinite bond case).

The reinforcement lengths shown in Fig. 12 are an absolute minimum for the appropriate soil friction angles and reinforcement designed to carry active stresses. The important omission is that they only represent force equilibrium and not moment equilibrium. The purpose of these results is to illustrate ways that reinforcement might maintain equilibrium in reinforced soil. This is useful for assessing possible reinforcement layouts.

Equilibrium could be maintained with shorter reinforcement lengths than those given in Fig. 12 for the **truncated length** case if the allowable reinforcement force was higher than that needed to support the active soil stresses. In this case there would be load shedding to lower reinforcement layers even on the **most critical plane**.

2.6 Stress transfer between reinforcement and soil in a wall

The stress distributions in Figs. 8 and 11 are for the **reinforced soil**, that is they represent the stresses in the soil in equilibrium with the mobilised reinforcement forces. The reinforcement is (conceptually) distributed through the soil rather than in discrete layers. However it is still possible to identify the overall pattern of stress transfer from the reinforcement to the soil.

The central feature of both the derived equilibrium states is the increasing horizontal outward shear stress in the soil behind the most critical plane and this causes the principal stress directions to rotate.

In the uniform zone between the wall face and the most critical plane the reinforcement force is constant and there is no transfer of stress between the soil and the reinforcement. The transfer happens at the wall face where the required horizontal stress is applied to the soil by the wall face, which in turn is supported by the reinforcement, Fig. 13a. This is an **ideal facing boundary**.

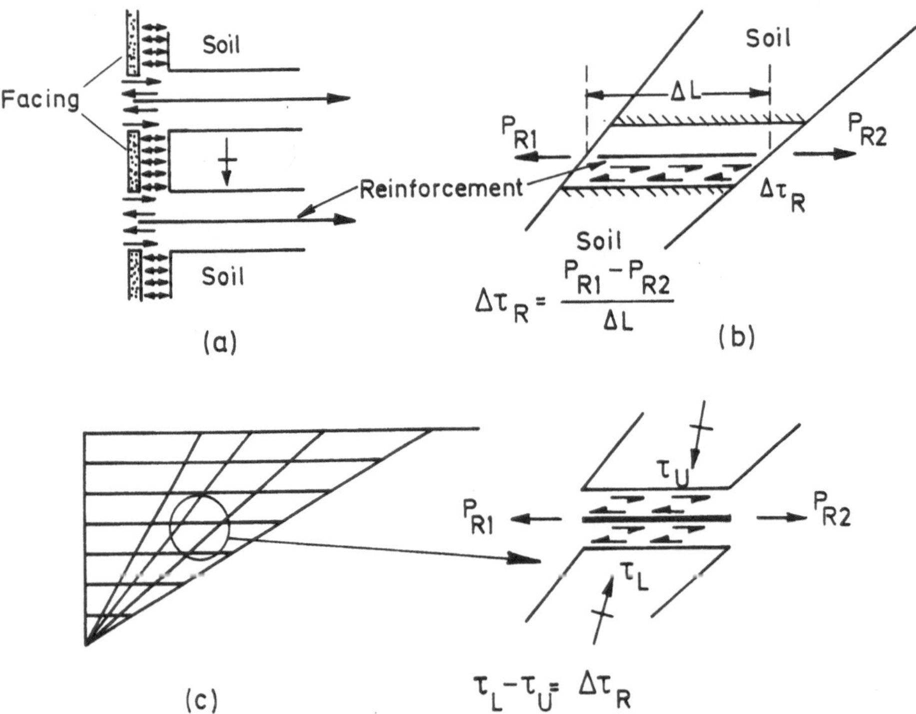

Fig. 13. Illustration of the load transfer mechanism between the reinforcement and the soil.

The reinforcement axial force is distributed back to the soil behind the most critical plane. The transfer of load occurs by shear stress developing between the reinforcement and the soil as illustrated in Fig. 13b, drawn for the case of wide width or sheet reinforcement. Overall, the axial force from the reinforcement layers are transmitted down into the soil. This is illustrated in Fig. 13c for the ideal length case.

At any level in the soil behind the **most critical plane** the principal stress directions in the reinforced soil are inclined from the vertical. A layer of wide width reinforcement would experience outward shear stress from the soil above which would be transmitted through the reinforcement to the soil below, see the detail in Fig. 13c. The reinforcement in turn also transmits some axial force to the soil by increasing the outward shear stress on the soil below.

The soil below the reinforcement layer is able to support the additional outward shear stress. Firstly, the principal stresses rotate further with depth in the soil which increases the ratio of shear stress to vertical stress. Secondly, the ability of the soil to carry shear stress increases with depth as the vertical self-weight soil stresses increase.

COLLAPSE LIMITS

3.1 Collapse without reinforcement rupture

Collapse will occur in a reinforced soil wall if on a potential failure mechanism there are insufficient available resisting forces. This may arise in two ways.

Firstly, if there is insufficient reinforcement length the **bond** can limit the mobilised reinforcement forces to below the value required to maintain equilibrium. Lack of equilibrium would lead to continued shear in the soil and collapse.

Secondly, the shearing resistance of soil across the surface of a layer of reinforcement in the ground can be less than the shearing resistance of the soil alone. In this case the layer of reinforcement provides a plane of reduced shearing resistance and if this does not balance the outward thrust of the unreinforced soil an **outward direct sliding** shear failure would develop.

3.2 Collapse with reinforcement rupture

Examples of equilibrium stress states in reinforced soil have been illustrated in Figs. 8 and 11. This pattern of stress equilibrium is likely to exist in a reinforced soil wall immediately prior to collapse caused by reinforcement **rupture**. If a reinforcement layer were to **rupture** (due to a local weakness, say) the local loss of equilibrium would cause the soil immediately above and below the rupture point to be overstressed and out of equilibrium. Local strains would develop as the soil locally redistributed the out of balance stresses. This may or may not cause rupture in an adjacent reinforcement layer before equilibrium is established once more.

Thus collapse caused by **reinforcement rupture** is started by local loss of equilibrium. Collapse caused by **bond** and **outward direct sliding** is started by general loss of equilibrium.

Collapse by **reinforcement rupture** propogates progressively from the local source of instability.

This is the reason why reinforcement layers should be distributed through the soil in a balanced way to everywhere satisfy local as well as overall equilibrium, (section 2.3).

WALL DEFORMATION

4.1 Sources of movement

Deformation of a reinforced soil wall should be expected from the three sources illustrated in Fig. 14.

Reinforced soil

With extensible polymer reinforcement materials the greatest deformation is likely to develop in the reinforced soil, **zones A** and **A'** Fig. 14. With reinforcement attached to the facing panels, the elongation in the reinforcement layers should be a good measure of the lateral deformation in the reinforced zone.

Between the most critical plane and the wall face, **zone A** in Fig. 14, the principal stresses in the reinforced soil are approximately vertical and horizontal. This zone is likely to deform relatively uniformly causing both horizontal and vertical movement at the face, Fig. 14b.

Behind the most critical mechanism, **zone A'** in Fig. 14, the principal stresses in the reinforced soil are inclined from the vertical, very markedly towards the base of the wall. The vertical movement in this zone will be much reduced compared with the horizontal movement.

Backfill and foundation

With truncated reinforcement the soil behind the reinforced zone must come into an active state of equilibrium, **zone B** in Fig. 14. This requires strain in the soil which will push the reinforced zone outwards.

Finally, the reinforced zone, and the unreinforced soil behind, rest on the foundation, **zone C** in Fig. 14. Movement caused by overstressing the foundation beneath the reinforced zone, or due to overall settlement in the foundation will cause additional "rigid body" movements at the wall face.

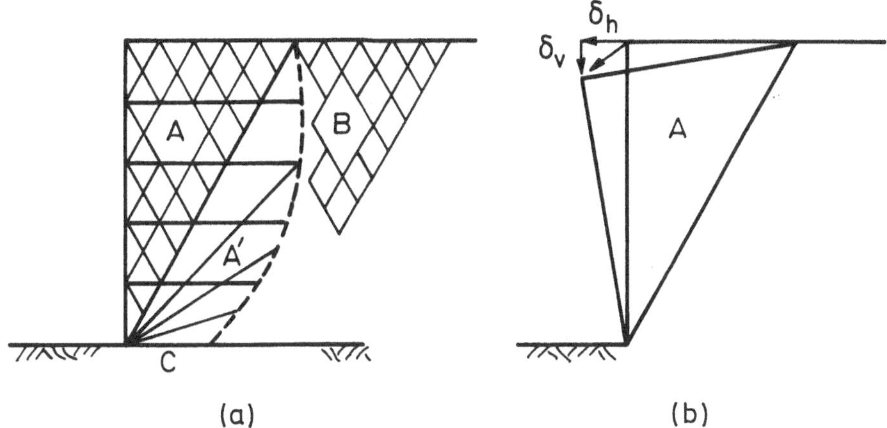

Fig. 14. Zones causing deformation at the face of a reinforced soil wall.

Care is needed with field observations of face movements in reinforced soil walls to separate the movement due to deformation of the reinforced soil, deformation behind the reinforced zone and the movement resulting from foundation deformation.

4.2 Deformation predictions for the reinforced zone

The two equilibrium stress fields for reinforced soil walls with **ideal length** and **truncated length** both directly give the magnitude and distribution of reinforcement force in the reinforcement layers. If the stiffness of the reinforcement is known, the strain along the reinforcement may be calculated and hence the elongation in the reinforcement layer can be derived analytically. With no relative slippage between the reinforcement layer and the soil, the elongation in the reinforcement would equal the horizontal component of movement at the wall face caused by deformation in the reinforced soil.

It is convenient to represent the horizontal outward movement δ as a proportion of the wall height **H**. The reinforcement stiffness K and axial force P govern the maximum reinforcement strain. The horizontal outward movement of the wall face due to deformation in the reinforced zone can be expressed by the non dimensional parameter

$$\left(\frac{\delta}{H}\frac{K}{P}\right)$$

The appropriate units for the load in geotextile and grid reinforcement are force per unit width (kN/m). The reinforcement **strain** should be expressed as a number (0.01) when evaluating parameters such as stiffness, rather than as a percentage (1%).

The reinforcement spacing arrangement influences the mobilised force in the reinforcement layers. With **ideal spacing** the reinforcement spacing is chosen so that the full allowable reinforcement force is mobilised in every layer. With **uniform spacing** the full allowable reinforcement force is only mobilised in the lowest reinforcement layer; progressively less force is required in higher layers in the wall to maintain equilibrium, which results in progressively less outward movement.

The non dimensional charts for outward movement presented below can be used for any chosen reinforcement spacing by using the mobilised reinforcement force P_{mob} calculated for local equilibrium in each layer, section .

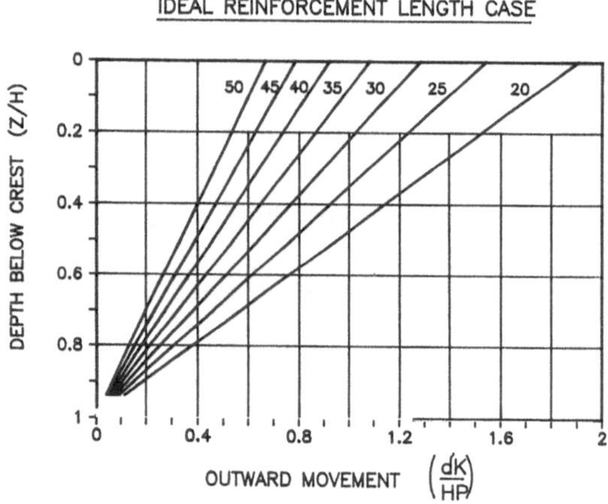

Fig. 15. Non dimensional outward movement at the face due to deformation in the reinforced zone. Reinforcement with ideal length.

4.3 Ideal length

The non dimensional chart for outward movement for reinforcement with **ideal length** is shown in Fig. 15. The variation of outward movement with depth in the wall closely reflects the variation of reinforcement length with depth,

Fig. 7. The results are for a range of angles of friction for the soil and, as would be expected, the soil with lower shearing resistance requires a greater zone of soil to be reinforced which results in greater deformation in the reinforced zone.

4.4 Truncated reinforcement length

The non dimensional chart for outward movement in the case of reinforcement with **truncated length** is given in Fig. 16. Again, the variation of outward movement with depth closely reflects the variation of the reinforcement length with depth, Fig. 9.

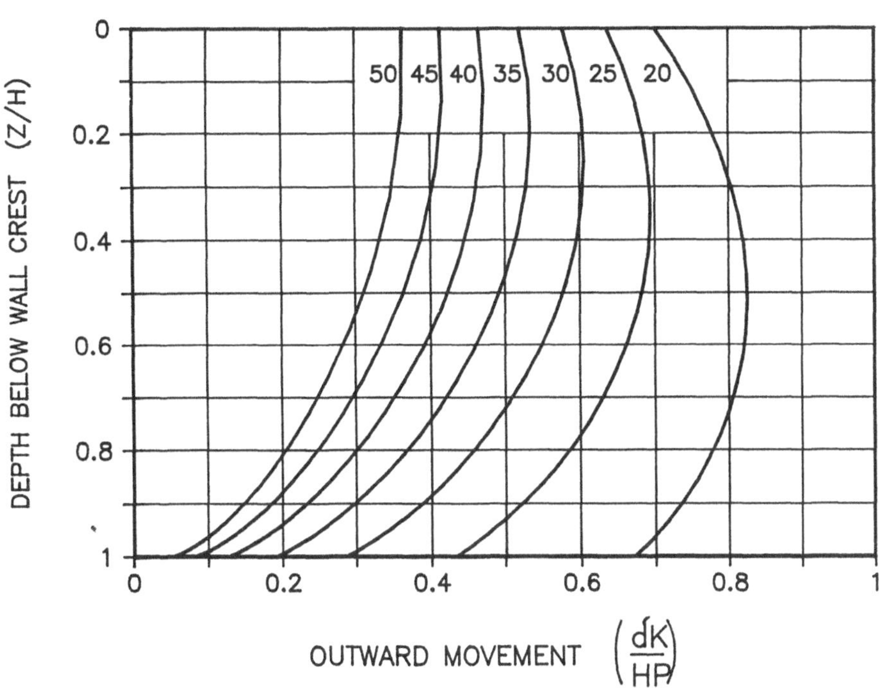

Fig. 16. Non dimensional outward movement at the face due to deformation in the reinforced zone. Reinforcement with truncated length.

The two charts in Figs. 15 and 16 can be used to estimate the outward movement caused by elongation of the reinforcement to maintain equilibrium. The movement depends only on the mobilised soil shearing resistance ϕ and the mobilised reinforcement tensile strain P/K. The distribution of outward movement with depth below the wall face should be calculated from both charts. The expected movement should be within the range calculated.

4.5 Example application

To illustrate the use of the non dimensional charts, consider an **8 metre** high wall, with no surcharge, soil unit weight **19 kN/m3** and designed for a mobilised angle of friction 30°. The reinforcement design allowable load is **26 kN/m** and the stiffness is **1000 kN/m** (representing material properties in the ground, at the end of the design life, at the design temperature). This stiffness gives an strain 0.026 (ie 2.6%) when the allowable load is mobilised in the reinforcement.

The gross required force for equilibrium on the **most critical plane** is

$$\left(P_R\right)_{required} = \frac{K_a \gamma H^2}{2} = 202 \quad kN/m$$

which can be provided by **8** reinforcement layers if **ideal spacing** is used.

The minimum required spacing at the base of the wall for a **uniform spacing** layout is given by the expression (for a wall without surcharge)

$$\frac{s_v}{H} = 1 - \sqrt{1 - \frac{P}{K_a \gamma \frac{H^2}{2}}}$$

which for the example above gives $s_v = 0.53\ m$, indicating that **15** layers of reinforcement are required.

If the **truncated length** arrangement is used, the face movement for the two spacing arrangements can be calculated from Fig. 16. The curve for the outward movement is the same as in the Fig. 16 for the **ideal spacing** case, except the actual value for P/K for the wall should be used. With **uniform spacing** the mobilised reinforcement force P decreases as a proportion of the depth below the wall crest, directly reducing P/K and hence the outward movement.

The results for the example are shown in Fig. 17. Two scales for the outward movement are indicated

(1) outward movement as a function of wall height

(2) the magnitude of the outward movement in mm.

The first scale is probably the most general representation of outward movement and for the design cases in Fig. 17 the maximum outward movement is approximately 1.6% and 1% of the wall height for the designs with 8 and 15 reinforcement layers respectively.

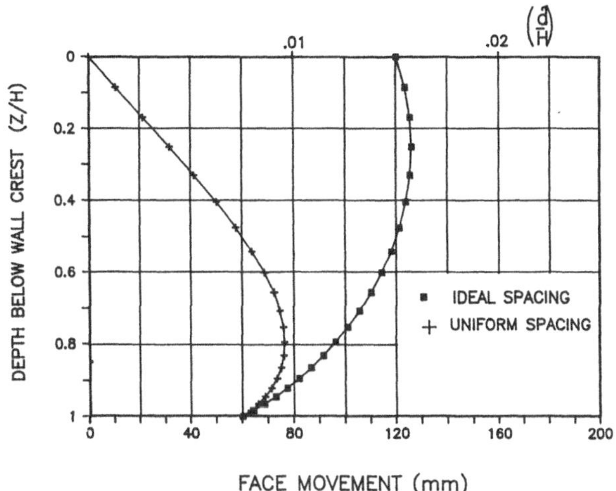

Fig. 17. Illustrative results for an 8 metre high wall. Reinforcement layout with truncated length; movement calculated for ideal spacing and uniform spacing arrangements.

4.6 Use with uniform surcharge loading

The non dimensional movement charts may be applied to walls subject to uniform surcharge load. All that is required is for the mobilised reinforcement force in each layer to be calculated from a local equilibrium balance allowing for the surcharge load (section 2.3).

The procedure is illustrated in detail in the *companion paper* describing the predictions for the RMC trial wall.

INFLUENCE OF THE CONSTRUCTION SEQUENCE

5.1 Introduction

The analysis for reinforced soil wall behaviour has been examined in stages. Firstly, equilibrium in a reinforced soil wall has been examined for soil with a fixed and constant angle of friction. Secondly, the state of equilibrium in the "constructed" wall has been used to deduce the magnitude and distribution of the reinforcement forces and hence the elongation in the reinforcement layers.

At the very least, the lateral deformation in the reinforced soil must equal the reinforcement elongations, and the distribution of reinforcement layer elongations were plotted in Figs. 15 and 16 to represent the outward movements at the wall face caused by deformation in the reinforced zone.

However, walls are not "created" but are built, and in this section the affect of **incremental construction** on outward movements will be discussed.

5.2 Incremental construction

The self-weight loading experienced by the soil during incremental construction progresses in a relatively uniform way as illustrated in Fig. 18. The position of the most critical mechanism, for example, stays constant in the soil throughout construction. The vertical stresses in the soil increase with construction and so must the horizontal stresses to provide equilibrium. Thus the reinforcement force and elongation builds up throughout construction.

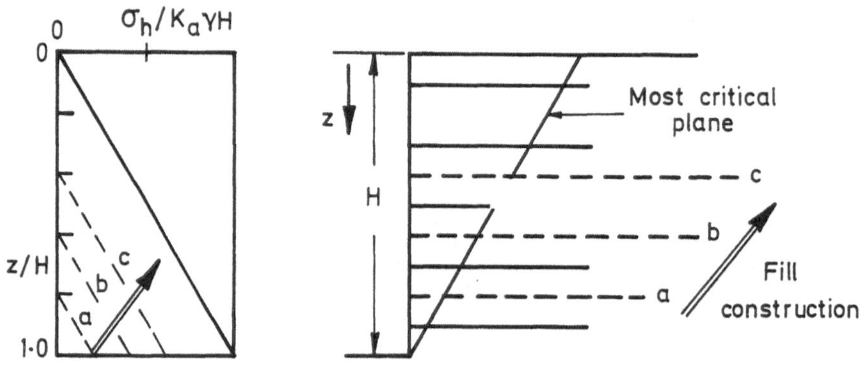

Fig. 18 Progressive increase in stress during incremental construction.

When it comes to examining outward movement at the face due to deformation in the reinforced zone, it is apparent that there is a **moving datum** that must be taken into account. This occurs because after constructing the first incremental lift of a wall, which deforms and moves outward, the subsequent facing is aligned over the already deformed lower facing.

For each additional fill layer, it is the outward movement developed in the layer immediately below, over which the facing is aligned, that causes the move in the datum.

In principle, a cumulative adjustment should be made at each incremental level to allow for the slightly moved starting point of each unstressed layer of reinforcement. This acts to **increase** the outward face movements higher in the wall, as will be illustrated.

Any calculation for this effect can only be approximate because of the many factors which can significantly alter the magnitude of the deformation in each increment of construction. For example,

the **compaction loading** is significant but this is likely to be quite variable from layer to layer

the **connection between the facings** is likely to significantly alter the amount of movement and the mode of movement - from translation to rotation

if the panel **rotates** rather than **translates** the movement at the top of the panel could change by a factor of at least two - and the alignment for the next facing panel is to the top of the previous panel.

Approximate calculation for incremental construction displacement

An **approximate** calculation for the average **incremental movement** caused by the construction of each layer is given below. Depending on whether the panel **rotates** or **translates** the movement at the top of the panel may be a factor of two larger. Thus only an approximate assessment is possible for what will actually happen in the field.

The calculation examines each increment of construction as if it were a small, independent reinforced soil wall. It is the deformation in the soil zone defined by the panel toe that causes the moving datum.

In the calculation, the reinforcement force required to maintain equilibrium for the panel under the soil weight and the compaction surcharge loading is estimated. The length of reinforcement over which the force will act is calculated based on the equilibrium stress states described earlier.

The incremental reinforcement force P_{INC} required to maintain equilibrium in a layer of construction of height H_{INC} is given by

$$P_{INC} = \frac{K_a \gamma H_{INC}^2}{2n} + \frac{K_a q_s H_{INC}}{n} \qquad (11)$$

where q_s is a uniform surcharge representing the load from the compaction, and n is the number of reinforcement layers connected to the panel.

The reinforcement force acts over a length of reinforcement given by the equation

$$L_{INC} = \left(H_{INC} + \frac{q_s}{\gamma} \right) \frac{\tan(90-\phi)}{2} \tag{12}$$

based on the **ideal length** equilibrium state.

The **incremental displacement** or outward movement caused by the construction of each layer is

$$\text{incremental displacement} = \frac{P_{INC} L_{INC}}{K} \tag{13}$$

where K is the stiffness of the reinforcement, taking account of the rather short loading period for an increment of construction compared with the long term design life for the wall.

The above analysis is for a translating panel; the incremental displacement at the top of the panel given by equation (13) should be **doubled** if the panel **rotates** around the base rather than **translates**.

Example

For the 8 metre high wall described earlier, and for the **uniform spacing** reinforcement layout, consider incremental construction in 8 layers (ie $H_{INC} = 1m$. There are two reinforcement layers in each increment $n = 2$). Use the same mobilised soil shearing resistance and reinforcement stiffness as before and assume that the compaction is equivalent to a uniform surcharge load $q_s = 5\ kN/m^2$.

The result when these values are substituted into equations (11) and (12) are

$$P_{INC} = 2.4\, kN/m$$

allowing for **two layers** of reinforcement per panel, and

$$L_{INC} = 1.1\, m$$

For reinforcement with a stiffness $K = 1000\, kN/m$ the **incremental displacement** is 2.6 mm from equation (13), or 5.2 mm if panel rotation rather than translation occurs.

The overall outward movement due to deformation in the reinforced zone for **uniform spacing** for the **truncated length** arrangement can now be illustrated, Fig. 19. The cumulative additional outward movement due to incremental construction amounts to 18 mm or 36 mm depending on the fixing between panels.

The effect of construction on the outward movement is much more significant in the upper part of the wall.

The **incremental displacement** depends on the number of reinforcement layers in each layer of construction. For **uniform spacing** this is the same up the wall and the incremental displacement for each layer is the same. For other spacing arrangements, such as **ideal spacing**, the number of reinforcement layers reduces higher in the wall, requiring calculation of the incremental displacement for each layer.

Because with ideal spacing the reinforcement is more widely spaced higher in the wall the incremental displacement will also be greater higher in the wall. This further increases the affect of incremental construction on the outward movement towards the top of the wall.

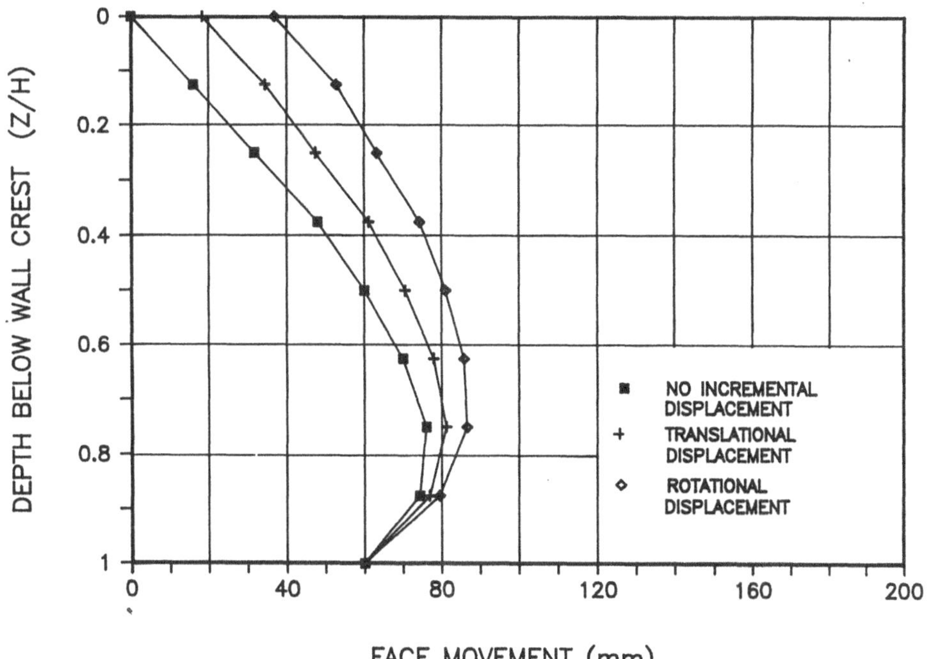

Fig. 19. The effect of incremental displacement due to incremental construction on the outward movement of the face due to deformation of the reinforced zone.

6.1 Introduction

There are almost any number of distributions of mobilised soil frictional resistance and mobilised reinforcement force which will provide equilibrium in a reinforced soil wall. For analysis, the problem has to be simplified by **assuming** that the mobilised frictional resistance is constant (at least in specified zones), and this technique has been used to investigate reinforcement force distributions which would provide equilibrium.

The two components providing equilibrium in the reinforced soil, the soil and the reinforcement, are not independent of one another, however, but linked through **strain compatibility.** The mobilised frictional resistance in the soil depends importantly on the strain which develops in the soil. Likewise, the mobilised reinforcement force depends on the strain in the reinforcement. If the unstressed reinforcement is placed in soil, then the **maximum possible** strain in the reinforcement equals the tensile strain in the soil in the direction of the reinforcement.

Between the most critical plane and the wall face the principal stress directions are approximately vertical and horizontal, as indicated in Figs. 8 and 11. In this region the reinforcement is aligned closely with the direction of principal tensile strain in the soil.

6.2 Compatibility curve

The link betweeen the mobilised soil frictional resistance and the mobilised reinforcement force can be shown on a **compatibility curve**, Fig. 20 (Jewell, 1985).

In terms of **required** and **available forces,** equilibrium can occur when the reinforcement mobilises sufficient **available force** to satisfy the requirements for equilibrium, the **required force** in the soil. This is where the two curves intersect, Fig. 20c.

A smaller unit of reinforced soil could also be examined with a compatibility curve in terms of the **local** required and available **stresses**, rather than overall forces.

The compatibility curve relating overall required and available forces in Fig. 20c is a direct illustration for **propped wall** construction. The starting point of zero tensile strain and unstressed reinforcement on the left axis is where the fill and the reinforcement have been placed, but the prop is providing all the required force. The required force is transmitted and distributed to the soil through the stiff facing.

On removal of the prop force there is a lack of equilibrium and the soil must strain to allow the reinforcement forces to be generated. As the soil strains it mobilises greater frictional resistance, thus reducing the required forces. The available reinforcement force increases as the reinforcement strains with the soil, until equilibrium is established. The overall required force is that which occurs on the most critical surface through the toe.

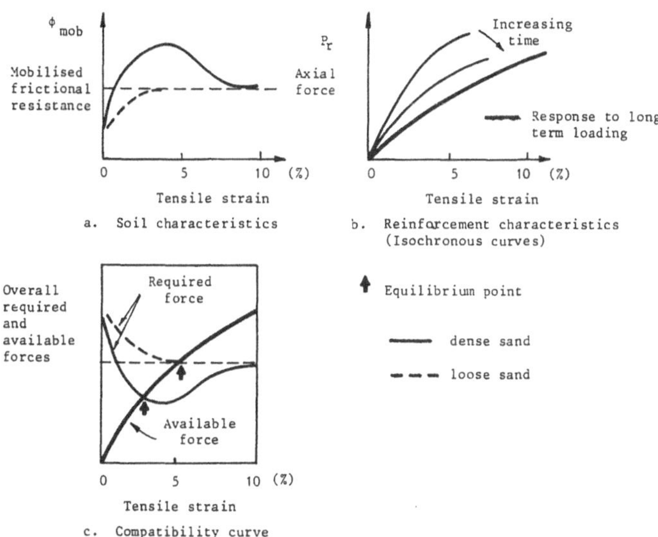

Fig. 20. Soil and reinforcement characteristics which may be used to indicate strain compatibility. (Jewell, 1985)

Stress paths

Another difference between **propped** and **incremental** wall construction is illustrated in Fig. 21. As indicated above, for a propped wall the soil will experience increasing vertical stress during construction and the horizontal stress is provided by the stiff, propped facing. Lateral strains will be small if the propping is stiff, and the stress path **RP** will lie close to the K_o line, Fig. 21a. On removal of the prop, the horizontal stress in the soil will reduce as the reinforced soil moves into equilibrium straining laterally to a higher stress ratio, **PQ** Fig. 21a.

With incremental construction the **reinforcement** must provide equilibrium at each stage. As the vertical stress increases so does the horizontal stress required to maintain equilibrium. Since the reinforcement is providing the horizontal stresses, the soil will experience increasing lateral strain throughout the construction process, therby mobilising increasing frictional resistance. The stress path in the soil for incremental construction is indicated by the line **RQ** in Fig. 21a.

What is required for an assessment of compatibility is the relationship between the mobilised frictional resistance and the principal tensile strain for the soil following stress paths similar to **RPQ** and **RQ**, under **plane strain** conditions, Fig. 21a.

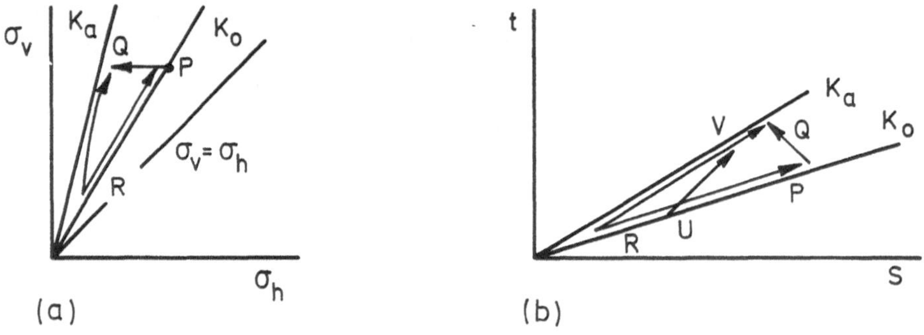

Fig. 21. Stress paths for reinforced soil for propped and incremental wall construction.

6.3 Stress path in direct shear tests

The direct shear test is the most common **plane strain** test for measuring the shearing resistance of granular fills. The stress path in a direct shear test increases both the mean and shear stresses in the soil up to the peak shearing resistance. Because the principal axes in the soil rotate during a direct shear test it is convenient to represent the test in terms of the mean normal stress **s** and mean shear stress **t**. The stress path **UV** (up to peak stress ratio) in a conventional direct shear test is shown in Fig. 21b.

Although the stress path is steeper than that experienced by the soil in an incremental wall, the direct shear test is as close as could be hoped for in a standard laboratory test.

The direct shear test is unsuited to modelling the **unloading** stress path **RPQ** for a propped wall, Fig. 21. The most suitable plane strain test would be in a biaxial apparatus, with K_0 consolidation of the sample followed by reduction in the minor principal stress.

The unloading stress path is, of course, almost identical to that in laboratory model tests on "active" retaining walls, where the wall is relaxed away from a bed of soil. The results of "active" wall experiments show that on an **unloading** stress path the soil shearing resistance can be mobilised with less lateral strain than on equivalent **loading** stress paths; see the well known results in Lambe and Whitman (1968), for example.

6.4 Strain softening

The stress paths in Fig. 21 are shown up to the peak stress ratio only. Continued shearing beyond the peak stress ratio in compact granular soils causes a gradual reduction in shearing resistance, which levels off only when the critical state has been reached at large strain.

In unreinforced dense granular soil the strain "post peak" will concentrate along discrete shear surfaces if at all possible. This is natural because once the shearing resistance has begun to reduce locally in the deforming soil, there is less resistance than in the adjacent soil so further shearing will occur preferentially in the already weakened soil. This mechanism of concentrated straining in a relatively thin or discrete band through the soil allows the overall shear strength to reduce with relatively **small deformations** at the soil boundaries.

This is the traditional concern in the design of soil slopes and walls where there may be non-uniform mobilised shearing resistance through the soil. In this case the designer anticipates that soil in a zone of relatively high strain could be rapidly losing shearing resistance with **small deformations** on **discrete shear surfaces** before the soil elsewhere has strained sufficiently to mobilise the design shear strength. The solution is to adopt relatively low values of mobilised shearing resistance for the soil, or high factors of safety.

Local strain softening would also occur in reinforced soil in the unreinforced soil behind the reinforced zone, or between the soil and the reinforcement during **outward sliding** or when the **bond limit** is reached between reinforcement and the adjacent soil.

However, in the main body of a reinforced soil wall (where **bond** is not limited) it seems unlikely that the mechanism of local strain softening would occur. The reason is that as long as there is available bond between the reinforcement and the soil any local strain along a discrete shear surface through the reinforced soil would locally mobilise additional reinforcement force which would re-establish equilibrium.

Between the most critical mechanism and the wall face there are a family of parallel potential slip surfaces which are all equally critically loaded by the soil self-weight, as was shown in Fig. 4. If the deformation mechanism is considered as slip along these parallel surfaces, slip along one particular surface would locally mobilise additional reinforcement force and hence increased shearing resistance; the adjacent parallel surfaces would then be more critical and deformation would occur next along one of them in preference. The process would then be repeated.

When designing for relatively extensible reinforcement, therefore, it is the detail of the **strain** in the soil during **strain softening** that is of interest, rather than the **deformation along a discrete shear surface** through the soil. It is the latter measurement that is recorded at the boundaries of most laboratory tests "post-peak".

6.5 Data from the simple shear test

The Cambridge University Simple Shear Apparatus is a **plane strain** test in which the central portion of the soil sample is kept uniform thereby allowing detailed direct measurement of the stress and the strain in the soil even during strain softening, Stroud (1971). The stress path in a simple shear test under a constant vertical stress is similar to the direct shear test.

The finding from simple shear tests on dense sand is that the stress ratio first increases rapidly with relatively small strain, and then stays at a high level while considerable strains develop in the soil. This is illustrated by Stroud's simple shear test results on dense Leighton Buzzard sand in Fig. 22. (Stroud, 1971). The stress ratio only gradually reduces towards the critical state stress ratio.

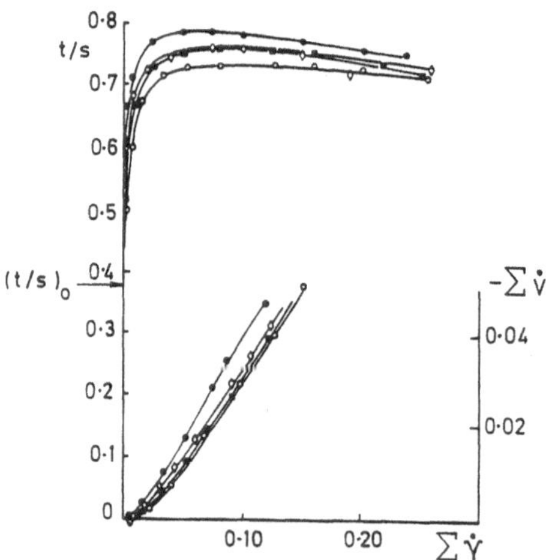

Fig. 22 Simple shear test results on dense Leighton Buzzard sand (Sroud, 1971).

The reason for this behaviour is that the **critical state shearing resistance** is reached only at the **critical state specific volume** in the sand (Schofield and Wroth, 1968). Very considerable volume expansion is required in initially dense sand to reach the critical state specific volume. This is what causes the dilation when dense sand is sheared. Thus large tensile strains develop in dense soil as it shears towards the critical state.

This feature of behaviour for granular soils is well illustrated by Stroud's simple shear data plotted in Fig. 23 (Stroud, 1971). The data are for three dense and three loose samples, and the mobilised shearing resistance (t/s) is shown plotted against the specific volume in the soil. The specific volume V_λ has been normalised to allow for different mean stress levels in the tests. What is clear from the data is the very rapid increase in shearing resistance in the dense sand which occurs with little volume change, followed by a rather gentle reduction in shearing resistance accompanied by considerable volume expansion (and tensile strain) as the soil strains towards the critical state.

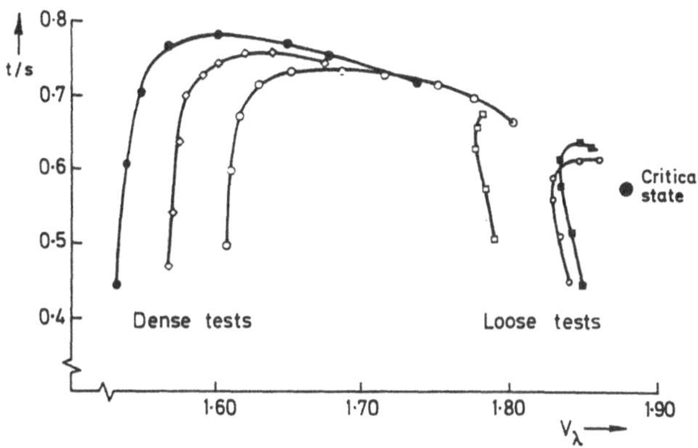

Fig. 23. The results of simple shear tests on sand plotted in terms of stress ratio versus (normalised) specific volume. (Stroud, 1971).

Similar detailed internal observation of the stress strain behaviour of sand has been made in the Imperial College Hollow Cylinder Apparatus (HCA) (Symes, 1983). The tests show the same pattern of behaviour for dense sand on similar stress paths; rapid mobilisation of shearing resistance at relatively low strain, followed by continued straining with an only gradual reduction in stress ratio.

Tests on dense sand in the HCA were also carried out on **unloading** stress paths relevant to **propped walls,** and these show, as anticipated, that less strain develops to mobilise the shearing resistance of the soil on **unloading** stress paths than on **loading** stress paths (at similar mean stress and specific volume).

6.6 Mobilised shearing resistance versus soil tensile strain

For **plane strain** design most of the soils data comes from unsophisticated direct shear tests. The triaxial test does not represent the correct loading conditions. The plane strain equivalent to the triaxial test, the biaxial test, which could provide plane strain data is not widely used.

For design against a collapse limit state the difficulty of an unsophisticated plane strain test is overcome by making a conservative estimate for the shearing resistance of the soil, and allowing suitable safety margins.

A calculation of serviceability, however, requires stress-strain properties for the soil. If there are no stress-strain data on relevant stress paths, then the only alternative for a prediction of compatibility is a simple model that captures the main features of the soil behaviour.

The simple model for sand described below allows direct shear test data to provide an approximate relationship between the mobilised soil shearing resistance and the soil tensile strain.

Simple model for sand

The soil surrounding an expanding cavity experiences large strain. An analytical solution for cavity expansion requires a relationship between the radial displacement of the cavity and the expansion pressure applied to the soil. Hughes et al (1977) solved the problem of cylindrical cavity expansion in frictional, dilating granular materials by modelling the behaviour of the sand as an elastic plastic material with the following characteristics.

The elastic behaviour is given by the shear modulus G and the drained value of Poisson's ratio ν. When the sand reaches the peak stress ratio $(t/s)_p$ it continues to deform with a constant angle of dilation ψ and at a constant stress ratio $(t/s)_p$.

For application to reinforced soil compatibility, the model can be simplified one step further. Stroud (1971) observed that the elastic volume change in dense sand was negligibly small unless the mean stresses were changing substantially along the stress path. If elastic volume change is ignored the material response to shear from an isotropic stress state can be summarised as shown in Fig. 24a, in terms of

$$(t/s)_p = \sin\phi_p \tag{14}$$

and

$$\sin\psi = \frac{d\epsilon_1 + d\epsilon_3}{d\epsilon_1 - d\epsilon_3} \tag{15}$$

where $d\epsilon_1$ and $d\epsilon_3$ are the incremental major and minor principal strains.

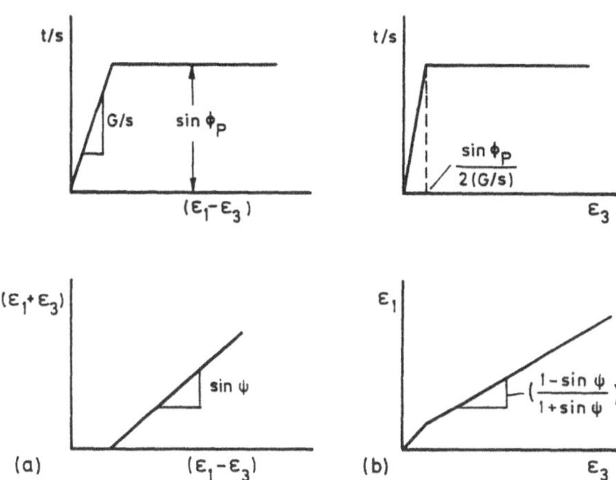

Fig. 24 Simple elastic plastic model for dense sand (after Hughes et al 1977).

The material behaviour is shown plotted as a function of the tensile strain in the soil in Fig. 24b. The peak stress ratio in the soil is reached with a mobilised tensile strain

$$\left(\epsilon_3\right)_{peak} = \frac{\sin\phi_p}{2(G/s)} \tag{16}$$

For soil sheared from an initial stress ratio, the tensile strain can be expressed in terms of the change in stress ratio $\Delta(t/s)$ to the required mobilised strength, so that

$$\left(\epsilon_3\right) = \frac{\Delta(t/s)}{2(G/s)} \tag{17}$$

Shear modulus for sand

The elastic shear modulus for sand depends on the sand density, the mean pressure and the strain amplitude. There are no well accepted relationships available for the evaluation of the shear modulus for sand in terms of other standard material properties. (See Wroth and Houlsby, 1985, for example).

An approximate estimate for the shear modulus could be made from a direct shear test. One set of assumptions to determine (G/s) would be:

(1) assume that there is uniform shear strain through the full depth of the sample for the elastic deformation so that the shear strain γ is

$$\gamma = \delta(x)/H \tag{18}$$

where $\delta(x)$ is the measured shear displacement and H is the thickness of the sand sample

(2) estimate the change in stress ratio between the start of the test and the mobilised stress ratio to which a secant shear modulus is to be calculated;

(a) if the stress path is assumed to start from an isotropic stress state and end at the peak shearing resistance then the change in the stress ratio would be

$$\Delta(t/s) = \sin\phi_p \tag{19}$$

so that the shear modulus is

$$G/s = \frac{\sin\phi_p}{(\delta(x)/H)} \tag{20}$$

(b) an alternative, which allows for the rotation of principal axes at the beginning of a direct shear test, would be to use the shear displacement and the change in stress ratio from the point of zero rate of change of volume to the desired mobilised stress ratio. The zero rate of volume change occurs at about the critical state stress ratio which would give

$$G/s = \frac{(\sin\phi_p - \sin\phi_{cv})}{(\delta(x)'/H)} \tag{21}$$

where $\delta(x)'$ is the relevant shear displacement, and ϕ_p the mobilised angle of friction.

The use of this soil model and the equations presented above is illustrated in the *companion paper* for the prediction of strain compatibility in the RMC trial wall.

REINFORCEMENT MATERIAL PROPERTIES

7.1 Relationship between force and extension

Isochronous load extension curves for the reinforcement material are a suitable way to represent the influence of time on the material properties of the reinforcement. The material properties were presented in this way for the RMC trial, see Bathurst and Jarrett (1986).

One set of isochronous curves applies at one temperature and for one chemical environment and state of mechanical damage for the reinforcement. For design purposes it is possible to make conservative assumptions - taking the highest likely temperature and the worst likely combination of soil environment and mechanical damage that might occur in the reinforced soil wall.

Isochronous curves alone may be insufficient to allow accurate estimates for stress relaxation, which might be important for long lived structures. This area of material behaviour still warrants more attention.

7.2 Stiffness at low strain

The working condition in most geotextile reinforced slopes and walls to date is one with unexpectedly low reinforcement force and deformation (see Yako and Christopher (1987) for recent examples). There are additional factors which can affect accurate predictions of working conditions with these low reinforcement loads.

Firstly, there is often **insufficient test data** for the reinforcement at low load.

Secondly, the **pre-conditioning** of the sample before testing (initial load-unload cycles when setting up the sample) and the **gripping arrangements** can influence the initial load-extension properties.

Construction also inevitably introduces factors which are more significant with relatively low reinforcement load, and hence extension - **slack in connections, misaligned reinforcement**, for example.

Another factor that is important at low loads is early changes in the stiffness of the reinforcement with load level. The "S" shaped load extension curve for some polyester materials is a well known example - relatively low initial stiffness which <u>increases</u> with increasing load (see Zanten (1986), for example). Change in stiffness in the early part of tests can become obscured by corrections to laboratory data to determine the <u>origin</u> for strain.

BOUNDARY CONDITIONS

8.1 Introduction

The boundary conditions for a reinforced soil wall are likely to affect the forces and displacements in the soil and the reinforcement. In general the boundary conditions for walls built on competent foundations are likely to reduce the forces and displacements in the reinforced soil compared with the assumed ideal boundary conditions.

In this paper it has been assumed that only the reinforcement provides additional stabilising forces in the soil, and that at the wall face the reinforcement tensile force is neatly transferred into a uniformly varying compressive stress on the soil surface. This may be thought of as **ideal facing**, as the face adds no kinematic restraint and is perfectly free from interaction with the foundation.

8.2 Base boundary

The height of a wall is determined by the **base boundary**. The base boundary defines the position of the **most critical surface** through the toe, and divides the soil into the uniform zone at the face and the zone of continuously varying stresses behind the most critical surface.

The rigidity and roughness of the base boundary affects the stability conditions in the zone behind the most critical surface. The base boundary can be considered to have **roughness** and **strength** provided by the foundation soil.

If the base boundary is not fully **rough** then potential failure mechanisms with a horizontal portion along the surface of the foundation may increase the required reinforcement forces behind the most critical surface.

A base foundation with inadequate shearing **strength** to support the vertical and shear loading from the wall at the foundation level would inevitably directly affect the equilibrium in a reinforced soil wall. This case of inadequate foundation bearing capacity is not considered further.

If the base boundary is both rough and strong (which is most common) it acts similarly to a plane of rough inextensible reinforcement. The additional "reinforcing" reduces the tensile strain in the soil near the base of the wall and hence reduces the mobilised reinforcement force in the lowest layers of reinforcement.

8.3 Face boundary and connection with the base

The influence of the face boundary can be thought of in terms of the face **continuity** and **stiffness,** and the **connection with the base**.

The continuity of the face, the connection between panels, and the stiffness of the face will provide kinematic restraint at the face on the overall deformation in the reinforced soil. An example was given earlier where the connection between panels was anticipated to influence the magnitude of construction induced movements by either allowing the panel to translate or forcing it to rotate about the base, (section 5.2).

A very stiff (unbending) continuous face, as in most **propped wall** constructions, could act to locally <u>increase</u> the mobilised reinforcement force close to the face in the upper part of a wall. This would occur if additional outward deflection higher in the wall caused by the stiff face pulled the reinforcement layers outwards. Local overstressing of the reinforcement adjacent to the face caused by a stiff, continuous face is probably not a problem in practice as practical reinforcement spacing arrangements typically provide more reinforcement layers than needed near the top of a wall.

The connection between the face and the base can also introduce large stabilising forces into a reinforced soil wall thereby reducing the reinforcement forces. Both vertical and horizontal force can be transmitted from the foundation to the wall.

The horizontal force would be transferred to the reinforced soil as a normal stress at the face and would reduce the required reinforcement force proportionally at that elevation. The vertical force at the face would provide an upward shear stress on the reinforced soil effectively making the front boundary "rougher", in the sense of conventional active earth pressure theory. This would locally reduce the vertical stresses in the reinforced soil and hence the required reinforcement force.

8.4 Summary on boundary conditions for practical walls

Most reinforced soil walls are built incrementally with connected facing panels and on a competent foundation. These boundary conditions reduce the reinforcement forces in the lower reinforcement layers close to the base of the wall.

This comes from two main sources. Firstly, the **rough, strong base boundary** acts like an additional substantial reinforcement layer at the base of the wall. Secondly, the **connection between the face and the base** can exert horizontal stabilising stresses directly to the surface of the soil through the facing. These stresses directly reduce the local required reinforcement forces for equilibrium.

8.5 Sidewall friction

When small sections of wall are tested in the field or laboratory the lateral boundary can also be important. Usually a test section is confined between rigid parallel boundaries and any movement in the soil results in shear deformation between the soil and the stationary side boundary. Any shearing resistance between the soil and the side boundary mobilises boundary forces directly resisting the deformation in the soil. So that in a Coulomb wedge calculation, for example, a net side wall force acting in the opposite direction to the implied soil movement should be included in the equilibrium force balance.

Bransby and Smith (1975) used a detailed numerical method to investigate the influence of side wall friction and wall geometry (the ratio of wall height to width H/w) on the active and passive earth pressure coefficients K_a and K_p for unsurcharged vertical walls. They found that the influence of side wall friction on the active earth pressure coefficient was quite small, a 14% reduction for a wall geometry H/w = 2 and a boundary friction 5.7°.

For self-weight loading the influence of the boundary roughness calculated by Bransby and Smith (1975) is approximately uniform with depth. This is not the case for surcharge loading where the side wall friction cumulatively reduces the vertical stress in the soil. Side wall friction can be more important for surcharge loading than for self weight loading. Simple closed form analysis for side wall friction is presented in the *companion paper* for self weight and surcharge loading, where it is suggested that it has a significant influence on the RMC trial wall.

CONCLUSIONS

A theoretical analysis for the behaviour of reinforced soil walls has been described, and the derivation of key soil and reinforcement material properties has been considered. The ideas are illustrated by practical application in a companion paper where they are used to make an analysis of the behaviour of the RMC trial walls.

The main points and conclusions from this paper are summarised below.

9.1 Equilibrium states

Two possible states of equilibrium have been presented for reinforced soil walls, together with the analytical equations describing them.

The **ideal length** equilibrium case is where the reinforcement locally provides the required stress in the soil everywhere that it is needed. The unreinforced soil behind the reinforced zone does not load the reinforced zone.

The **truncated length** equilibrium case is where equilibrium in the soil is maintained with the reinforcement everywhere carrying the maximum allowable force. In this case there are unreinforced zones (behind the reinforced soil) that require reinforcement stresses but do not contain reinforcement. These come into an active state of equilibrium loading the back of the reinforced zone, and the required reinforcement force is thereby shed to the lower reinforcement layers.

It has been suggested that the two equilibrium cases are likely to "bound" the actual state of equilibrium in a reinforced soil wall.

9.2 Reinforcement force and deformation

The distribution of the reinforcement force in equilibrium with the soil is described analytically for the two equilibrium cases. The magnitude of the reinforcement force depends on the reinforcement spacing. The concept of a balanced reinforcement layout satisfying local and overall equilibrium has been reviewed, and this determines the **ideal spacing** arrangement.

The elongation in the reinforcement is an important component of the outward movement in a reinforced soil wall. The analytical calculation for the reinforcement elongations in the two equilibrium cases is provided in **non dimensional charts** which allow the deformation in the reinforced zone to be directly calculated. The movement depends only on the mobilised soil strength, the mobilised reinforcement force and the reinforcement stiffness. The charts are applicable to any reinforcement spacing arrangement.

The important finding (which is particularly emphasised in the *companion paper)* is that the difference in the movement calculated from the two "bounding" equilibrium cases is rather small, and the analysis therefore predicts the deformation in the reinforced zone to within close limits.

9.3 Incremental construction movement

Aligning facing panels during incremental construction above the previously built facing which has already deformed causes a **moving datum** for the position of the initially unstressed reinforcement layers. This leads to cumulative additional outward movement that must be taken into account for each layer. The additional movement can be thought of as the deformation that would arise if the incremental layer were built as an independent wall.

An analysis for incremental movement is presented. The incremental movement should be added to the movement caused by the reinforcement elongation to give the overall movement at the face. Because the incremental movement is cumulative it can significantly affect the outward movement in the upper half of a wall.

Perhaps <u>the</u> most important distinction between an **incremental wall** and a **propped wall** is that propping during filling eliminates the **moving datum** and hence the incremental construction movement. For the propped wall the outward face movement is due only to the reinforcement elongation.

9.4 Strain compatibility

The tensile strain in the soil and the reinforcement provides the link between the mobilised values of soil shearing resistance and reinforcement axial force. This determines the conditions for equilibrium in the reinforced soil. The tensile strain in initially unstressed reinforcement cannot exceed the tensile strain in the adjacent soil.

A **compatibility curve** can be used to estimate the equilibrium state, and this is most conveniently carried out in terms of the overall equilibrium of forces on the most critical surface.

Biaxial tests, which would be the the most appropriate source of **plane strain** data for the soil stress-strain characteristics, are not widely used. In the more common **plane strain** direct shear test the relationship between the mobilised shearing resistance and the soil tensile strain is not measured.

A simple elastic plastic soil model is proposed for use with the compatibility curve which allows an estimate to be made of the compatible equilibrium strain and mobilised shearing resistance in a reinforced soil wall. A method is given for estimating the relationship between the soil strength and tensile strain from a direct shear test on sand. The application of the method is illustrated in the *companion paper*.

9.5 Summary

The theory and analysis in this paper provides a complete method for estimating the serviceability of a reinforced soil wall. Two solutions can be derived which are expected to "bound" the actual equilibrium. **Non dimensional charts** and equations are presented which allow the reinforcement force distribution along the layers (the maximum force distribution in the layers is the same in both cases) and the face deformation due to the reinforcement elongations to be calculated. In practice, the difference between the predicted movements for the two "bounding" equilibrium cases is relatively small giving a rather precise prediction for the likely movement of a reinforced soil wall.

Separate and additional movement due to incremental construction of a reinforced soil wall has been identified. This provides the major distinction between an **incremental wall** where it occurs and a **propped wall** where it does not occur. The calculation for the construction induced movement is very much less precise, although the magnitude of movement is shown to be significant, particularly higher in the wall. An analysis is given for incremental construction movement.

Acknowledgements

Aspects of the reported study were completed while the author was supported by the Royal Society/SERC industrial fellowship scheme. The author is grateful to the Soil Mechanics Group at the Department of Engineering Science, University of Oxford, for stimulating support, to Guy Houlsby for the many discussions on the behaviour of sand, and to George Milligan for thoroughly reviewing the manuscript and making many useful suggestions.

References

Bathurst, R.J. & Jarrett, P.M. (1986). Class A prediction exercise for reinforced earth walls. *Bulletin No. 1 for NATO Advanced Research Workshop, Application of Polymeric Reinforcement in Soil Retaining Structures*, Royal Military College, Kingston.

Bransby, P.L. & Smith, I.A.A. (1975). Side friction in model retaining wall experiments. *Journal of Geotechnical Engineering,* **ASCE GT7,** July, 615-632.

Hughes, J.M.O., Wroth, C.P. & Windle, D. (1977). Pressuremeter tests in sand. *Geotechnique* **27,** 455-477.

Jewell, R.A. (1985). Limit equilibrium analysis of reinforced soil walls. *Proc. 11 Int. Conf. Soil Mech. Fdn Engng, San Francisco,* **Vol 3,** 1705-1708.

Jewell, R.A. (1987). Analysis and predicted behaviour for the Royal Military College trial wall. *Proc. NATO Advanced Research Workshop, Application of Polymeric Reinforcement in Soil Retaining Structures,* Martinus Nijhoff.

Jewell, R.A., Paine N.P. & Woods R.I. (1984). Design methods for steep reinforced slopes. *Proc. Int. Conf. Polymer Grid Reinforcement, London,* 70-81.

Jewell, R.A. and Wroth, C.P. (1987). Direct shear tests on reinforced sand. *Geotechnique* **37,** No. 1, 53-68.

Lambe, T.W. & Whitman, R.V. (1968). *Soil Mechanics.* John Wiley, New York.

Stroud, M.A. (1971). *The behaviour of sand at low stress levels in the simple shear apparatus.* PhD Thesis, University of Cambridge.

Symes, M.J.P.R. (1983). *Rotation of principal stresses in sand.* PhD Thesis, Imperial College, London.

Schofield, A.N. & Wroth, C.P. (1968). *Critical State Soil Mechanics,* McGraw Hill, London.

Wroth, C.P. & Houlsby, G.T. (1985). Soil mechanics - property characterisation and analysis procedures. *Proc. 11 Int. Conf. Soil Mech. Fdn Engng, San Francisco,* **Vol 1,** 1-55.

Yako, M.A. & Christopher B.R. (1987). Polymerically reinforced retaining walls and slopes in North America. *Proc. NATO Advanced Research Workshop, Application of Polymeric Reinforcement in Soil Retaining Structures,* Martinus Nijhoff.

Zanten, R.V. Van (ed.) (1986). *Geotextiles and Geomembranes in Civil Engineering,* Balkema, Rotterdam.

REINFORCEMENT EXTENSIBILITY IN REINFORCED SOIL WALL DESIGN

Rudolph Bonaparte, GeoServices Inc. Consulting Engineers, USA
Gary R. Schmertmann, Graduate Student, University of California, Berkeley, USA

1. INTRODUCTION

There are many types of reinforcing materials and systems available for the construction of reinforced soil walls. Of the many types, the Reinforced Earth system developed by Vidal (1966) in France has predominated and has provided the basis for most theoretical and empirical knowledge of the behavior of reinforced soil walls. The Reinforced Earth system has a number of distinguishing characteristics that include:

- steel reinforcing elements that have tensile moduli on the order of 2 x 10^8 kPa (3 x 10^7 lbs/in^2);
- reinforcing elements that are discrete strips, approximately 50 mm (2 in.) wide and 5 mm (0.2 in.) thick; and
- concrete facing (skin) elements that can individually undergo limited translation and rotation in response to movements in the reinforced fill or settlements of the foundation soils.

More recently, reinforced soil walls have been constructed with geosynthetic reinforcement and various facing elements. The two types of geosynthetics commonly used in reinforced soil wall construction are geogrids and geotextiles. Geosynthetic soil reinforcement systems have distinguishing characteristics that include:

- polymer reinforcing elements that have tensile moduli on the order of 1 x 10^6 kPa (1.5 x 10^4 lbs/in^2);
- reinforcing elements that are continuous or semi-continuous sheets with thicknesses in the range of 1 to 5 mm (0.04 to 0.20 in.); and
- a variety of possible facing (skin) elements including concrete panels, timbers, and geosynthetics.

It is clear that the Reinforced Earth system and geosynthetic reinforcement systems incorporate fundamentally different types of reinforcing elements: steel strips versus polymer sheets. While the nature of the reinforcing elements are different, the most commonly

P. M. Jarrett and A. McGown (eds.), The Application of Polymeric Reinforcement in Soil Retaining Structures, 409–457.

used procedures for analysis and design are similar. These procedures derive largely from research on Reinforced Earth and other steel reinforcement systems. There prevails an inherent assumption that analysis and design procedures developed for steel reinforcement are also applicable to geosynthetic reinforcement. This assumption is questionable.

This paper will examine differences in the behavior of soil walls reinforced with steel strips or other relatively "inextensible" materials (such as steel grids) and soil walls reinforced with relatively "extensible" geosynthetic materials (geotextiles or geogrids). This comparison will be made for the ideal case of a frictionless wall facing, braced during construction, and free to rotate about its toe after construction. Wall systems and construction procedures causing deviations from this "ideal" case will be discussed. Analysis and design procedures for reinforced soil walls will be reviewed and conclusions will be drawn on the procedures most appropriate for use with geosynthetic reinforcement. These procedures will be used to conduct parametric studies and to predict the behavior of the instrumented geogrid-reinforced soil walls constructed at the Royal Military College (RMC) in Canada.

2. COMPARISON OF INEXTENSIBLE AND EXTENSIBLE REINFORCEMENT

There are two basic differences between steel and geosynthetic reinforcing elements:

- as stated in the introduction, the tensile moduli of steel reinforcement and geosynthetic reinforcement are vastly different: the ratio, E_s/E_g, is on the order of 100 to 1000 (E_s = modulus of elasticity of steel and E_g = low-strain tensile modulus of polymers used in geosynthetics); and
- as shown by design experience, the volume of steel in a steel-reinforced soil wall is much smaller than the volume of polymer in a comparable geosynthetic-reinforced wall (0.02% to 0.05% for steel reinforcement compared to 0.2% to 0.5% for geosynthetic reinforcement).

The consequences of these differences are discussed below.

2.1 Definitions

McGown et al. (1978) originally defined inextensible and extensible reinforcements (inclusions) as follows:

- inextensible reinforcements (inclusions) are those that "have rupture strains which are less than the maximum tensile strains in the soil without inclusions, under the same operational conditions"; and
- "extensible reinforcements (inclusions) are those that have rupture strains larger than the maximum tensile strains in the soil without inclusions, under the same operational conditions".

These definitions are difficult to interpret since the word extensible relates to the stress-strain response of a material rather than to the material's rupture strain. A better understanding of the meaning of the two terms, extensible and inextensible, can be obtained by comparing the horizontal strain in an element of reinforced soil subjected to a given load, to the strain required to develop an active plastic state in an element of the same soil without reinforcement:

- inextensible reinforcement is reinforcement used in such a way that the tensile strain in the reinforcement is significantly less than the horizontal extension required to develop an active plastic state in the soil; and

- extensible reinforcement is reinforcement used in such a way that the tensile strain in the reinforcement is equal to or larger than the horizontal extension required to develop an active plastic state in the soil.

As will be shown subsequently, steel reinforcement meets the criterion for inextensible reinforcement in most practical applications. It will also be shown that currently available geosynthetic reinforcing materials meet the criterion for extensible reinforcement in almost all practical applications.

Two extreme cases of extensible and inextensible reinforcement can be considered: (i) an "absolutely" inextensible reinforcement which is so stiff that equilibrium is achieved at virtually zero horizontal extension (K_0 conditions theoretically prevail in the reinforced soil mass if compaction induced stresses, soil arching, and other secondary effects are ignored); and (ii) an "absolutely" extensible reinforcement which has such a low modulus that virtually no tensile forces are introduced to the soil mass at the strain required to develop an active plastic state (K_a conditions theoretically prevail).

2.2 Stresses and Strains in Unreinforced Soil Element

A small element of compacted granular soil is considered at some intermediate point in the construction of a retaining wall with a frictionless, rigid facing and no reinforcement (Figure 1a). Initially, the element of soil is in an at-rest state of stress (point A in Figures 2a and 2b). As successive fill levels are placed, the major and minor principal stresses acting on the soil element increase. Several possible stress increments associated with wall construction are shown in Figure 1b and corresponding stress paths are shown in Figure 2a. Stress increments include: (i) compression loading; (ii) compression unloading; (iii) proportional loading; and (iv) general incremental loading. It is instructive to relate these stress increments to the soil element "A" in Figure 1a. If the wall facing is braced during fill placement, the minor and major principal stress increments (ignoring compaction stresses) will be proportional and their ratio will be equal to K_0 (line segment AB in Figure 2a). If the braces are removed after construction and the facing is free to rotate about its base, σ_3 will decrease by $\sigma_1 \cdot (K_0 - K_a)$ along a compression unloading stress path (line segment BC in Figure 2a) as the fill yields into an active plastic state.

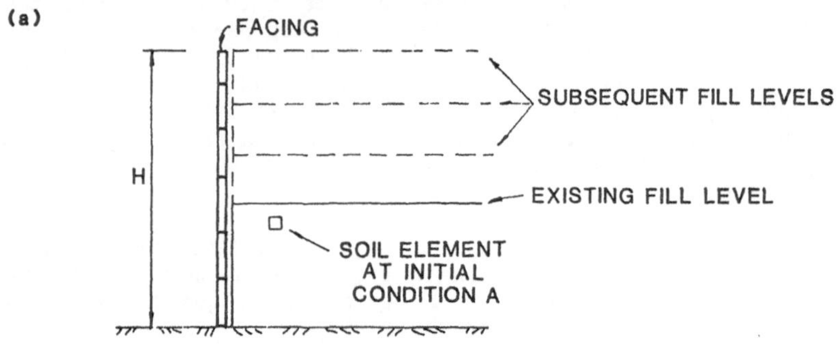

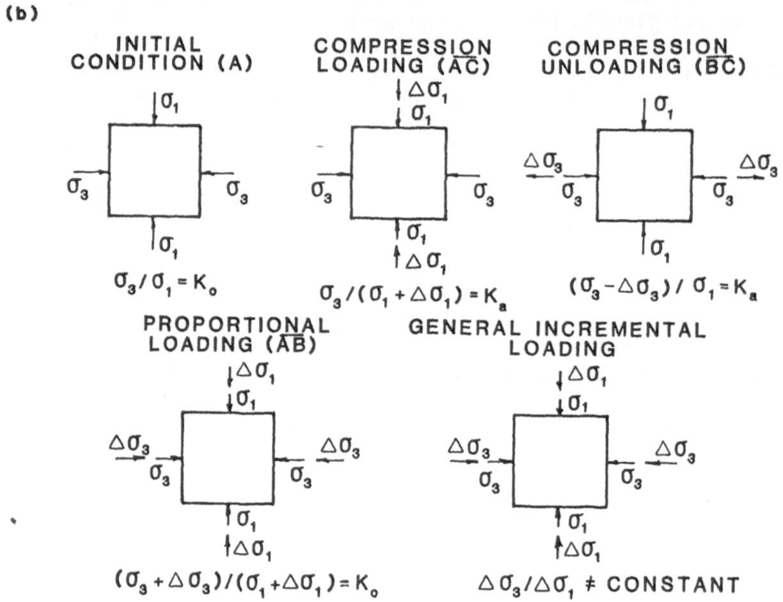

Figure 1. Possible stress states for soil element A in a reinforced soil wall: (a) soil element "A"; (b) stress states.

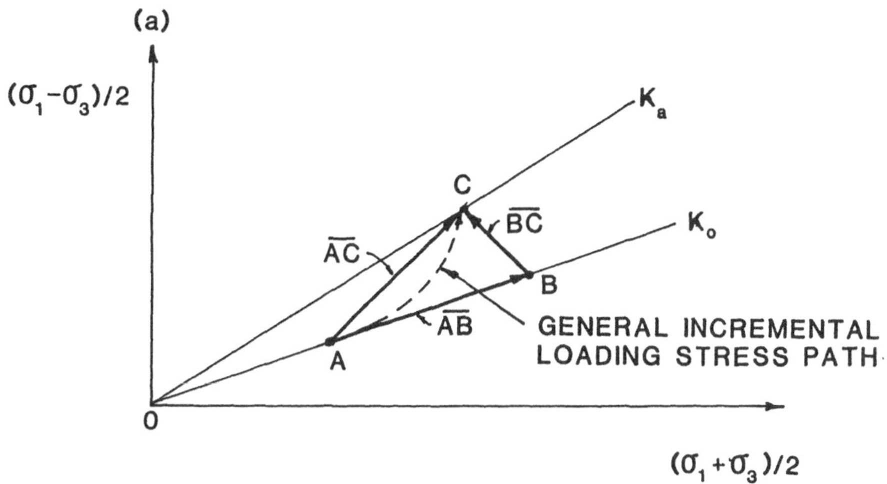

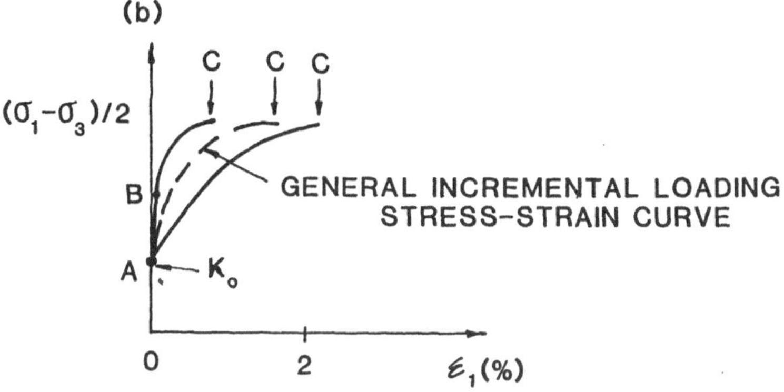

Figure 2. Idealized stress-paths and stress-strain curves for stress states in Figure 1: (a) stress paths; (b) stress-strain curves.

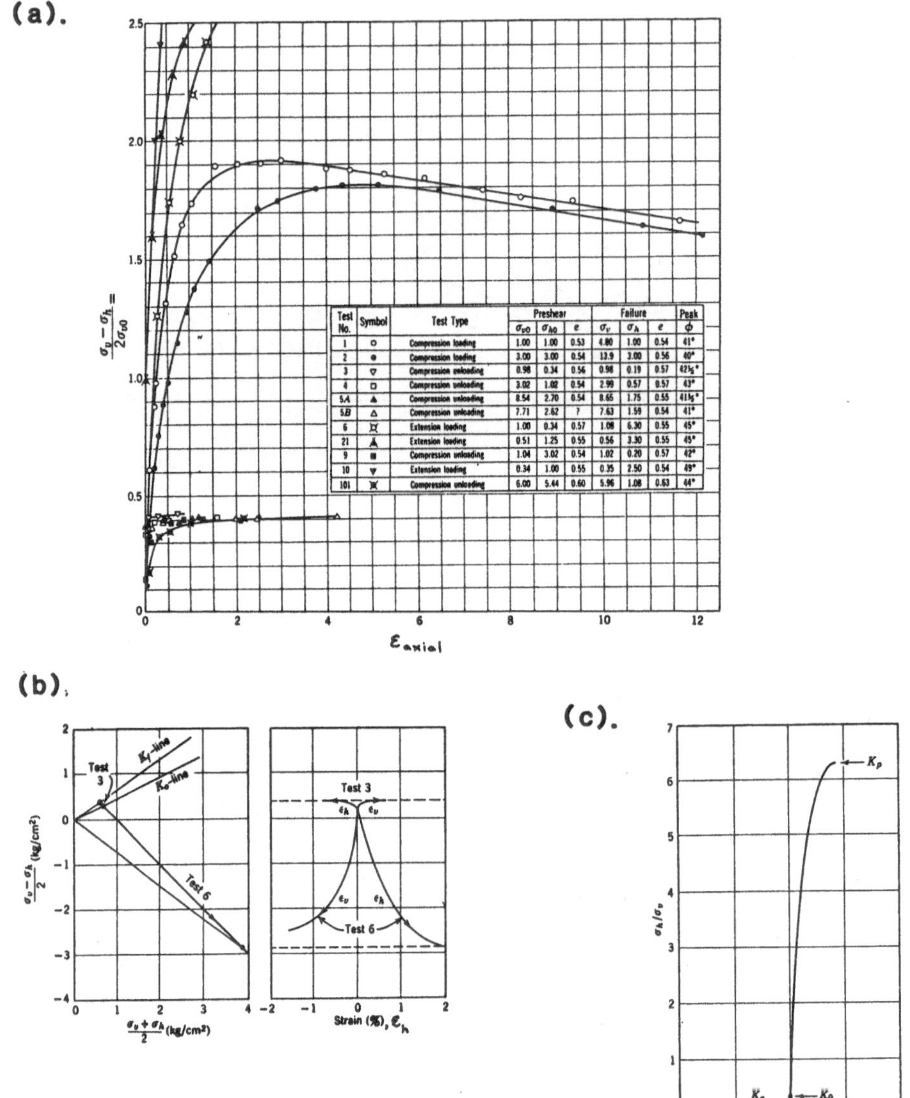

Test No.	Symbol	Test Type	Preshear σ_{v0}	Preshear σ_{h0}	Preshear e	Failure σ_v	Failure σ_h	Failure e	Peak ϕ
1	○	Compression loading	1.00	1.00	0.53	4.80	1.00	0.54	41°
2	●	Compression loading	3.00	3.00	0.54	13.9	3.00	0.56	40°
3	▽	Compression unloading	0.98	0.34	0.56	0.98	0.19	0.57	42½°
4	□	Compression unloading	3.02	1.02	0.54	2.99	0.57	0.57	43°
5A	▲	Compression unloading	8.54	2.70	0.54	8.65	1.75	0.55	41½°
5B	△	Compression unloading	7.71	2.62	?	7.63	1.59	0.54	41°
6	⌑	Extension loading	1.00	0.34	0.57	1.08	6.30	0.55	45°
21	⩓	Extension loading	0.51	1.25	0.55	0.56	3.30	0.55	45°
9	■	Compression unloading	1.04	3.02	0.54	1.02	0.20	0.57	42°
10	▼	Extension loading	0.34	1.00	0.55	0.35	2.50	0.54	49°
101	⌘	Compression unloading	6.00	5.44	0.60	5.96	1.08	0.63	44°

Figure 3. Stress-paths and stress-strain curves for dense sand specimens in a triaxial tests: (a) stress-strain curves; (b) stress and strain paths; (c) plot of lateral earth pressure coefficient, K, versus horizontal strain derived from test results for compression unloading and extension loading. (From Lambe and Whitman, 1968.)

In contrast to the above case, most reinforced soil walls are built with facing elements that can deform incrementally as the wall is built. Consider soil element "A" for this case. Initially, the soil element will follow the K_0 stress path (again ignoring compaction stresses). As lateral stresses induce deformation of the wall face, however, $\Delta\sigma_3 < K_0\Delta\sigma_1$. Eventually, $0 < \Delta\sigma_3 < K_a\Delta\sigma_1$. The stress path for this general incremental loading must fall between the stress paths in Figure 2a for compression loading (line segment AC in Figure 2a) and K_0 loading followed by compression unloading.

Stress-strain curves associated with the stress paths in Figure 2a are shown in Figure 2b. The curves in Figure 2b have been idealized from data such as that shown in Figure 3. Starting from K_0 conditions, the major principal strain for soil element "A" corresponding to the stress path AB + BC is 0.5% or less (Lambe and Whitman, 1968). Assuming $\epsilon_3 \approx -\epsilon_1$ (i.e, $\epsilon_2 = 0$), $\epsilon_h = \epsilon_3 < 0.5\%$ for an active plastic stress state. For compression loading, $\epsilon_h \approx 2\%$ to 3%. By deduction, for general incremental loading $\epsilon_h \approx 1\%$ to 2%.

2.3 Stresses and Strains in Reinforced Soil Element

It was just shown that for an element of unreinforced compacted granular soil, $\epsilon_h < 0.5\%$ for a compression unloading stress path, and $\epsilon_h \approx 1\%$ to 2% for a general incremental loading stress path. In this section, the mobilized reinforcement forces at these strain levels are investigated.

An element of compacted granular soil is considered to be initially in an at-rest (K_0) state of stress. Plane-strain deformations are assumed and compaction induced stresses are neglected. Principal stress rotation due to the horizontal tensile stresses is also neglected and thus the ratio of minor to major principal stress is taken as:

$$K = \sigma_h/\sigma_v \tag{1}$$

where σ_v is the vertical effective stress and σ_h = horizontal effective stress.

To evaluate the influence of reinforcement extensibility, the relationship between K and horizontal strain, ϵ_h, will be defined in the subsequent analyses using a hyperbola. (Hyperbolas have been widely used in soil mechanics to describe the shape of stress-strain curves (e.g., Kondner, 1963; Duncan and Chang, 1970).) The hyperbola used to define the K-ϵ_h curve in Figure 4a is given by:

$$K_0 - K = \frac{K_0 - K_a}{1 + (\lambda/\epsilon_h)} \tag{2}$$

where K_0 is assumed to be given by Jaky's (1944) equation:

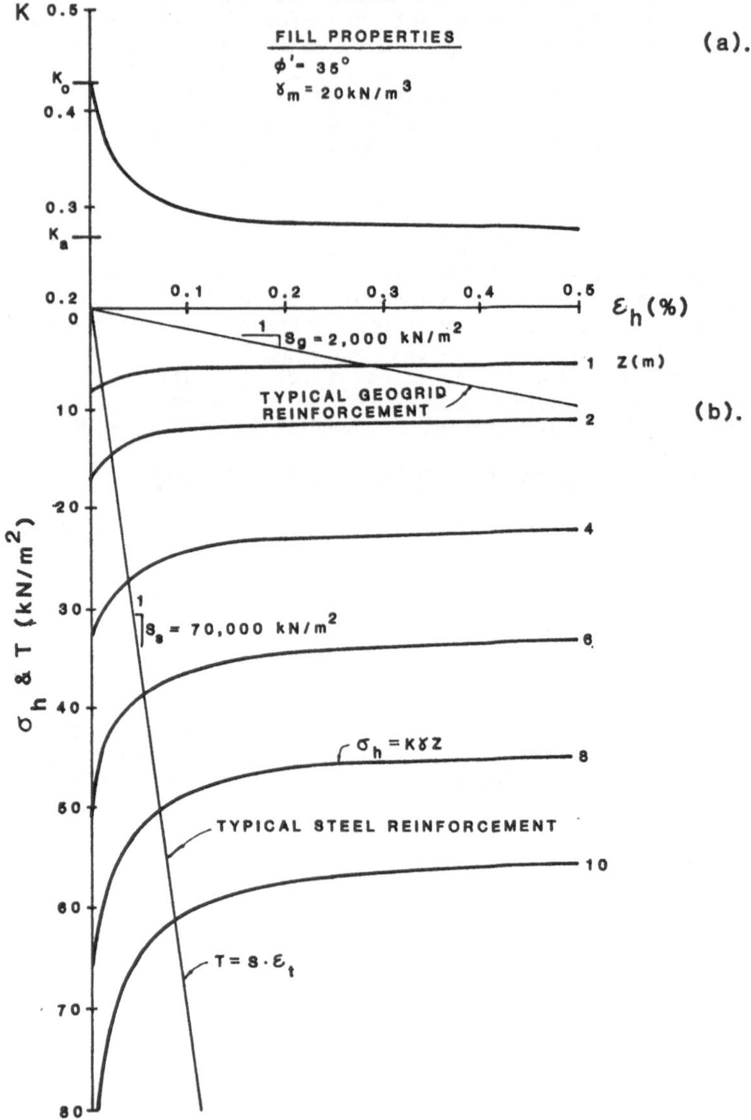

Figure 4. Graph for estimating equilibrium strains in an ideal element of reinforced soil subject to a compression unloading stress path: (a) hyperbolic curve of coefficient of lateral earth pressure, K, versus horizontal strain ϵ_h; and, (b) relationship between horizontal effective stress, σ_h, and ϵ_h. (See Figure 4 for definitions of S_s and S_g.)

$$K_o = 1 - \sin\phi \tag{3}$$

and K_a corresponds to the active Rankine state of stress, given by:

$$K_a = \frac{1 - \sin\phi}{1 + \sin\phi} \tag{4}$$

In Equation 2, $K = K_a$ at infinite strain. The parameter "λ" controls the relationship between $(K_o - K)$ and ϵ_h. The value of "λ" can be estimated from test results or backcalculated from a set of assumed conditions. To backcalculate "λ" it is assumed that $(K_o - K) = 0.95\ (K_o - K_a)$ at 0.5% strain for the compression unloading stress path. In other words, at 0.5% strain, σ_h is assumed to have decreased from $K_o\sigma_v$ to a value just slightly larger than $K_a\sigma_v$. The hyperbolic curve shown in Figure 4a is related to a granular soil with $\phi = 35°$ and $\gamma = 20$ kN/m³ (127 lb/ft³). For this soil, $K_o = 0.426$, $K_a = 0.271$, $(K_o - K_a) = 0.155$ and $\lambda = 2.6 \times 10^{-4}$.

Figure 4b shows the relationship between σ_h and ϵ_h obtained using K values, calculated as indicated above, in the equation $\sigma_h = K\gamma z$, with $\gamma = 20$ kN/m³. If it is assumed that the reinforcement tension just balances σ_h, the curves in Figure 4b represent the reinforcement force per unit soil area required to maintain equilibrium in the soil element. With this interpretation, Figure 4b can be used to estimate the strains in a soil element with reinforcements having various tensile stiffnesses.

To lend practical value to these estimates the steel and geosynthetic reinforcement configurations shown in Figure 5 are considered. Figure 5a represents a standard Reinforced Earth panel with four steel reinforcing elements (strips). Typical dimensions of the steel strips are given in the figure. From the information provided in Figure 5 a reinforcement stiffness per unit soil area (with the soil area taken normal to the direction of the reinforcing element) can be defined for steel strips as follows:

$$S_s = E_s \cdot N \cdot a \cdot b/A \tag{5}$$

where: S_s = steel reinforcement stiffness per unit soil area (kPa); E_s = tensile modulus of the reinforcement (kPa); N = number of reinforcing elements (strips or layers) per wall facing panel; a and b = dimensions of the reinforcing elements (strips or layers) as shown in Figure 5. For a Reinforced Earth structure with $E = 2 \times 10^8$ kPa (29×10^6 lbs/in²), N = 4, a = 4 mm (0.16 in.), b = 50 mm (2 in.), and A = 2.25 m² (24.5 ft²), S_s = 70,000 kPa (1.5×10^6 lbs/ft²).

For geosynthetic reinforcement, the tensile modulus, E_g, and thickness, a, are usually combined into a reinforcement stiffness per unit width, J_g (kN/m). In this case, Equation 5 becomes:

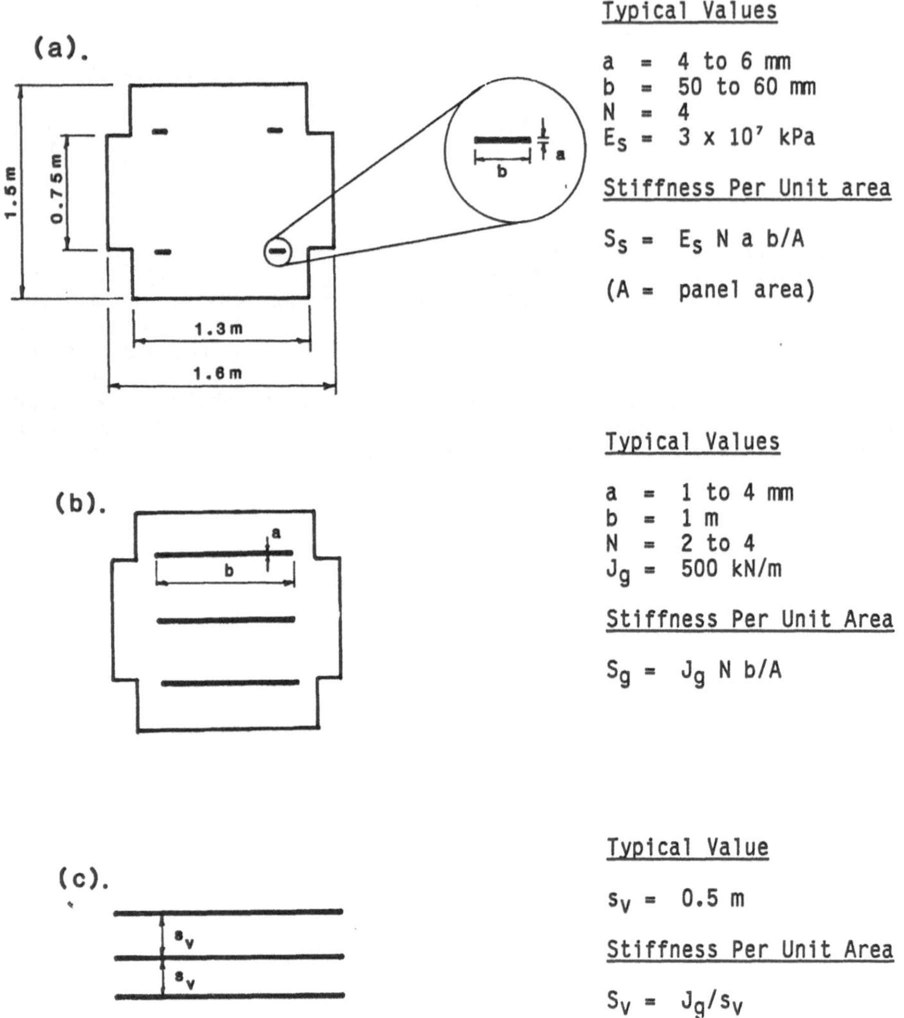

Figure 5. Equation for estimating the reinforcement stiffness per unit soil area, S (kPa): (a) steel strip reinforcement (S_s) with a segmental facing panel; (b) geosynthetic reinforcement with a segmental facing panel (S_g); and (c) continuous layers of geosynthetic reinforcement. The unit soil area is normal to the direction of the reinforcing elements.

$$S_g = J_g \cdot N \cdot b/A \quad (6)$$

Equation 6 is valid in the case of geosynthetic strips (Figure 5b). In the case of continuous layers, the reinforcement stiffness is given by:

$$S_g = J_g/s_v \quad (7)$$

where s_v is the vertical spacing between reinforcement layers.

To estimate S_g for a typical geosynthetic reinforcement application, the Tucson, Arizona retaining wall project presented by Berg et al. (1986) is considered. In this project, continuous layers of geogrid reinforcement were used with vertical spacings between layers ranging from 0.30 m to 0.75 m (12 and 30 in.). The low-strain secant J_g of the geogrid used on the Tucson project is on the order of 600 kN/m (40,000 lbs/ft). For the case of s_v = 0.3 m (12 in.), $S_g \approx$ 2,000 kPa (40,000 lbs/ft²).

Once S_s or S_g is obtained, the required reinforcement force per unit soil area, T (kN/m²), can be related to the reinforcement strain as follows:

$$T = S \cdot \epsilon_t \quad (8)$$

where: S = S_s (for steel) or S_g (for geosynthetic) as defined by Equations 5, 6, or 7; and ϵ_t = tensile strain in the reinforcement. In Figure 4b, the line for S_s = 70,000 kPa is related to steel reinforcement and the line for S_g = 2,000 kPa is related to geosynthetic reinforcement. Equation 8 is represented by straight lines in Figure 4b.

In this paper it is assumed that $\epsilon_t = \epsilon_h$. The use of $\epsilon_h = \epsilon_t$ implies that: (i) the presence of reinforcement does not locally affect soil strains (which is not true for steel and may not be true for geosynthetics); and (ii) there is no slippage between the soil and reinforcement. (This is a good assumption for geogrids reinforcing granular fill since the fill penetrates the apertures of the grid. The assumption may be less appropriate for geotextiles where there is a greater likelihood of localized slip between the fill and geotextile). The intersections of the T-ϵ_t lines with the σ_h-ϵ_h lines in Figure 4b represent equilibrium states for the reinforced soil element. These two sets of lines are analogous to the strain compatibility curves presented by Jewell (1985b).

The strains obtained from Figure 4b have been replotted in Figure 6. This figure includes results for a range of steel reinforcement stiffnesses (S_s = 70,000 to 300,000 kPa) and geosynthetic reinforcement stiffnesses (S_g = 500 to 3,300 kPa). It can be seen that for steel reinforcement, the equilibrium strains are on the order of 0.01% to 0.1%. The range of strains in Figure 6 for geosynthetic reinforcement is large due to the wide range of tensile stiffnesses associated with currently available products. Even for

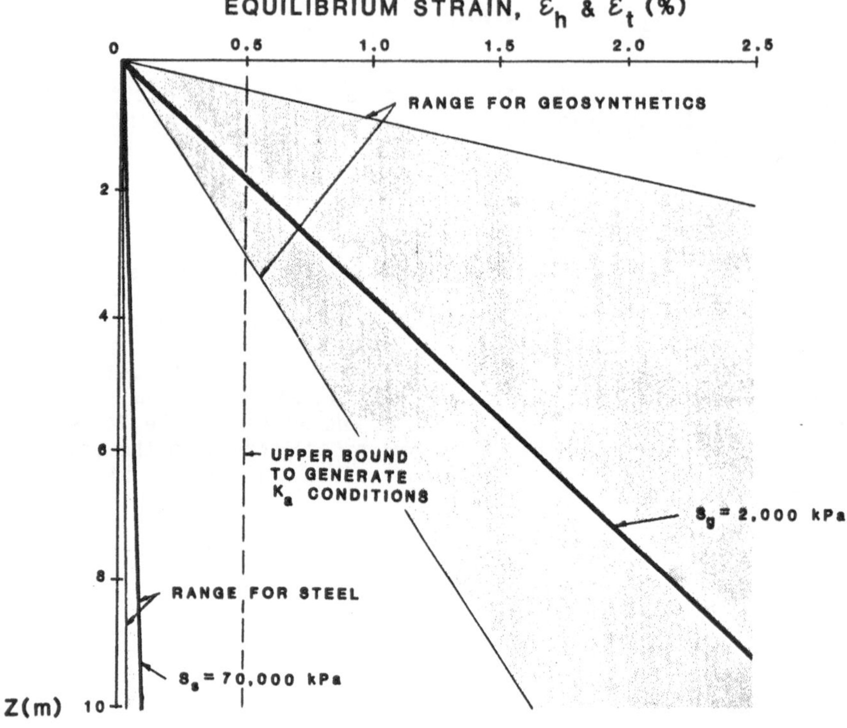

Figure 6. Range of equilibrium strains in an element of reinforced soil incorporating either steel reinforcement (S_s) or geosynthetic (S_g) reinforcement and following a compression unloading stress path. (S_s and S_g are defined in Figure 4.)

the stiffest geosynthetics currently used, however, strains are on the order of 20 times greater than those for steel reinforcement. For the case of S_g = 2000 kPa (40,000 lbs/ft²), strains range from about 1% up to about 2.5%.

2.4 Lateral Stresses and Mobilized Tensions in Reinforced Soil

The equilibrium strains obtained from Figure 4b have been used with the hyperbolic K-ε curve of Figure 4a to produce Figure 7. The derivation of the curves in this figure is as follows:

- Lateral stresses in the soil are given by:

$$\sigma_h = K\gamma z \tag{9}$$

- Lateral stresses are transferred to reinforcement:

$$T = \sigma_h \tag{10}$$

- Since it is assumed that $\epsilon_t = \epsilon_h$:

$$T = S\,\epsilon_t = S\,\epsilon_h \tag{11}$$

- Combining Equations 2, 9, 10, and 11 gives the following equation which was used to generate the K - z curves presented in Figure 7:

$$z = \frac{\lambda S}{\gamma}\left(\frac{K_0 - K}{K\,(K - K_a)}\right) \tag{12}$$

  where: z = depth; λ = parameter used in Equation 2 and linked to the strain required for the soil to reach the Rankine active state of stress; S = reinforcement stiffness per unit area; γ = unit weight of soil; K_0 = at-rest lateral earth pressure coefficient; K_a = active lateral earth pressure coefficient; and K = lateral earth pressure coefficient at strain ϵ_h.

Figure 7 presents a plot of the mobilized reinforcement force per unit area (normalized in terms of the lateral earth pressure coefficient, K) required to keep a soil element at depth z in equilibrium. Results are presented for both geosynthetic and steel reinforcement for a soil with $\lambda = 2.6 \times 10^{-4}$. Figure 7 is revealing because it illustrates the role of reinforcement tensile stiffness on the horizontal stresses that must be resisted by the reinforcement. With geosynthetic reinforcement, the calculated lateral stresses are virtually those corresponding to the active Rankine state, even for relatively stiff geosynthetic reinforcement. Clearly, geosynthetics fulfill the criterion given previously for extensible reinforcement.

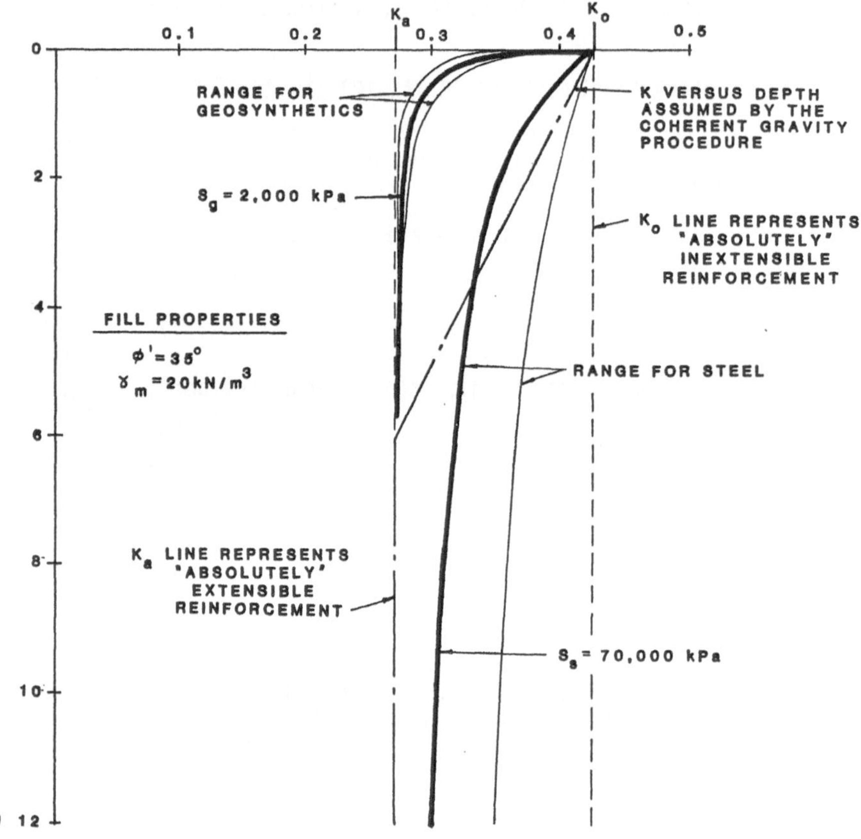

Figure 7. Calculated lateral earth pressure coefficient, K, for an element of reinforced soil incorporating either steel (S_s) or geosynthetic (S_g) reinforcement and following a compression unloading stress path. Calculated values are based on equilibrium strains obtained from Figure 4. (S_s and S_g are defined in Figure 5.)

According to Figure 7, mobilized reinforcement forces and lateral earth pressures are larger for steel than for geosynthetics. Figure 7 indicates that for steel, lateral earth pressures near the top of a wall should be close to those associated with at-rest conditions. In this case, steel reinforcement fulfills the criterion for inextensible reinforcement. At depth, the lateral earth pressures tend toward those associated with active conditions. Based on the hyperbolic curve in Figure 4a, steel reinforcement would need to undergo a tensile strain on the order of its initial yield strain ($\epsilon_g \approx 0.2\%$) in order for the soil to reach an active plastic state. Since the steel reinforcement used in Reinforced Earth walls is designed so as not to exceed a tensile stress of about $0.55f_y$, where f_y is the initial yield stress of the steel, the reinforced fill should theoretically never reach an active plastic state.

This last observation is interesting in light of the assumed distribution of lateral earth pressure versus depth used in the coherent gravity design procedure recommended by LCPC - SETRA (1979) for the design of Reinforced Earth walls (also see Ingold, 1984 or Jones, 1985 for a description of the procedure). In the coherent gravity procedure, the lateral earth pressure coefficient, K, is assumed to decrease from a value of K_o at the ground surface to K_a at (and below) a depth of 6 m (20 ft). Based on the above discussion, it is concluded that K_a stresses can only develop in a steel reinforced soil structure if there is slip between the soil and reinforcement. If there is no slip, a larger value of K should be considered for the coherent gravity procedure for depths greater than 6 m (20 ft).

Figure 8 is similar to Figure 7 except that it shows the influence of the K-ϵ_h relationship on the calculated K - z curves. Results are shown for $\epsilon_t = \epsilon_h = 0.5\%$, 1%, and 2%. As previously noted $\epsilon_h = 0.5\%$ is assumed to be correspond to a compression unloading stress path and $\epsilon_h = 1\%$ to 2% is assumed to correspond to a general incremental loading stress path. It is clear that while the effect of stress path (and thus ϵ_h) is significant, it does not alter the conclusions of the previous paragraphs.

Mitchell (1987) has presented results from incremental, plane-strain finite element analyses conducted by Collin (1986) consistent with the results presented above. The finite element model used by Collin accounts for nonlinear soil stress-strain properties, including dilatancy and hysteresis, independent facing units, soil-reinforcement interface behavior, incremental construction processes, and compaction induced stresses. The model was used to predict the behavior of 4.3 m (14 ft) and 6.1 m (20 ft) high reinforced soil walls. The walls were intended to model a steel bar-mat reinforced soil wall constructed in Hayward, California. The geometries and properties of the walls are shown in Figure 9. Numerical simulations were conducted for three types of reinforcement: steel bar-mat (S_s = 112,000 kPa); welded wire mesh (S_s = 310,000 kPa) and geogrid (S_g = 1400 kPa). Results are shown in Figure 9 along with measurements obtained from the Hayward Wall. Inspection of this figure shows that for the steel reinforcement both the numerical simulations and measured data indicate σ_h magnitudes in the range of $K_o\sigma_v$ or greater, with σ_v calculated using the "Meyerhof (1953)" type stress distribution to account for the lateral thrust of the retained backfill. Similar results were

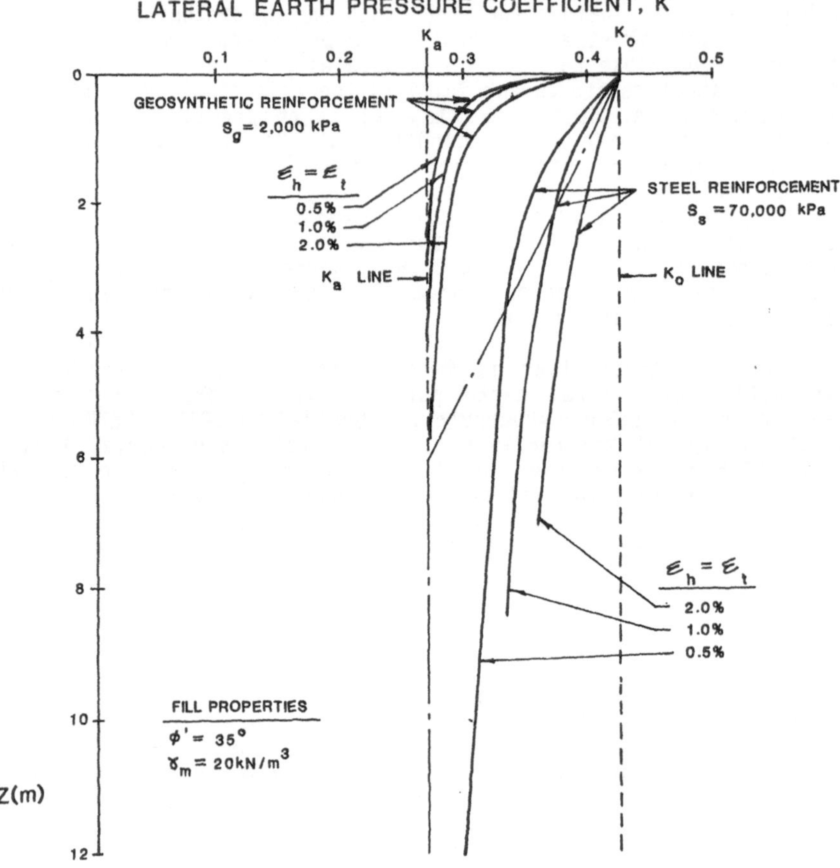

Figure 8. Influence of stress path on the calculated lateral earth pressure coefficient, K, for an element of reinforced soil incorporating either steel (S_s) or geosynthetic (S_g) reinforcement: ϵ_h = 0.5% corresponds to compression unloading (Figure 2a) and ϵ_h = 1.0% to 2.0% corresponds to general incremental loading.

obtained from the numerical simulations of the welded wire mesh wall. The results from the numerical simulations incorporating geogrid reinforcement indicate $\sigma_h = K_a\sigma_v$, where σ_v is the overburden pressure ($\sigma_v = \gamma z$). This result is consistent with Figures 7 and 8.

2.5 Influence of Construction

The preceding analyses were for single elements of reinforced soil. These analyses can be extended to the behavior of a geosynthetic-reinforced wall if it is assumed that the reinforcement does not affect the stress state in the soil (i.e., the soil stresses and reinforcement tensions are uncoupled).

For the case of an unreinforced wall that is braced during construction, but free to rotate about its toe after the braces are removed, the wall fill undergoes K_0 loading followed by compression unloading (stress path AB + BC in Figure 2a). If the wall face were frictionless, an active Rankine state would theoretically develop. This latter case corresponds to that of a flexible, frictionless cantilever wall rotating about a fixed toe (Figure 10a). If the wall contains inextensible reinforcement that makes a good bond with the soil, the large tensile stiffness of the reinforcement will suppress horizontal soil strains. In the limiting case of an "absolutely" inextensible reinforcement, the horizontal direction becomes a zero extension direction and the failure surface is as indicated in Figure 10b (Bassett and Last, 1978). For the case of extensible reinforcement, horizontal soil strains should be only partially suppressed, as indicated by Figures 7 and 8. It is hypothesized that the failure surface for this case will be close to that shown in Figure 10a.

For the case of incremental wall construction the situation is more complex. At depth, the soil will follow some general incremental loading stress path (Figure 2a), eventually reaching an active plastic condition. Near the top of the wall, the soil should be in a subfailure state, with the soil stress path being closer to one of proportional loading (AB in Figure 2a) than unloading (BC in figure 2a). The failure mechanism for this case is more complex than given by Figure 10a. Suggestions have been put forth that the failure mechanism is similar to one for a retaining wall deforming about a hinged crest. There is some experimental support for this suggestion for inextensible reinforcement systems (Juran and Schlosser, 1978) but not for extensible reinforcement systems. Additional research is needed in this area.

2.6 Conclusions for Geosynthetic Reinforcement

It is concluded that currently available geosynthetic materials are extensible forms of reinforcement. This conclusion is based on an analysis of the amount of horizontal extension required to induce an active plastic state in an element of unreinforced soil. The amount of extension is stress path dependent and, for a compacted granular fill, ranges from less than 0.5% for compression unloading up to 1% to 2% for general incremental loading. The amount of mobilized geosynthetic reinforcement tension at these low extensions is small. Theoretically, local yielding of the soil element will occur before

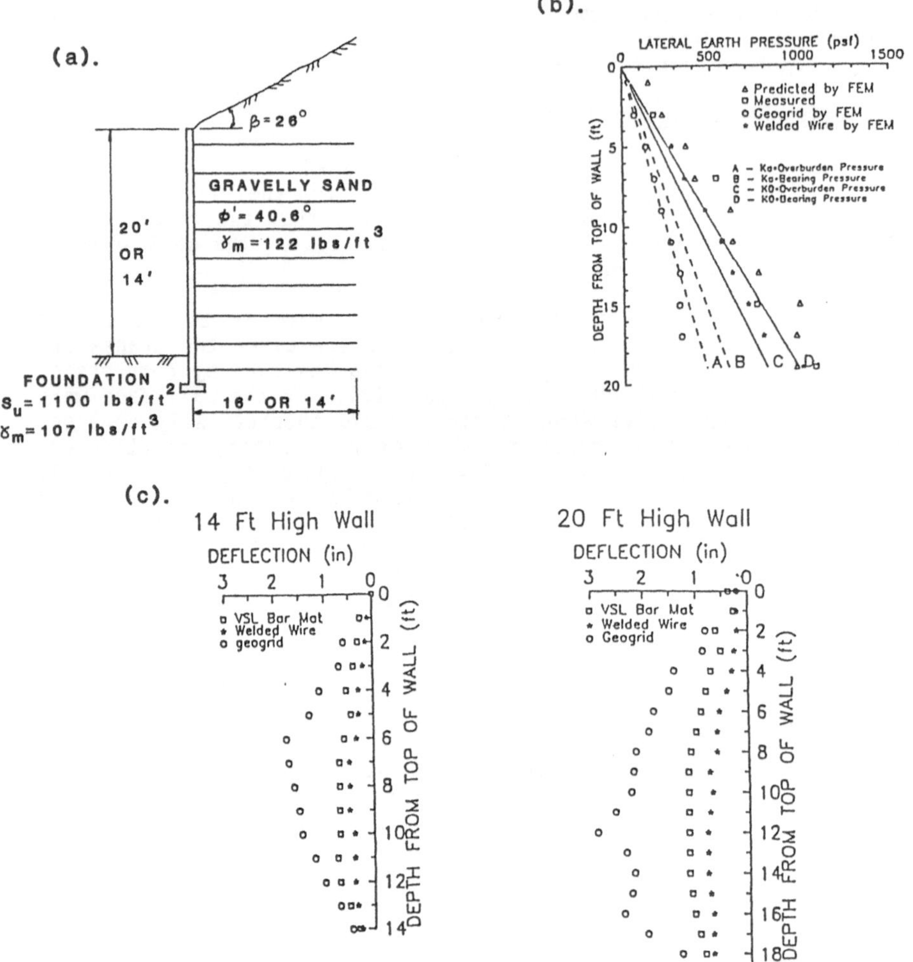

Figure 9. Predicted and measured performance of Hayward reinforced soil walls: (a) wall cross-section and steel bar-mat reinforcement layout; (b) measured and predicted lateral earth pressures for different types of reinforcement; (c) predicted wall deformations. Wall facing consisted of segmental concrete panels (from Mitchell, 1987).

enough geosynthetic tension is mobilized to reach equilibrium. This is in contrast to steel reinforcement where the reinforcement tension will be mobilized at comparatively small values of horizontal extension (i.e., $\epsilon_h \approx 0.01$ to 0.1%).

Geosynthetic tensile stiffnesses are much lower than steel tensile stiffnesses (i.e., $S_s \approx 25$ to 600 S_g). This fact was used as justification for uncoupling the stress-strain response of an element of soil from the tension-strain response of the reinforcement. Calculation of horizontal force equilibrium for the soil and reinforcement yielded the strains and stresses in the soil and reinforcement (Figure 6). Extension of the analysis to full-scale retaining wall structures was discussed and the influence of construction procedures were qualitatively considered. It was concluded that due to their low tensile stiffnesses, geosynthetics will be less effective than steel reinforcement in suppressing horizontal soil strains. As a result larger soil shear stresses will be mobilized in geosynthetic reinforced soil walls than in steel reinforced walls. The wall fill in geosynthetic reinforced soil walls will be closer to active plastic conditions. For these reasons, limit state analysis procedures should be more applicable to geosynthetic reinforced soil walls than to steel reinforced soil walls. In the next section of this paper, limit state analysis procedures will be reviewed.

3. LIMIT STATE ANALYSIS METHODS

Two approaches to limit state analysis are reviewed: (i) approaches based on force or moment equilibrium of a wedge (or block of soil with some other shape such as a circle, log-spiral, etc.) with one side of the wedge coincident with the potential failure surface; and (ii) approaches based on plasticity solutions.

3.1 Limit Equilibrium Solutions

Approaches based on limit equilibrium calculations all involve the separation of the reinforced soil into two zones (Juran and Schlosser, 1978): an active zone, limited on one side by the potential failure surface which is also assumed to be the locus of maximum reinforcement tensions; and a resistant zone which provides reinforcement anchorage. The stress and strain states within the active and resistant zones are indeterminate and only the shear and normal stresses on the potential failure surface can be defined. Limit equilibrium solutions vary mainly by the shape of the assumed failure surface and by the way that the reinforcement forces are introduced into the equilibrium equations. The most commonly assumed shapes for the failure surface are shown in Figure 11 and include a straight wedge, two-part wedge, circle, and log-spiral.

The simplest assumption regarding the shape of the potential failure surface is that it is a straight wedge. This assumption was proposed by Coulomb in 1776 and has traditionally been used to determine active pressures against yielding retaining walls with smooth or rough surfaces. Graphical solutions have been developed for varying values of wall friction (e.g., Terzaghi and Peck, 1967) and comparisons with more accurate solutions show that the error

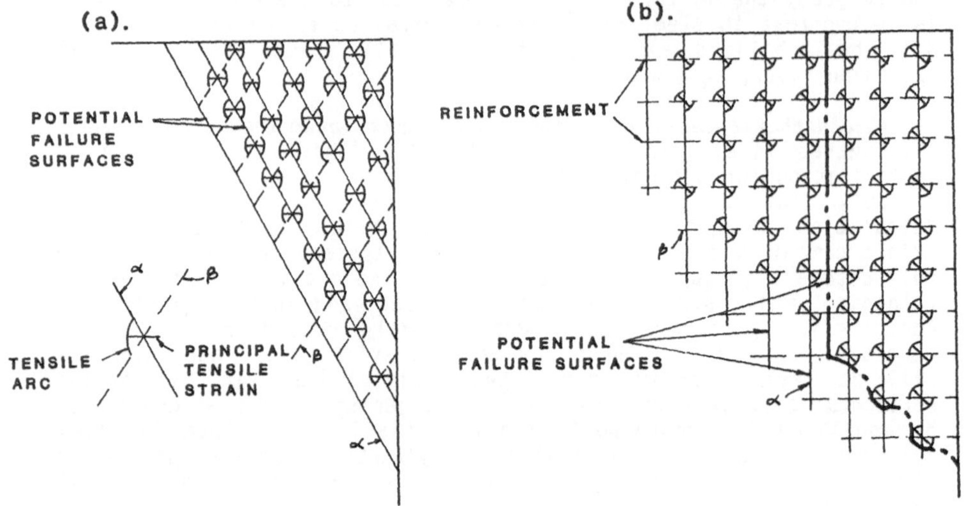

Figure 10. Idealized zero-extension trajectories (α and β trajectories) and potential failure surfaces presented by Bassett and Last (1978) for: (a) a flexible cantilever wall bending about its toe (data from Milligan, 1974); (b) a reinforced soil wall with very stiff (e.g.,metallic) reinforcement.

associated with the use of a straight wedge for active pressure calculations are small. For the simple case of zero wall friction and horizontal backfill with uniform surcharge, the straight wedge solution is identical to that obtained with Rankine theory (a plasticity solution), both predicting a critical wedge passing through the wall toe at an angle of (45°-ϕ/2) from the vertical. Application of the straight wedge to reinforced soil consists of introducing reinforcement forces into the equilibrium equations. Early theoretical formulations for the analysis of Reinforced Earth walls used Coulomb wedges (e.g., Schlosser and Vidal, 1969; Lee et al., 1973). Implicit in the use of this method is the assumption that the reinforcing elements have a negligible effect on the stresses and strains within the soil mass. Subsequent investigations (e.g., Baguelin, 1978; Juran and Schlosser, 1978) have shown that this assumption may not be appropriate for soils reinforced with steel.

The straight wedge failure surface (Figure 11a) is a special case of the more general two-part wedge failure surface (Figure 11b). The straight wedge can be used to evaluate the critical potential failure surface for simple loading configurations and for calculating the reinforcement tension along that surface. However, straight wedges cannot be used to investigate failure surfaces involving sliding of the reinforced soil over a layer of reinforcement or failure surfaces that may be critical for external stability. In addition, failure surfaces observed in small-scale tests, numerical studies, and reinforcement tensions in stable prototype structures, all suggest that the critical failure surface in steel reinforced structures is better modeled with a two-part wedge (or a curved failure surface) than a straight wedge. Two-part wedge limit equilibrium models have been proposed by Romstad et al. (1978) and Jewell (1985b) for reinforced soil walls, and by Stocker et al. (1979) for nailed soil walls. Two-part wedge models have been used for the analysis of geosynthetic reinforced slopes by Murray (1984), Jewell et al. (1984), Schmertmann et al. (1987), and others.

Circular and logarithmic-spiral failure surfaces (Figures 11c and 11d) were suggested by Juran and Schlosser (1978) in conjunction with the requirement for the failure surface to be vertical at the top of the wall. This requirement was derived from their observation that steel reinforcement suppresses horizontal soil strains in the upper portions of the active zone, causing K_0 conditions to prevail. The kinematic condition resulting from Juran's and Schlosser's requirement is one of wall rotation about a hinged crest. Juran and Schlosser hypothesized that failure resulted from shear along a very thin surface between rigid active and resistant blocks. They found that both the log spiral and circular failure surfaces resulted in reasonably good predictions of the critical heights of small-scale model walls. These predictions were better than those based on Coulomb or Rankine theory. However, only the log-spiral surface provided a reasonable estimation of the shape and location of the failure surface. Terzaghi (1943), and more recently Milligan (1983) and Fang and Ishibashi (1985), have shown that the lateral pressure distribution associated with walls having a fixed crest is significantly different than the triangular distribution associated with rotation about a hinged toe.

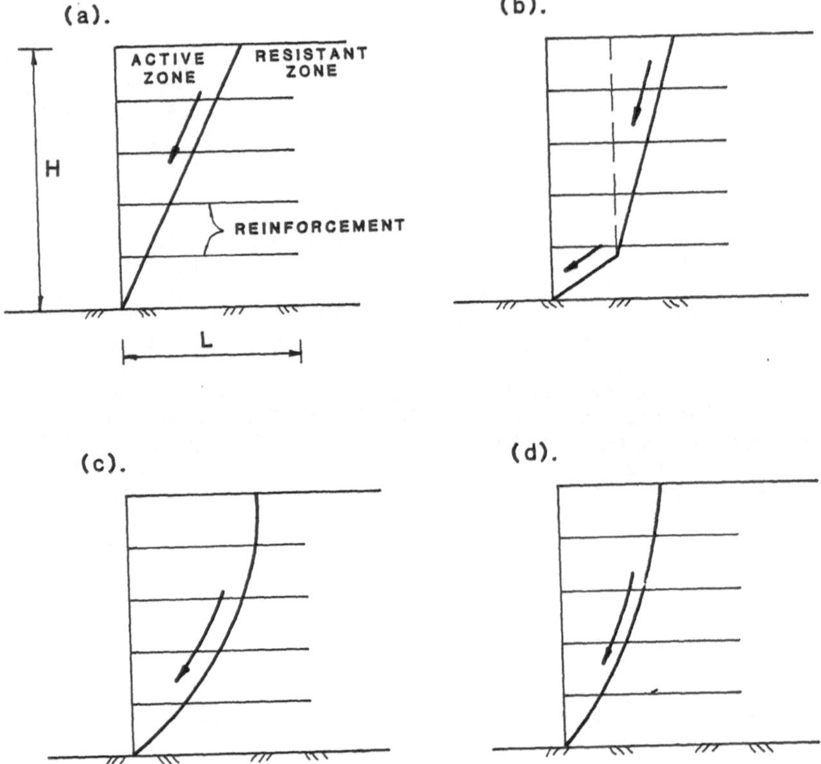

Figure 11. Common shapes for potential failure surfaces for limit equilibrium stability analysis: (a) straight wedge; (b) two-part wedge; (c) circle; and (d) logarithmic spiral.

A logarithmic-spiral failure surface has been proposed by Leshchinsky and Perry (1987) for the analysis and design of geosynthetic reinforced soil walls. Leshchinsky and Perry also assume that, at incipient failure, the geosynthetic deforms at its intersection with the failure surface so that it is orthogonal to the radius of the logarithmic-spiral. Gourc et al. (1986) also consider a "reorientation" of the reinforcement at its intersection with the failure surface in their "displacement method" of analysis. Gourc et al. assume a circular failure surface separating active and resistant zones. Using assumptions about the deflected shape of the geosynthetic in the vicinity of the failure surface, they calculate reinforcement deformations based on a soil-geosynthetic elasto-plastic interface deformation model. An additional feature of Gourc et al.'s model is that it allows slip between the soil and reinforcement. Limit equilibrium analysis models utilizing circular slip surfaces have also been proposed for geosynthetic reinforced embankments by Studer and Meier (1986) and Ruegger (1986).

3.2 Plasticity Solutions

Classical plasticity methods can be used to develop ultimate loads for simple rigid-plastic soils under plane strain conditions. Reinforcement tensions required for equilibrium can be deduced from the results. Stresses and strains outside of the plastic zone are indeterminate. Rigorous plasticity solutions have only been developed for a few simple cases due to their complexity. Approximate solutions using lower bound stress fields and upper bound kinematic analyses extend the range of available solutions. Incremental elasto-plastic numerical formulations incorporating anisotropic, strain-hardening constitutive relationships vastly increase computational possibilities. However, the complexity of these formulations and the time involved in using them preclude their use in all but research applications.

Rankine's solution for a cohesionless, rigid-plastic soil represents the simplest state of plastic equilibrium for a retaining wall. Solutions for the more complex cases of walls with surface friction and sloping backfills were solved using lower bound stress fields by Caquot and Kerisel (1949) and Sokolovski (1965). Meyerhof (1980) provides a brief overview of recent work related to the use of plasticity theories to determine failure mechanisms and stress fields behind retaining walls with various boundary stress and deformation conditions, including walls rotating around a hinged top. Scott (1985) recently presented a comprehensive overview of the history of the use of plasticity solutions in soil mechanics.

3.3 Comparison of Solutions

Comparisons of several different limit state analysis procedures were presented by Morgenstern and Eisenstein (1970), Figure 12. These comparisons are for a retaining wall free to rotate about its toe, backfilled with a cohesionless, rigid-plastic soil with a horizontal top surface and various amounts of wall friction. It can be seen that for active plastic conditions, the various limit equilibrium and plasticity solutions agree to within about ±10% for equal values of wall friction. For the case of zero wall friction, the predicted (a).

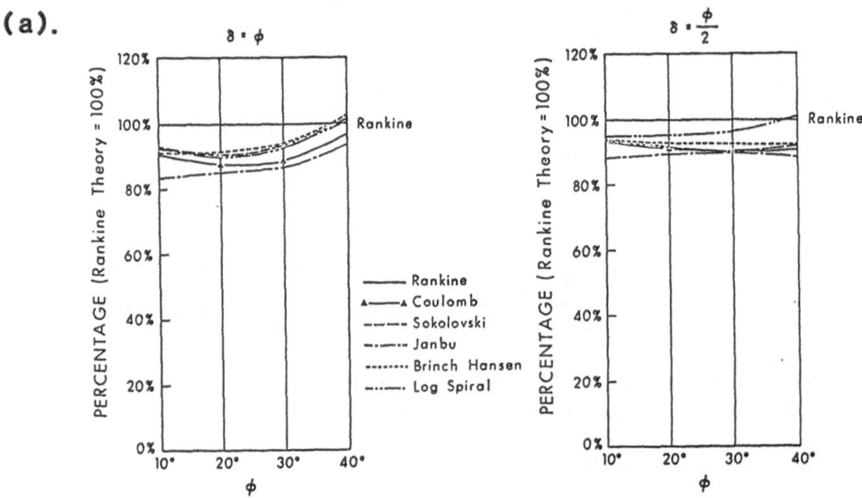

(b).

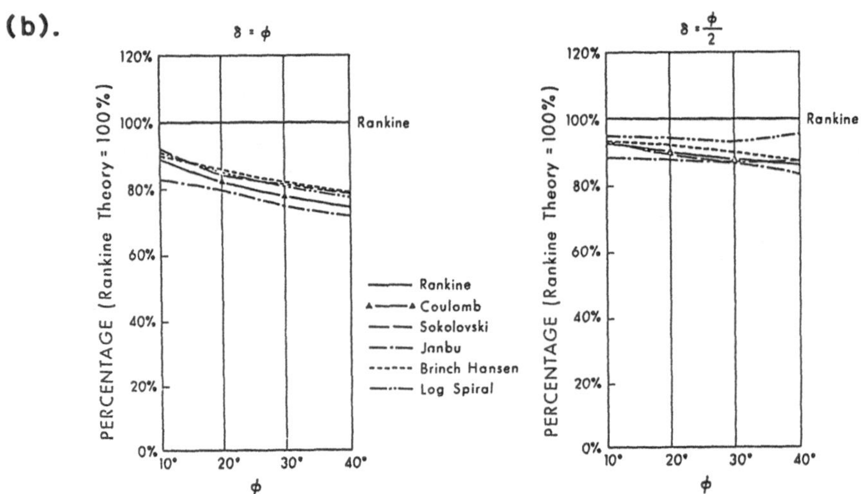

Figure 12. Comparisons of active pressures calculated using various limit equilibrium and plasticity analysis methods for a flexible retaining wall yielding about its toe: (a) total active earth pressures for various theories; and (b) horizontal components of total active earth pressures (from Morgenstern and Eisenstein, 1970). δ represents the angle of friction between the retaining wall and soil.

failure surface and horizontal stresses required for equilibrium are those given by Rankine's solution. From these results, it is concluded that if the boundary conditions (surcharges and wall friction) are properly accounted for, the selected limit equilibrium or plasticity analysis method will not have a large effect on the computed equilibrium reinforcement tensions (unless the considered failure surfaces are constrained as is the case with Juran and Schlosser's logarithmic spiral procedure). Therefore, in the remainder of this paper, calculations will be based on a simple two-part wedge limit equilibrium model.

4. ANALYSIS OF GEOSYNTHETIC-REINFORCED SOIL WALLS

In this section, geosynthetic-reinforced soil walls are analyzed using the two-part wedge model shown in Figure 13. The walls are assumed to rest on competent foundations. The influence of the facing unit and construction procedures on wall behavior are neglected. The important variables considered are the strength of the soil, tensile stiffness of the geosynthetic reinforcement, and wall geometry. The two-part wedge analysis will be used to investigate geosynthetic forces and strains under working conditions. This technique for evaluating reinforcement tensions was used by Jewell (1985b). In general, the application of limit equilibrium procedures to working conditions is questionable. The authors suggest however, that these procedures may be applicable to the extent that a plastic zone develops in the geosynthetic-reinforced fill under working conditions.

An IBM PC computer program was written to analyze two-part wedges with failure surfaces emerging behind the wall crest and at points up the front face of the wall. The program, written in nondimensional form, models interslice friction, wall face friction and surcharge loads. The program automatically analyzes a very large number of potential failure surfaces and calculates the required reinforcement force to maintain equilibrium on each surface. The soil is assumed to be a uniform frictional material complying with the Mohr-Coulomb failure criterion. Two-part wedge failure surfaces involving horizontal sliding of soil over reinforcement account for reduced shear strength along the soil-reinforcement interface through use of a coefficient of interaction, μ, defined as the ratio of the soil-reinforcement interface shear strength to the soil shear strength. A value of $\mu = 1.0$ was used for the analysis presented herein. The interwedge friction angle, ψ, was found to have only a very minor influence on the analysis results ($\psi = 0$ was used herein).

4.1 Global vs. Local Analysis

Jones (1985) has pointed out that internal stability calculations for reinforced soil walls fall into one of two categories: (i) those in which local stability is considered in the vicinity of every single strip or element of reinforcement; and (ii) those in which the overall stability of wedges or blocks are considered. Bolton et al. (1978) considered Reinforced Earth walls as an assemblage of individual anchors, with each anchor holding back a small area of facing. They concluded that local equilibrium must be evaluated, since, if any one layer of reinforcement breaks, the other layers will successively break because of sudden load transfer. Both Bolton et al. and

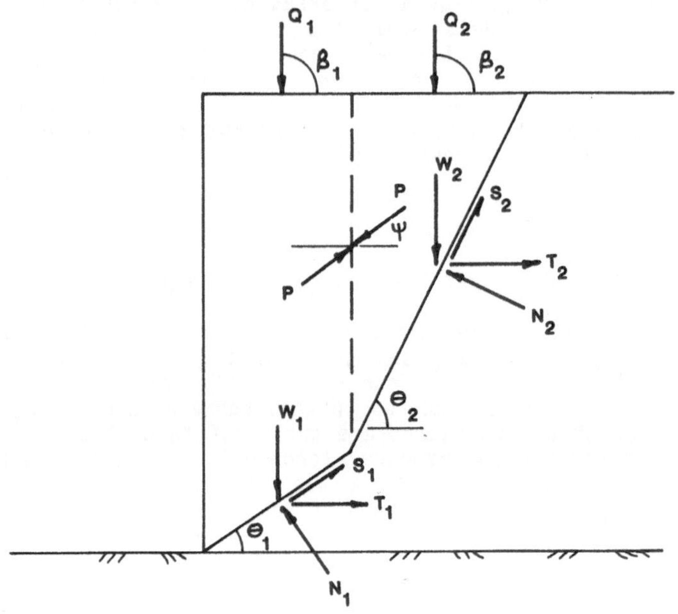

DEFINITIONS:

W = Soil weight
N = Normal force on potential failure surface
S = Shear resistance on potential failure surface
P = Interwedge force
T = Horizontal reinforcement force
Q = Surcharge loads
θ = Angle of potential failure surfaces
β = Angle of surcharge loads
ψ = Angle of interwedge force

Figure 13. Two-part wedge model used for reinforced soil wall stability analysis.

Segrestin (1979) point out that the use of a wedge analysis assumes that all reinforcing elements crossing the failure surface yield simultaneously. They note that calculated factors of safety based on Coulomb wedge calculations are larger than those based on local equilibrium of individual reinforcing elements. As a result, some widely used design procedures (e.g., British Dept. of Transport (1978)) incorporate both local and global stability calculations.

Strict adherence to local stability requirements may be overly restrictive for geosynthetics because: (i) they are much more ductile than steel; and (ii) they occupy a substantially larger volume in the reinforced zone (they are in contact with many more soil particles) than steel. Geosynthetics pick up increasing tensile loads with increasing tensile strains up to and past the point at which the soil is yielding. It is reasoned that if local overstressing occurs during the loading process, the soil and reinforcement will deform locally, increasing the tensile force carried by the reinforcement. Soil shear stresses generated as a result of differential soil strains should result in load transfer between reinforcement layers. The relatively large volume of geosynthetic reinforcement should enhance load transfer.

Strict adherence to local stability requirements may not be necessary with geosynthetic reinforcement. However, local stability can be easily achieved by doing a thorough analysis that assesses the required reinforcing forces on a wide range of potential failure surfaces, and then ensuring that the reinforcement distribution provides the required forces on all surfaces. The procedure described below for determining the reinforcement force distribution satisfies local stability requirements.

4.2 Reinforcement Force Distribution

In keeping with the first part of this paper, it is assumed that the stress-strain response of the soil can be uncoupled from the tension-strain response of the geosynthetic and that locally $\epsilon_h = \epsilon_t$. With these assumptions, the two-part wedge model shown in Figure 13 can be used to estimate the distribution of reinforcement tensions required to maintain wedge equilibrium. For any considered wedge, the total required horizontal reinforcement force is assumed to have two components: a resultant due to the weight of the soil fill acting at the one-third height of the wedge (i.e., resulting from a triangular distribution of required reinforcement tensions); and a resultant due to surcharge loads acting at the one-half height of the wedge (i.e., resulting from a uniform distribution of required reinforcement tensions).

In order to investigate the distribution of reinforcement tension with horizontal (x) and vertical (z) distance from the wall crest the following multi-step analysis was carried out:

(i) The geometry of the soil fill behind the wall face was nondimensionalized by dividing the vertical and horizontal axes by the wall height. The fill was then subdivided into square elements, the side length of each being H/20.

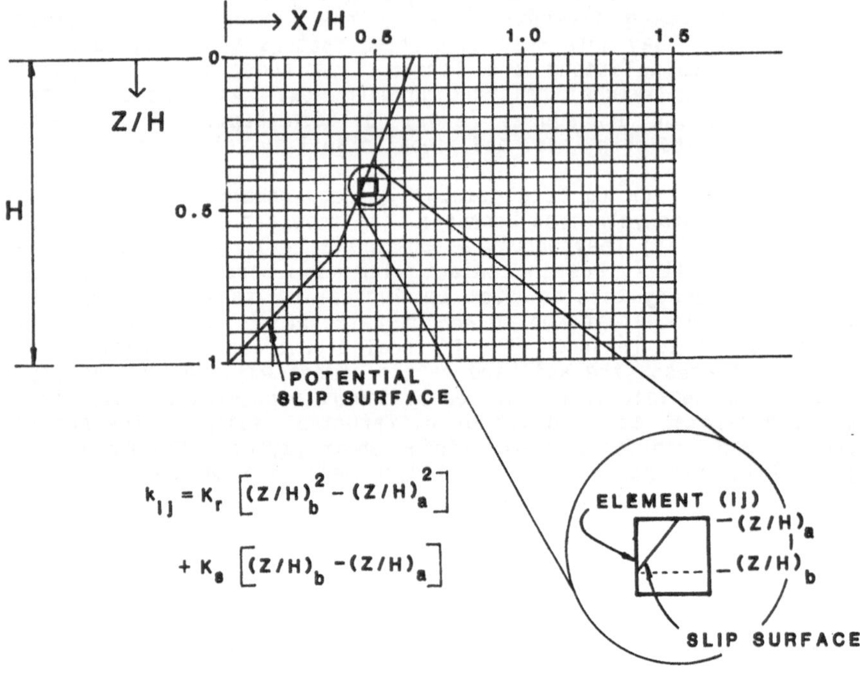

Figure 14. Interpretation of two-part wedge analysis results to estimate the distribution of required reinforcement forces per unit area, T_{ij} (kPa), in reinforced soil walls.

Each soil element was assigned to an element of a 20 by 30 matrix (Figure 14).

(ii) A trial wedge was selected and equilibrium calculations were carried out to determine the normalized total reinforcement force, K_r, due to the weight of the soil fill alone, and the normalized reinforcement force, K_s, due to surcharge loads and zero fill weight. Trial wedges with their toes at all levels of the facing were considered.

(iii) K_r and K_s for each trial wedge were apportioned to the elements of the matrix intersecting the considered wedge (Figure 14); the portion of K_r apportioned to each element increased proportionally with the depth of the element below the crest and with the portion of the potential failure surface intercepted by the element (as shown by the equation given in Figure 14). The portion of K_s apportioned to each element increased proportionally with the portion of the potential failure surface intercepted by the element.

(iv) The above steps were repeated for a very wide variety of wedges (several thousand wedges for each set of the input variables, ϕ, ψ, and $q/\gamma H$); the maximum normalized tensile force for each element, k_{ij}, was stored in the matrix. When the search was completed, the array of maximum k_{ij} values was output.

(v) The distribution of localized reinforcement forces per unit area (normal to the wall face), T_{ij} (kPa) required to maintain equilibrium at the location of any matrix element was determined by using the k_{ij} matrix in the equation:

$$T_{ij} = \gamma H (k_{ij}) \qquad (13)$$

where γ (kN/m³) is the unit weight of the wall fill and H is the wall height.

The above procedure for estimating T_{ij} is conservative (i.e, gives a high value of T_{ij}) because it implicitly assumes that the localized reinforcement forces are mobilized everywhere at the same time (i.e., the soil shear strength is simultaneously mobilized on all of the trial wedges).

Figure 15 shows the reinforcement force distribution obtained for the case H = 10 m, ϕ = 35°, γ = 20 kN/m³ and q = 0. The distinguishing features of the T_{ij} distribution include: reinforcement tension per unit area that increases linearly with depth, but is constant along horizontal planes located between the wall face and a line at an angle of 45° - $\phi/2$ from the wall face (active Rankine state); decreasing reinforcement tension with increasing distance behind the active zone; and a zero force line extending from the base of the wall back into the fill. This plot is analogous to one presented previously by Jewell (1985b). An alternate way to illustrate the required force distribution is through contour

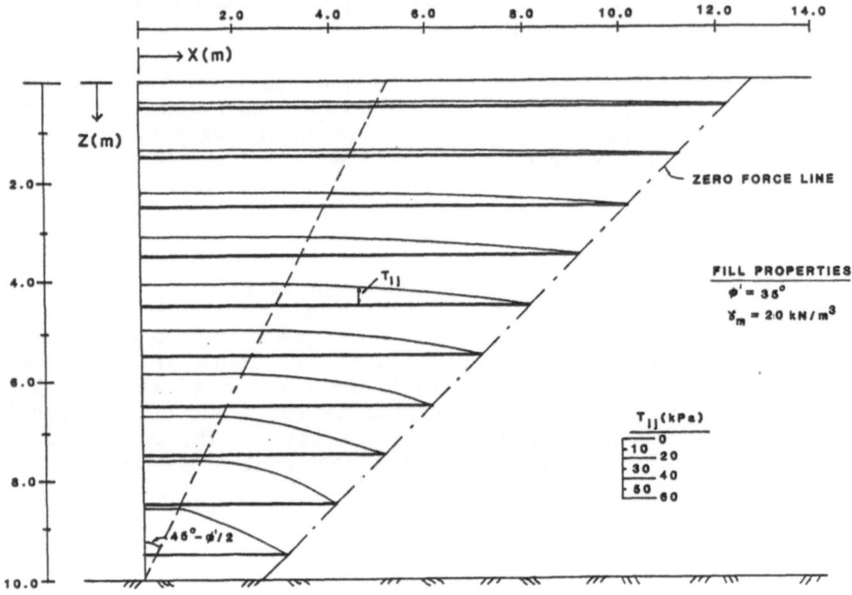

Figure 15. Distribution of required reinforcement forces per unit area, T_{ij} (kPa), required to maintain equilibrium in a 10 m (33 ft) high reinforced soil wall.

plots such as those shown in Figure 16. These plots show all the same features as Figure 15. In addition, they allow rapid evaluation of T_{ij} anywhere in the fill. The effect on T_{ij} of soil strength (ϕ) and uniform vertical surcharges is shown in Figure 16.

Figures 15 and 16 show that to satisfy local equilibrium requirements, tensile forces are required to a distance back from the wall crest equal to 1.0 to 1.5 times the wall height ($X/H = 1.0$ to 1.5). In practice, reinforcement lengths are typically on the order of $L/H = 0.8$. The wedge of unreinforced soil between the reinforced zone and zero force line exerts a thrust and overturning moment on the reinforced soil mass. Calculations show that for $L/H = 0.8$, $\phi > 35°$, and $q/\gamma H < 0.2$, the overturning moment due to the thrust of the retained fill is insignificant, as is the calculated vertical stress increase within the reinforced soil mass due to the overturning moment. Therefore, calculation of vertical stresses using the "trapezoidal" or "Meyerhof" distributions is unnecessary. Further, Jewell (1985b) noted that reinforcement truncation short of the zero force line causes the locus of maximum reinforcement tensions to move toward the wall face. While this effect may be significant for steel strip reinforcement, the effect was found to be insignificant for geosynthetic reinforcement (due to the short required bond lengths for geosynthetics compared to steel strips).

4.3 Reinforcement Deformations

Reinforcement tensile strains, ϵ_{ij}, were calculated from the local reinforcement tensions as follows:

$$\epsilon_{ij} = T_{ij}/S_g \tag{14}$$

where T_{ij} (kN/m) was obtained from Equation 13 and S_g was calculated as shown in Figure 4. In Equation 14, T_{ij} and S_g are independent variables (i.e, the reinforcement tensile stiffness does not influence the state of stress). The total reinforcement elongation at any elevation $\delta_r(z)$ was obtained from:

$$\delta_r(z) = \int_0^L \epsilon_{ij} dx \tag{15}$$

where L (m) is the length of reinforcement.

Figure 17 presents ϵ_{max} and δ_r values for a reinforced soil wall with $H = 10$ m, $\phi = 35°$, $\gamma_m = 20$ kN/m³, and $q = 0$. Results are presented for a range of S_g values and for $S_s = 70{,}000$ kPa. S_g and S_s are assumed to be constant with depth (uniform reinforcement layout). For these conditions, the maximum calculated reinforcement strain occurs at the bottom of the wall and the maximum calculated reinforcement elongation occurs at $(z/H) \approx 0.6$. For $S_g = 2{,}000$ kPa, $\delta_{r,max} = 80$ mm (3.2 in) and $\delta_{r,max}/H = 8 \times 10^{-3}$. In contrast, for $S_s = 70{,}000$ kPa, $\delta_{r,max}/H = 4 \times 10^{-6}$. The form of these results are consistent with those reported by Mitchell (1987) and shown in Figure 9.

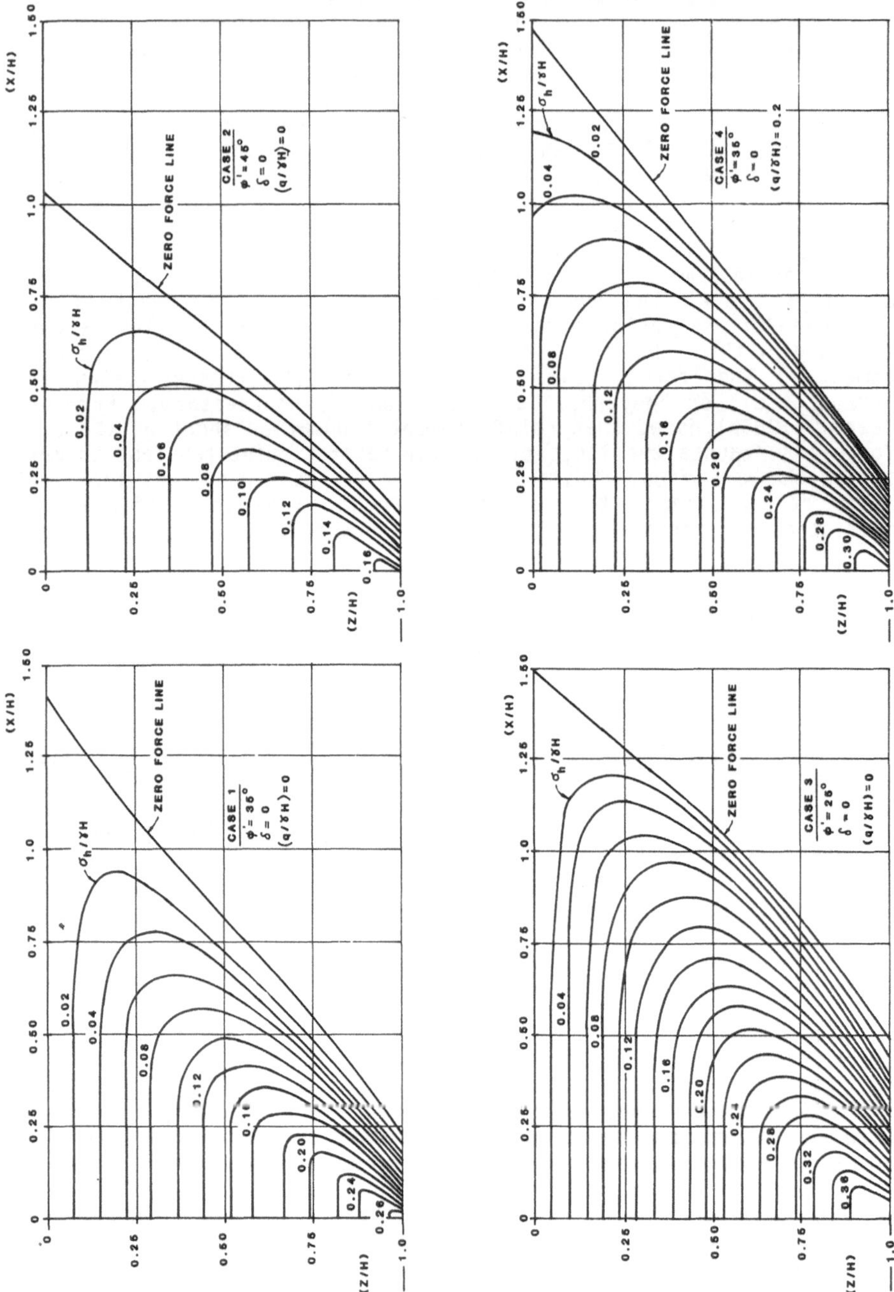

Figure 16. Contours of Reinforcement force per unit area, $(T_{ij}/\gamma H)$, required to maintain equilibrium in a reinforced soil wall. Contours obtain using two-part wedge analysis results.

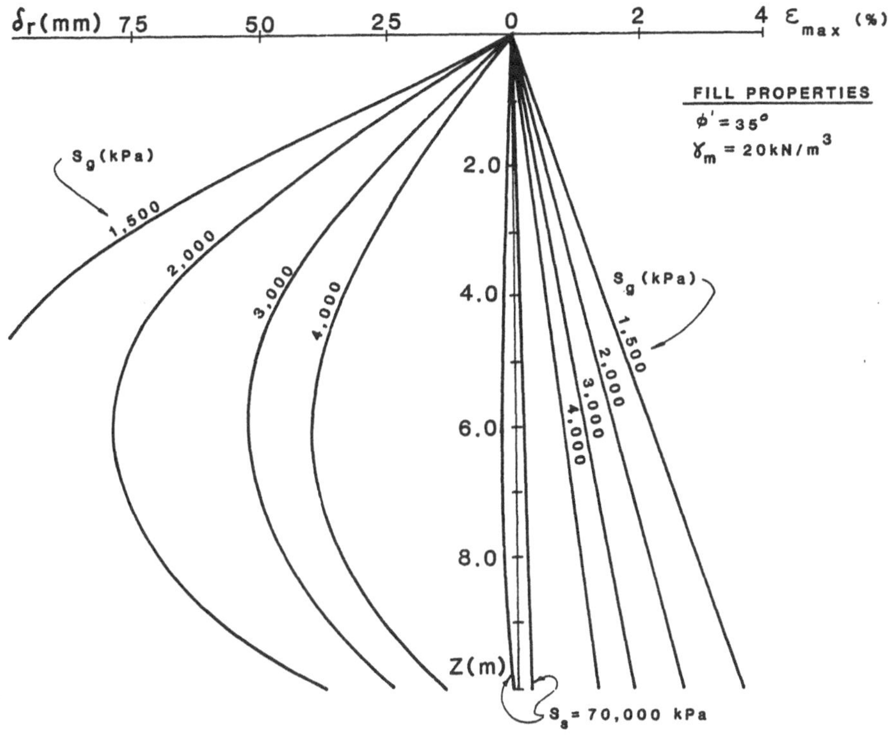

Figure 17. Reinforcement elongation (δ_r) and maximum tensile strain required to maintain equilibrium in a reinforced soil wall.

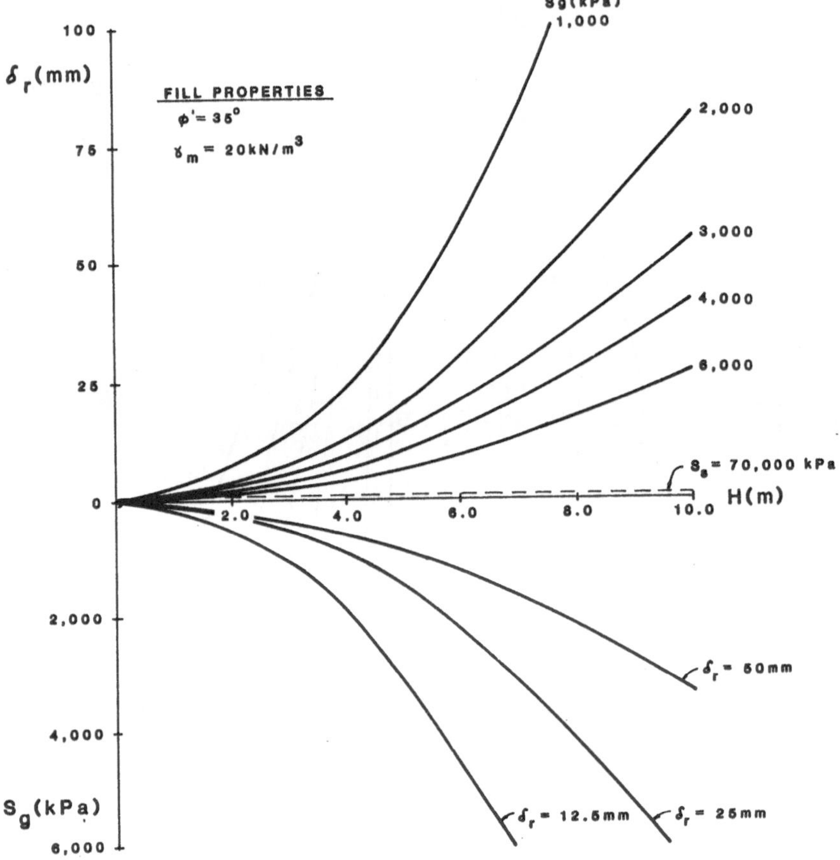

Figure 18. Relationship between wall height (H), reinforcement stiffness per unit soil area (S_g and S_s), and reinforcement elongation (δ_r) in a reinforced soil wall.

Figure 18 shows the relationship between wall height, H, reinforcement stiffness per unit area (S_g and S_s), and reinforcement elongation (δ_r). For a given allowable reinforcement elongation, the maximum wall height is dependent on the reinforcement stiffness. It is important to note that δ_r is not the total wall face displacement, as measured from a vertical plane through the wall toe at the start of construction. The total displacement will increase with distance above the wall base (neglecting initial wall batter). The total displacement can only be obtained through an analysis that takes into account the incremental nature of wall construction. δ_r should be interpreted strictly as the reinforcement elongation at any level. Further, δ_r will be affected by the boundary conditions imposed on the soil and reinforcement.

5. DESIGN OF GEOSYNTHETIC REINFORCES SOIL WALL

5.1 Method of Analysis

The analysis procedures described in Section 4 can be used to design geosynthetic-reinforced soil walls. The design process must take into account all potential failure mechanisms through, under, and behind the reinforced soil mass. Various shapes for the potential failure surface can be assumed (Figure 11). Two-part wedge failure surfaces (Figure 11b) may offer the best combination of accuracy, versatility and simplicity for widespread application as a design tool. Direct sliding of soil on reinforcement can be modeled using two-part wedges with a horizontal lower failure surface. The soil shear strength along this lower surface is reduced by the soil-reinforcement coefficient of interaction, μ, obtained from the results of direct shear tests. Two-part wedges can be used to investigate global stability. The design process should also include an analysis for adequate foundation bearing capacity.

In most geosynthetic-reinforced soil walls, the length of reinforcement is controlled by external stability considerations (resistance to direct sliding, bearing capacity, or global stability) rather than by reinforcement pullout from the resistant zone behind the potential failure surface. Due to the comparatively high geosynthetic reinforcement volume (0.2 to 0.5% of the reinforced fill volume), stress levels in the reinforcement are low compared to those in steel. This fact, coupled with geosynthetics' large embedded surface areas, and their generally good bond with granular soils, preclude concern over the length required to prevent pullout for all but the uppermost layer(s) of reinforcement. With geosynthetics, there is more concern over reinforcement rupture than pullout.

5.2 Soil and Reinforcement Properties for Design

The design process necessitates selection of soil and reinforcement properties for design and the selection of appropriate factors of safety.

For design, soil strengths and geosynthetic tensions should be selected at compatible values of strain. Since geosynthetics are extensible and ductile, it appears both logical and conservative to define a limit state for design corresponding to mobilization of the large-strain, constant-volume strength of the soil. Recommendations for selection of reinforcement tensions compatible with the use of ϕ_{cv} were first discussed by McGown et al. (1984a,b) and subsequently by Jewell (1985) and Bonaparte et al. (1985, 1987) and are not repeated here.

The selection of soil properties to determine working strains and deformations is more complex due to progressive soil strength mobilization (e.g., Murray, 1987) and progressive strain softening (e.g., Rowe, 1969b). Practically, the use of ϕ_p is acceptable for analyses of working conditions, as experience with existing geosynthetic reinforced soil walls indicates relatively small reinforcement strains under these conditions.

In most practical applications, both ϕ_{cv} and ϕ_p can be based on plane strain deformation conditions. Jewell and Wroth (1987) have recently presented a comprehensive review of the relationship between plane strain (ϕ_{ps}) and direct shear angles of friction (ϕ_{ds}). They make recommendations for the determination of ϕ_{ps} from direct shear test results.

5.3 Factors of Safety

Design of reinforced soil walls involves the application of an overall factor of safety against reinforcement rupture. There are two ways to incorporate the factor of safety (FS) and both are commonly used: factor the shear strength of the soil by FS and use this factored strength in design calculations; or, use the actual shear strength in design calculations and increase the calculated reinforcement tension by FS. These two procedures give roughly the same reinforcement design tension (α_d) at $\phi = 35°$. For $\phi = 25°$, the first procedure results in α_d about 10% smaller than the value of α_d obtained with the second procedure. At $\phi = 45°$, the reverse is true. At more extreme values of ϕ, the results based on the two methods diverge significantly. The design engineer should be aware of this difference, particularly when dealing with unusual problems or comparing different designs for the same application (i.e., designs from different proprietary wall suppliers).

5.4 Design Details

Design of geosynthetic-reinforced soil walls includes many details that can significantly influence wall behavior. Most important of these are the type of facing unit and the construction procedures. As indicated in Section 2.5, the type of facing unit and construction procedures strongly influence the wall failure mechanism and soil strain field.

Design of the reinforcement-to-panel connection is important not only with respect to the connection strength, but also with respect to deformation of the wall facing. For instance, Berg et al. (1987) attributed 6 to 9 mm of measured face movement in a 5.3 m (17.4 ft) high geogrid-reinforced soil wall to insufficient pretensioning of the geogrid-to-panel connection. Compaction may result in movement of the wall facing as well as compaction-induced soil stresses and reinforcement tensions (Collin, 1986; Mitchell, 1987). Finally, foundation preparations and drainage provisions will both affect wall behavior.

6. ANALYSIS OF RMC REINFORCED SOIL WALLS

In this section, the analysis procedures described in Section 4 are applied to two geogrid-reinforced soil walls constructed and loaded in 1986/1987 at the Dolphin Structures Laboratory, Royal Military College (RMC), Kingston, Ontario. For completeness, a brief description of the RMC test walls is included. A more thorough discussion of the materials, design, and construction of the walls can be found in other papers presented at the NATO Workshop.

6.1 Description

Cross-sections and front views of the two RMC reinforced soil walls are shown in Figure 19. Both walls were 3.0 m (9.8 ft) high and were constructed with medium dense (relative density ≈ 55%) coarse sand and four layers of 3.0 m (9.8 ft) long TENSAR SR2 geogrid reinforcement. Geogrid strips were fastened to the wall facing panels prior to panel erection. Each geogrid layer was pretensioned to 0.4 kN/m (27 lbs/ft) to remove slack prior to backfilling over the layer. The fill was compacted using a small vibrating plate tamper. One wall was constructed with incremental timber facing panels. The 0.75-m (2.5-ft) high panels were separated by horizontal layers of foam rubber filler. The bottom-most course of panels rested directly on a concrete leveling pad that was restrained from forward movement. A second wall was constructed using a propped vertical timber facing which also rested on a horizontally restrained leveling pad. Three levels of props prevented movement of the timber facing during fill placement and compaction. The props were removed when the fill height reached the top of the wall. A uniform vertical surcharge of 12 kPa (250 psf) was placed on top of each of the walls immediately after construction. The surcharge was achieved by placing a 750 mm (2.5 ft) layer of uncompacted loose sand over the entire area of sand backfill. Additional vertical surcharges were applied using pneumatically inflated neoprene bags.

6.2 Selection of Reinforcement and Soil Parameters

The reinforcement input parameters are the tensile stiffness, J_g (kN/m), and the tensile force per unit width at failure, α_f (kN/m). Both J_g and α_f are strain-rate and temperature dependent. Normally, information on soil-reinforcement bond and reinforcement pullout is also required. Based on the work of Sarsby (1985), the coefficient of soil-reinforcement interaction, μ, should be on the order of 0.9 or more. Simple calculations quickly demonstrate that the reinforcement lengths are sufficient to prevent reinforcement pullout prior to rupture.

The required soil input parameters are the plane-strain friction angle, ϕ_{ps}, and the unit weight γ (kN/m³). The dry density (ρ) of the sand fill at D_r = 55% was reported to be about 1.79 Mg/m³ and the compaction moisture content was about 3%; therefore, the moist unit

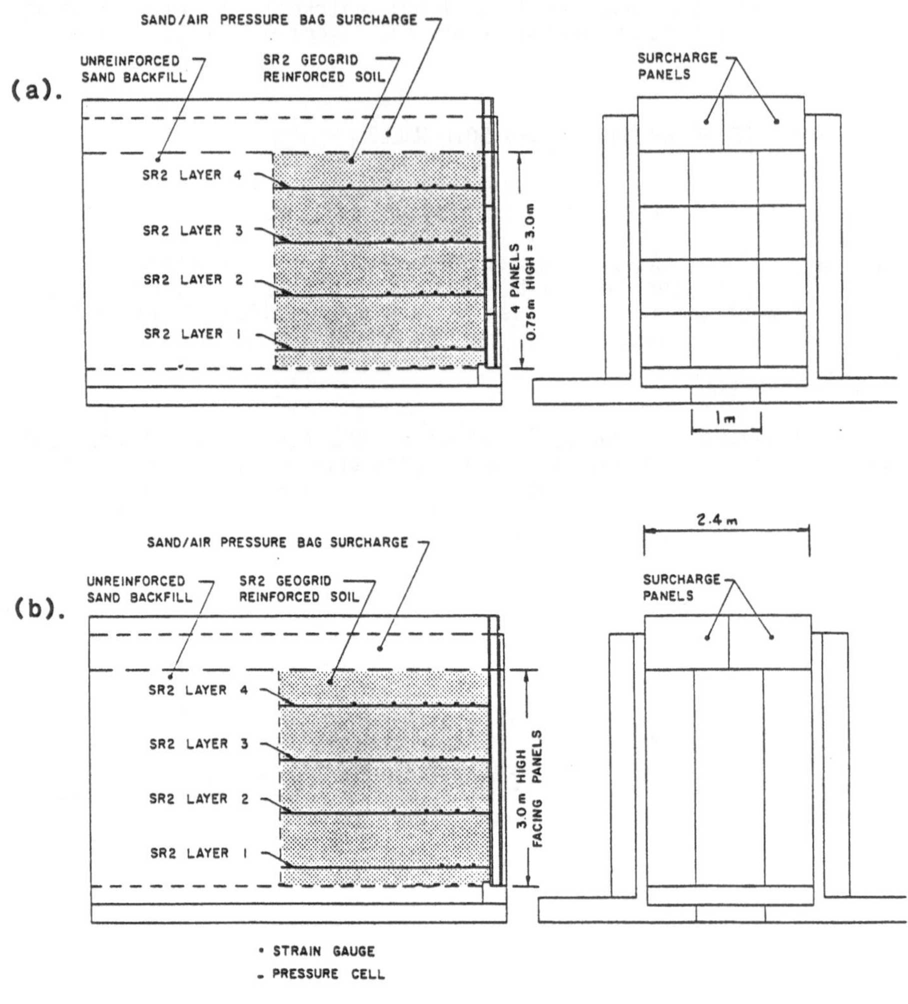

Figure 19. Cross-sections and front views of RMC reinforced soil walls: (a) incremental timber facing panels; (b) propped vertical timber facing panels.

weight of the fill was about 18.1 kN/m³. The sand shear strength was measured in 60 mm square direct shear tests and in consolidated-undrained triaxial tests. Direct shear tests carried out on fill specimens with an initial dry density of ρ_0 = 1.78 Mg/m³ gave a peak friction angle $(\phi_p)_{ds}$ = 43°. Direct shear tests also gave a value for the large-strain, constant-volume friction angle of $(\phi_{cv})_{ds}$ = 40°. Using the equation by Rowe (1969a) relating plane strain and direct shear friction angles:

$$\tan\phi_{ps} = (\tan\phi_{ds}) / \cos(\phi_{cv})_{ds} \tag{16}$$

yields $(\phi_p)_{ps}$ = 51° for ϕ_{ds} = 43° and $(\phi_{cv})_{ps}$ = 48° for ϕ_{ds} = 40°. Therefore, ϕ = 48° represents a lower bound of ϕ and ϕ = 51° represents an upper bound. The authors have elected to use ϕ = 45° and ϕ = 50° as approximate lower and upper bounds in subsequent calculations.

The properties of the high density polyethylene (HDPE) geogrid in the RMC test wall have been described by McGown et al. (1984a). Specific test data on the reinforcement materials used to build the test walls were provided by the workshop organizers. No temperature corrections to the test data are required since both geogrid physical property testing and test wall construction were at 20°C (68°F).

The following secant tensile stiffness values (J_g in kN/m) have been selected by the authors from the results of constant - load creep tests provided by the workshop organizers:

Tensile Strain (%)	Load duration (hr)	
	100	1000
0 - 2	600	500
2 - 5	500	400

For purposes of calculating the critical wall height and factors of safety, the reinforcement tensile force per unit width at failure is required. The authors have chosen to select α_f based on a 10% limiting strain criterion. This results in α_f = 35.3 kN/m for a load duration of 100 hours and α_f = 34.0 kN/m for a load duration of 1000 hours.

6.3 Analysis Results

Analysis of the RMC reinforced soil walls was carried out using the two-part wedge computer program described in Section 4. Unfortunately, it is not possible to use the two-part wedge model to differentiate the behavior of the incremental panel and propped panel walls. This is a significant limitation of limit equilibrium models.

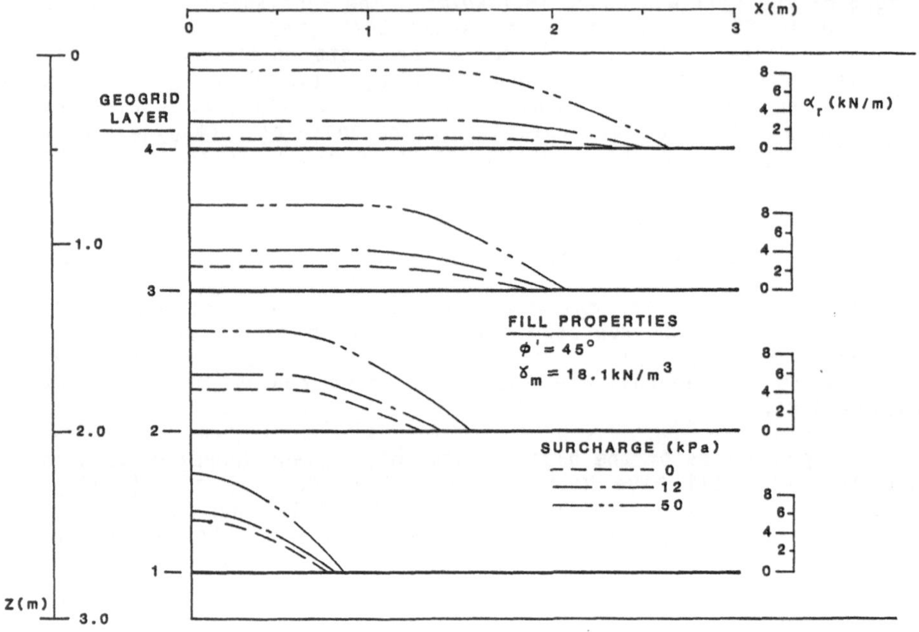

Figure 20. Reinforcement forces per unit width, α_r (kN/m) in geogrid layers. Results for RMC reinforced soil walls obtained using two-part wedge analysis.

Frictional forces on the sidewalls of the reinforced soil mass were also neglected.

Results of the analysis are shown in Figures 20 and 21. Figure 20 shows the distribution of required reinforcement forces in each of the four layers of geogrid reinforcement for $\phi = 45°$. The calculated maximum reinforcement tensions, $\alpha_{r,max}$ (kN/m), in each geogrid layer are:

	q (kPa)			q (kPa)		
Geogrid Layer	12	30	50	12	30	50
4	3.0	5.7	8.5	2.3	4.5	6.7
3	4.3	6.6	9.1	3.4	5.1	7.1
2	6.1	8.4	10.9	4.7	6.5	8.5
1	6.5	8.5	10.5	5.0	6.6	8.2
	$\phi = 45°$			$\phi = 50°$		

The analysis results indicate a maximum reinforcement tension, $\alpha_{r,max}$, at $q = 50$ kN/m² (1040 psf) on the order of 8 kN/m (550 lbs/ft) for $\phi = 50°$ and 11 kN/m (750 lbs/ft) for $\phi = 45°$. If the factor of safety (FS) of the reinforced walls is taken as the ratio of $\alpha_f/\alpha_{r,max}$, then FS ≈ 3 for $\alpha_{r,max} = 11$ kN/m (750 lbs/ft).

Reinforcement strains are calculated from reinforcement tensions using Equation 14. The maximum calculated strains at each geogrid elevation are shown in Figure 21.

For the case of a 100 hour load duration and q = 12 kPa (250 psf), $\epsilon_{max} = 1.1\%$ for $\phi = 45°$ and 0.8% for $\phi = 50°$. For q = 50 kPa (1040 psf), $\epsilon_{max} = 1.8\%$ for $\phi = 45°$ and 1.4% for $\phi = 50°$. Calculated strains increase 20% as the surcharge durations increase from 100 to 1000 hours. The calculated maximum strain occurs in the lowest geogrid layer when q = 12 kPa and in the next-to-lowest layer when q = 50 kPa.

Calculated reinforcement elongations are shown in Figure 21. For q = 0 to 12 kPa (0 to 250 psf), the maximum elongation takes place around the mid-height of the wall. For q = 30 to 50 kPa (620 to 1040 psf), however, the maximum elongation occurs in the top geogrid layer. Maximum geogrid elongations for various surcharges and a 100 hour load duration are:

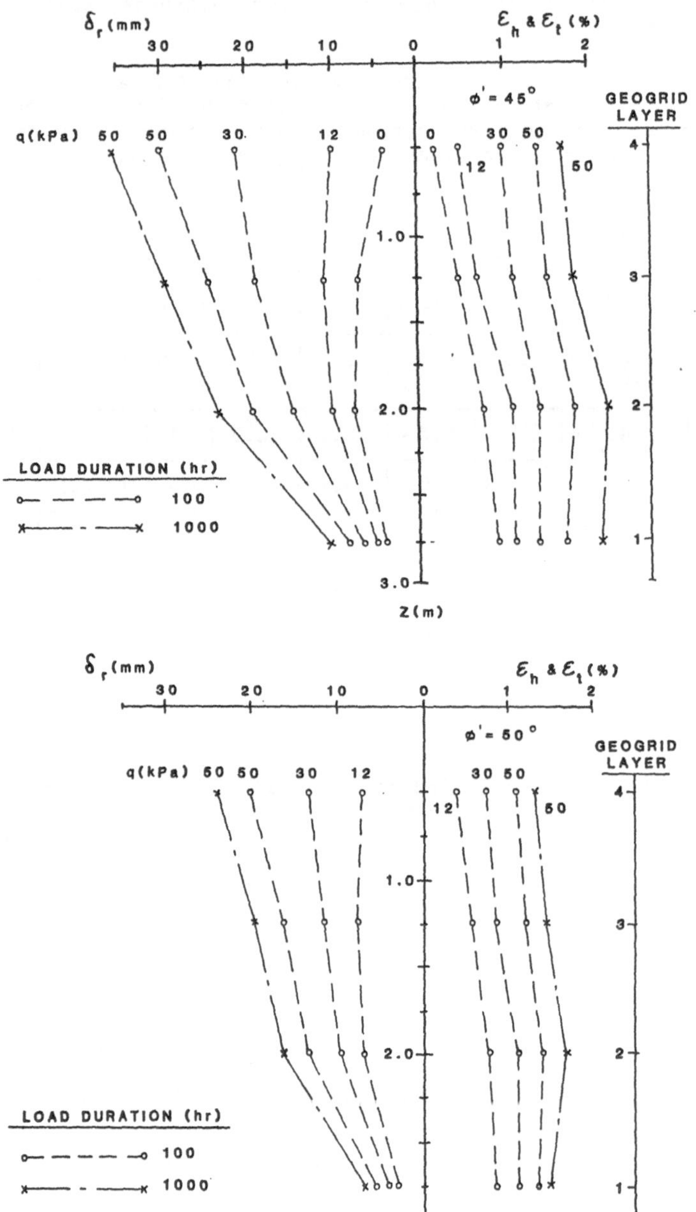

Figure 21. Geogrid elongation (δ_r) and maximum strain (ϵ_t) for RMC reinforced soil walls. Results obtained using two-part wedge analysis.

		q (kPa)		
		12	30	50
$\phi = 45°$	$\delta_{r,max}$ (mm)	11	21	30
	$\delta_{r,max}/H$	0.004	0.007	0.01
$\phi = 50°$	$\delta_{r,max}$ (mm)	7	13	20
	$\delta_{r,max}/H$	0.002	0.004	0.007

6.4 Discussion of Results

The analysis results shown in Figures 20 and 21 appear reasonable, particularly given the uncertainties inherent in attempting to use limit equilibrium procedures to predict working strains and deformations. The analysis results should only correlate with the actual behavior of the RMC walls if the actual soil and reinforcement properties and applied boundary conditions match those used in the analysis. The properties of the soil and reinforcement used in the RMC walls are reasonably well defined. Most of the uncertainty rests with the boundary conditions.

The boundary condition on top of the walls was modeled as a uniform vertical surcharge. This model is incomplete because it neglects the fact that the actual surcharge consisted of 0.75 m (2.5 ft) of loose sand. The sand strength was neglected in the model as were shear stresses developed at the interface between the top of the wall and the loose sand surcharge. The net effect of neglecting these factors is unknown. Plane strain conditions were assumed in the model, requiring frictionless sides to the RMC test wall facility. The actual polyethylene-soil interface friction angle along the side walls will be on the order of 12°. This friction will result in soil shear stresses that will reduce geogrid strains and wall movements. If a sidewall friction angle of 12° were incorporated into the analysis, average geogrid strains and tensions would be reduced by about 10%.

Two different sets of boundary conditions must be considered for the front of the wall: those associated with the incremental timber facing panels and those associated with the propped vertical timber facing panels.

With respect to the incremental panel wall, each panel is attached to only one geogrid layer. This causes the panels to be statically unstable unless sufficient normal and shear forces develop between panels (and between the lowest panel and the leveling pad) to prevent panel rotation. Either individual panels will rotate significantly out of alignment or large shear and normal forces will be transmitted between panels. In this latter case, the foam rubber filler will compress. The filler may become ineffective in minimizing shear stresses at the soil-timber interface. If soil-timber interface friction is fully mobilized, lateral earth pressures would be reduced by about 10%, according to Figure 12. Reinforcement elongations and strains would also be reduced. Interface shear stresses between the timber and reinforcement may also reduce σ_v in the vicinity of the wall face to values less than $(\gamma z + q)$.

For the tilt-up panel wall, reinforcement strains in the vicinity of the wall face will be controlled by the amount of wall rotation. This may lead to a geogrid strain distribution near the wall face different than that predicted from the two-part wedge solutions. Frictional forces should develop between the tilt-up panels and the reinforced soil resulting in reduced horizontal soil stresses and geogrid strains. The frictional forces, coupled with the deformation constraints of the tilt-up panel, may result in significant deviations from the conditions assumed in the two-part wedge model.

In conclusion, the authors believe that the boundary conditions imposed on the RMC reinforced soil walls may significantly influence the measured soil and reinforcement stresses and strains, particularly in the vicinity of the facing units. The influence of these boundary conditions cannot be adequately accounted for using the two-part wedge analysis. In general, however, the imposed boundary conditions should reduce geogrid strains and elongations below the values calculated using the two-part wedge model, possibly by as much as 20 to 30%.

7. CONCLUSIONS

An analysis of the influence of reinforcement tensile stiffness on the behavior of an element of soil was carried out. The analysis indicated that the tensile stiffnesses of steel reinforcement are sufficient to suppress horizontal soil strains and thereby prevent full mobilization of the shear strength of the soil. As a result, the calculated lateral stresses in a reinforced soil mass are close to those associated with an at-rest (K_0) state of stress. The tensile stiffnesses of polymer materials used in geosynthetic-reinforced soil walls are much lower than those of steel reinforcement. The calculation results indicated that geosynthetic reinforcement will not reach equilibrium until sufficient strain has occurred to fully mobilize the shear strength of the soil. As a consequence, an active (K_a) plastic state of stress should develop in geosynthetic-reinforced soil walls.

It was suggested that classical limit state analysis procedures, modified to take into account the presence of reinforcement forces, are appropriate for analysis and design of geosynthetic-reinforced soil walls. A limitation of these procedures as they apply to reinforced soil is that they involve an uncoupling of the soil stresses and reinforcement forces. It was argued that this uncoupling is more acceptable for extensible geosynthetic reinforcement than for inextensible steel reinforcement. Available limit equilibrium and plasticity analysis methods were reviewed and it was noted that the selected method of analysis has little influence on the calculated required tensile forces for active earth pressure problems.

The parametric analyses used in the latter part of paper were based on a two-part wedge limit equilibrium model. The model is conservative and results in reinforcement tensions consistent with an active (K_a) Rankine state of stress. The model was used to investigate the influences of reinforcement tensile stiffness, soil shear strength and wall geometry on the distribution of reinforcement tensile strains, forces and elongations in reinforced soil walls. The analysis results show reinforcement strains and elongations inversely proportional to reinforcement stiffness. Conversely, for a given reinforcement stiffness, calculated reinforcement strains and elongations increase rapidly with increasing wall height and decreasing soil strength. It was suggested that for geosynthetic reinforcement, overturning moments and vertical pressures due to the thrust of the retained fill can be neglected. Therefore, it is unnecessary to incorporate the "Meyerhof" or "trapezoidal" vertical stress distributions into design calculations. Vertical stresses should be based on the fill weight and the applied surcharge loads.

In the last part of the paper, the RMC reinforced soil walls were analyzed using the two-part wedge model. It was noted that limit equilibrium models cannot be used to differentiate walls built using different facing units and construction procedures. Results from the analysis were presented. The results indicate a maximum reinforcement strain on the order of 1% for a surcharge load of 12 kPa (250 psf) and 1.5% for a surcharge load of 50 kPa (1040 psf). A review was carried out to assess whether the boundary conditions used in the limit equilibrium model matched those in the RMC walls. It was concluded that the facing units used for construction of both walls impose significant boundary stresses on the front of the reinforced soil mass. Boundary stresses also result from the dead load surcharge on top of the reinforced soil and from friction along the sides of the test facility. These boundary stresses should reduce geogrid strains and elongations below the values calculated using the two-part wedge model, possibly by as much as 20 to 30%.

REFERENCES

Bacguelin, F. (1978), "Construction and Instrumentation of Reinforced Earth Walls in French Highway Administration", Proceedings, Symposium on Earth Reinforcement, American Society of Civil Engineers, Pittsburgh, pp. 186-201.

Bassett, R.H. and Last, N.C. (1978), "Reinforcing Earth Below Footings and Embankments", Proceedings, Symposium on Earth Reinforcement, American Society of Civil Engineers, Pittsburgh, pp. 202-231.

Berg, R.B., La Rochelle, P., Bonaparte, R. and Tanguay, L. (1987), "Gaspe Peninsula Reinforced Soil Seawall-Case History", Proceedings, Symposium on Soil Improvement, ASCE Geotechnical Special Publication No. 12, Atlantic City, pp. 309-328.

Berg, R.R., Bonaparte, R., Anderson, R.A., and Chouery, V.E. (1986), "Design, Construction and Performance of Two Geogrid Reinforced Soil Retaining Walls", Proceedings, Third International Conference on Geotextiles, Vienna, Vol. 2, pp. 401-406.

Bolton, M.D., Choudhury, S.P. and Pang, P.L.R. (1978), "Reinforced Earth Walls: A Centrifugal Model Study", Proceedings, Symposium on Earth Reinforcement, American Society of Civil Engineers, Pittsburgh, pp. 252-281.

Bonaparte, R. and Berg, R.R. (1987), "Long-Term Allowable Tension for Geosynthetic Reinforcement", Proceedings, Geosynthetics '87, New Orleans, Vol. 1, pp. 181-192.

Bonaparte, R., Holtz, R.D. and Giroud, J.P. (1985), "Soil Reinforcement Design Using Geotextiles and Geogrids", Geotextile Testing and the Design Engineer, American Society for Testing and Materials, Philadelphia, pp. 69-115.

British Dept. of Transport (1978), "Reinforced Earth Retaining Walls and Bridge Abutments for Embankments", Technical Memo BE3/78.

Caquot, A. and Kerisel, J. (1956), "Traite de Mecanique des Sols", Gauthier-Villars, Paris.

Collin, J.G. (1986), Earth Wall Design, Ph.D. Dissertation, University of California, Berkeley, 440 p.

Duncan, J.M. and Chang, C.Y (1970) "Nonlinear Analysis of Stress and Strain in Soils", Journal of the Soil Mechanics and Foundations Division, ASCE, Vol. 96, No. SM5, Sep, 1629-1653.

Fang, Y.S. and Ishibashi, I. (1986), "Static Earth Pressures with Various Wall Movements", Journal of Geotechnical Engineering, Vol. 112, No. 3, Mar, pp. 317-333.

Gourc, J.P., Ratel, A. and Delmas, P. (1986), "Design of Fabric Retaining Walls: The 'Displacement' Method", Proceedings, Third International Conference on Geotextiles, Vienna, Vol. 2, pp. 289-294.

Handy, R.L. (1985), "The Arch in Soil Arching", Journal of Geotechnical Engineering, ASCE, Vol. 111, No. 3, Mar, pp. 302-318.

Ingold, T.S. (1982), Reinforced Earth, Thomas Telford Ltd., London, 141 p.

Jaky, J. (1944), "The Coefficient of Earth Pressure at Rest", Journal of the Society of Hungarian Architects and Engineers, pp. 355-358.

Jewell, R.A. (1985a), "Material Properties for the Design of Geotextile Reinforced Slopes", Geotextiles and Geomembranes Journal, Vol. 2, No. 2, pp. 83-109.

Jewell, R.A. (1985b), "Limit Equilibrium Analysis of Reinforced Soil Walls", Proceedings, Eleventh International Conference on Soil Mechanics and Foundation Engineering, San Francisco, Vol. 2, pp. 1705-1708.

Jewell, R.A. and Wroth, C.P. (1987), "Direct Shear Tests on Reinforced Sand", Geotechnique, Vol. 37, No. 1, pp. 53-68.

Jewell, R.A., Paine, N. and Woods, R.I. (1984), "Design Methods for Steep Reinforced Embankments", Proceedings, Symposium on Polymer Grid Reinforcement in Civil Engineering, The Institution of Civil Engineers, London, pp. 70-81.

Jones, C.J.F.P. (1985), Earth Reinforcement and Soil Structures, Butterworths Advanced Series in Geotechnical Engineering, London, 183 p.

Juran, I. and Schlosser, F. (1978), "Theoretical Analysis of Failure in Reinforced Earth Structures", Proceedings, Symposium on Earth Reinforcement, American Society of Civil Engineers, Pittsburgh, pp. 528-555.

Kondner, R.L. (1963), "Hyperbolic Stress-Strain Response: Cohesive Soils", Journal of the Soil Mechanics and Foundations Division, ASCE, Vol. 89, No. SM1, Feb, p. 115.

LCPC-SETRA (1979), "Les Ouvrages en Terre Armée", Recommandations et Règles de l'Art, 195 p.

Lambe, T.W. and Whitman, R.V. (1968), Soil Mechanics, John Wiley and Sons, Inc., New York, 553 p.

Lee, K.L., Adams, B.D. and Vagneron, J.J. (1972), "Reinforced Earth Retaining Walls", Journal of the Soil Mechanics and Foundations Division, ASCE, Vol. 99, No. 10, Oct, pp. 745-764.

Leshchinsky, D. and Perry, E.B. (1987), "A Design Procedure for Geotextile Reinforced Walls", Proceedings, Geosynthetics '87, New Orleans, Vol. 1, pp. 95-107.

McGown, A., Andrawes, K.Z., and Al-Hasani, M.M. (1978), "Effect of Inclusion Properties on The Behavior of Sand", Geotechnique, Vol. 28, No. 3, pp. 327-346.

McGown, A., Andrawes, K.Z., Yeo, K.C. and DuBois, D.D. (1984a), "The Load-Strain-Time Behavior of Tensar Geogrids", Proceedings, Symposium on Polymer Grid Reinforcement in Civil Engineering, The Institution of Civil Engineers, London, pp. 11-17.

McGown, A., Paine, N. and Dubois, D.D. (1984b), "Use of Geogrids in Limit Equilibrium Analysis", Proceedings, Symposium on Polymer Grid Reinforcement in Civil Engineering, Institution of Civil Engineers, London, pp. 31-36.

Meyerhof, G.G. (1953), "The Bearing Capacity of Foundations Under Eccentric and Inclined Loads", Proceedings, Third International Conference on Soil Mechanics and Foundation Engineering, Zurich, Vol. 1, pp. 440-445

Meyerhof, G.G. (1980), "Limit Equilibrium Plasticity in Soil Mechanics", Proceedings, Symposium on Applications of Plasticity and Generalized Stress-Strain in Geotechnical Engineering, ASCE, Florida, pp. 7-24.

Milligan, G.W.E. (1974), "The Behavior of Rigid and Flexible Retaining Walls in Sand", Ph.D. Dissertation, University of Cambridge, England.

Milligan, G.W.E. (1983), "Soil Deformations Near Anchored Sheet Pile Walls", Geotechnique, Vol. 33, No. 1, pp. 41-55.

Mitchell, J.K. (1987), "Reinforcement for Earthwork Construction and Ground Stabilization", Theme Lecture, Preprint, VIII Pan American Conference on Soil Mechanics and Foundation Engineering, Cartagena, Aug.

Morgenstern, N.R. and Eisenstein, Z. (1970), "Methods for Estimating Lateral Loads and Deformations", Proceedings, Specialty Conference on Lateral Stresses and Earth Retaining Structures, American Society of Civil Engineers, Ithaca, pp. 51-102.

Murray, R.T. (1984), "Reinforcement Techniques in Repairing Slope Failures", Proceedings, Symposium on Polymer Grid Reinforcement in Civil Engineering, London, pp. 47-53.

Murray, R.T. (1987), "Factor of Safety Considerations Relating to Reinforced Soil Structures", Preprint, NATO Advanced Research Workshop on Polymeric Reinforcement in Soil Retaining Structures.

Romstad, K.M., Al-Yassin, A., Hermann, L.R. and Shen, C.K. (1978), "Stability Analysis of Reinforced Earth Retaining Structures", Proceedings, Symposium on Earth Reinforcement, Pittsburgh, pp. 685-713.

Rowe, P.W., (1969a), "The Relation Between the Shear Strength of Sands in Triaxial Compression, Plane Strain and Direct Shear", Geotechnique, Vol. 19, No. 1, pp. 75-86.

Rowe, P.W. (1969b), "Progressive Failure and Strength of a Sand Mass", Proceedings, Seventh International Conference on Soil Mechanics and Foundations Engineering, Mexico City, Vol. 1, pp. 341-349.

Ruegger, R. (1986), "Geotextile Reinforced Structures on Which Vegetation Can Be Established", Proceedings, Third International Conference on Geotextiles, Vienna, pp. 453-459.

Sarsby, R.W. (1985), "The Influence of Aperture Size/Particle Size on the Efficiency of Grid Reinforcement", Proceedings, Second Canadian Conference on Geotextiles and Geomembranes, Edmonton, pp. 7-12.

Schlosser, F. (1978) "La Terre Armée, historique development actuel et Futur", Proceedings, Symposium on Soil Reinforcing and Stabilizing Techniques, NWSIT/NSW University, pp. 5-28.

Schlosser, F. and Vidal, H. (1969), "La Terre Armée", Bulletin de Liason du Laboratoire des Ponts et Chaussées, No. 41, Nov, pp. 101-144.

Schmertmann, G.R., Bonaparte, R., Chouery, V.C. and Johnson, R. (1987), "Design Charts for Geogrid Reinforced Soil Slopes", Proceedings, Geosynthetics '87, New Orleans, Vol. 1, pp. 108-120.

Scott, R.F. (1985), "Plasticity and Constitutive Relations in Soil Mechanics", Journal of Geotechnical Engineering, ASCE, Vol. 111, No. 5, May, pp. 563-605.

Segrestin, P. (1979), "Design of Reinforced Earth Structures Assuming Failure Wedges", Proceedings, International Conference on Soil Reinforcement, Paris, Vol. 2, pp. 163-168.

Sokolovski, V.V. (1965), Statics of Granular Materials, Pergamon Press, London.

Stocker, M.F., Korber, G.W., Gassler, G. and Gudehus, G. (1979), "Soil Nailing", Proceedings, International Conference on Soil Reinforcement, Paris, Vol. 2, pp. 469-474.

Studer, J.A. and Meier, P. (1986), "Earth Reinforcement with Nonwoven Fabrics: Problems and Computational Possibilities", Proceedings, Third International Conference on Geotextiles, Vienna, Vol 2, pp. 361-365.

Terzaghi, K. (1943), Theoretical Soil Mechanics, John Wiley and Sons, New York, 729 p.

Terzaghi, K. and Peck, R.B. (1967), Soil Mechanics in Engineering Practice, John Wiley and Sons, New York, 510 p.

Vidal, H. (1966), "La Terre Armée", Annales de l'Institut Technique du Bâtiment et des Travaux Publics, No. 223-234, Jul-Aug.

DESIGN OF REINFORCED SOIL RETAINING WALLS:
ANALYSIS AND COMPARISON OF EXISTING DESIGN METHODS AND PROPOSAL FOR A NEW APPROACH

J.P. GOURC, A. RATEL, Ph. GOTTELAND
IRIGM, UNIVERSITE DE GRENOBLE, FRANCE

1. INTRODUCTION

The study presented here is a comparative study of several methods for designing earth retaining walls with geosynthetic reinforcement. In this study, only the methods considering limit equilibrium analysis will be considered. Other methods using numerical models for simulation (finite difference method and finite element method) are excluded, because they are generally too laborious to design an economical job. Moreover, it appears difficult to numerically simulate the very variable technological conditions during construction (however, it is one of the essential advantages of finite element method which can simulate the construction stages such as layering and fixing the fabric layers to the face of the wall, compacting the soil, etc.).

Since the first presence of the method "Terre Armée" (Schlosser, et al. -(1)-), then that of bolting and geotextiles, numerous methods for calculation at limiting state have been proposed. Without stating all of them, they can be classified according to their hypothesis. These consider principally the following points:

- external equilibrium
- critical slip line
- local equilibrium of an isolated reinforcement

At the IRIGM Laboratory of Grenoble University, we carried out a comparison of these methods. In addition, IRIGM, in cooperation with LCPC, developed a specific method, called "displacements method", for designing geosynthetic walls.

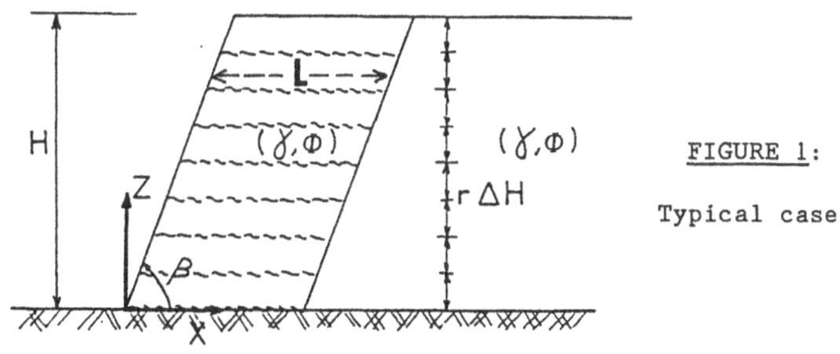

FIGURE 1: Typical case

P. M. Jarrett and A. McGown (eds.), The Application of Polymeric Reinforcement in Soil Retaining Structures, 459–506.

In order to simplify the problem and the comparative calculations, it is necessary to limit the number of variable parameters. Thus, the typical case considered is shown in Fig. 1.

- The problem is considered as a plane strain problem.
- The reinforcement layers are of the same length L, regularly spaced (ΔH), and horizontal.
- The reinforcement layers are composed of the same geosynthetics (tensile modulus, J, tensile strength, T_F, soil - geosynthetic friction angle: $\tan \phi_g = 2/3 \tan \phi$).
- The soil is homogeneous for reinforced mass and the backfilling material: cohesionless, unsaturated (γ, ϕ).
- The foundation soil has a sufficient bearing capacity in all cases studied (deep failure passing under the base of the wall will not be examined).
- The conditions of the wall face are not specified.

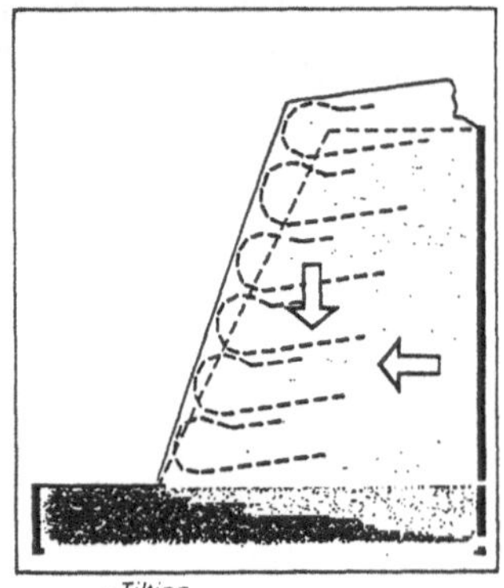

- Tilting.

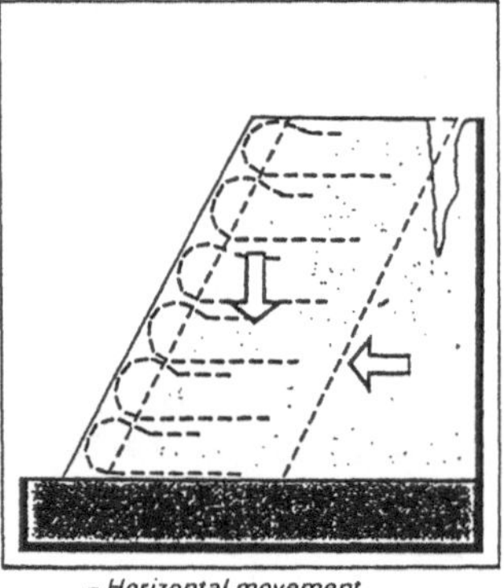

- Horizontal movement.

FIGURE 2: External equilibrium

2. EXTERNAL EQUILIBRIUM

The observations of movements of whole reinforced soil masses, either on reduced models or on large scale models, justify the overall study of a reinforced mass. Hence the classical procedure for analyzing retaining walls will be used (Figure 2):

- Overturning stability: factor of safety F_R
- Sliding stability: factor of safety F_G
- Bearing stability: factor of safety F_P
- Condition of no tension induced beneath the base of the wall.

This study does not bring in the characteristics of geosynthetic reinforcement, except the interface condition at the base of the mass ($\phi_b = \phi$ or ϕ_g). The calculation obtains a predesign of the wall:

$$L = L\ \min\ (H, \beta)$$

This traditional procedure is not consistent with that for studying internal stability of walls. Instead, the latter uses the concept of slip line associated with a factor of safety F_s on the shear strength of the soil ($F_s = \tan \phi / \tan / \phi_c$) as that in limit equilibrium analysis of slopes. While for retaining walls, the factor of safety focuses on the loads applied (F_G, F_R).

The stability of sliding can thus be examined from slip lines such as those shown in Figure 3.

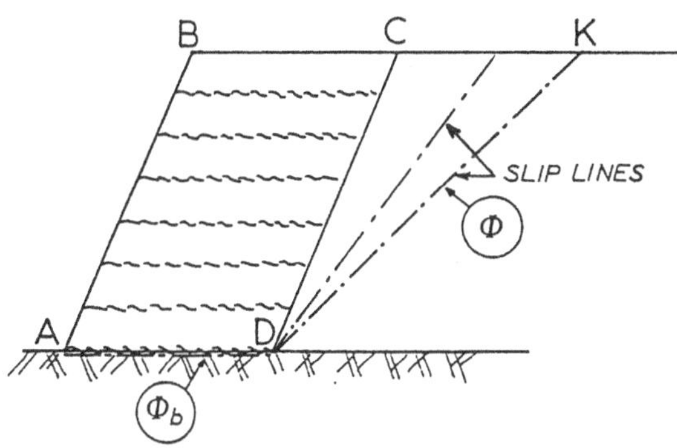

FIGURE 3: Proposal for sliding stability study

This procedure obtains continuous results between internal and external stabilities if, for example, the same mechanism is considered for both cases. Besides, this calculation allows a direct estimation of the force on DC without laborious use of the tables of acting force. However, this procedure generally slightly underestimates the active force.

It should be noted that this kinematic of failure is equivalent to that of sliding of a wall on its base, but is not equivalent to that of overturning. In reality, the condition of overturning is practically satisfied at the base ($F_R \geq 1.5$) when $F_G \geq 1.5$ except for vertical facing. Therefore, its systematic verification is not generally necessary.

3. PRINCIPLE OF GEOSYNTHETIC REINFORCEMENT

A retaining wall resists the force P exerted by the backfill (active block ABK) as shown in Figure 4.

$$\vec{\Sigma}\,(-\vec{P},\ \vec{W},\ \vec{R}) = \vec{0}$$

$$\vec{\Sigma}\,(\vec{P},\ \vec{P^*}) = \vec{0}$$

Similarly, if the retaining wall is replaced by a reinforced mass, the whole reinforcement will have to resist the thrust of the soil P (active block ABK) (Figure 5).

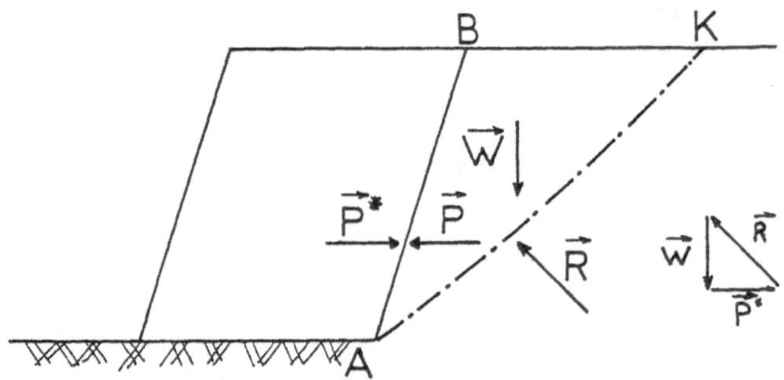

FIGURE 4: Influence of a retaining wall on the slope stability

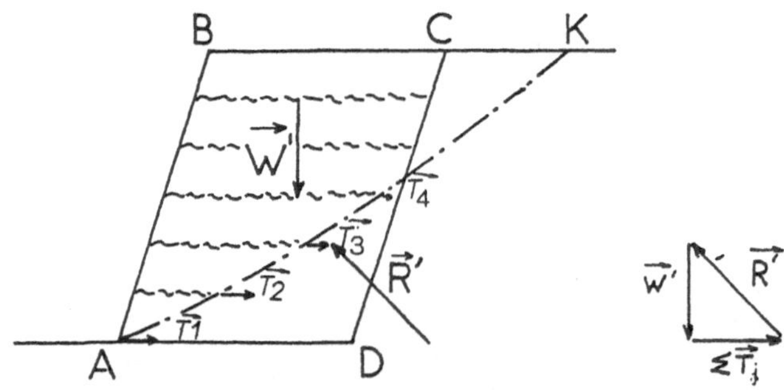

FIGURE 5: Influence of reinforcement layers on the slope stability

$$\vec{\zeta}(-\vec{P'}, \vec{W'}, \vec{R'}) = \vec{0}$$

$$\vec{\zeta}(\vec{P'}, \vec{\sum T_j}) = \vec{0}$$

All the present approaches of design calculation of reinforced soil structures are thus based on the principle of limit equilibrium:

* All the methods assume a slip line which is the position of the maximum tensile stresses in the reinforcement (Juran et al. -(2)-). This corresponds to the inversion of the directions of tangential stresses along the soil-geotextile interfaces at the points of kinematic discontinuity. T_j is the maximum tensile force in the reinforcement j.

* Each layer of reinforcement is considered as having no flexural rigidity (like a membrane) and only mobilizes tensile forces. In most methods, it is assumed that the displacement velocity along the slip line does not influence the orientation of

tensile forces. Their directions are assumed to follow the initial horizontal direction of reinforcement ($\beta_j = 0$, see 5.2.2).

4. THE PRINCIPLES OF LIMIT EQUILIBRIUM METHODS

In one of several respects, the various limit equilibrium methods are different in the shapes of the failure surfaces (slip line) assumed.

4.1 Polygonal slip line

The most general slip line is a concave polygonal slip line. The classical procedure of analysis consists of cutting the sliding mass into n slices. The number of the reinforcing layers cut by the slip line is represented by r (Figure 6).

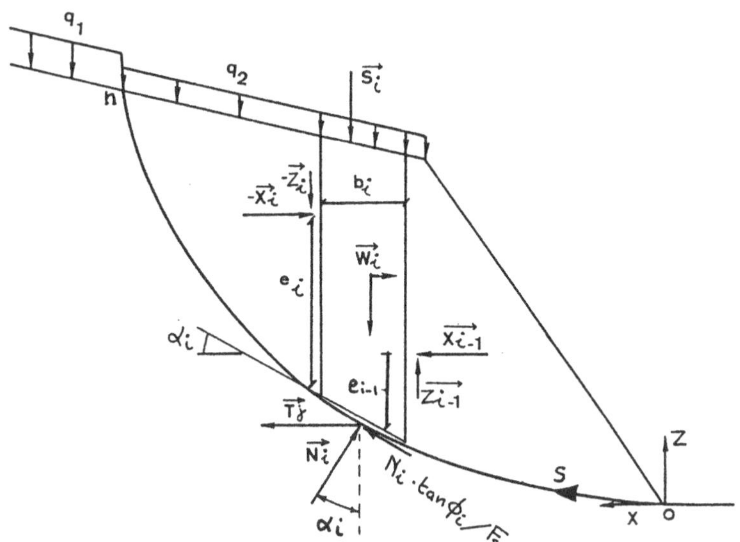

FIGURE 6: Analysis of reinforced slope stability by the method of slices

Unknowns		Equations	
X_i, Z_i, e_i	$3(n+1)$	$X_0 = Z_0 = e_0 = 0$	6
N_i	n	$X_n = Z_n = e_n = 0$	
F_s	1	equilibrium equations	3n
T_j	r		

In the stability analysis for slopes without reinforcement, there is a total of 3n equations obtained from equilibrium conditions of n slices. Hence, it lacks (n-2) equations (number of unknowns - number of equations).

In order to solve the problem, it thus needs (n-2) supplementary assumptions. A method called "perturbations method" (Raulin et al. -(3)-) will be used here. It consists of choosing a distribution of normal stress along the slip line:

$$N_i = N_{io} (\mu_1 f(s_i) + \mu_2 g(s_i))$$

The above assumption supplies n supplementary equations plus two unknowns μ_1 and μ_2 to be determined, i.e. (n-2) equations given. The example presented in the following text uses:

$$N_i = N_{io} (\text{Fellenius}) \cdot (\mu_1 + \mu_2 \cdot \tan\alpha_i)$$

In the case of non-reinforced slopes, this method can obtain the factor of safety of the shear strength of the soil, F_s. It can also obtain a realistic distribution of normal forces N_i at the base of each slice. However, it can not solve the problem with r tensile forces T_j of reinforcements. Therefore, T_j must be determined separately.

4.2 Circular slip line

For a non-reinforced homogeneous soil mass (c, ϕ, γ constant), a method of overall equilibrium can be used (without cutting the mass into slices). A complementary assumption, eg., on the resultant force R at the base of the slip line (Biarez -(4)-), allows F_s to be obtained (Phan et al. -(5)-).

The forces in the reinforcements T_j need to be determined separately, as previously, if the factor of safety for the reinforced soil mass F_s is to be determined. The use of the method of slices in this case can also be examined (cf 4.1). However, there are some points which need to be noted in using the method of slices, say "inexact".

Bishop's simplified method -(6)- only considers vertical force equilibrium of slices and overall moment equilibrium with respect to the center of the circle. The shape of the slip line considered must be circular.

Janbu's method -(7)- can analyze both circular and polygonal slip lines: vertical, horizontal, and moment equilibrium of each slice are satisfied, with the exception of overall moment equilibrium (Ratel -(8)-). It can thus be used to analyse a polygonal slip line.

In addition, there are several other different circular slip lines such as a slip line composed of an arc and a straight line corresponding to the sliding plane either along a geosynthetic layer (when $\phi_g < \phi$) or along a preferential slip line (for instance along the wall base, when $\phi_b < \phi$) (Figure 7). Werner et al. -(9)- proposed a particular slip line considering local instability between reinforcement layers (Figure 8).

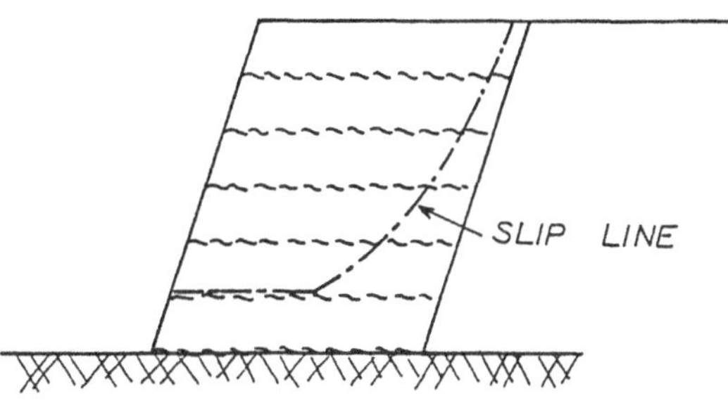

FIGURE 7: Bi-component slip line

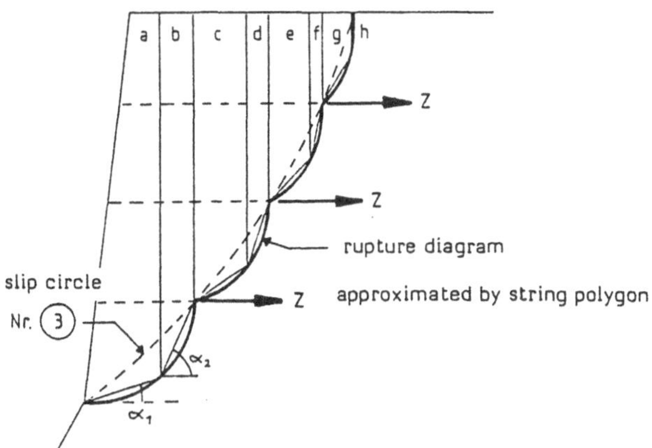

FIGURE 8: Modified slip line -(9)-

4.3 Log-spiral slip line

This method assumes a logarithmic spiral shape using the equation below:

$$R = R_o \cdot e^{\alpha \tan\phi_c}$$

(Juran et al. -(2)-, Figure 9; Baker et al. -(10)-; Leshchinsky et al. -(11)-, Figure 10). It can obtain analytically a distribution of normal stresses along the slip line. In the absence of reinforcement, F_s can be obtained without assumptions. For the case of a reinforced mass, T_j must be predetermined as in the previous cases for obtaining F_s.

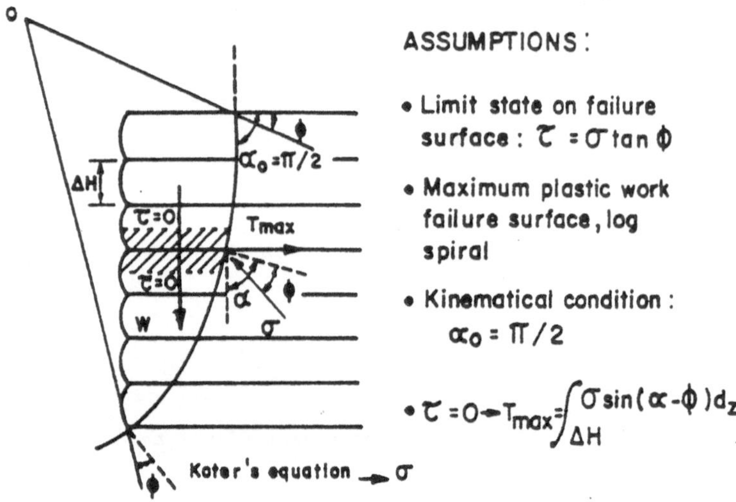

ASSUMPTIONS:
• Limit state on failure surface: $\tau = \sigma \tan \phi$
• Maximum plastic work failure surface, log spiral
• Kinematical condition: $\alpha_0 = \pi/2$
• $\tau = 0 \rightarrow T_{max} = \int_{\Delta H} \sigma \sin(\alpha - \phi) d_z$
Kotter's equation → σ
Tmax
W
ΔH

FIGURE 9: Log-spiral slip line -(2)-

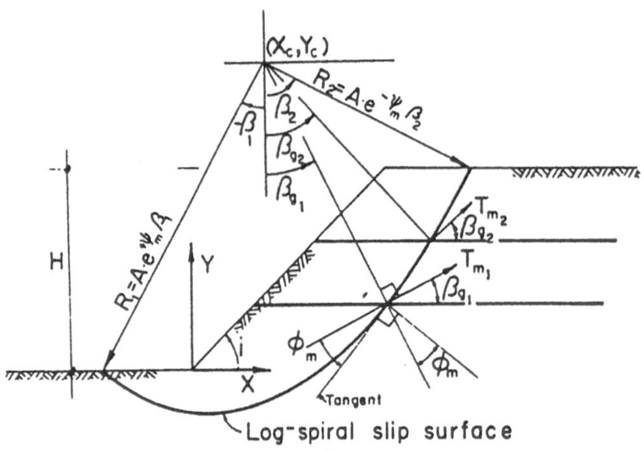

FIGURE 10: Log-spiral slip line -(11)-

4.4 Two-block

This method is very useful for analysing reinforced soil structures. The active zone is composed of two blocks. The interblock face is either vertical (Bordairon -(2)-, Figure 11; Jewell -(13)-), as in the method of slices, or inclined (Hamilton -(14)-, Figure 12).

In the method of slices the number of equations required is (n-2) (cf 4.1). Even in the two-block case (n=2), the problem still can not be solved, i.e. the position of N_a and N_b at the base of blocks (Figure 11) must be specified.

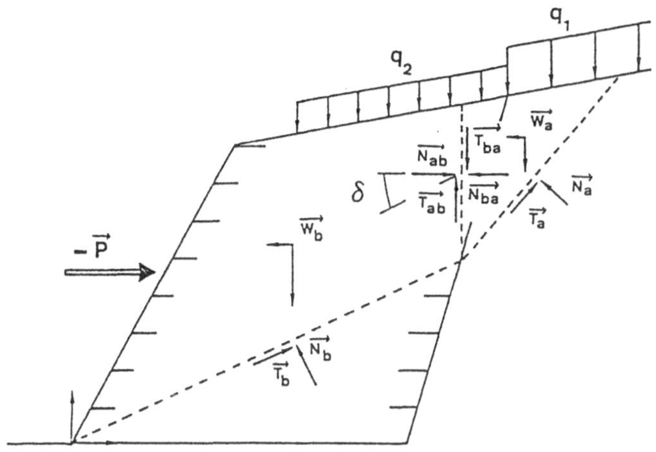

FIGURE 11: Two-block stability analysis -(12)-

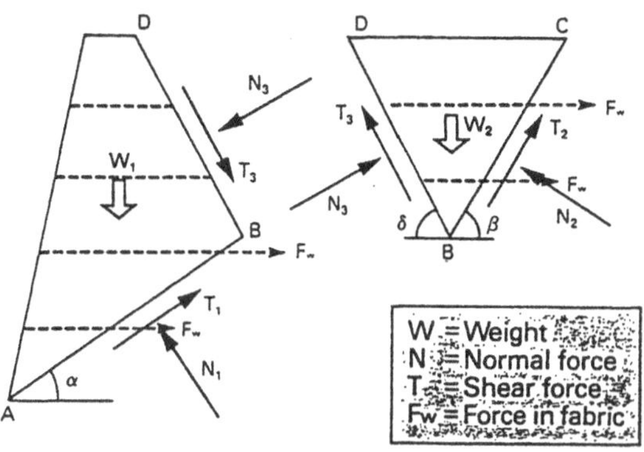

FIGURE 12: Two-block stability analysis - modified cutting of the active zone -(14)-

The "two-block method" is thus generally used without verifying the moment equilibrium.

Unknowns		Equations	
N_a, N_b	2	equilibrium equations	4
N_{ab}, Tab	2		
F_s	1		
$\sum T_j$	1		

The usual procedure consists of considering the case when inclination of interblock face δ is most unfavourable.

Therefore, in the absence of reinforcement, the problem can be solved by fixing δ, then F_s will be determined. The reinforcement only intervenes by its sum $\sum T_j$. If $\sum T_j$ is determined separately, then F_s can be obtained. Inversely, F_s can be fixed to calculate $\sum T_j$.

4.5 Coulomb wedge

The slip line assumed is a straight line inclined at α with respect to the horizontal plane.

Unknowns		Equations	
N_a	1	equilibrium equations	2
F_s	1		
$\sum T_j$	1		

In this method, moment equilibrium is not satisfied either. As previously, F_s can be determined according to $\sum T_j$. Broms -(15)- proposed a modified shape of active block to take into account a soil-geosynthetic unilateral sliding (Figure 13).

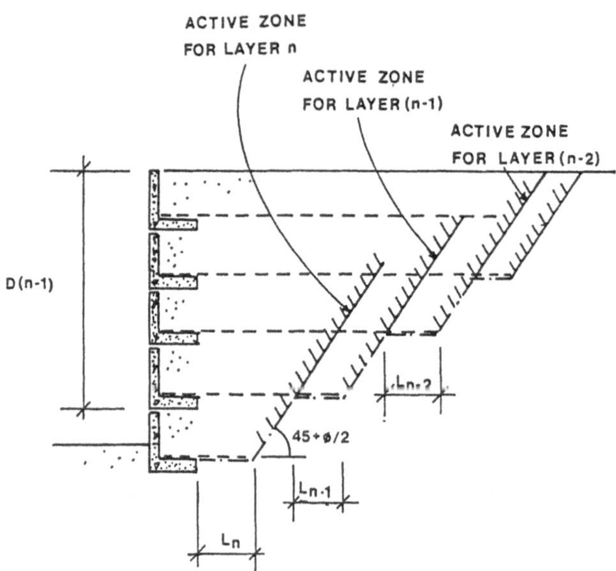

FIGURE 13: Proposal of modified Coulomb wedge for reinforced soil -(15)-

5. TAKING INTO ACCOUNT THE FORCE OF REINFORCEMENT

As it has been mentioned in Chapter 4, the stability analysis of reinforced soil mass does not allow to determine simultaneously F_s and the forces T_j, or their sum $\sum T_j$. In order to solve this problem, two different approaches, global approach and local approach, will be distinguished.

5.1 Global approach

It estimates the mobilisable strength in each reinforcement:

$$T_j \leq T_{jmob} = \min \left(\frac{T_{fj}}{F_T}, \frac{T_{aj}}{F_f}\right)$$

where T_{fj} = failure tensile strength of the jth geosynthetic layer

F_T = factor of safety with respect to tensile failure

$T_f/F_T = T_I$: allowable tensile strength of geosynthetics

T_{aj} = maximum anchorage force of the jth reinforcement

$$T_{aj} = 2 \cdot \int_o^{L_{aj}} (\sigma_n \cdot \tan \phi_g) \cdot dL$$

F_f: factor of safety with respect to soil-geosynthetic friction

L_{aj}: anchor length (to the slip surface) L_{aj} in active zone (L^a_{aj}) and passive zone (L^P_{aj}) will be distinguished

σ_n: normal stress acting on the reinforcement in the zone of anchorage

In order to control a design, the following procedure is followed.

Assume that a reinforced mass of fixed length L and of predetermined reinforcement, identical for all the reinforcement (T_f fixed), the choicė of F_T will depend on the conditions of the use of geotextiles and of its nature (Mir Arabchari -(16)-, Andrawes et al. -(17)-).

Once the slip line has been selected T_{aj} can be calculated. With F_f fixed, the mobilisable strength in each reinforcement, $T_{j\ mob}$, can be obtained:

$$T_j = T_{j\ mob}$$

The next step is the determination of F_s, by limit equilibrium method.

However, some variations exist. Another less physical viewpoint such as Leschinsky -(11)- does not consider any limitation in anchorage strength (reinforcement of large length) and assigns the same factor of safety to both shear strength of soil and the tensile strength of the reinforcement:

$$F_T = F_s \quad \rightarrow\rightarrow \quad T_{j\ mob} = \frac{T_f}{F_s}$$

The limit equilibrium method associated with variationally extremisation can obtain $(F_s = F_T)_{min}$.

In the most general case, $T_{j\ mob}$ varies with the reinforcement, and also with the anchor length L_{aj}, and therefore with the position of the slip line. The critical slip line is the one corresponding to $(F_s)_{min}$ for a given reinforcement $(L,\ T_f/F_T)$. Hence, the calculation of limit equilibrium must be performed for a great number of potential slip lines.

One important remark: in order to facilitate the calculation, many other authors follow a more rapid simplified process. It consists of choosing independently the slip line corresponding to maximal horizontal force P (Figure 4) and the length L of the reinforcement layers. To equate P_{max} and the tensile forces in the reinforcements:

$$\vec{P}_{max} + \sum \vec{T}_j = \vec{0}$$

In the case of vertical wall ($\beta = \pi/2$), the maximal horizontal force corresponding to the Rankine block is:

$$P = \frac{1}{2} K_a \cdot \gamma \cdot H^2 \qquad \text{where } K_a = \tan^2 \left(\frac{\pi}{4} - \frac{\phi_c}{2}\right)$$

This kind of slip line is most traditionally admitted ("Standard" method - Gourc et al. -(18)-; Murray -(19)-).

For inclined walls, Jewell et al. -(13)- do the same thing: the slip line corresponding to the force P_{max}.

However, Segrestin -(20)- demonstrated analytically that the inclination α of the critical block was not identical for non-reinforced soil and reinforced soil.

Non-reinforced soil -(Figure 14)-

$$P_{max} \text{ for } \sin 2\alpha = \sin 2(\alpha - \phi_c)$$

$$\rightarrow\rightarrow \quad \alpha = \frac{\pi}{4} + \frac{\phi_c}{2}$$

Reinforced soil -(Figure 15)-

$$(\text{same } T_I = \frac{T_f}{F_T} \text{ for each reinforcement layer})$$

The limit value of T_I to equilibrate the block for:

$$\sin 2\alpha - \lambda \frac{\Delta H}{L} \cos^2\alpha = \sin 2(\alpha - \phi_c)$$

$$\rightarrow\rightarrow \quad \alpha < \frac{\pi}{4} + \frac{\phi_c}{2}$$

Note that the critical angle α depends on L, i.e., on the length of embedment.

Therefore, it should be remembered that the critical slip line can not be determined without taking into account the reinforcement.

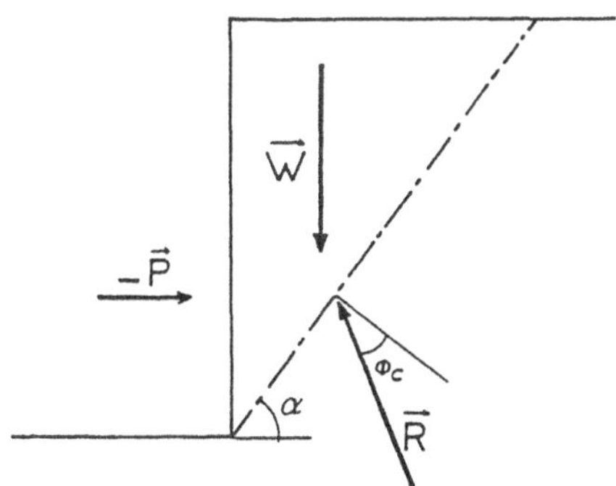

FIGURE 14: Static equilibrium for unreinforced soil

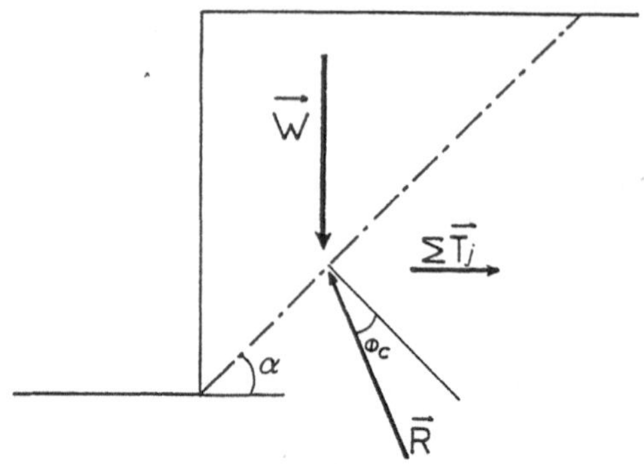

FIGURE 15: Static equilibrium for reinforced soil

Moreover, even in the case where the global approach is rigorously performed, without precedent approximation, this approach is not conservative because equilibrium is obtained from maximal mobilisable strength ($T_{j\ mob}$ with factors of safety F_T and F_f):

$$\sum_j T_j = \sum_j T_{j\ mob} = \sum_j \min\left(\frac{T_{fj}}{F_T}, \frac{T_{aj}}{F_f}\right)$$

However, nothing guaranties a simultaneous mobilisation of r forces $T_{j\ mob}$ as this is examined in the calculation. It's all the more true for a relatively deformable structure.

5.2 Local approach

Most other design methods, aware of the inherent uncertainty in the global approach and the approximations made on the critical slip line (the shape of the line or line of maximal tension), have completed the precedent equilibrium condition by a local equilibrium condition for each layer of reinforcement. Two types of approaches are distinguished:

- equilibrium of a horizontal layer of soil-reinforcement composite;
- equilibrium of a reinforcement considered as an anchored membrane.

5.2.1 Equilibrium of a composite soil-reinforcement layer

Juran et al. (Figure 9 -(2)-) showed that the vertical stress σ_z and horizontal stress σ_x are principal stresses on a central plane between two reinforced layers because of symmetry. Consequently, the shear stress τ on these planes is zero -(21)- (Figure 16).

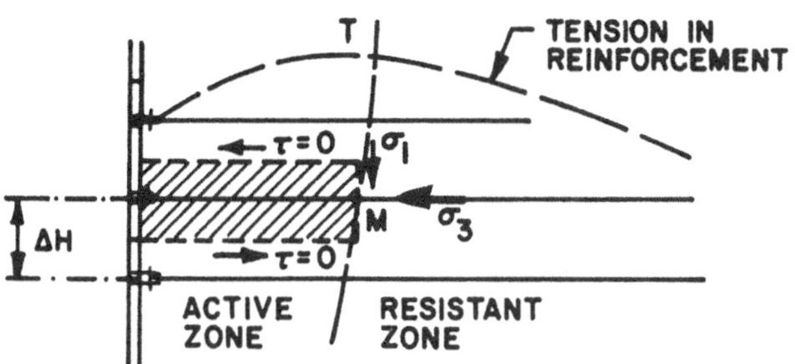

FIGURE 16: Local equilibrium -(21)-

We can then consider that the active zone is composed of horizontal composite layers maintained horizontally by the forces exerted along the slip line (Figure 17).

Each composite element is subjected to a force ($\sigma_x \cdot \Delta H$) and to a tensile force T_j. From local equilibrium,

$$T_j = -\sigma_x \cdot \Delta H$$

This is the classical principal of local equilibrium.

Next, the various methods considering the values of σ_x take into account:

the compatibility which overall equilibrium imposes:

$$\int_o^H \sigma_x \cdot dZ = P = -\sum T_j$$

This equation is checked in "Standard" method (method C) applied to vertical walls -(18)-

$$\text{since } \sigma_x = K_a \cdot y \cdot (H-z) \text{ where } K_a = \tan^2 \left(\frac{\pi}{4} - \frac{\phi_c}{2}\right)$$

$$\int_a^H \sigma_x \cdot dZ = \frac{1}{2} K_a \cdot y \cdot H^2$$

in agreement with

$$P = P_{max} = \frac{1}{2} \cdot K_a \cdot y \cdot H^2 \text{ for } \alpha = \pi/4 + \phi/2$$

(refer to 5.1)

This equation is also checked in Jewell's block method -(13)- applied to inclined walls:

$$\sigma_x = K \cdot \gamma \cdot (H - z) \text{ where } K = \frac{P_{max}}{\frac{1}{2}\gamma H^2}$$

$$P_{max} = \frac{1}{2} K \cdot \gamma \cdot H^2$$

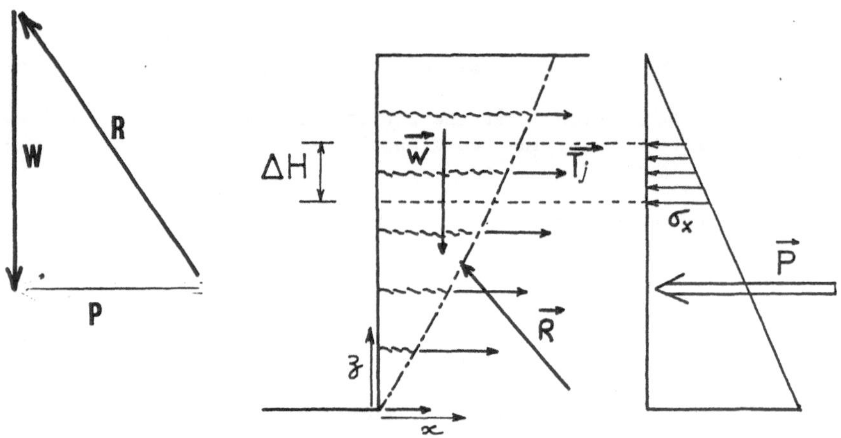

FIGURE 17: Compatibility between local equilibrium and overall equilibrium for active zone

It is clear that these r conditions of equilibrium for the r layers of reinforcement impose safer conditions than only using overall equilibrium condition (5.1)

$$\text{local equilibrium:} \quad \sigma_{xj} \cdot \Delta H \leq \min\left(T_I, \frac{T_{aj}}{F_s}\right) \rightarrow T_{I2}$$

$$\text{global equilibrium:} \quad P \leq \sum \min\left(T_I, \frac{T_{aj}}{F_s}\right) \rightarrow T_{I1} \leq T_{I2}$$

This can compensate partially or totally the simplification on the critical slip line identified from the line of maximal thrust seen in 5.1. Moreover, the condition of overall equilibrium seen in 5.1 is usually not used, the slip line goes out of the reinforced soil mass, at the top of the wall.

Other authors obtained a result even more conservative in overestimating the force P obtained from overall equilibrium:

-(22)- (Steward et al., - method [D]) $\sigma_x = K_o \cdot \gamma \cdot (H - z)$

where $K_o = 1 - \sin \phi_c$ (soil at rest condition)

-(15)- (Broms, - method [E]) $\sigma_x = 0.65 \cdot K_a \cdot \gamma \cdot H$

stress distribution used in tieback walls

and $K_a = \tan^2 (\frac{\pi}{4} - \frac{\phi_c}{2})$

In the case of "reinforced-earth" walls with vertical facing the reinforced backfill is considered as monolithic: one obtains a horizontal distribution of nonhomogeneous vertical stress σ_z, owing to the moment of overturning (Figure 18).

-(19)- (Murray, - method [B]) trapezoidal distribution of σ_z

The thurst is increased in considering maximal stress $(\sigma_z)_{max}$

$$\sigma_x = K_a \cdot (\sigma_z)_{max} = K_a \left[\gamma(H-z) + \frac{6M_z}{(L)^2}\right]$$

where $K_a = \tan^2 (\frac{\pi}{4} - \frac{\phi_c}{2})$ and M_z the resulting moment.

-(21)- (LCPC - method [A]) distribution of σ_z from Meyerhof

$$\sigma_x = K_a \cdot \frac{\gamma(H-z)}{1 - \frac{2M_z}{\gamma(H-z)\cdot(L)^2}}$$

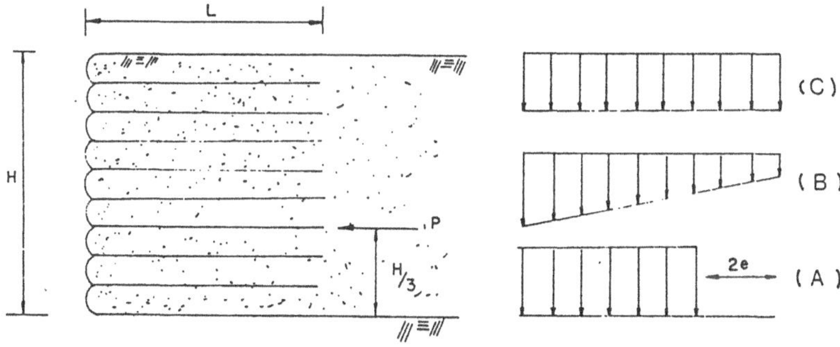

FIGURE 18: Distribution of vertical stress at geosynthetics level

5.2.2 Equilibrium of geosynthetic considered as an anchored membrane - the "displacements method" -

5.2.2.1 Basic principle

The notable difference existing between a metallic reinforcement ("Terre Armée": Reinforced Earth) and a usual geosynthetic reinforcement is due to the larger deformability of geosynthetic with respect to the soil strains. In this circumstance, mobilisation of tensile forces in the reinforcement will depend on the displacement that it is imposed along the slip line.

The proposed mechanism is shown in Figure 19 (Gourc et al. -(23)-).

The displacement of the active block correspond with displacement $\Delta\theta$ at top of the wall. A displacement field is imposed in soil in the vicinity of the slip line, i.e., $\Delta\theta_j = f(\Delta\theta)$ at the jth reinforcement.

Once the behaviour of each reinforcement is defined for a membrane anchored at both sides, in the passive and active zones, $T_j = g(\Delta\theta_j)$, the relation $\Delta\theta_j = f(\Delta\theta)$ allows a relation of compatibility for tensile forces T_j mobilized in each reinforcement to be found. Thus, during the displacement of the block ($\Delta\theta$ increasing), the inclination β_j and the tensile force T_j enlarge.

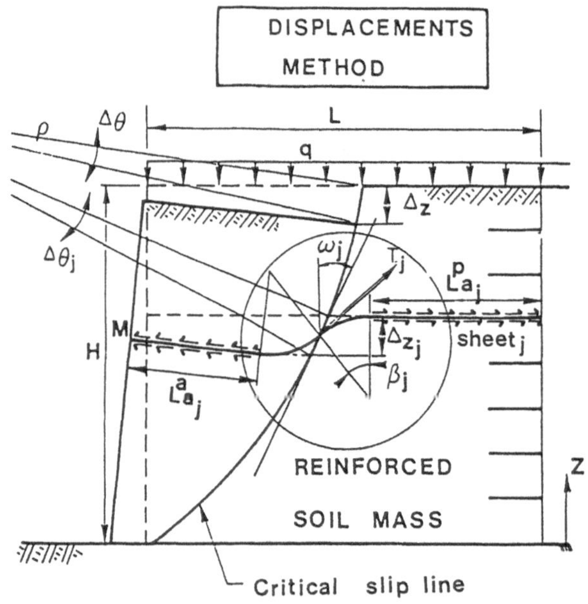

FIGURE 19: "Displacements method" -(22)-, -(25-)

For each value of $\Delta\theta$, there are r couples (T_j, β_j) obtained from local equilibrium of each reinforcement. These r couples are then substituted into the general equilibrium equations of the block (here, the "perturbations method" -(3)-). The process is repeated by increments of $\Delta\theta$ until a system of r couples (T_j, β_j) stabilizing the sliding block is obtained.

5.2.2.2 <u>Local behaviour of reinforcement</u>

The experiments performed at Grenoble University (Gourc et al. -(23)-) on a small scale model of temporary road on weak soil, reinforced by a fabric, brought the proposition of the mechanism of an anchored membrane.

* <u>Anchorage</u>: The two sides of shear zone, the reinforcement is anchored in active (L_{aj}^{a}) and passive (L_{aj}^{p}) zones. For a geo-

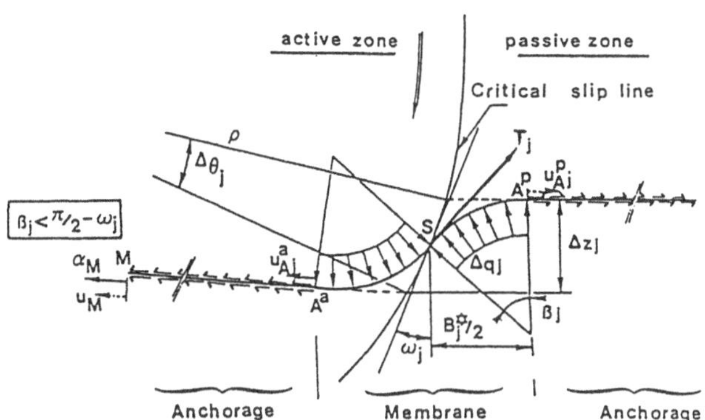

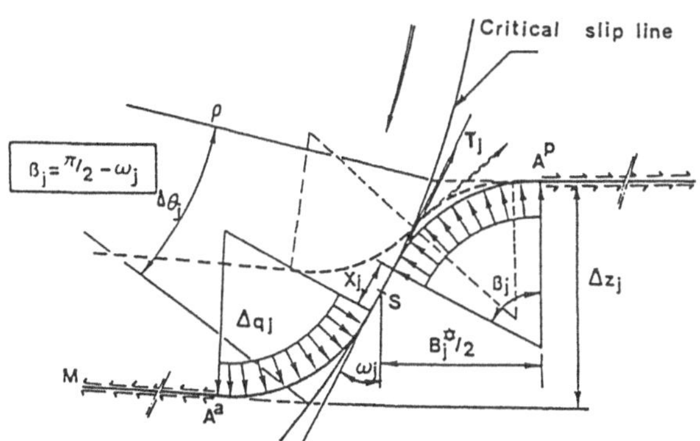

<u>FIGURE 20</u>: Local behaviour of the geosynthetic layer near the slip line: anchored membrane -(22)-

synthetic of linear elastic behaviour under tension ($T=J\cdot\varepsilon$, where ε represents the relative elongation) and elasto-plastic behaviour under friction, one can determine, at the soil-reinforcement interface, the laws of anchorage behaviour

$u^a_{Aj} = v\ (T_{aj},\ L^a_{aj})$ active zone

$u^p_{Aj} = v\ (T_{aj},\ L^p_{aj})$ passive zone

which relate the displacement of points A^a_j and A^p_j to the tensile force T_j in the central zone of membrane (Figure 20).

* Membrane: This concept has been clarified from the shear test on a soil-geosynthetic composite, the direction of shear being perpendicular to the fabric layer (Gourc et al. -(24)-) (Figure 21).

The geosynthetic takes its shape like a membrane in the vicinity of the shear surface, due to its negligible flexural rigidity. It supports an increase in stress Δ_{qj} normal to its plane (Figure 20). In order to simplify the problem, Δ_{qj} was taken to be uniform. This implies a bi-circular deformed shape in the vicinity of the slip line, at least for $\Delta\theta_j$ is sufficiently small ($\beta_j < \alpha_j$). For larger displacements (with regard to soil stiffness K_s), the membrane becomes tangential to the sliding surface ($\beta_j = \alpha_j$).

FIGURE 21: Shear test on a soil-geosynthetic composite sample -(24)-

It is considered: $\Delta q_j = K_s \cdot \Delta Z_j/4$

equilibrium of the membrane: $T_j = \Delta q_j \cdot \dfrac{B^*_j}{2 \cdot \sin \beta_j}$

elastic behaviour of the membrane:

$$\frac{T_j}{J} + 1 + \frac{(u^a_{Aj} + u^P_{Aj}) - B^*_j \cdot \dfrac{\beta_j}{\sin\beta_j} - \dfrac{X_j}{\sin\alpha_j}}{B^*_j - (\Delta Z_j - X_j) \cdot \cot \alpha_j} = 0$$

The above equation, which integrates the displacements at the head of anchorage (u^a_{Aj}, u^P_{Aj}), assures continuity between the two zones of anchorage and the zone of membrane.

This method is very versatile because it can envisage an anchorage in nonhomogeneous soils, or particular conditions such as the geosynthetic fixed at the wall face ($u_M = 0$ for a motionless facing - Figure 19). This method was used as an operational design program, perfected jointly with LCPC (Cartage Program - Delmas et al. -(26)-). However, at the present stage, it appears difficult to determine experimentally the soil stiffness K_s. Hence, the behaviour of the membrane is simplified as follows (Figure 22):

"Small displacements" approach:

$$\Delta Z_j = (u^a_{Aj} + u^P_{Aj})/\cot \alpha_j$$

$$\beta_j = 0$$

"Large displacements" approach:

$$\Delta Z_j = (u^a_{Aj} + u^P_{Aj}) \cdot \sin \alpha_j$$

5.2.2.3 Limit equilibrium study

In the general case, i.e., a polygonal slip line with the application of method of slices (Perturbations), the number of equations and unknowns are as follows:

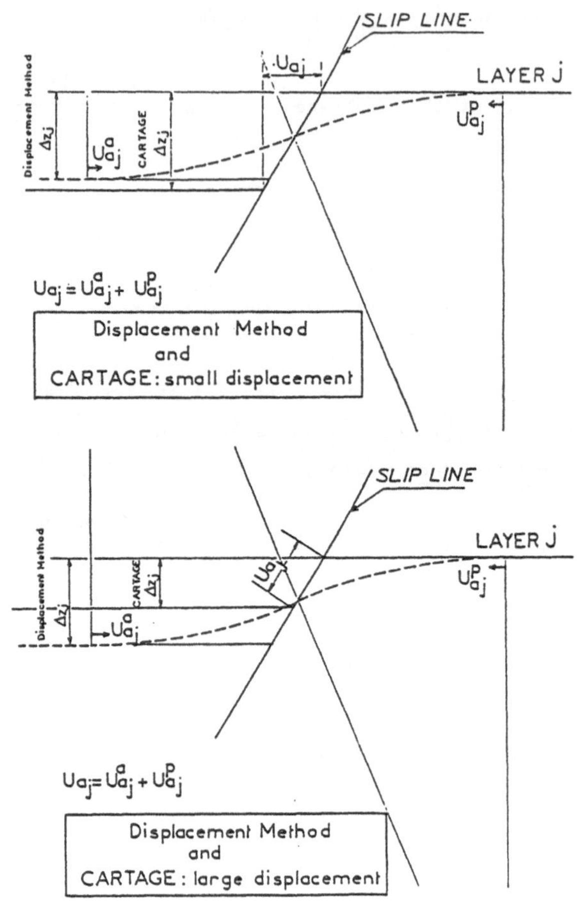

FIGURE 22: "Displacements method": CARTAGE program, specific hypothesis

Unknowns		Equations	
X_i, Z_i, e_i	$3(n + 1)$	equilibrium of slices	$3n$
F_o	1	$X_o = Z_o = e_o = 0$	3
N_i	n	$X_n = Z_n = e_n = 0$	3
μ_1, μ_2	2	$N_i = N_{io}\ (\mu_1 + \mu_2 \cdot f(s_i))$	n
T_j, β_j	2r	equilibrium of anchored membrane	2r
$\Delta\theta$	1	$F_s = F_1$	1
	$4n+2r+7$		$4n+2r+7$

The solution of the system is obtained by fixing first the factor of safety $F_S = F_1$ and determining the $\Delta\theta$ (or ΔZ), which brings the active block into equilibrium (one $\Delta\theta$ for every slip line). The critical slip line is that corresponding to maximum displacement in equilibrium.

In demonstrative case, this procedure is applied in the case of the critical slip line such as the case of Coulomb wedge (angle of wedge is: $(\pi/2 - \alpha)$ (Figure 23). The displacements of equilibrium obtained are compared for method [A] -(21)- and method [E] -(15)-. It is concluded that there exists in function of the angle of the block a maximum ΔZ (or $\Delta\theta$) for all cases.

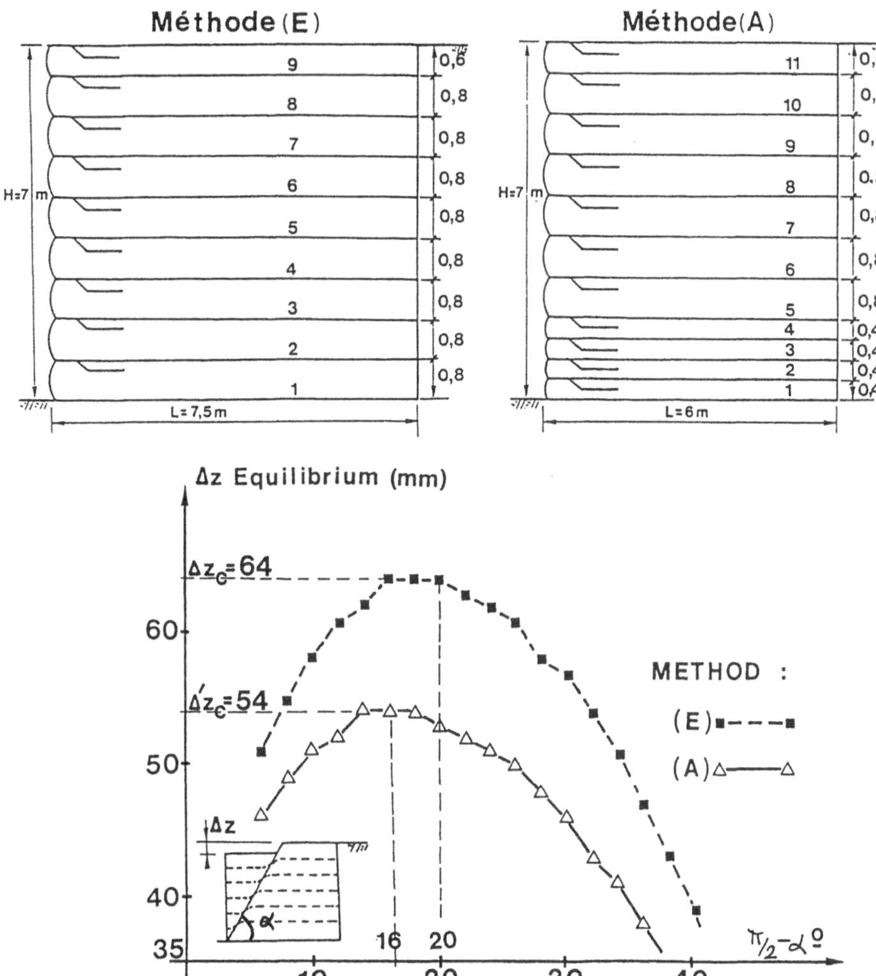

FIGURE 23: Comparison of two design methods (A -(21)-) and (E -(15)-) by "Displacements method"

In the case shown in Figure 23, the design according to method [E] obtained an equilibrium displacement larger than that according to method [A]. Thus, a method of evaluation for one design method in relation to another is disposed.

If the "displacements method" is used as a design method, the process is different: Delmas et al. -(26)- proposed in the program Cartage to set a limit for maximal equilibrium displacement ΔZ (or $\Delta\theta$).

6. INFLUENCE OF DIFFERENT HYPOTHESIS OF DESIGN CALCULATION

As it has been mentioned in Chapter 5, there are important differences in the hypotheses of design calculation. The problem is to know if this involves differences in design calculation. Without studying all cases, we will emphasize some significant points which will let the designing engineers know the influence of a particular hypothesis of calculation on the design of a reinforced retaining wall.

6.1 Choice of a reference method

The simple method, "two-block" (Bordairon -(12)-, will be chosen as a reference method. It satisfies overall equilibrium (cf 5.1) but does not satisfy local equilibrium (cf 5.2).

In general, the design calculation depends on the following parameters:

γ, β, H, ΔH, ϕ, ϕ_b, ϕ_g, δ (Figure 24)

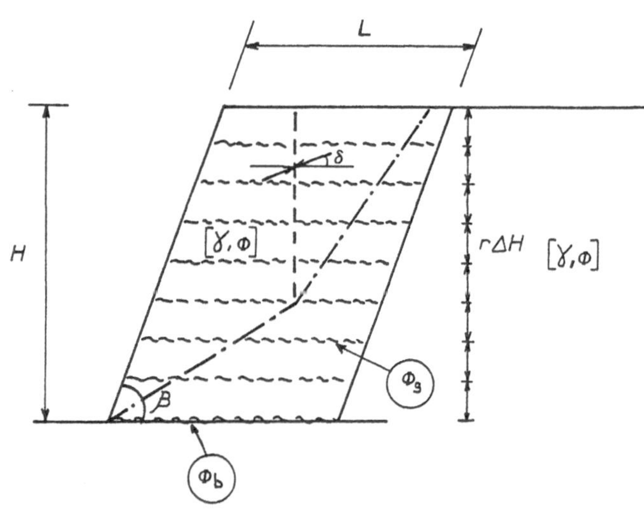

FIGURE 24: "Two-block" stability: design parameters

The soil is considered homogeneous, cohesionless, and $\tan \phi_g = 2/3 \tan \phi$. The factors of safety for external stability are F_G and F_R, and for internal stability is $F_s = \tan \phi / \tan \phi_c = F_f$. Once the limiting values of F_G and F_R have been chosen (here $F_G = F_R = 1$), the charts of the following type are obtained, i.e., L is the length of the mass in function of the height H and $T_I = T_f/F_T$ is the allowable tensile strength (Figure 25).

The diagrams can be divided into three zones (with T_I fixed):

- The first zone is for L varies linearly with H ($L = L_{min}$). In this case, it is the external stability that influences the design.
- The level zone is where the choice of an allowable tensile strength (T_I) imposes a limiting height of the retaining wall. This is the case that $P = -\sum T_j = rT_I$ with overestimated length of fabric anchorage.
- The intermediate zone is where $T_{aj}/F_f < T_I$ for some layers.

6.2 Interdependence of different factors of safety

6.2.1 F_G and F_R, $f(F_s)$ (Figure 26)

It is noted in Section 2 that it seems more logical for external stability to maintain the same principle of study (the type of slope stability analysis method) than for internal stability. From this point of view, $F_s = 1.5$ with $F_R = F_G = 1$ were chosen. Then, from the minimal lengths obtained, i.e., L_{RC} (for $F = 1$, $F_R = 1.5$), L_{GC} (for $F_G = 1$, ($Fs = 1.5$), and L_{pc} (for $\sigma_n > 0$, no tensile stress at the base of the wall, $Fs = 1.5$), the required minimal external length $L_{ec} = \max (L_{RC}, L_{GC}, L_{pc})$ is determined. This design is then verified for a system of factors of safety $F^*_s = 1$, $F^*_R > 1.5$, $F^*_G > 1.5$ - retaining walls design method). It is certain that the external length calculated L_{ec} is overabundant towards these coefficient F* values. Only L_{RC} calculated in the case of vertical walls can not obtain a coefficient $F^*_R > 1.5$ (for $F^*_s = 1$).

6.2.2 T_I and Lmin f(Fs) (Figure 27)

In Figure 27, the variation of T_I and L are presented in function of the factor of safety F_s fixed for soil. Because there is no relationship existing, a balanced choice of the factor of safety for soil (F_s) and geosynthetic ($F_T = T_f/T_I$) remains difficult.

The minimum width Lmin and the intrinsic tensile force T_I (minimum tensile force for equilibrium) are increasing with factor of safety Fs.

To set, for designing, large factors of safety Fs, induce a more conservative value of T_I.

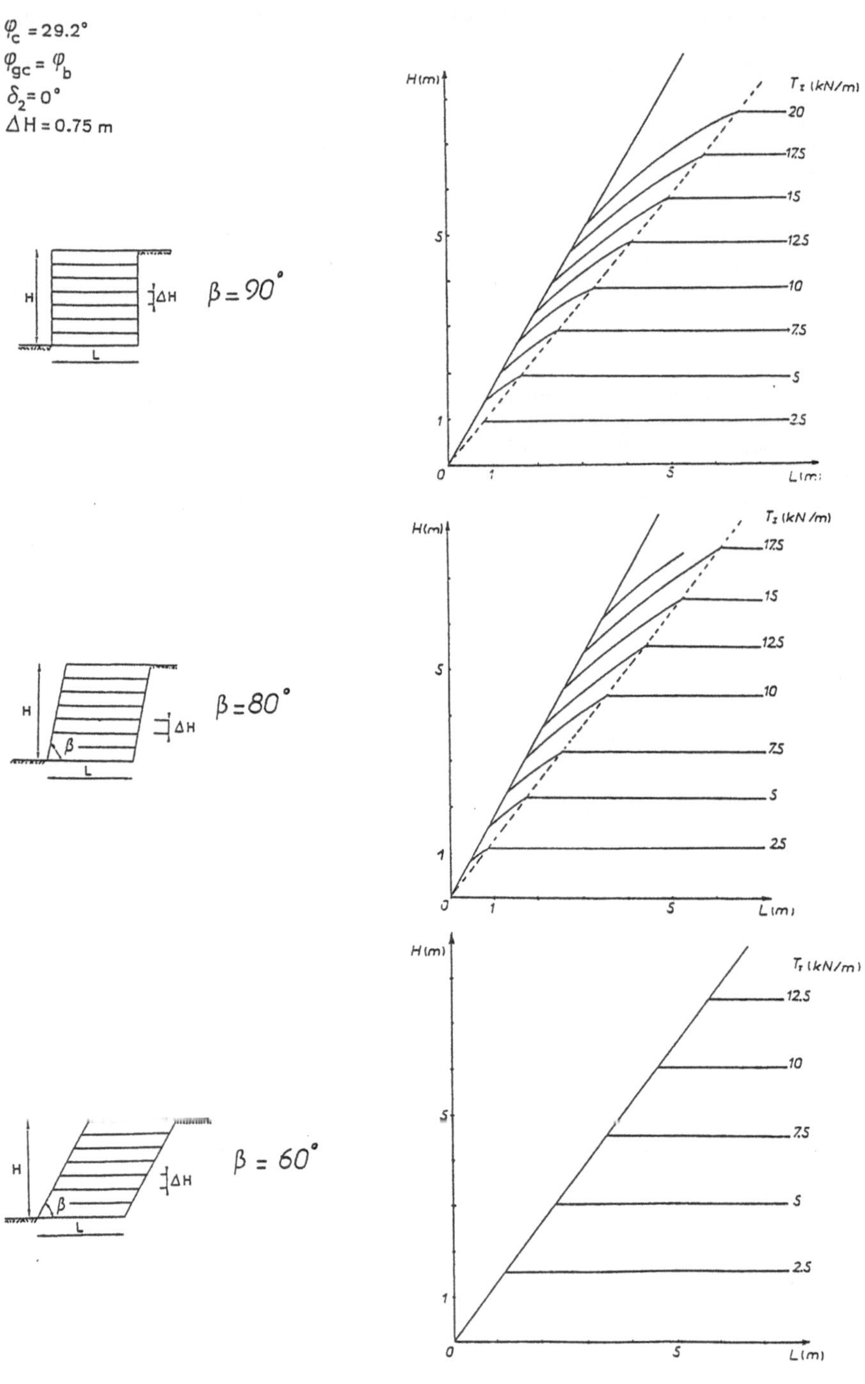

FIGURE 25: "Two-block method" -(12)- charts for designing

retaining walls design method: $F_G^* = 1.5$, $F_R^* = 1.5$, $F_S^* = 1$

reference method: $F_G = 1$, $F_R = 1$, $F_S = 1.5$

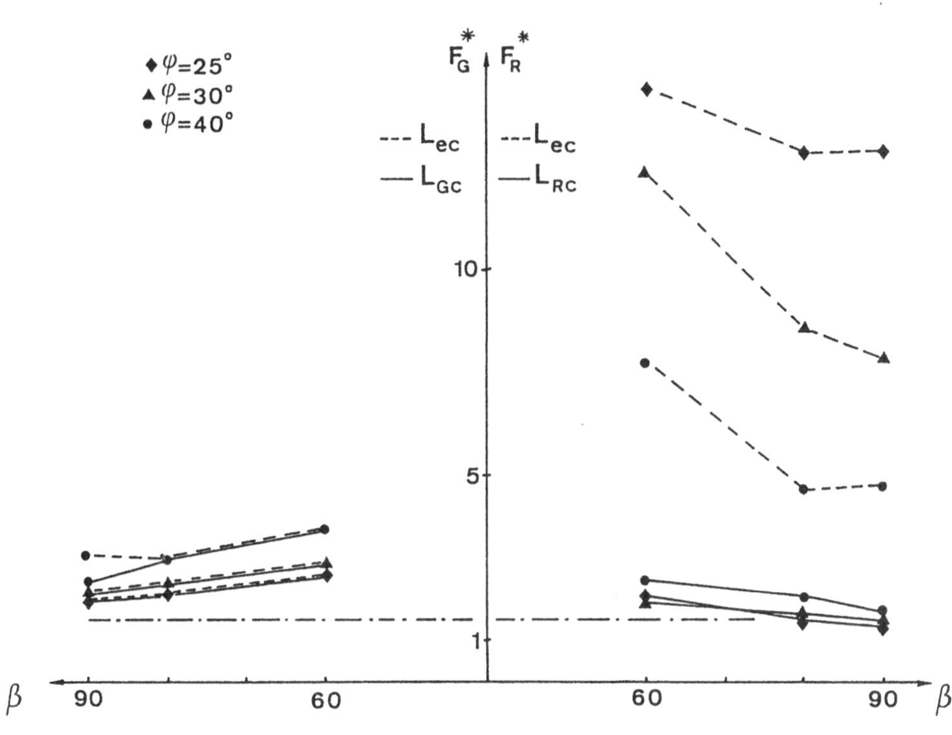

FIGURE 26: "Two-block method": choice of the factors of safety

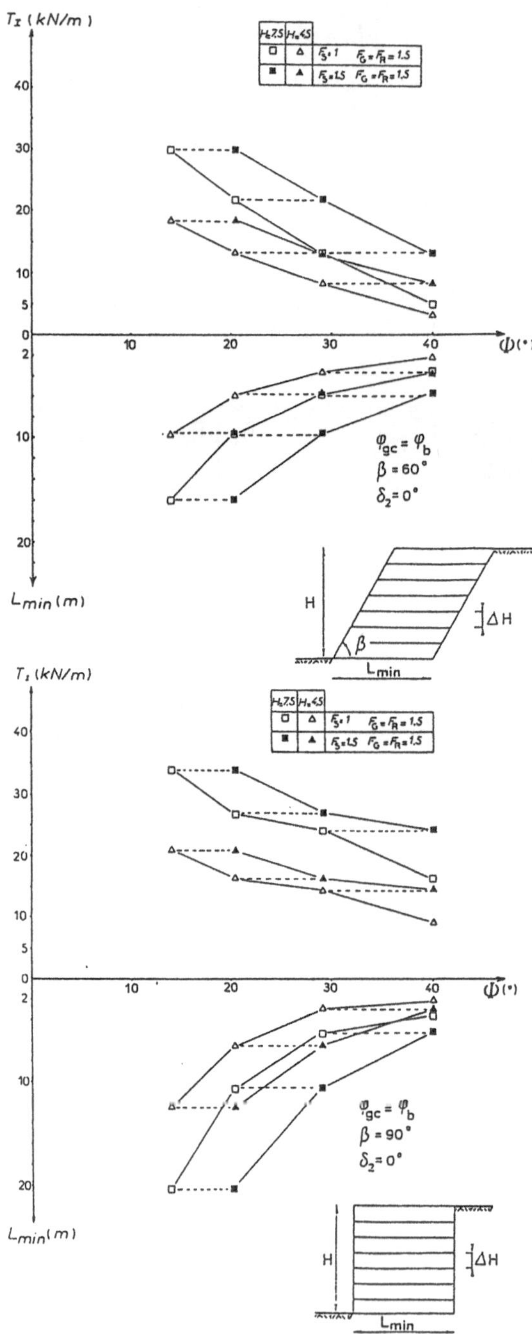

FIGURE 27 -a-: "Two-block method": influence of the factor of safety F_S on the T_I and L_{min} values

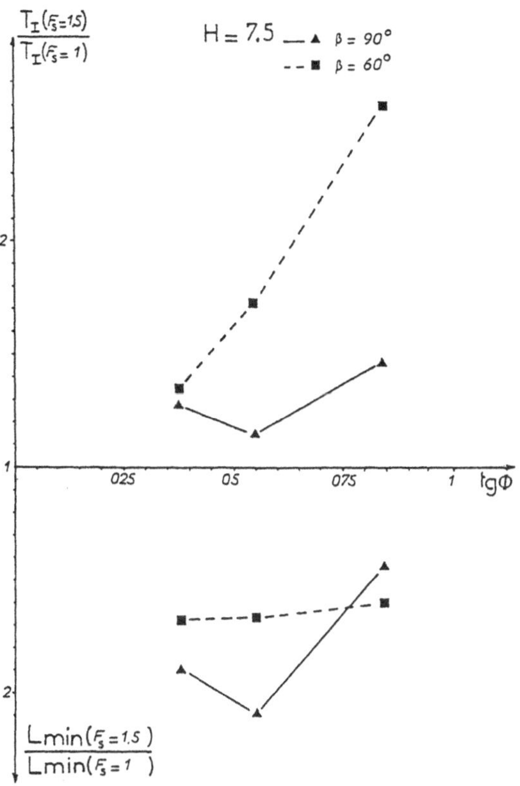

FIGURE 27 -b-: "Two-block method": security on Lmin and T_I induced by F_S

6.3 Method of blocks: influence of interblock conditions (Figure 28)

In paragraph 4.4, it is seen that the inclination δ (Figure 24) must be fixed first.

The most critical case ($T_{I\ max}$) corresponds to $\delta = 0$. A similar result has been obtained by Jewell -(13)-.

6.4 Comparison of "two-block method" with circular sliding method

The comparison is made for the following three methods: "two-block method" with $F_s = 1$; "perturbations method" (calculation of corresponding F_p); and "Bishop's simplified method" (calculation of corresponding F_b). The results are compatible (Figure 29). Note that the "perturbations method" is an exact method which satisfies all the equilibrium equations, however, "two-block method" and "Bishop's simplified method" are inexact.

In the case where $\phi_b = \phi_{gc} = \tan^{-1}(2/3 \tan \phi_c)$, the method of circular sliding overestimates F_s. This is because the slip line of "two-block method" can follow the weak plane of the wall base, but not for circular slip line.

In Figure 30, the critical slip lines corresponding to each method are presented.

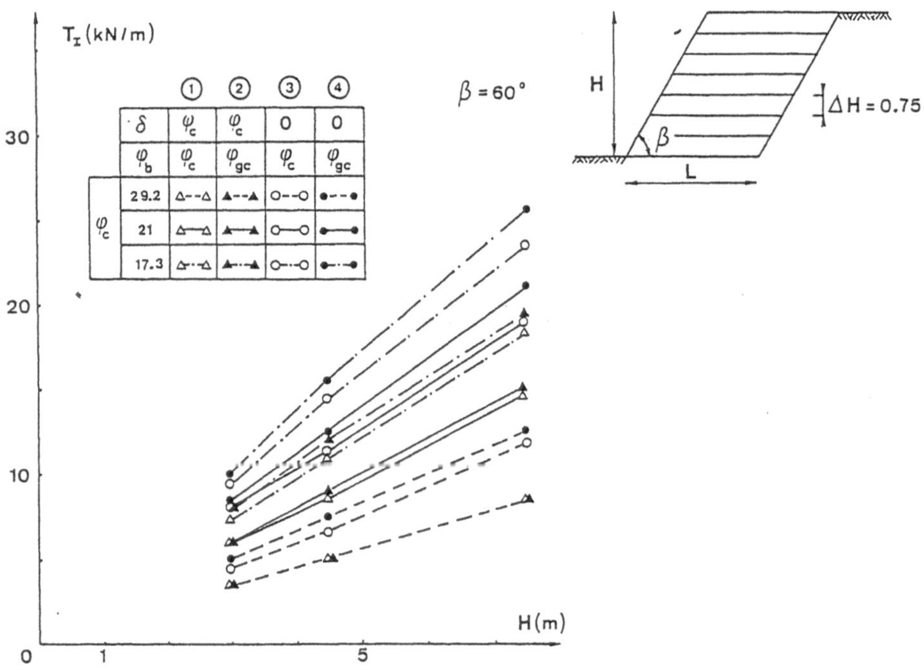

FIGURE 28: "Two-block method": influence of δ and ϕ_b values on T_I

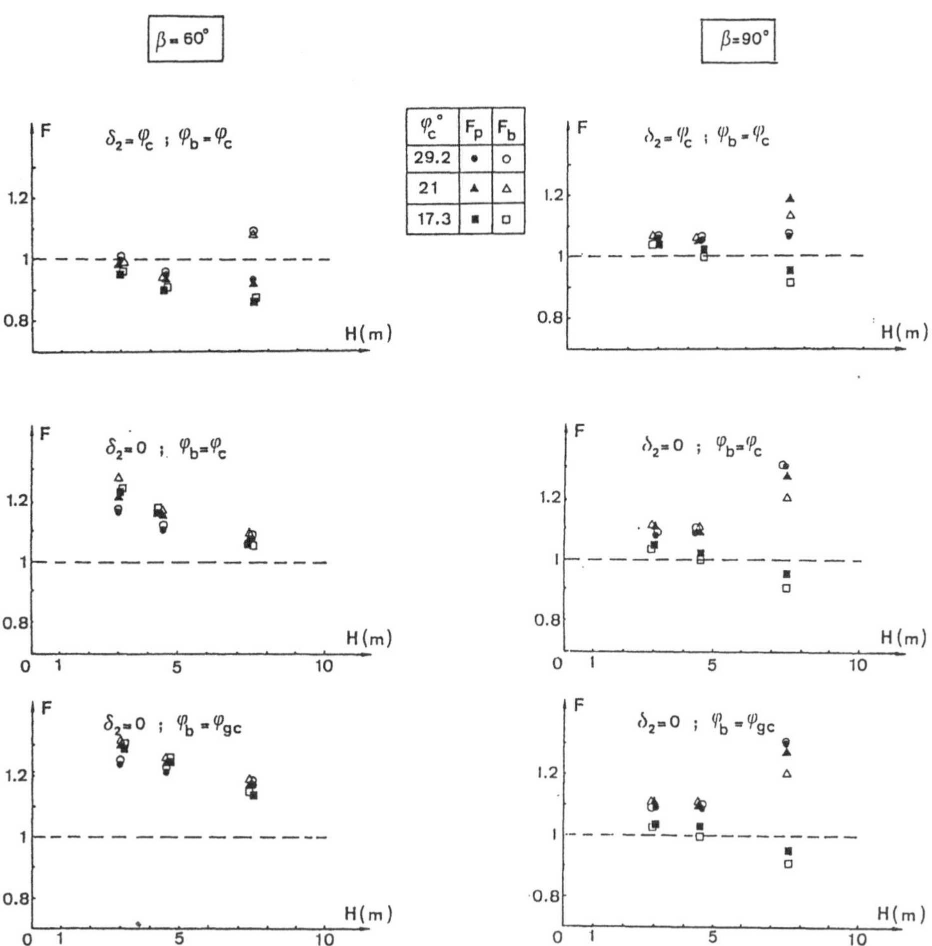

FIGURE 29: Comparison of critical slip lines: "slice method" with circular slip line and "two-block method"

	X (m)	Z (m)	F
C_b	- 3	9	1.08
C_p	- 3	9	1.07

	P(kN/m)	T_I(kN/m)
S_c	107	11
S_m	107	11

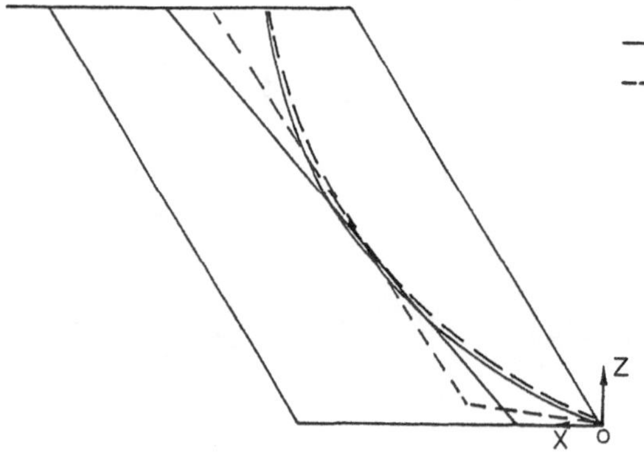

$\delta_2 = 0$; $\varphi_b = \varphi_c$

F_p : "Perturbations method"

F_b : "Bishop's simplified method"

$F_s = 1$: "Two-block method"

FIGURE 30: Comparison of "slice method" for circular slip line and "two-block method": deviation of factors of safety

6.5 "Two-block method": choice of critical sliding surface

In 5.1, it is mentioned that an arbitrary choice of the slip line corresponding to maximal thrust P_m (S_m), instead of choosing the critical slip line (S_c), may induce an underestimation of minimal T_I.

For vertical walls, Figure 31 shows (Figure 31 -a-: H = 4.5 m and Figure 31 -b-: H = 7.5 m):

- the slip line Sm for the maximal thrust Pmax = Pm;

- the slip line Sc for the critical thrust Pc corresponding to the maximal value of the tensile force T_I.

For a vertical wall, the Sm slip line (for the maximal value of the thrust P_m) agrees with Rankine wedge. Sm is independant of the reinforced retaining wall width L.

On the other hand, the Sc slip line (for the critical value of the thrust P_c is a function of the width L. As a matter of fact, the computer calculation provides two slip lines for every value of the width L (Sci, "inferior", and Scs, "superior") and for the same maximal value of T_I.

The larger the L value, the larger is the critical thrust value until the maximal thrust Pm is reached (superabundant anchorage length Laj). Also the larger the L value, the smaller is the T_I value until its minimum for $P_c = P_m$ is reached.

However, it should be noted that in the method of reference (13), the unconservative hypothesis slip line Sm corresponding to $P = P_m$ instead of the slip line Sc, is compensated by the added condition related to local equilibrium.

6.6 Choice of local equilibrium conditions

For vertical walls, we compare the values of the tensile forces T_I computed by methods A,B,C,D (see 5.2) and also method G, the "two-block method". To allow a practical comparison, the same factor of safety is used for all the five methods.

In Figure 32 -a-, $F_G = F_R = 1$, $F_s = 1.5$. The selected width of the reinforced retaining wall is the minimum width L = Lmin, again identical for all the five methods.

T_I values are related to local equilibrium conditions and increase with the value of the horizontal stress σ_x.

In Figure 32 -b-, the conventional design calculations for retaining walls choose the following factors of safety:

$$F_G = F_R = 1.5 \qquad F_s = 1 \quad (6.2.1)$$

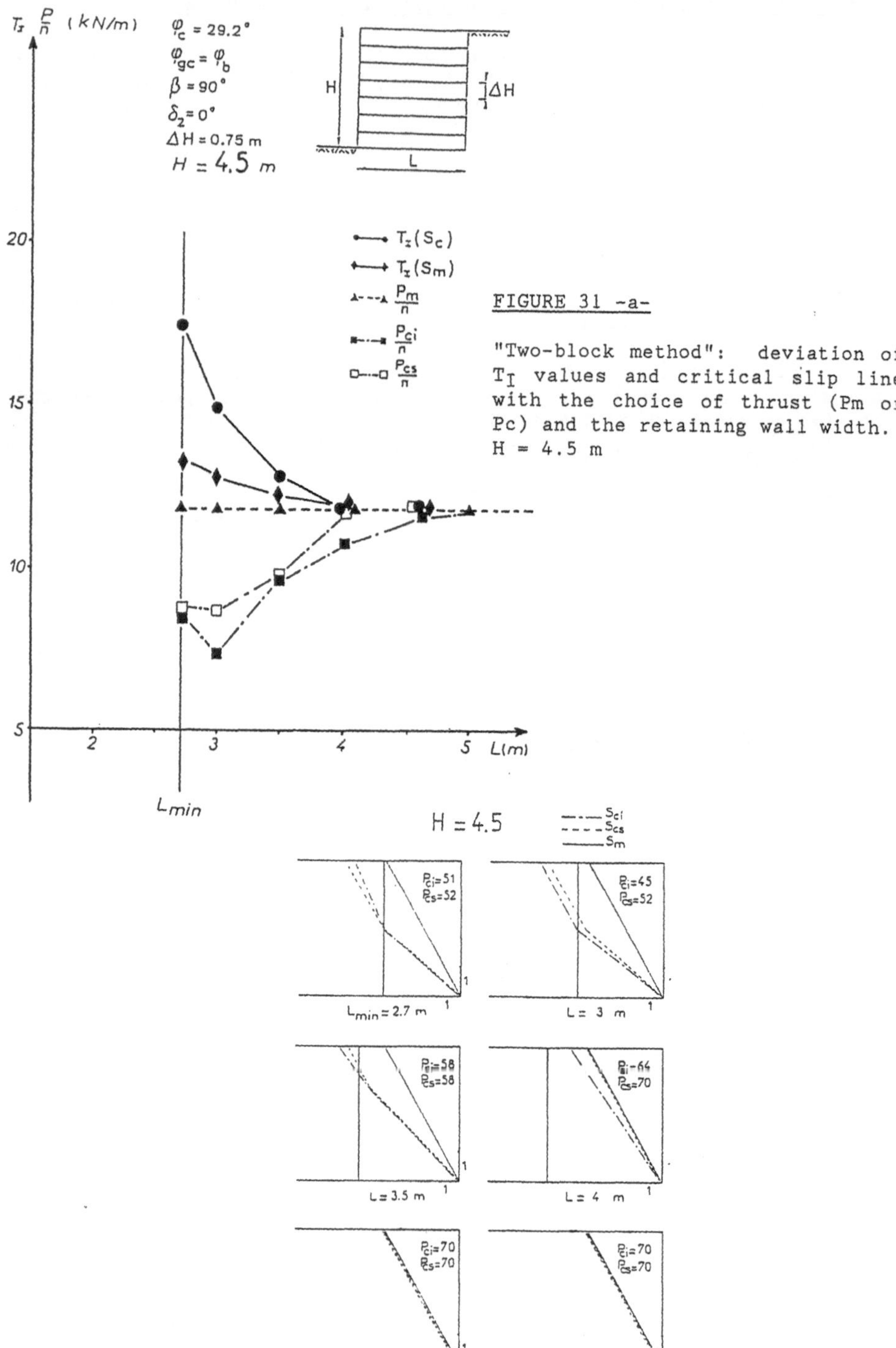

FIGURE 31 -a-

"Two-block method": deviation of T_I values and critical slip line with the choice of thrust (Pm or Pc) and the retaining wall width. H = 4.5 m

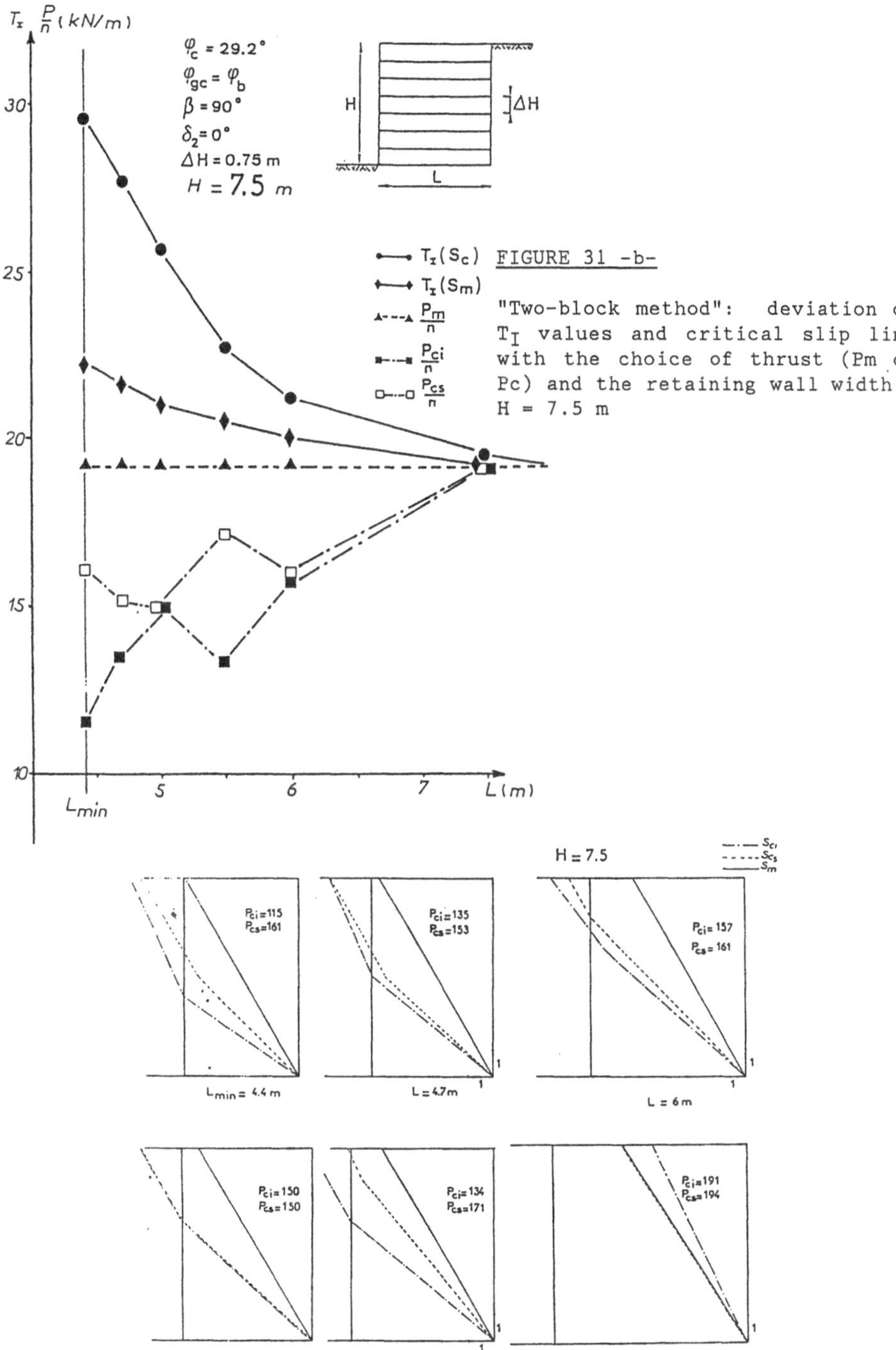

FIGURE 31 -b-

"Two-block method": deviation of T_I values and critical slip line with the choice of thrust (Pm or Pc) and the retaining wall width. H = 7.5 m

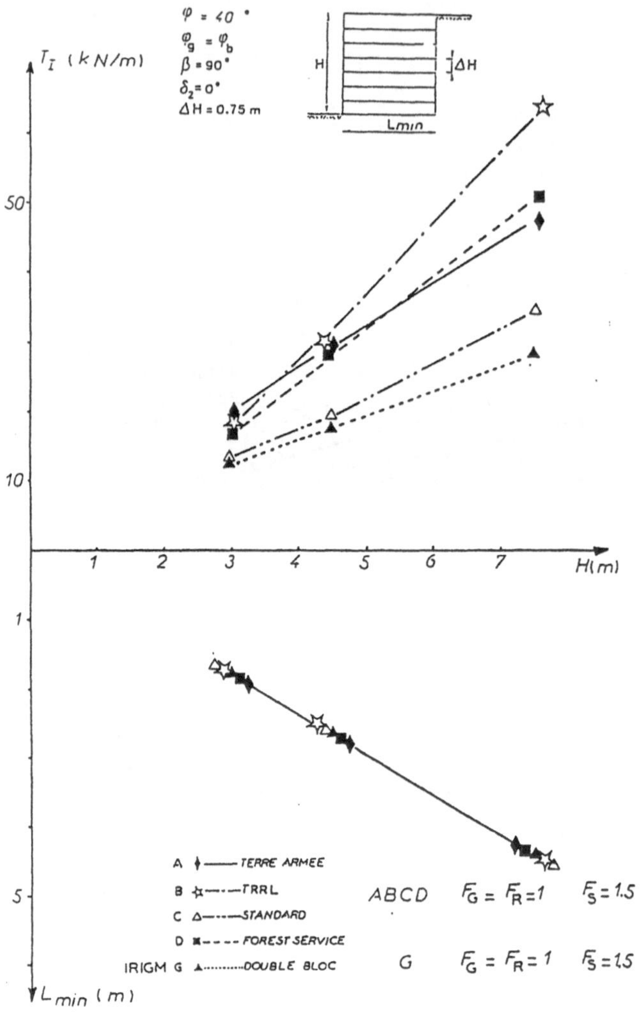

FIGURE 32 -a-: Comparison of traditional design methods with "two-block method"

32 (a): methods A,B,C,D

$F_S = 1.5 \quad F_G = F_R = 1$

-G, "two-block method"

$F_S = 1.5 \quad F_G = F_R = 1$

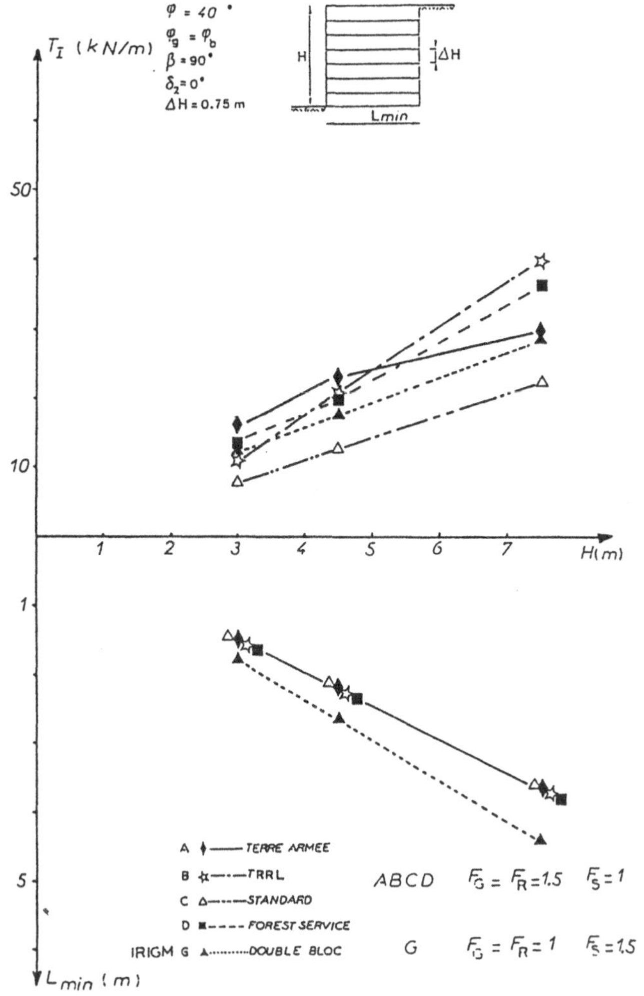

FIGURE 32 -b-: Comparison of traditional design methods with "two-block method"

32 (b): methods A,B,C,D

$F_s = 1.5$ $F_G = F_R = 1.5$

-G, "two-block method"

$F_s = 1.5$ $F_G = F_R = 1$

We take the example of last Figure 32 -a- in reference the "two-block method" (for $F_G = F_R = 1.5$ $F_s = 1$). If Figure 32 -b- is compared with Figure 32 -a-, then for methods A,B,C,D, Lmin and T_I are smaller than those in the previous case (Figure 32 -a-).

T_I values are closer to those by method G.

A further comparison for the case of Figure 33 shows that a deviation on σ_x values introduces a deviation on T_I values. The parameters used for Figure 33 are those of the RMC Trial Wall:

$$\beta = \frac{\pi}{2} \quad H = 3\ m \quad L = 3\ m \quad \gamma = 17.9\ kN/m$$

$$\phi = 43° \quad \phi_g = 43° \quad c = 0 \quad c_g = 0$$

$F_s = 1$ (except for "method of slices": T_I fixed)

6.7 "Displacements method": influence of complementary parameters

6.7.1 Anchorage behaviour

For the case shown in Figure 34, the influence of the parameters u_p, L_a, and J on $T_a = f(u_A)$ is presented three-dimensionally in Figures 35-37.

(Figure 35) u_p has little influence.

(Figure 36) L_a influences little away from the maximal anchorage strength.

(Figure 37) On the other hand, J has a notable influence.

6.7.2 Membrane behaviour

The variation of soil stiffness K_s induces a modification of equilibrium conditions.

While K_s increases, the equilibrium displacement ΔZ_j decreases and β_j (the inclination of tensile force T_j) increases (see Figure 38).

Note also that simplification of the geometry of membrane zone (Cartage program - see 5.2.2.2) in tensile force (Figure 39: case of Figure 33 - Q = 50 kPa, "small displacements" approach).

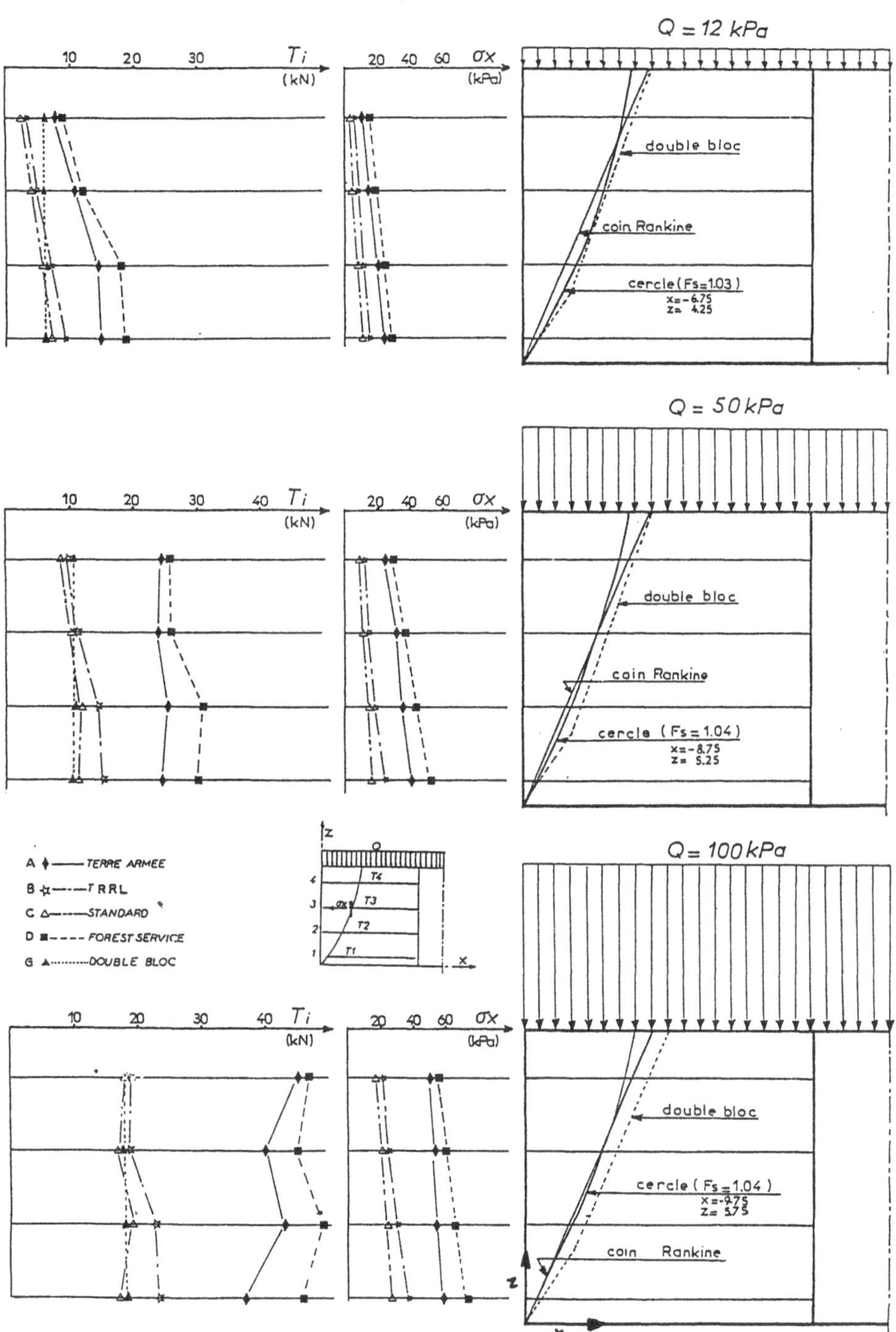

FIGURE 33: Comparison of traditional design methods with "two-block method" and "method of slices" associated with circular slip line (RMC-Kingston geogrid reinforced earth wall)

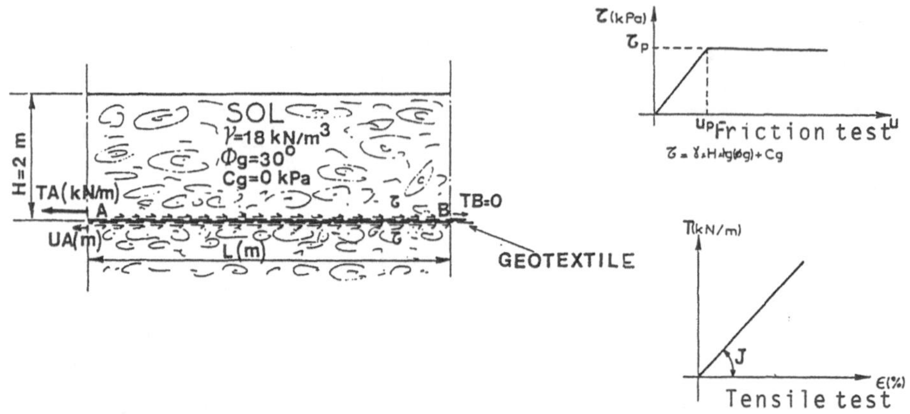

FIGURE 34: Anchorage behaviour parameters for a geosynthetic buried in soil mass

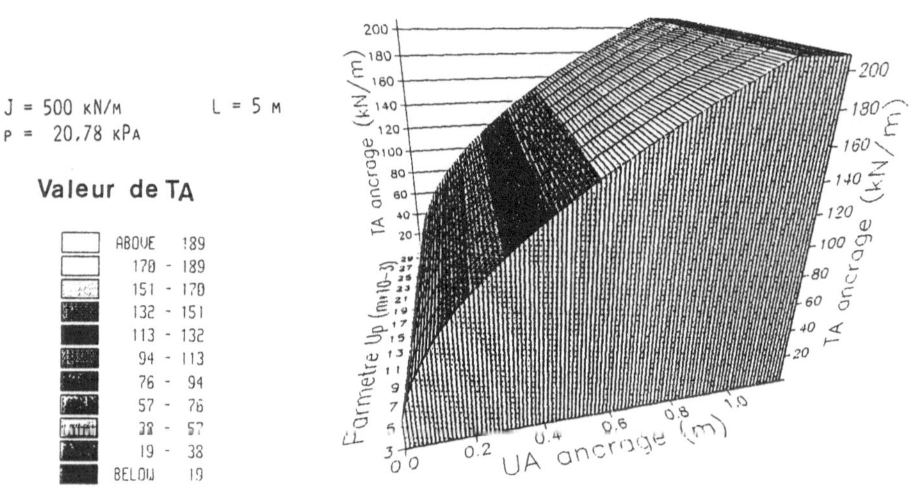

FIGURE 35: Anchorage behaviour of a geosynthetic buried in soil mass: influence of Up

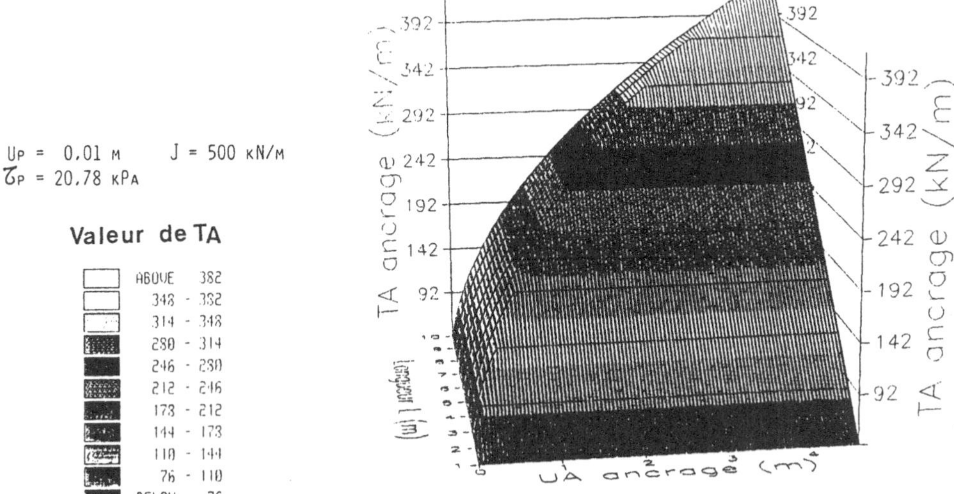

FIGURE 36: Anchorage behaviour of a geosynthetic buried in soil mass: influence of La

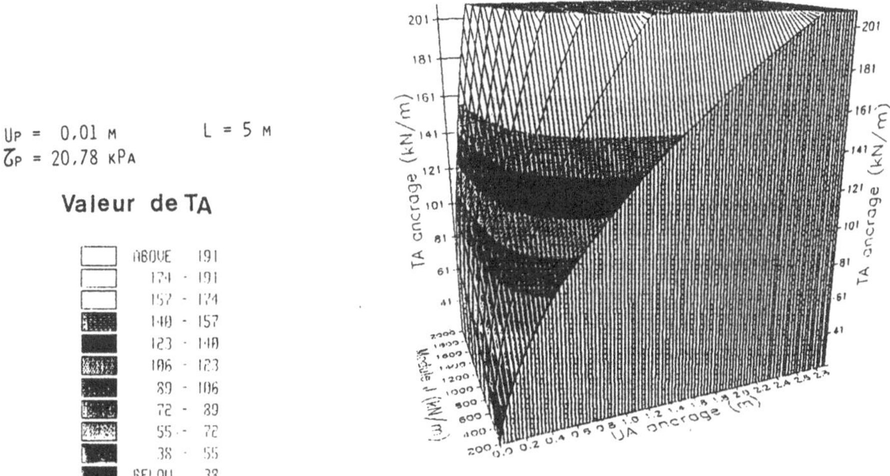

FIGURE 37: Anchorage behaviour of a geosynthetic buried in soil mass: influence of J

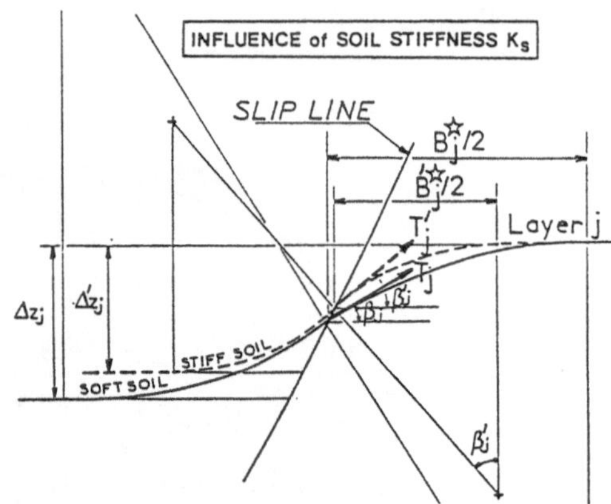

FIGURE 38: Membrane behaviour: influence of soil stiffness K_s

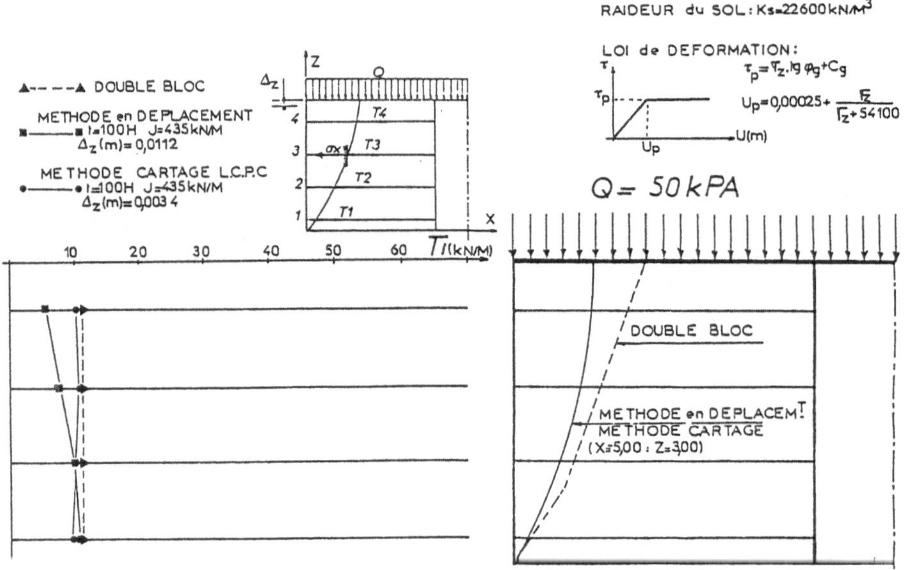

FIGURE 39: Comparison between "displacements method" and "two-block method": case of Figure 33
(RMC-Kingston geogrid reinforced earth wall)

The following is a design example by the program Cartage, with different assumptions shown in the table (Figure 40) - (Delmas et al. -(26)-).

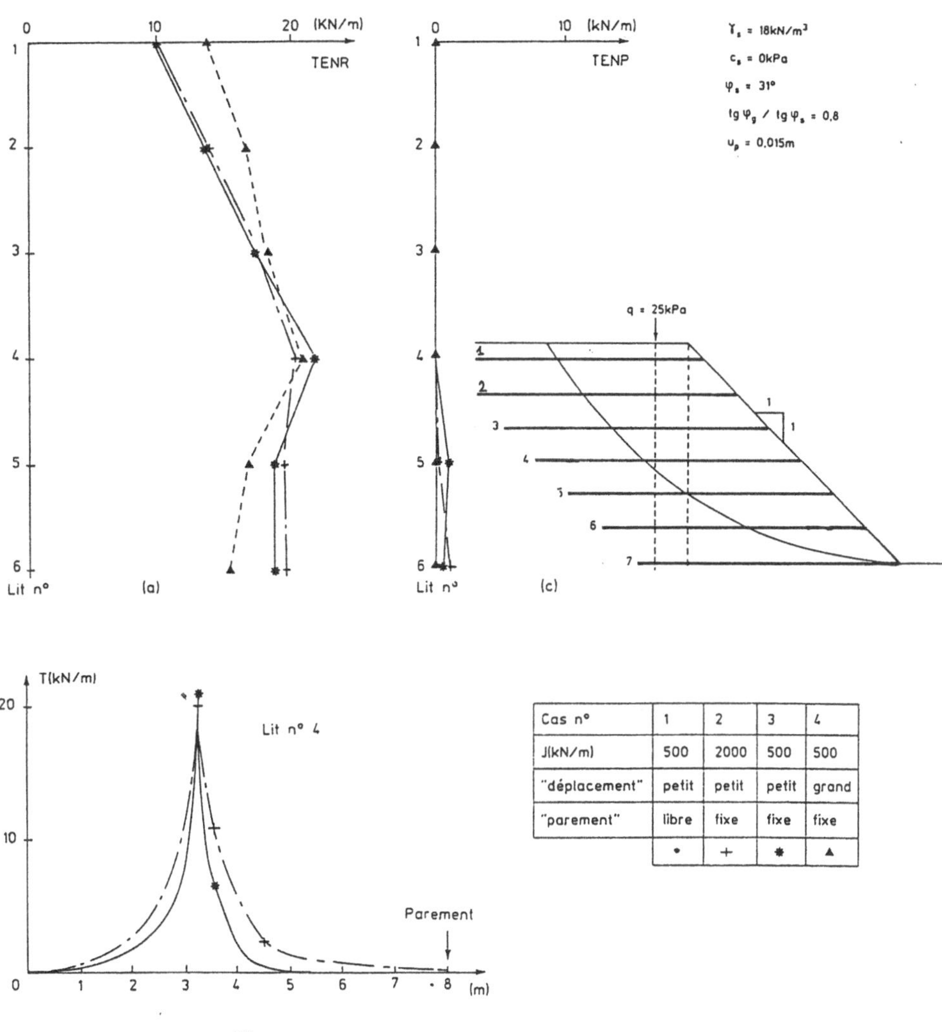

Cas n°	1	2	3	4
J(kN/m)	500	2000	500	500
"déplacement"	petit	petit	petit	grand
"parement"	libre	fixe	fixe	fixe
	•	+	✱	▲

FIGURE 40: Design of CARTAGE program -(26)-

6.8 Seismic effect

The "two-block method" is used again and the seismic load is simulated by a psudo-static load (with a seismic coefficient k). The internal seismic action is taken into account in the form of a horizontal force acting towards the slope, $k \cdot \vec{W}$. Similarly, an external seismic action is assumed existing on the active block behind the reinforced wall. This thrust is a uniformly distributed load acting on DC (Figure 41) (Monobe et al. -(27)-). This consideration is conservative because both maximal values of the two effects (internal and external) are assumed acting simultaneously (Bastick et al. -(28)-).

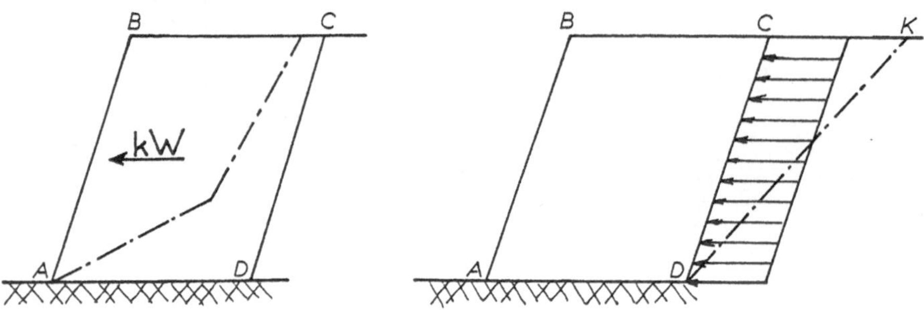

FIGURE 41: Internal and external seismic effect on a reinforced retaining wall

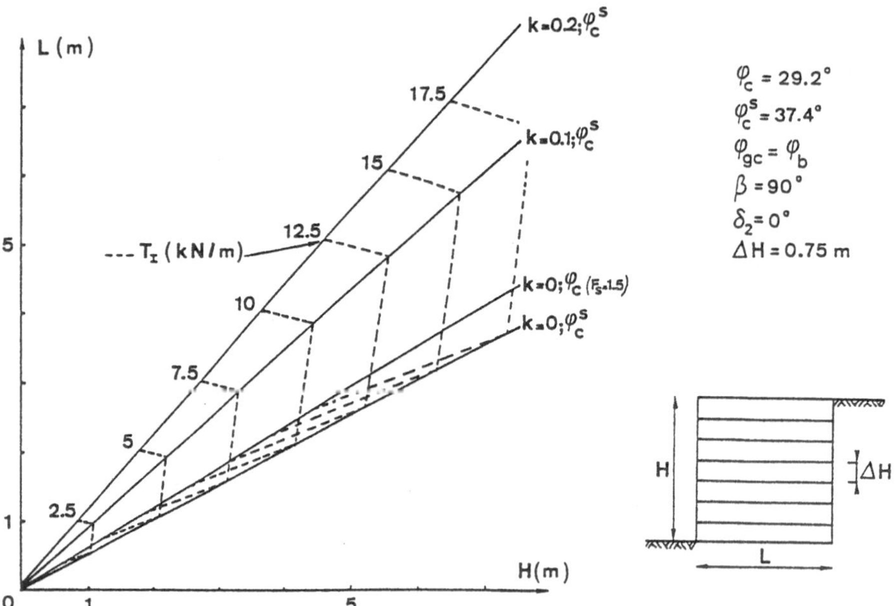

FIGURE 42: "Two-block method": seismic effect

If our results are compared to those obtained from Bonaparte et al. -(29)-, it is noted that for the same factor of safety ${}^{s}F_{s} = 1.1$, we obtain the values of the width L distinctly high (the increase of the external thrust not being taken into account in reference (29)).

By comparing with the results for static case ($F_s = 1.5$), it is seen in Figure 42 that the protection brought by a static design is weak in relation to an earthquake ($F_s = 1.5 \rightarrow F_s^s = 1.1$ for $k < 0.1$).

6.9 Cohesion of soil of reinforced mass (Figure 43)

In the design of a reinforced soil mass, the weak cohesion of the material is usually neglected ($c_c = 0$ instead of $C_c = C/F_s$). This induces a distinct overestimation of the minimal value of T_I (Figure 43).

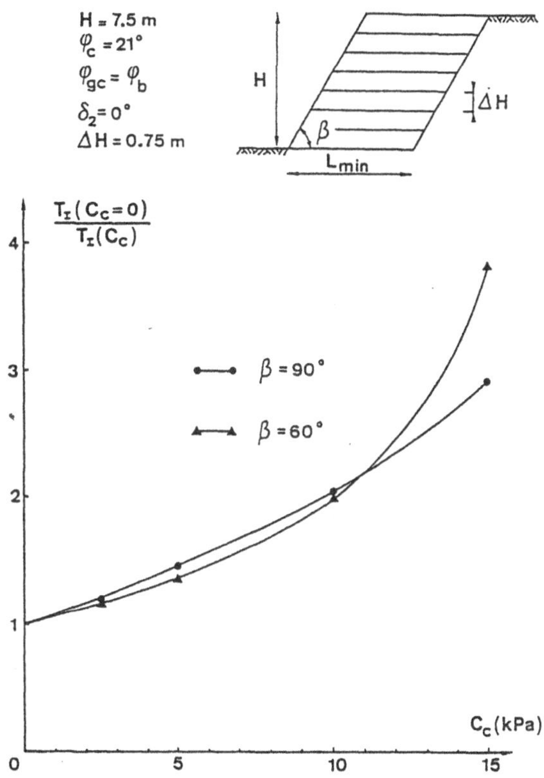

FIGURE 43: "T· ›-block method": influence of the soil cohesion

7. CONCLUSION

The numerical and graphic methods will provide the engineers with explicit charts for rapid predesign. However, it is also necessary to know well the fundamental assumptions in each method to estimate the design scattering.

Nonetheless, the design methods still need to be improved for taking into account such effects as, eg., a point load acting on the top of the wall, the folding of the fabric layers to make the wall facing, etc. These remain to be developed. However, the recent methods have indeed made great progress in design calculation.

ACKNOWLEDGEMENT

The authors sincerely thank Dr. Rong-Her CHEN, Associate Professor of National Taiwan University in Taipei, for translating this text during his stay in our research group of IRIGM, Grenoble University, FRANCE.

REFERENCES

1. Schlosser, F., Vidal, H. La Terre Armée. Bulletin de liaison des Laboratoires des Ponts et Chaussées no 41. November 1969.

2. Juran, I., Schlosser, F. Theoretical Analysis of Failure in Reinforced Earth Structures. Convention ASCE. Pittsburg 1978.

3. Raulin, P., Rouquès, G., Toubol, A. Calcul de la stabilité des pentes en rupture non circulaire. Rapport de recherche des Laboratoires des Ponts et Chaussées no. 36. 1974.

4. Biarez, J. Equilibre limite des talus et barrages en terre. Annales de l'ITBTP. Sols et Fondations no. 51. 1965.

5. Phan, T.L., Segrestin, P., Schlosser, F., Long, N.T. Etude de la stabilité interne des ouvrages en terre armée par deux méthodes de cercle de rupture. Colloque International sur le Renforcement des Sols. Paris, 1979.

6. Bishop. The Use of the Slip Circle in the Stability Analysis of Slopes. Geotechnique, Vol. V.

7. Jambu, N. Application of Composite Slip Surfaces for Stability Analysis. European Conference on Stability of Earth Slopes, Discussion, Vol. III. Stockholm, 1954.

8. Ratel, A. La "méthode en déplacements" appliquée aux massifs en sol renforcé par géosynthétiques. Thèse D.I., IRIGM, Université de Grenoble I, 1987.

9. Werner, G., Resl, S. Stability Mechanisms in Geotextile Reinforced Earth Structures. 3rd International Conference on Geotextiles. Vienna, 1986.

10. Baker, R., Garber, M. Theoretical Analysis of the Stability of Slopes. Geotechnique 28 no. 4. 1978.

11. Leshchinsky, D., Reinschmidt, A. Stability of Membrane Reinforced Slopes. ASCE Journal of Geotechnical Engineering, Vol. III no 11. November 1985.

12. Bordairon, M. Dimensionnement des massifs en sol renforcé par géosynthétiques. Thèse 3e C., IRIGM, Université de Grenoble I, November 1986.

13. Jewell, K.A., Paine, N., Wood, R.I. Design Methods for Steep Reinforced Embankments. Symposium on Polymer Grid Reinforcement in Civil Engineering. 1984.

14. Hamilton, M. Calculation Method for the Stability of Reinforced Embankments. Delft, 1984.

15. Broms, B.B. Polyester Fabric as Reinforcement in Soil. 1st International Conference Soils, Textiles. Paris 1977.

16. Mir Arabchari, N. Fluage des matériaux textiles utilisés dans les ouvrages de génie civil. Thèse D.I., Ecole Centrale. Paris, 1985.

17. Andrawes, K.Z., McGown, A., Murray, R.T. The Load-Strain-Time-Temperature Behaviour of Geotextiles and Geogrids. 3rd International Conference on Geotextiles and Geomembranes. Vienna 1986.

18. Gourc, J.P., Bordairon, M. Rembalis renforcés par géotextiles. Comparaison des méthodes de calcul. Journée sur le Renforcement des sols par géotextiles. Rapport interne Comité Français des géotextiles et Géomembranes. 1984.

19. Murray, R.T. Design of Reinforced Earth Walls. TRRL Int. Report. 1978.

20. Segrestin, P. Calcul d'un massif en terre armée par les coins de rupture. International Conference on Reinforced Earth. Paris, 1979.

21. Ministère des Transports Français, DGTI. Les ouvrages en Terre Armée Recommandations et règles de l'art. LCPC, SETRA. 1979.

22. Steward, J.E., Williamson, R., Mohney, J. Guidelines for Use of Fabrics in Construction and Maintenance of Low Volume Roads. Chapter 5. U.S. Forest Service. Portland, Oregon, June 1977.

23. Gourc, J.P., Ratel, A., Delmas, Ph. Design of Fabric Retaining Walls: The "Displacements Method". 3rd International Conference on Geotextiles and Geomembranes. Vienna, 1986.

24. Gourc, J.P., Matichard, Y., Perrier, H. Delmas, Ph. Capacité portante d'un bicouche, sable sur sol mou, renforcé par géotextile. 2nd International Conference on Geotextiles. Las Vegas, 1982.

25. Gourc, J.P., Mommessin, M., Monnet, J. Geotextile Reinforced Embankment Over Weak Soil: Different Theoretical Approaches. 3rd International Conference on Geotextiles and Geomembranes. Vienna, 1986.

26. Delmas, Ph., Berche, J.C., Gourc, J.P. Le dimensionnement des ouvrages renforcés par géotextile: Programme Cartage. Bulletin de liaison des Laboratoires des Ponts et Chaussées, 142. Mars, avril 1986.

27. Monobe, N., Matsuo, H. On the Determination of Earth Pressure During Earthquakes. Work Engineering Congress, Vol. 9, no 388. Tokyo.

28. Bastick, M., Schlosser, F. Comportement et dimensionnement dynamiques des ouvrages en Terre Armée.

29. Bonaparte, R., Schmertman, G.R., William, N.D. Seismic Design of Slopes Reinforced with Geogrids and Geotextiles. 3rd International Conference on Geotextiles and Geomembranes. Vienna, 1986.

FACTOR OF SAFETY CONSIDERATIONS IN REINFORCED SOIL STRUCTURES

R.T. MURRAY, Transport and Road Research Laboratory, England.

ABSTRACT

A comparison is given of the conventional method of design of reinforced soil structures, involving the use of overall or 'lump' factors of safety, with an alternative method which allows a more realistic assessment of the contribution made by the soil. The results of the comparisons for both internal and external stability are presented with respect to the specified factor of safety determined in the usual way. The study has shown that relative to the specified factors of safety, the values calculated by the proposed method are generally smaller, apart from at the limiting condition when they are in agreement, provided that the assumed movement of the wall is outward rotation about the base. For the case of outward rotation about the top, even at the limiting condition the results generally disagree. The study has also shown that, regardless of the fact that only one safety factor may be specified, there are at least two different safety factors influencing behaviour. This result thus provides support for the use of partial factors in design as has been recently proposed.

The importance of considering strain behaviour in design is discussed in the paper. Although the method referred to above does not truly consider strain, it provides much greater consistency and compatibility of soil-reinforcement design while at the same time preserving the simplicity of existing methods. The method seems best suited to design involving "stiff" reinforcements.

The increasing use of polymeric reinforcements has highlighted the need for an alternative approach which considers strain behaviour directly. One such method has been described in the paper and the equations for internal local stability are presented. The analysis has been limited to the simple case of linear variation of strain of the soil and reinforcement but could be readily extended to more complex situations.

1. INTRODUCTION

The subject of factor of safety is one which generates considerable controversy in civil engineering (Burland et al,1981;Symons,1983). In this paper the conventional approach to the

P. M. Jarrett and A. McGown (eds.), The Application of Polymeric Reinforcement in Soil Retaining Structures, 507–540.

design of reinforced soil structures employed in the United Kingdom (Dept. of Transport,1978) is compared with an alternative deterministic procedure which is considered to provide a more realistic basis for design. To allow the conventional and proposed solutions to be directly comparable, similar forms of design equations are employed in the two cases but this is not a requirement of the proposed method. An important feature of the method is that it permits the influence of the constraints imposed by the reinforcement on the soil properties to be taken into account.

The calculations involve the usual assumption in design that the wall movements will involve outward rotation about the base. However, as recent research has indicated that reinforced soil retaining structures constructed with part-height panels tend to rotate about the top of the structure (McGown et al,1987) further analysis of internal stability behaviour has been also carried out for this case.

As described in the paper,the behaviour of a reinforced soil structure is greatly influenced by the strains developed in the soil and reinforcement. This is of particular relevance to soil structures with polymeric reinforcement although at present the design of reinforced soil structures employing geotextiles or related materials as reinforcement generally follows the procedures laid down for metallic and other types of high modulus reinforcement. However, polymeric reinforcements are often more strain susceptible than metallic types and to avoid excessive strains being induced it is usual practice to increase the factor of safety such that the polymeric reinforcement has a working load which is a much smaller proportion of its ultimate (long term) load capacity than its metallic counterpart.

As far as is known this approach generally offers a viable solution to coping with the greater strain susceptibility of geotextiles and related materials but their increasing use for reinforced soil applications warrants the development of a more appropriate method of design which takes due consideration of the material properties. The paper, therefore, gives consideration to the development of a design procedure which is based on an assessment of strain in the soil and reinforcement and may thus have better applicability to structures employing geotextiles or related materials.

2.0 ASSESSMENT OF SAFETY FACTORS BASED ON OUTWARD WALL ROTATION ABOUT THE BASE.

In the design of retaining structures it is usual practice to assume that the limiting condition will correspond to the active state produced by outward rotation about the base. The analysis given in this Section is also based on this assumption while an alternative, and often more realistic approach is considered in the following Section.

2.1 Adherence Resistance.

The conventional procedure for determining the required length of reinforcement to prevent adherence failure is based on a limit equilibrium approach.The method of assessing local stability will be considered in relation to self-weight of the fill only but the techniques discussed are equally applicable to other more complex loading conditions.

At limiting equilibrium the force (D) applied to a single element is given by :

$$D = K_a \gamma z S_V S_H \quad \ldots\ldots (1)$$

The resisting force (R) is given by :

$$R=2B\mu_L \gamma z L_L \quad \ldots\ldots (2)$$

At limiting equilibrium these two equations are equal

$$\text{i.e } R=D$$

$$2B\mu_L \gamma z L_L = K_a \gamma z S_V S_H$$

$$2B\mu_L L_L = K_a S_V S_H$$

Assuming that the objective of the design is to establish a safe length of reinforcement having previously selected values of B, S_V and S_H for known values of μ_L and K_a, then the above equation can be re-arranged to give the limiting length required :

$$\text{i.e. } L_L = \frac{K_a S_V S_H}{2\mu_L B} \quad \ldots\ldots (3)$$

Now the length to be employed in the final design must have an adequate factor of safety against adherence failure and the required design length is therefore given by :

$$L_D = F L_L$$

It is apparent that for values of F other than unity, there is no longer an equality with the right hand side of Equation 3. As S_V, S_H and 2B are unchanged, the effect of increasing the length is to alter the properties K_a and μ_L, previously defined as limiting values, and which is no longer the case. The manner in which these two values are altered with increasing specified factor of safety will be governed by their relations with soil strain. From a theoretical standpoint the extreme range of the possible values that can occur is given by :

$$\mu\ (\text{minimum}) = \frac{\mu_L}{F} \ \longleftarrow\ \mu, K = \text{function}(F)\ \longrightarrow\ K(\text{maximum}) = K_a . F$$

This relation states that in one extreme only the interface friction is influenced by the safety factor while the other extreme is the full effect of safety factor being taken only by the soil friction. Between these extremes both μ and K are affected to some extent.

For purposes of the following analysis it is assumed that as both values are a function of the soil friction, the interface friction coefficient μ_L can be related to the peak friction angle by a coefficient α:

$$\text{i.e } \mu_L = \tan\phi_\mu = \alpha.\tan\phi_P$$

Moreover, the mobilised friction angle ($\tan\phi_M$) may be related to the peak friction angle by the well-known equation employed in slope stability analysis :

$$\text{i.e } \mu_M = \alpha\tan\phi_M = \frac{\alpha\tan\phi_P}{F_S} \quad \text{.........................(4)}$$

Thus the design length can be expressed by the relation :

$$L_D = \frac{K_M S_V S_H}{2\mu_M B} \quad \text{...(5)}$$

$$\text{Now } K_M = \tan^2(45 - \phi_M/2) \quad \text{...........................(6)}$$

Expanding Equation 6 and substituting Equation 4 gives the mobilised lateral earth pressure coefficient in terms of the safety factor and peak friction angle :

$$\text{i.e } K_m = \frac{\sqrt{(F_S^2 + \tan^2\phi_P)} - \tan\phi_P}{\sqrt{(F_S^2 + \tan^2\phi_P)} + \tan\phi_P} \quad \text{....................(7)}$$

$$\text{now as } F = \frac{L_D}{L_L}$$

$$F = \frac{K_M.\mu_L}{K_a.\mu_M} = \frac{K_M.\alpha\tan\phi_P}{K_a\alpha\tan\phi_P/F_S} \quad \text{.........................(8)}$$

$$\text{and } F_S = \frac{F.K_a}{K_M} \quad \text{....................................(9a)}$$

or in suitable iterative form :

$$F_{S,i+1} = \frac{F.K_a[\sqrt{(F_{S}^2{}_{,i} + \tan^2\phi_P)} + \tan\phi_P]}{[\sqrt{(F_{S}^2{}_{,i} + \tan^2\phi_P)} - \tan\phi_P]} \quad \text{................}(10)$$

thus enabling the factor of safety of the soil (F_S) to be determined as defined by Equation 4. Moreover, as the strength properties of the soil are not fully mobilised, the factor of safety against adherence or bond resistance (F_A) does not correspond to the specified value F.

i.e. $$F_A = \frac{R_M}{D_M} = \frac{2BL_D\mu_M}{S_VS_HK_M}$$

and $$\frac{2BL_L}{S_VS_H} = \frac{K_a}{\mu_L} \qquad \text{where } L_L = L_D/F$$

$$\frac{2BL_D}{S_VS_H} = \frac{K_aF}{\mu_L}$$

$$F_A = \frac{K_a\mu_MF}{\mu_LK_M} = \frac{K_aF}{K_MF_S} \quad \text{...........................}(11)$$

It should be noted that the numerical value of F_A in Equation 11 is unity (apart from the effect of rounding errors) for all values of friction angle and specified factor of safety. This occurs because with increasing specified factor of safety the ratio μ_M/K_M reduces at the same rate to exactly compensate and preserve the equilibrium of the equation. However, this result should not be interpreted as indicating that the structure is at the point of failure as any tendency for bond failure to occur would induce soil strain and an increasing value of mobilised friction angle. Thus to attain the limiting values of μ and K, the pull-out force required would be equal to F_A times the working load which would correspond to the ultimate force based on the specified factor of safety F. It could well be that the strains to achieve such a condition would be unacceptably large although this is most likely to arise with soils in a loose condition, or with highly extensible geotextile reinforcement.

In essence the analysis makes the assumption that with a factor of safety of unity, sufficient yielding of the wall takes place to fully mobilise active earth pressure conditions and interface friction. As the factor of safety is increased the amount of yielding reduces proportionately with a corresponding reduction in mobilised friction angle. Thus greater force is developed which must be resisted by the element. This behaviour is analogous to a rigid retaining wall or rigid tunnel lining which inevitably needs to support greater forces because of the constraints they apply to the soil. Neglecting situations where passive pressures might develop, it would seem reasonable to assume that a lower bound friction angle consistent with zero strain and corresponding to K_0 conditions would apply.

Figure 1 shows the relation between the specified safety factor and the value of F_S derived from Equation 9a for a peak friction angle of 30°. Similar forms of relation are produced for other values of peak friction angle. The figure also shows the relation between specified factor of safety and the factor of safety in adherence based on Equation 11. In the analysis the mobilised friction angle was not permitted to fall below the zero strain or K_0 value. It was previously stated that the factor of safety against adherence failure will remain sensibly constant at about unity but an anomalous situation occurs when the specified safety factor is further increased after the zero strain condition has been attained.

It is important to consider how this factor of safety in adherence is developed in order to preserve the equality between the restoring force and disturbing force. Because of the assumption that both soil friction and interface friction are modified identically by soil strain, the implication of the analysis is that any additional length of reinforcement, corresponding to an enhancement of the specified safety factor beyond the zero strain condition, does not contribute to the resisting force but nonetheless provides additional strength reserve in adherence. The increase in adherence safety factor shown in Figure 1, after the safety factor of the soil attains a constant value, arises from this length in reserve.

An alternative method of analysis which has been carried out is based on the assumption that the interface friction angle continues to reduce with increasing length of reinforcement so that the same resistance will be developed over a longer length. This analysis therefore requires that the relations between soil strain and both soil friction and interface friction, which were assumed to be the same initially, diverge after the zero strain condition is reached. It would be possible to employ different soil strain relations for these two modes of friction but this would add further complexity to the analysis. It should be noted that, although implied otherwise by the theory, even with this latter method of analysis a stage must be reached with large values of L_D when further increases will not contribute to resistance. The extra length would thus be redundant but could still be considered an additional safety factor in reserve as for the former analysis. The factor of safety of the soil on the basis of this latter analysis is given by:

$$\text{for } \epsilon = 0 \text{ ;} \qquad F_S = \frac{F.K_a}{K_0} \qquad \text{(9b)}$$

Figure 1 also shows the alternative relation between safety factor of the soil based on Equation 9b and the specified safety factor. The two relations showing the factor of safety of the soil in Figure 1 may be contrasted with the normal assumption that the soil strength will be fully mobilised at all stages which corresponds to a constant safety factor of unity.

The above considerations of safety factor are not necessarily indicative of unsafe design but highlight inconsistencies in the present

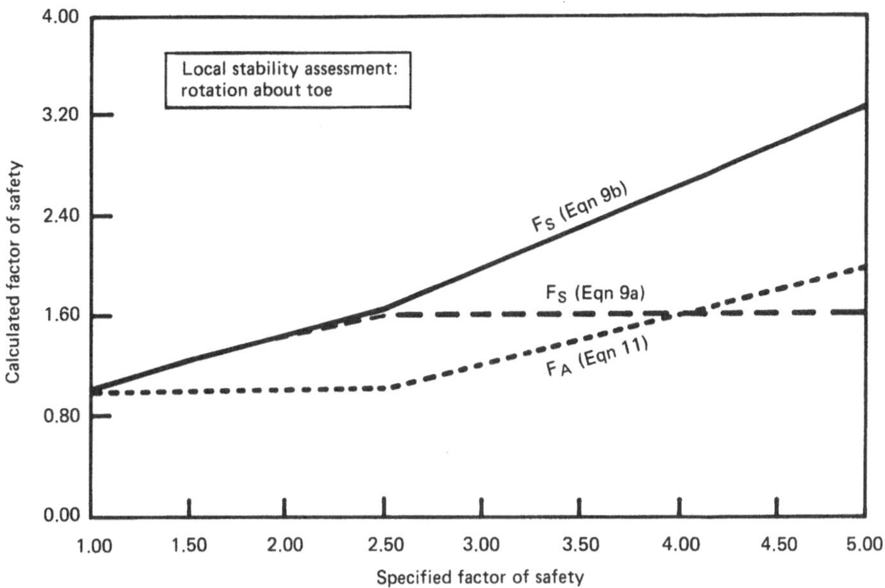

Fig. 1 Relations between specified and calculated factors of safety

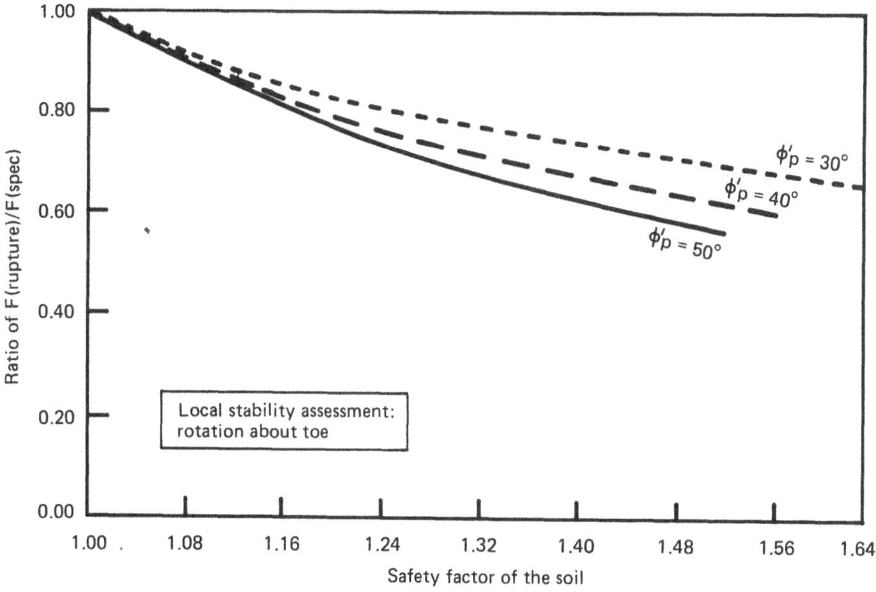

Fig. 2 Relations between safety factor of soil and F(rupture)/F(spec)

approach based on the use of overall or lump factors of safety. Nonetheless, there may well be situations where difficulties with local stability can arise. For example, where there is poor soil-reinforcement interaction as a result of inadequate compaction, overlarge fill material or other cause, the larger forces may induce excessive deformation. However, the analyses has more serious implications in regard to the design of local stability against tensile rupture as described below.

An important result from the above analysis is that although only a single factor of safety was specified, it is apparent that two factors of safety, F_S and F actually determine behaviour. This result seems to vindicate the use of the partial factor of safety approach proposed by Murray and McGown (1987).

2.2 Local tensile stability.

The approach adopted in the United Kingdom for assessing the tensile strength requirements of reinforcing elements (Dept. of Transport,1978) is based on a permissable stress method in which a permitted value of stress for a particular material (p_M) multiplied by the minimum area of cross-section (a) is equated to the horizontal force to be supported by the element. It is assumed for purposes of the design calculations that the shear strength of the soil is fully mobilised.

Although the design calculations for bond resistance and tensile resistance are treated independently, it is apparent that the same disturbing force applies in the two cases and there is therefore a direct interrelation between the two possible modes of failure. Thus the imposition of a factor of safety to prevent bond failure constrains the soil and increases the forces which also have to be resisted in tension. However, whereas bond failure would be preceded by soil strain and hence an increase in mobilised friction, with rupture this is not necessarily the case, particularly for high modulus reinforcement which would strain very little prior to failure. It is thus important to ensure that the factor of safety against bond failure is always smaller than against tensile rupture if sudden or very rapid modes of collapse are to be avoided, especially where the possibility of corrosion or degradation exists.

In the following determination of the safety factor against tensile rupture, the mobilised friction angles for the soil are assumed to correspond to those values determined previously in assessing bond stability. In some circumstances this may be considered an optimistic assumption as the relation between force and strain will be different for the two cases and with metallic and other forms of high modulus reinforcement, the strains induced as a consequence of tensile force are likely to be very small. The procedure is as follows :

The tensile resistance at the working condition (R_M) is given by

$$R_M = a.p_M = a.p_L/F$$

where the specified factor of safety $$F = \frac{a.p_L}{K_a \gamma z S_V S_H} \qquad \ldots\ldots\ldots\ldots(12)$$

defining the actual safety against rupture (F_R) as

$$F_R = \frac{a.p_L}{K_M \gamma z S_V S_H} \qquad (13)$$

Thus the ratio of specified to actual safety factor against rupture is given by :

$$\frac{F}{F_R} = \frac{K_M}{K_a}$$

Now F is specified and K_M is given by Equations 6 or 7. As F_S is assumed to correspond to the same relation with K_M as previously determined for local adherence stability, the actual factor of safety against rupture can be determined in terms of F_S.

i.e. $$F_R = \frac{K_a.F}{K_M}$$

Or expressing F_R/F as a ratio and replacing K_M by the relation given in Equation 7 then:

$$\frac{F_R}{F} = \frac{K_a[\sqrt{(F_S^2 + \tan^2\phi_p)} + \tan\phi_p]}{[\sqrt{(F_S^2 + \tan^2\phi_p)} - \tan\phi_p]} \qquad (14)$$

The ratio of actual factor of safety to specified factor of safety against rupture is shown plotted against safety factor of the soil in Figure 2 for peak friction angles of 30°, 40° and 50°. It should be noted that although the local safety factors in adherence given by Equations 9a and 9b are different, this does not affect the results produced by Equation 14 as K_M has the same values over the entire range for the two methods. The relation shown in Figure 2 thus applies to both methods of analysis.
A point to note is that the results have been normalised with respect to the limiting condition. Thus as the specified factor of safety increases, the gain in F_S or F_R may appear smaller for soils with a larger angle of friction.

In contrast to the previous analysis involving adherence behaviour, with high modulus reinforcement there will not necessarily be any further increase in mobilised shear strength of the soil up to failure of the reinforcement.

2.3 Overall adherence resistance.

A similar approach to that described above for local stability may be employed for the assessment of overall bond resistance. Referring to Figure 3 then the bond resistance of the ith element assuming a plane surface as shown is given by :

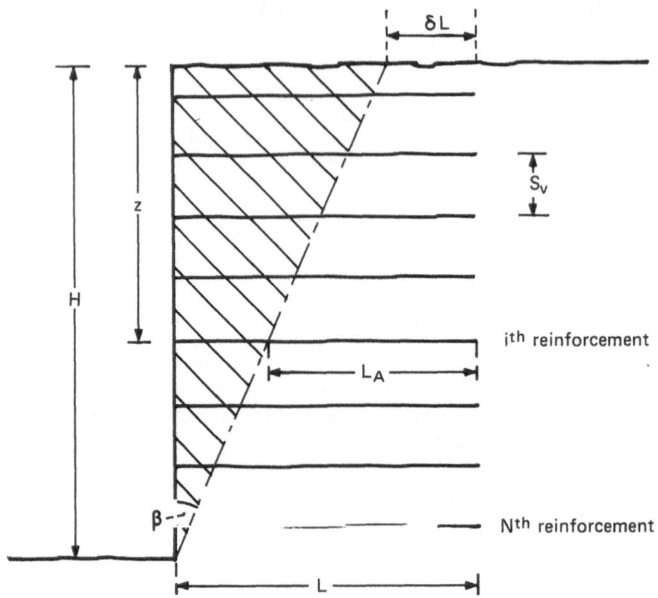

Fig. 3 Geometrical considerations in assessment of overall stability

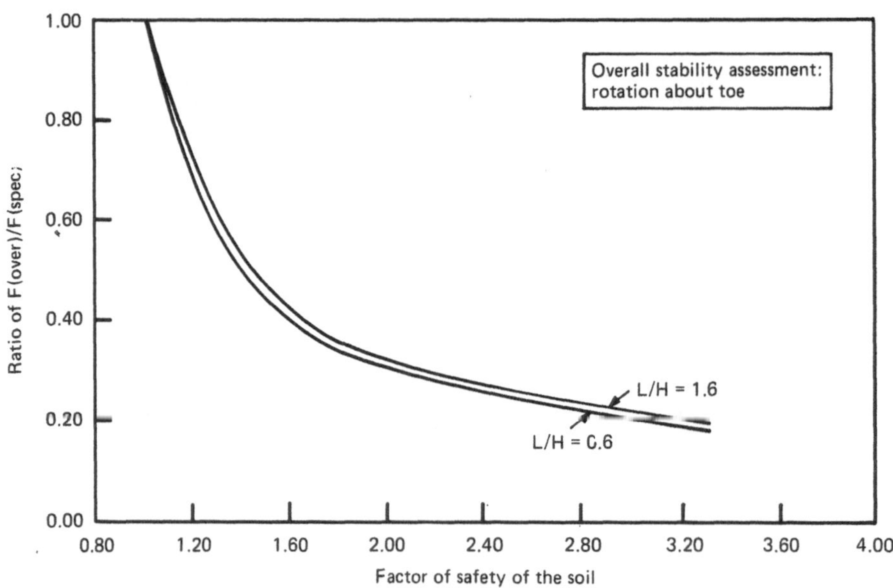

Fig. 4 Relations between F(soil) and ratio of F(over)/F(spec)

$$T_i = 2Bu_L\gamma z(\delta L + z\tan\beta) \qquad (15)$$

Now $\delta L = L - H.\tan\beta$; $z = i.S_V$

$$T_i = 2B\mu_L\gamma(iS_V\delta L + i^2S_V{}^2\tan\beta) \qquad (16)$$

Summating Equation 16 from 1 to N can be shown to produce the following equation :

$$T_T = \sum_{i=1}^{i=N} T = \tfrac{1}{2}\gamma B\mu_L H^2.2(N + 1).[\frac{\delta L}{H} + \frac{1}{3}.\tan\beta(2 + \frac{1}{N})] \qquad (17)$$

$$\text{i.e. } T_T = \tfrac{1}{2}\gamma B\mu_L H^2.2(N + 1)[\frac{(L - H\tan\beta)}{H} + \tan\beta(\frac{2}{3} + \frac{1}{6N})] \qquad (18)$$

Now the disturbing force $D = \tfrac{1}{2}K\gamma H^2$, where K takes the value of K_a at the limiting condition or K_M according to the mobilised friction angle. The specified factor of safety (F) is therefore given by :

$$F = \frac{B\mu_L}{K_a}.2(N + 1)[(\frac{L}{H} - \surd K_a) + \surd K_a(\frac{2}{3} + \frac{1}{6N})]$$

$$\simeq \frac{2B\mu_L}{K_a}.(N + 1)[\frac{L}{H} - \frac{\surd K_a}{3}] \qquad (19)$$

and actual safety factor (F_{OVER}) by :

$$F_{OVER} = \frac{2B\mu_M}{K_M}(N + 1)[\frac{L}{H} - \frac{\surd K_M}{3}] \qquad (20)$$

Thus the ratio of F_{OVER}/F is given by :

$$\frac{F_{OVER}}{F} = \frac{K_a[L/H - \surd K_M/3]}{F_S.K_M[L/H - \surd K_a/3]} \qquad (21)$$

The ratio of F_{OVER}/F has been determined from the values of K_M and F_S calculated in the local stability analyses employing Equations 7 and 10 respectively. The results are presented for peak friction angles of 30°, 40° and 50° in Figure 4 for geometries of L/H of 0.6 and 1.6. As can be seen the results are very similar for these two cases in view of the fact that the data have been normalised with respect to the limiting conditions.

It may be noted that the bracketed terms in Equation 21 have very similar numerical values and a close approximation to the safety factor against bond failure is given by :

$$F_{OVER} \simeq \frac{F.K_a}{F_S K_M}$$

The above equation corresponds to that derived for local bond stability (Equation 11) and the agreement between the two possible modes of instability can be explained by the fact that the same relation between specified factor of safety and mobilised friction angle has been assumed for both cases. In practice, the mobilised friction angle may vary throughout the height of the wall according to the magnitude and form of yielding that takes place. However, the analysis which has been described above relates to the most commonly assumed case of outward rotation about the base.

3.0 SAFETY FACTORS FOR INTERNAL STABILITY ASSOCIATED WITH WALL ROTATION ABOUT THE TOP.

There are essentially two forms of reinforced soil walls; the most common form is constructed with part height panels which permit considerable articulation during construction. The other type involves the use of full height panels which are generally propped during construction. Because of the different construction techniques used for the two methods, the strains developed in the soil are best represented by rotation about the top in the former case and by rotation about the base in the latter (McGown et al, 1987). These differences in the pattern of strain distribution have a significant influence on the magnitude and distribution of the lateral pressures.

A method of assessing the influence of different forms of wall movement on the lateral pressures developed behind retaining walls has been proposed by Dubrova (Harr ,1966). For the usual assumption of rotation about the base, the method produces the conventional Rankine solution. To evaluate the lateral pressures produced by rotation about the top, it is assumed that the mobilised friction angle increases linearly from a minimum value at the top to the fully mobilised value at the base. In the analysis presented by Dubrova, the value at the top was assumed to be zero but it seems more appropriate to use the value corresponding to "at-rest" conditions. On this basis the variation in the maximum friction angle which can be mobilised (ϕ_{PZ}) is given by :

$$\phi_{PZ} = \phi_0 + \eta z \qquad (22)$$

where $\eta = (\phi_P - \phi_0)/H$

A further assumption involved in the method is that the total force (P) at the limiting condition conforms to that produced by the usual Coulomb analysis. The lateral earth pressure is determined by differentiation of the total force equation and treating K_{az} as a variable :

i.e. $P = \frac{1}{2} K_{az} \gamma z^2$

i.e. $$\frac{dP}{dz} = K_{az}\gamma z + \frac{1}{2}\gamma z^2 . \frac{d}{dz}(K_{az})$$

now $$\frac{d}{dz}(K_{az}) = \frac{d}{dz}\frac{[1 - \sin(\phi_0 + \eta z)]}{[1 + \sin(\phi_0 + \eta z)]}$$

$$\text{i.e.} \quad \sigma_h = \frac{dP}{dz} = \gamma z[K_{az} - \frac{\eta z.\cos(\phi_0 + \eta z)(1 + K_{az})}{2(1 + \sin(\phi_0 + \eta z))}] \quad .....(23)$$

For the case of wall translation, Dubrova proposed that the lateral pressure distribution was obtained from the average of the distributions determined for rotation about the base and the top.

The above method of analysis has been employed in the assessment of internal stability involving rotation about the top.

3.1 Local stability assessment.

At the overall limiting condition the mobilised friction angle is assumed to be given by Equation 22. It is apparent that in the upper part of the reinforced soil wall the friction angle will not attain its peak value even at the limiting condition. This may be contrasted with the case of rotation about the base when the strength of the soil was fully mobilised at the limiting condition. It should be noted, however, that if the outward movements of the facing continue beyond that required to achieve the limiting condition, further redistribution of lateral pressure would occur and eventually attain the full strength of the soil at all locations.

As for the case of rotation about the base, the influence of the safety factor will be to constrain the soil. Proceeding with the analysis as previously, an equivalent equation for local bond stability for rotation about the top is as follows :

$$F = \frac{L_D}{L_L} = \frac{K_{MZ}.\tan(\phi_P)}{K_a.\tan(\phi_{MZ})} \quad .....(24)$$

Where the subscript m indicates mobilised friction while the subscript z indicates variation with depth.

Defining $\theta_0 = \phi_0(1 - z/H)$; $\theta_P = \phi_P z/H$

$$\tan\phi_{MZ} = \tan(\theta_0 + \theta_{MZ})$$

$$\tan\phi_{MZ} = \frac{F_S\tan\theta_0 + \tan\theta_P}{F_S - \tan\theta_0\tan\theta_P} \quad .....(25)$$

$$\text{and } K_{MZ} = \tan^2(45^\circ - \phi_{MZ}/2) \quad .....(26a)$$

$$\text{i.e } K_{MZ} = \frac{\sqrt{(F_S{}^2 + \tan^2\phi_{MZ})} - \tan\phi_{MZ}}{\sqrt{(F_S2 + \tan^2\phi_{MZ})} + \tan\phi_{MZ}} \quad .....(26b)$$

$$\text{where } \phi_{MZ} = \tan^{-1}\left(\frac{F_S.\tan\theta_O + \tan\theta_P}{F_S - \tan\theta_O\tan\theta_P}\right) \quad \text{.................(27)}$$

Substituting Equation 25 in Equation 24 permits the factor of safety for the soil (F_S) to be determined :

$$\text{i.e. } F_S = \frac{\tan\theta_P(FK_a + \tan\theta_O K_{MZ}\tan\phi_P)}{(K_{MZ}\tan\phi_P - FK_a\tan\theta_O)} \quad \text{.................(28a)}$$

Note that as F_S also occurs in the term K_{MZ}, Equation 28a must be solved iteratively. More rapid convergence is obtained by casting the equation into the following form for iteration purposes :

$$F_{Si+1} = \sqrt{\frac{F_{Si}.\tan\theta_P(FK_a + \tan\theta_O K_{MZ}\tan\phi_P)}{(K_{MZ}\tan\phi_P - FK_a\tan\theta_O)}} \quad \text{..............(28b)}$$

Proceeding as previously for the assessment of local stability against rupture enables the following equation to be derived corresponding to Equation 14 :

$$\frac{F_R}{F} = \frac{K_a}{K_{MZ}} \quad \text{...(29)}$$

The values of K_{MZ} are obtained from Equation 26 while F_S is derived from Equation 28 based on the assessment of local bond stability. The variation of F_R and F_S over the depth of the structure for a specified value of F equal to unity is shown plotted against depth factor in Figure 5. Although the limiting condition corresponding to unit factor of safety has been specified, this only occurs at the base where sufficient movement has occurred. Figure 5 shows that the strength of the soil towards the upper part of the structure is not fully mobilised by virtue of the fact that F_S exceeds unity and also as F_R is less than unity. These results may be contrasted with the corresponding condition for rotation about the base when the factors of safety are the same. The influence of wall movement on mobilised friction angle may be seen more clearly in Figure 6 where the relations between depth factor and mobilised friction angle are presented for peak angles of 30°, 40° and 50°.

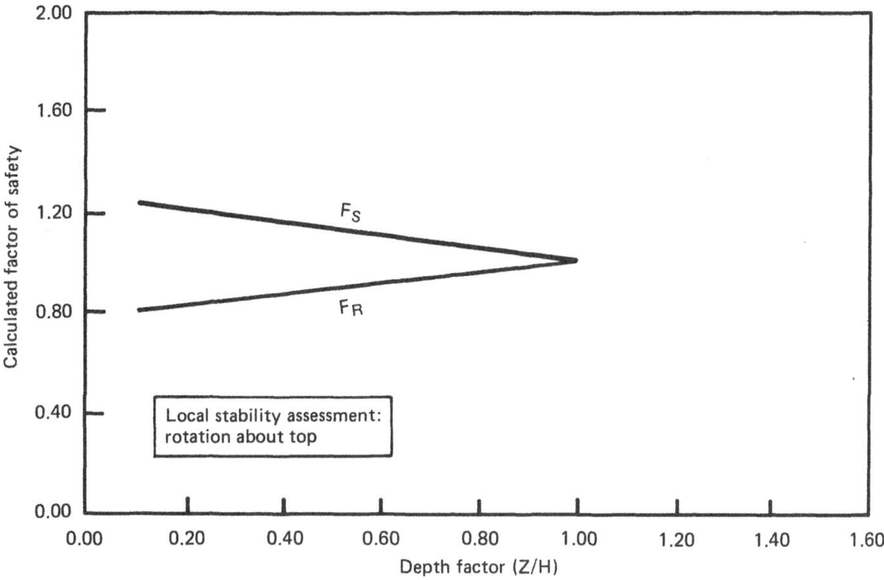

Fig. 5 Relations between depth factor and F(rupture) for F(spec) of unity

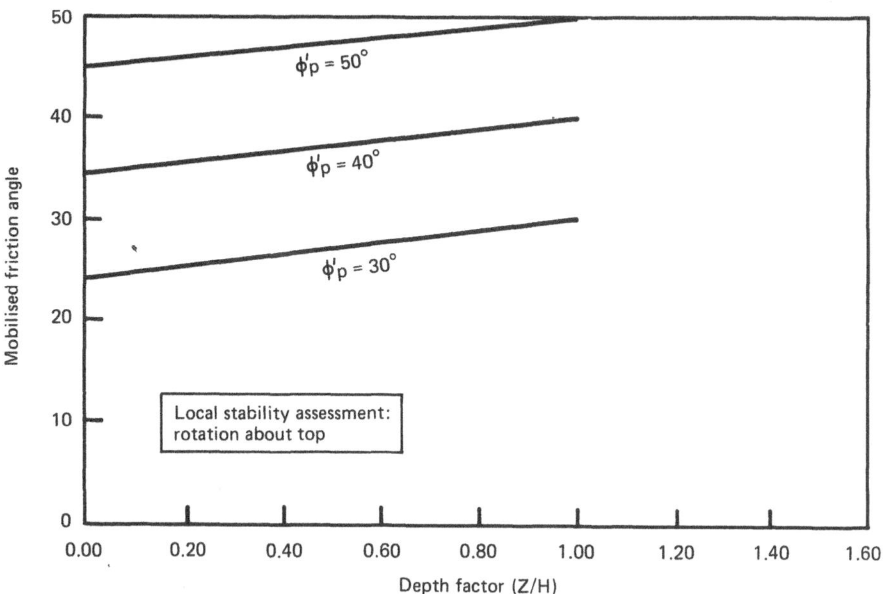

Fig. 6 Relations between depth factor and mobilised friction angle

3.2 Overall adherence resistance.

To simplify the analysis the assessment of overall adherence resistance for outward rotation about the top is based on the average values of safety factor of the soil, mobilised friction angle and earth pressure coefficient determined from the local bond analysis described in Section 3.1. The equation for overall adherence stability on this basis is as follows :

$$\frac{F_{OVER}}{F} = \frac{K_a}{F_{AS}K_{MA}}\left[\frac{L/H - \sqrt{K_{MA}}/3}{L/H - \sqrt{K_a}/3}\right] \qquad (30)$$

The results obtained from the analysis based on Equation 30 are shown in Figure 7 plotted against the average safety factor (F_{AS}) of the soil. Also shown on the figure is the relation between average lateral pressure coefficient (K_{MA}) and F_{AS}. A comparison of the results produced by Equations 21 and 30 corresponding to rotation about the base and top respectively is presented in Figure 8 in terms of the ratio of F_{OVER}/F versus F_{AS}. The close agreement between the two curves confirms the point made at the outset that the total force for both cases is the same and is assumed to correspond to $\frac{1}{2}K\gamma H^2$.

4. INTERNAL STABILITY CONSIDERATIONS WITH DIRECT ASSESSMENT OF STRAINS.

In the previous sections the influence of strain behaviour of the soil and reinforcement was assessed indirectly by assuming that the strains developed would be inversely proportional to factor of safety. An alternative approach will now be described in which the lateral strains are evaluated directly, thus allowing the mobilised friction angle to be calculated. The lateral pressures and tension in the reinforcement may then be determined on this basis. The method is essentially an extension of the conventional design whereby an evaluation of the strain permits the use of more realistic angles of friction in the design equations.

4.1 Basis of the method.

The method requires a knowledge of the relations between tensile load and strain for the reinforcement and between mobilised friction angle and lateral strain for the soil. The simplest approach is to assume a linear relation for the reinforcement and a bilineal relation for the soil (Figure 9).
In this latter case the ultimate friction angle (ϕ_{CV}) is used in preference to the peak friction angle as potential post-peak reduction in strength and rapid progressive failure are then avoided.

The relation between strain in the reinforcement (ϵ_R) and tensile load (T) is given by :

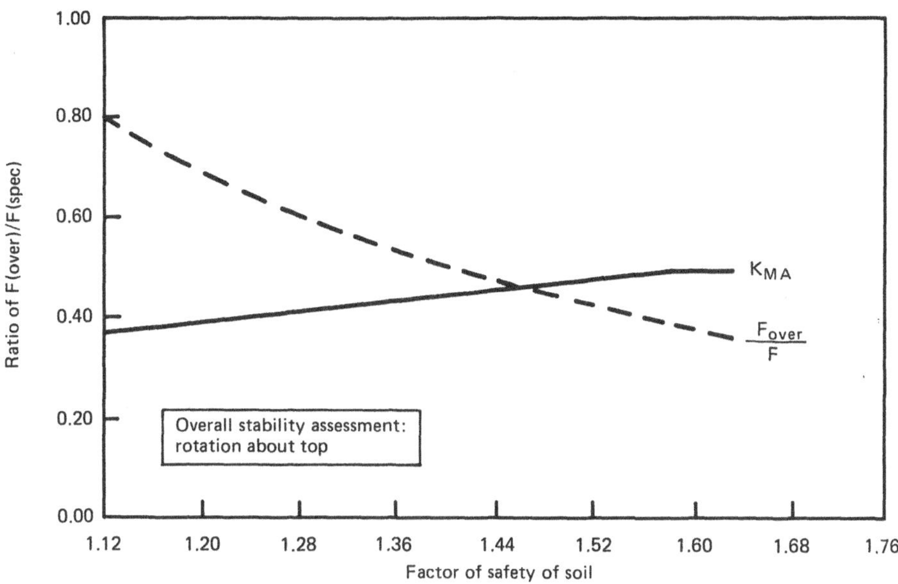

Fig. 7 Relations between F(soil) and ratio of F(over)/F(spec)

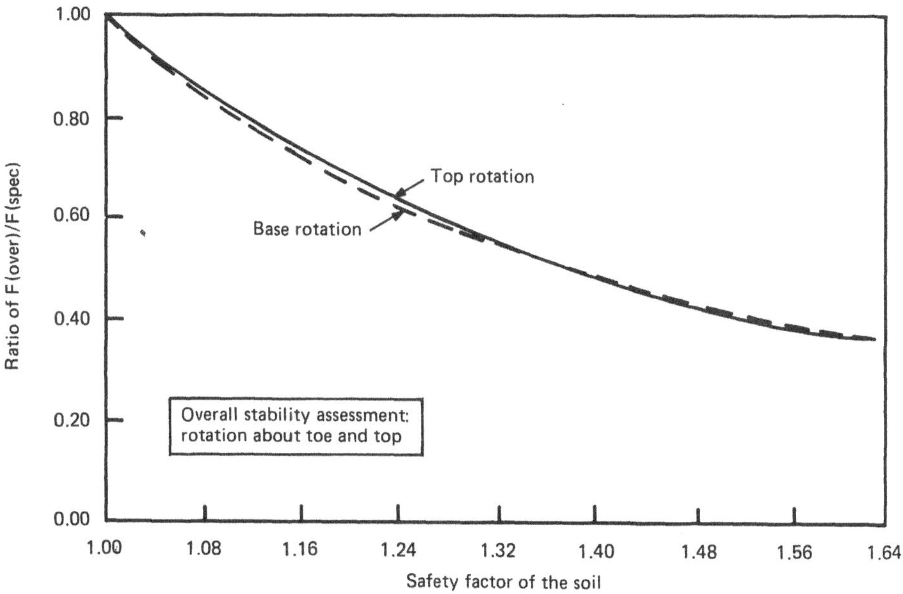

Fig. 8 Relations between F(soil) and ratio of F(over)/F(spec)

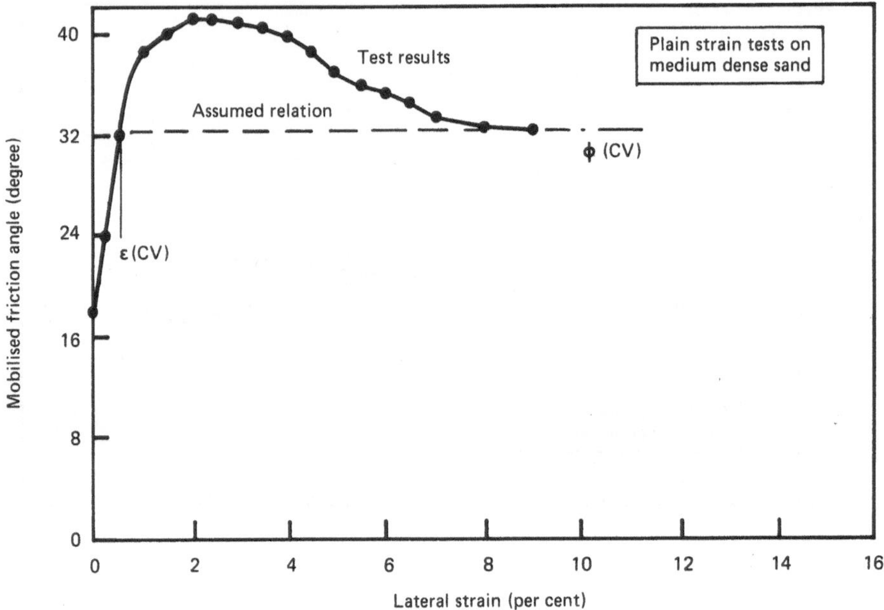

Fig. 9 Relations between lateral strain and mobilised friction angle

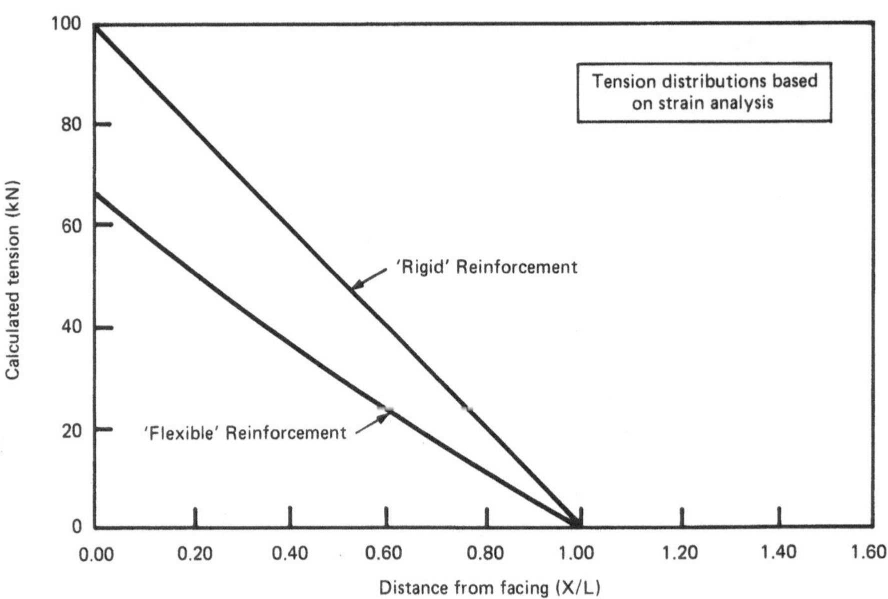

Fig. 10 Tension distributions for 'rigid' and 'flexible' reinforcement

$$T = m.\epsilon_R \quad \ldots\ldots(31)$$

The relation between lateral strain in the soil (ϵ_S) and mobilised friction angle is given by :

$$\phi_M = \phi(K_0) + n\epsilon_S \qquad \epsilon_S < \epsilon_{CV} \quad \ldots\ldots(32a)$$

$$\phi_M = \phi_{CV} \qquad \epsilon_S \geq \epsilon_{CV} \quad \ldots\ldots(32b)$$

Where ϵ_{cv} is as shown in Figure 9 and corresponds to the minimum lateral soil strain to mobilise ϕ_{CV}. Now for purposes of evaluating local stability the lateral force (D_M) is given by :

$$D_M = K_M \sigma_V S_V S_H = K_M \gamma z S_V S_H$$

The lateral force has to be resisted by the tensile force (T) in the reinforcement :

$$\text{i.e. } T = K_M \gamma z S_V S_H \quad \ldots\ldots(33)$$

$$m.\epsilon_R = K_M \gamma z S_V S_H$$

Where $K_M = (1 - \sin\phi_M)/(1 + \sin\phi_M)$, and $\phi_M = \phi_{K0} + n\epsilon_S$

Now for compatibility, the (average) strain in the reinforcement and soil must be equal :

$$\text{i.e. } \epsilon = \frac{\gamma z S_V S_H[1 - \sin(n\epsilon + \phi_{K0})]}{m[1 + \sin(n\epsilon + \phi_{K0})]} \quad \ldots\ldots(34)$$

It is apparent for consistency of the results that the argument of the sin terms in Equation 34 must not exceed the design value of friction angle. The above equation has been re-arranged to the following form to allow a more rapid iterative solution to be obtained :

$$\epsilon_{i+1} = \sqrt{\frac{\epsilon_i \gamma z S_V S_H[1 - \sin(n\epsilon_i + \phi_{K0})]}{m[1 + \sin(n\epsilon_i + \phi_{K0})]}} \quad \ldots\ldots(35)$$

The strain determined from Equation 35 is then used to evaluate ϕ_m from Equation 32. It should be noted, however, that the values obtained relate to the conditions directly behind the facing and it is of interest to consider how the strains may vary along the reinforcement.

For metallic or other types of high modulus reinforcement, the magnitude of the strains will be negligible and any outward displacement of the facing will produce a corresponding movement at every point on the reinforcement. Thus the application of a pull-out force would result in a uniform shear strain at the interface between the soil and reinforcement. Such a distribution of shear strain would produce a constant mobilised friction angle along the length of the reinforcement together with a linear variation in tension from a maximum value at the face to zero at the rear.

In contrast, the application of a pull-out force to an extensible reinforcement would induce the greatest strain at the face where the maximum tension occurs. If it is assumed that the strain varies linearly from the face, to zero strain at a distance L from the face where K_0 conditions apply, then the tension developed in the reinforcement at distance x can be obtained from the following differential equation :

$$\frac{dT}{dx} = 2A\mu_m\sigma_v C_1$$

Where $\mu_M = f(x)$
where A = surface area of reinforcement
where C_1 = constant of proportionality

Taking $\mu_M = \alpha.\tan\phi_M = \alpha.\tan(\phi_{K0} + n\epsilon)$ where α = an adherence factor between zero and unity.

Now ϵ is assumed to vary linearly from ϵ_M at the face as determined from Equation 35 to 0 at distance L :

$$\text{i.e. } \epsilon_x = \epsilon_M(1 - x/L)$$

Where $\epsilon_M \leq (\phi_{CV} - \phi_0)/n$ if ϕ_{CV} is used in design

$$\frac{dT}{dx} = -2A\sigma_v\alpha C_1.\tan[\phi_{K0} + n\epsilon_M(1 - x/L)]$$

$$T_x = \frac{2A\sigma_v\alpha L_D C_1}{n\epsilon_M}.\ln.[\cos(\phi_{K0} + n\epsilon_M(1-x/L))] + C_2$$

when x=0 ; $T_X = T_M$

$$C_1 = \frac{T_M}{\frac{(2A\sigma_v\alpha L_D)}{n\epsilon_M}.\ln\frac{[\cos(\phi_{K0} + n\epsilon_M)]}{[\cos\phi_{K0}]}}$$

when x = L ; $T_X = 0$

$$0 = \frac{2A\sigma_v\alpha L_D C_1}{n\epsilon_M}.\ln.[\cos(\phi_{K0})] + C_2$$

$$C_2 = -\frac{T_M.\ln[\cos\phi_{K0}]}{\ln\frac{[\cos(\phi_{K0} + n\epsilon_M)]}{[\cos\phi_{K0}]}}$$

$$T_X = \frac{T_M}{\ln.\frac{[\cos(\phi_{K0}+n\epsilon_M)]}{[\cos\phi_{K0}]}}.\ln.\left\{\frac{\cos(\phi_{K0} + n\epsilon_M(1-x/L))}{\cos(\phi_{K0})}\right\} \quad \text{......(36)}$$

Although the tension distribution produced by Equation 36 is not considered representative of an actual situation, it allows a comparison to be made between the calculated results for 'rigid' and 'flexible' reinforcement (Figure 10). As could be expected the "rigid" reinforcement has attracted greater tensile forces because the amount of strain of the soil is reduced with a corresponding reduction in mobilised friction. Moreover ,because of the uniformity of the shear strains at the interface, the tension distribution for the "rigid" reinforcement varies linearly over the reinforcement. In contrast, the "flexible" reinforcement shows a non-linear distribution because the soil strains reduce from a maximum value at the facing to zero at the rear.

An assessment of local bond stability requires that, for flexible reinforcement, the lateral strain distribution along the length of the reinforcement must be known or assumed. Making the same assumption as before that the lateral strain reduces linearly from the facing to zero at a distance L allows the resisting force (R_M) against pull-out to be determined :

$$\text{i.e. } R_M = \int 2B\mu_M \gamma z\, dx$$

where $\mu_M = \alpha.\tan\phi_M = \alpha.\tan(\phi_0 + n\epsilon)$

For the simple case assumed, ϵ varies linearly from the maximum value e_m at the facing to zero at L :

$$\text{i.e. } \epsilon = \epsilon_M(1 - x/L)$$

Thus R_M is determined from :

$$R_M = \int_0^L (2B\gamma z\alpha).\tan\{\phi_0 + n\epsilon_M(1 - x/L)\}dx$$

$$\text{i.e } R_m = \frac{2B\gamma z\alpha L_D}{n\epsilon_M} \ln.\left(\frac{\cos\phi_0}{\cos(\phi_0 + n\epsilon_M)}\right) \quad \text{.......................(37)}$$

The disturbing force corresponds to the lateral force acting over the area of facing supported by a single element :

$$\text{i.e. } D_M = K_M \gamma z S_V S_H$$

Hence the actual factor of safety is given by :

$$F_A = \frac{R_M}{D_M}$$

Now the specified factor of safety is equal to :

$$F = \frac{R}{D}$$

$$\frac{F_A}{F} = \frac{R_M.D}{D_M.R}$$

$$\text{i.e } \frac{F_A}{F} = \frac{K_a}{K_M.n.\epsilon_M.\tan\phi_L} \ln\left(\frac{\cos\phi_0}{\cos(\phi_0 + n.\epsilon_m)}\right) \qquad \text{.................(38)}$$

The calculated factor of safety is plotted versus the depth below the surface in Figure 11 for peak friction angles of 30°, 40° and 50°. The reasons for the relatively small factors of safety are as follows :

(1) The mobilised friction angle is less than the limiting value behind the facing so that the lateral force is greater than the minimum value corresponding to active earth pressure conditions.

(2) The magnitude of the lateral strain reduces along the length of the reinforcement with a corresponding reduction in interface friction so that less pull-out force is developed.

However, as the limiting conditions approach, greater strains will be induced and the available soil strength will increase until it is fully mobilised. The development of such strains will rarely be uniform and the maximum strength may be achieved at some locations while at others only a small proportion will have been attained. Thus with increasing strain the former locations will tend to reduce in strength as post-peak conditions are developed while elsewhere the strength may be approaching or at the peak value. It is unlikely, therefore, that with extensible reinforcement the peak strength will be attained simultaneously at all locations and although the available resistance in adherence will improve with increasing strain, it will be usually less than calculated on the basis of ideal limiting strength values. A prudent approach to design would thus be to assume that the maximum available friction angle of the soil corresponds to the ultimate friction angle of the soil ϕ_{CV}.

Notwithstanding that the ultimate pull-out resistance is likely to be less than calculated on the basis of fully mobilised strengths, the local adherence stability based on conventional design equations will generally be adequate although the factor of safety at the working condition will be usually smaller than specified. Of greater significance will be the effect of the smaller mobilised soil strengths in relation to potential rupture.

The assessment of local stability against tensile rupture was based on the same method as described in Section 2.1 whereby the lateral forces and mobilised friction angles determined for adherence stability are again employed. It should be noted, however, that such an approach is not necessary in this case as an independent calculation of local stability against rupture would produce the same strains and mobilised frictions and thus highlights the much greater consistency of a method based on a strain assessment. The relations between depth factor and calculated factor of safety, for the same values of peak friction angle

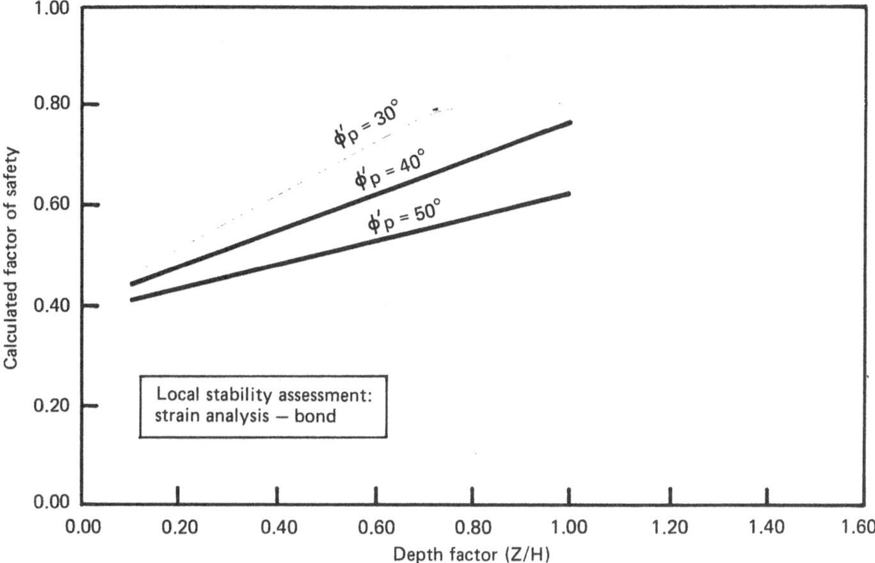

Fig. 11 Relations between depth and calculated factor of safety

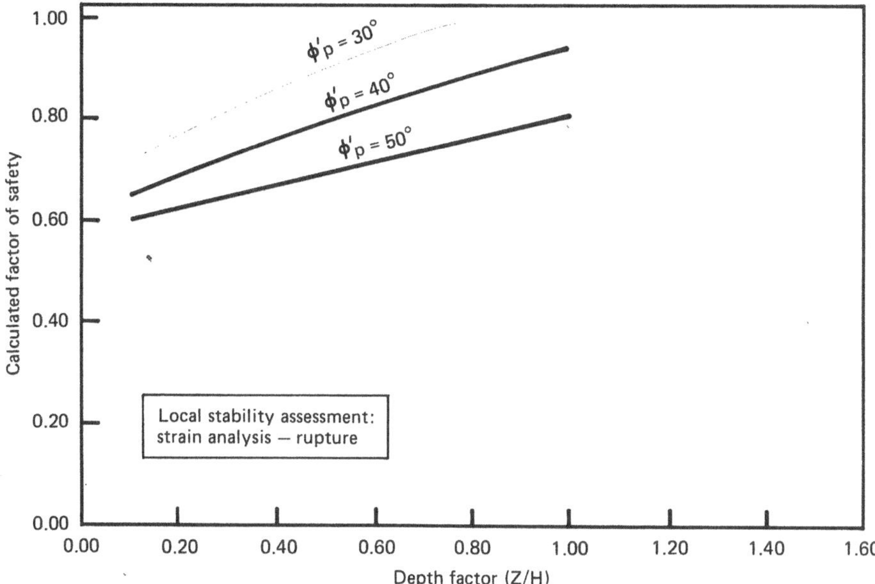

Fig. 12 Relations between depth and calculated factor of safety

as used previously, are shown in Figure 12.

The description of a possible design method based on the assessment of strain has been limited to considerations of local stability and for a very simple assumption concerning the distribution of internal strains. The extension of the method to overall stability would not appear to present any special difficulties and could be based on a similar approach to that described in Section 3.2 for the case of rotation about the top of the wall when average values were employed.

Alternatively, in preference to the development of an overall design equation involving the use of average values, it may prove more convenient to determine the total resistance by summation of the contributions made by individual elements incorporating the actual properties of the soil and reinforcement at each level. As far as the use of alternative patterns of strain behaviour is concerned, the main problem is that the basis for more complex assumptions may prove difficult to justify and will lead to more involved design equations. However, the latter difficulty may be overcome by the use of Southwell-type plots of the relations between tensile load and extension for a geotextile and between mobilised friction angle and lateral strain for soil which generally allow both forms of relation to be linearised (Murray,1980).

5.0 EXTERNAL STABILITY CONSIDERATIONS.

The assessment of the external stability of an earth retaining structure is carried out to ensure that the wall does not fail by sliding or overturning and, when cohesive soils are present behind or beneath the wall, to ensure that the wall is safe against a potential slip failure. This latter mode of possible failure is usually assessed by the Slices Method (Bishop,1955), or similar, and involves an iterative procedure for determining factor of safety and mobilised shear strength. Such an approach is consistent with the methods described earlier for internal stability and will not be considered further. The following sections describe procedures which attempt to also bring consistency into both sliding and overturning aspects of external stability.

5.1 Resistance to sliding.

The safety factor against forward sliding of the wall on the basis of limiting strengths is given by :

$$F = \frac{W.\tan\phi_p}{\frac{1}{2}K_a \gamma z^2} \quad \ldots\ldots (39)$$

The conditions for limiting equilibrium are :

$$2\tan\phi_p.L_L = K_a.z$$

and the limiting width of wall (L_L) at the point of sliding is obtained from :

$$L_L = \frac{K_a . z}{2\tan\phi_P} \qquad (40)$$

Thus the design length (L_D) is related to the limiting length by a factor of safety :

$$L_D = F . L_L \qquad (41)$$

As assumed previously, up to a condition of zero strain, the greater the factor of safety against sliding the more the soil will be constrained such that less of its shear strength may be mobilised. Thus the actual safety factor against sliding is given by :

$$F_{SLID} = \frac{2L_D . \tan\phi_M}{K_M . z} \quad : \epsilon_S > 0 \qquad (42a)$$

and for equilibrium $L_D = F . L_L = \dfrac{K_M . z}{2\tan\phi_M}$

$$\text{i.e.} \quad \frac{F . K_a . z}{2\tan\phi_P} = \frac{K_M . z}{2\tan\phi_M}$$

Assuming $\tan\phi_M = \tan\phi_P / F_S$

then $F_S = F . K_a / K_M$ .... (43)

$$\text{where } K_M = \frac{\sqrt{(F_S{}^2 + \tan^2\phi_P)} - \tan\phi_P}{\sqrt{(F_S{}^2 + \tan^2\phi_P)} + \tan\phi_P}$$

Thus Equation 43 can be solved iteratively for F_S .

To preserve the equality of Equation 42a the factor of safety against sliding will correspond to unity, until the zero strain condition is reached. An anomaly occurs, however, as to how the equality of Equation 42a is preserved when the specified factor of safety is increased beyond the state when zero strain is attained. One possible approach is to assume that although K_M remains constant at K_0, the value of $\tan\phi_M$ in the numerator of Equation 42a reduces to compensate for any further increase in the length L_D. This assumption implies that F_S will further increase to reduce the mobilised sliding friction over a greater length. The equation for the factor of safety against sliding is therefore given by:

$$F_{SLID} = \frac{2L_D.\tan\phi_M}{K_0.z} \quad : \epsilon_S = 0 \quad \text{.............................(42b)}$$

Clearly a stage may be attained with large values of specified safety factor when further increases in L_D will not produce any redistribution of the applied sliding force and any additional length will be redundant. At this stage the redundant length will need to be considered as a reserve of safety factor which does not contribute to the equilibrium of Equations 42.

In preference to the specified safety factor F which assumes fully mobilised strengths, or to the value of safety factor against sliding F_{SLID} which generally corresponds to unity, a more realistic measure of the safety factor against sliding is provided by the safety factor of the soil F_S. The relations between F_{SLID} based on Equations 42 and specified safety factor for peak friction angles of 30°, 40° and 50° are shown in Figure 13.

5.2 Stability against overturning or bearing failure.

The conventional method of designing against overturning assumes that the wall behaves as a rigid body when resisting the thrust from the backfill. It is further assumed that a trapezoidal pressure distribution will be developed at the base of the wall. The method is essentially a permissable stress approach in which the width of the wall is determined by an allowable bearing capacity for the foundations. The limiting strength of the backfill is employed in determining the earth pressure coefficients and a usual requirement is to avoid the development of tensile soil stresses in the foundations. The equation for the maximum foundation pressure on this basis is given by :

$$\sigma_f = \gamma H(1 + K_a.H^2/L^2) \quad \text{....................................(44)}$$

where σ_f must not exceed the allowable bearing pressure q_a. Thus one definition of safety factor against overturning may be given by :

$$F = \frac{q_L}{\sigma_f} \quad \text{...(45)}$$

where q_L is the ultimate bearing capacity of the soil.

As was the case for internal stability it is apparent that there are inconsistencies in such an approach as the greater the ultimate bearing capacity relative to the applied pressure ,the smaller the amount of yielding taking place with less scope for mobilising the shear strength of the backfill. An alternative method is described which involves an assumption that the mobilised shear strength is reduced in inverse proportion to the factor of safety against bearing failure. The procedure is therefore similar to that employed previously for

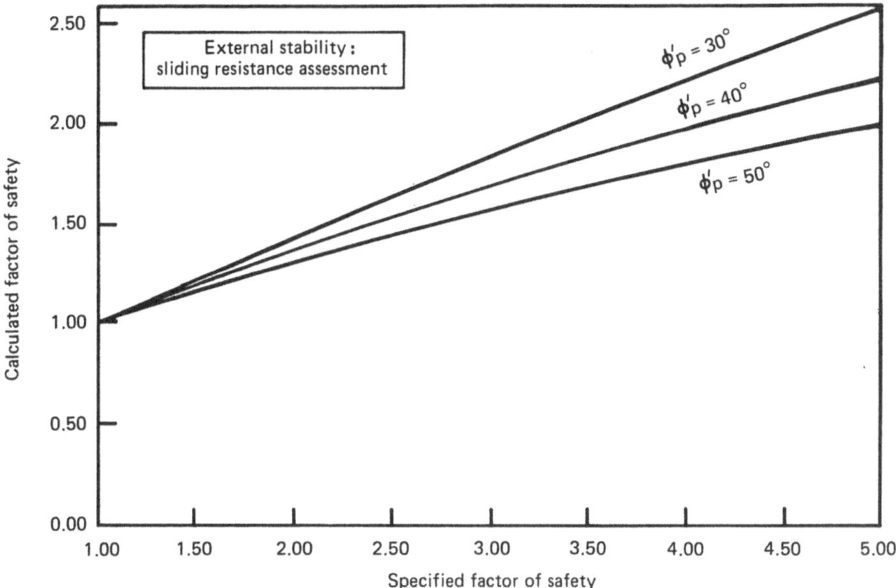

Fig. 13 Relations between specified and calculated factor of safety

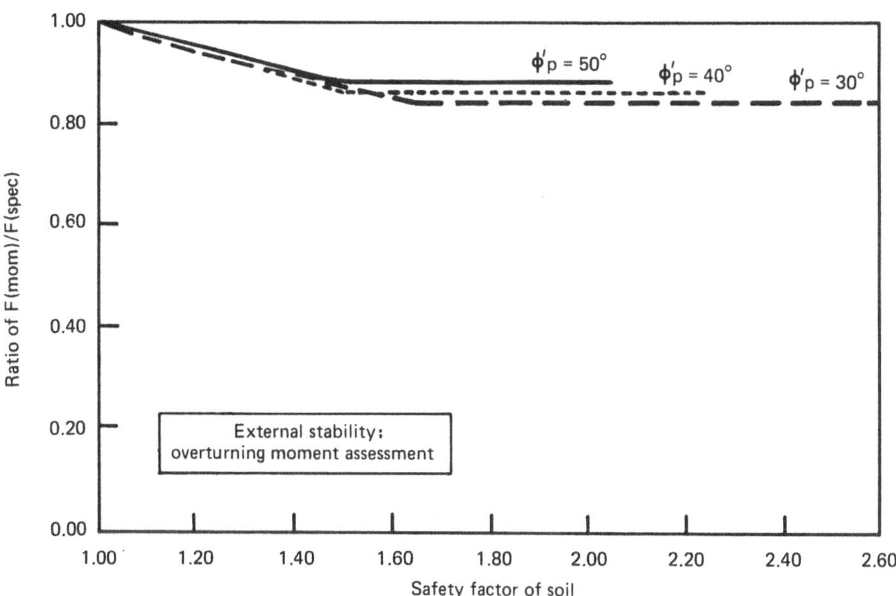

Fig. 14 Relations between F(soil) and ratio of F(mom)/F(spec)

evaluating the internal stability characteristics of reinforced soil walls.

From Equations 44 and 45 the limiting bearing capacity of the soil is given by :

$$q_L = F\gamma H(1 + K_a.H^2/L^2) \quad \ldots\ldots (46)$$

Now if the shear strength of the soil is not fully mobilised the lateral earth pressure coefficient increases to K_M and the actual factor of safety against overturning (F_{MOM}) is given by :

$$q_L = F_{MOM}.\gamma H(1 + K_M.H^2/L^2) \quad \ldots\ldots (47)$$

Equations 46 and 47 may be solved for F_{MOM} provided the factor of safety of the soil (F_S) is known :

$$\text{i.e } \frac{F_{MOM}}{F} = \frac{(1 + K_a.H^2/L^2)}{(1 + K_m.H^2/L^2)} \quad \ldots\ldots (48)$$

One technique is to employ the same value of F_S as was determined in the assessment of stability against sliding described in Section 5.1. Assuming that $\tan\phi_M = \tan\phi_P/F_S$, then the lateral earth pressure coefficient K_M can be determined employing the same equation as given previously allowing Equation 48 to be solved. The relations between the ratio of F_{MOM}/F determined from Equation 48 are shown in Figure 14 plotted versus the safety factor of the soil F_S.

It should be noted that friction between the backfill and rear of the wall has been ignored in the analysis .In some cases where the backfill is attempting to move down relative to the wall this will result in an underestimate of the factor of safety. However, it seems better to ignore "wall friction" unless reliable data on the possible movements are available as in other cases the wall may be attempting to move down relative to the backfill with an increased overturning effect and reduction in factor of safety. Thus the use of zero wall friction provides a reasonable compromise between these two extreme situations.

6. FURTHER ASPECTS OF SAFETY FACTOR

The considerations of safety factor have been limited, so far, to the influence of strain characteristics on behaviour and how the calculation procedures may be modified to take account of such behaviour in design. It is apparent from the foregoing analysis that the with greater constraints on soil strain larger forces have to be carried by the structure. Thus an important factor in reducing costs is the development of techniques which avoid inhibiting soil strains but still ensure stable and serviceable structures. Methods of achieving this have been recently proposed and are the subject of ongoing research (McGown et al,1987).

There are a number of other factors influencing the performance of reinforced soil structures which are normally accounted for by selecting an appropriate factor of safety. In particular there is the variability of material properties, construction and design tolerances, the influence of construction and deviations in the applied loads. The application of an overall factor of safety to account for these variables for structures reinforced with polymeric materials may prove uneconomic in view of the wide range of strength and strain characteristics of these materials. Partial factors of safety have been proposed, therefore, to enable the influence of each factor to be considered separately (Murray and McGown,1987).

The partial factors to be considered in the design of reinforced soil structures are listed in Table 1. The values $Ø_{m1}$ and $Ø_{m2}$ must be determined for the specific material in the particular application. The recommended procedure is to establish the variation in properties between control specimens and those obtained from strength tests on normal production material in controlled laboratory conditions tested under both constant rate of strain and sustained loading in both the short and long term. The partial factors $Ø_{m2}$ is also best evaluated from such strength tests and comparisons with control specimens to enable the influence of site damage and environmental effects to be assessed. Figure 15 shows schematically how the load versus strain relation of polymeric reinforcement may be influenced by site damage effects.

The partial factors γ_{fL} and γ_{f3} are normally provided in Codes of Practice relating to the more general aspects of civil engineering works (e.g. B.S. 5400,1980).

TABLE 1

PARTIAL FACTORS CONTRIBUTING TO THE OVERALL FACTOR OF SAFETY

PARTIAL FACTOR	PURPOSE
γ_{M1}	To cover the possible reductions in material properties compared to the the properties of control specimens.
γ_{M2}	To cover for site damage and construction or manufacturing tolerance on site, such as misalignment, undulations and mis-shaped products.
γ_{fL}	To cover unfavourable deviations in loads.
γ_{f3}	To cover errors or inaccuracies in the design method.
Overall factor of safety = $(\gamma_{M1} * \gamma_{M2} * \gamma_{fL} * \gamma_{f3})$	

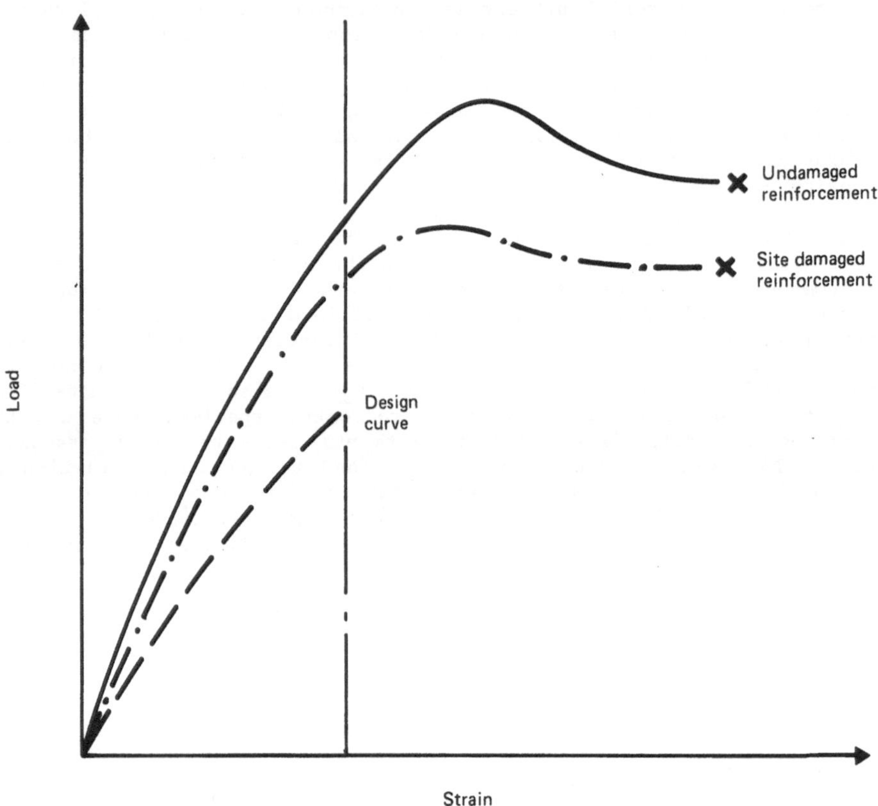

Fig. 15 Influence of site damage on load versus strain relation of polymer reinforcement shown schematically

7. CONCLUSIONS.

1. In the design of reinforced soil structures it is normally assumed that the full strength of the soil is mobilised and that the properties are unaffected by the presence of the reinforcement or boundary constraints. It has been demonstrated that such assumed behaviour is unrealistic and tends to inhibit progress towards a better understanding of how reinforced soil structures actually behave.

2. An alternative method of design has been described in which the mobilised soil strength is related to factor of safety. This method highlights the fact that with increasing support by other components in the reinforced soil system, e.g. reinforcement, facing and foundation, the work done by the soil reduces. In effect the soil inadvertently develops a "safety factor".

3. The alternative approach has been shown to be applicable to design in terms of both internal and external stability and appropriate equations have been developed for the normal assumption of outward rotation about the toe as well as for outward rotation about the top which is considered more relevant for designs involving part-height facing panels. Although the method does not truly represent strain behaviour, it provides much greater consistency and compatibility in design while at the same time preserving the simplicity of the existing methods and would seem to be particularly suited for use with "stiff" reinforcements.

4. The increasing use of geotextile reinforcements has highlighted the need for an alternative design approach which considers strain behaviour. One such method has been described and the equations for internal local stability presented. The analysis has been limited to the simple case of linear variation of strain in the soil and reinforcement but could readily be extended to more complex situations.

5. Because of the wide range of strength and strain characteristics of polymeric reinforcements, the most economic approach in the design of reinforced soil structures employing such materials is likely to involve the use of partial safety factors.

8. ACKNOWLEDGEMENTS.

The work described in this paper forms part of the research programme of the Transport and Road Research Laboratory and the paper is published by permission of the Director. The author is particularly grateful to Mr. I.F. Symons of TRRL for a number of helpful suggestions in the preparation of this paper.

9. REFERENCES.

Bishop,A.W. (1955) The use of the slip circle in the stability analysis of slopes.**Geotechnique 5**,pp 8 - 17 ,London

British Standards Institution (1980).BS5400 - Steel, concrete and composite bridges. British Standards Institution, London.

Burland,J.B. Potts,D.M. and Walsh,N.M. (1981). The overall stability of free and propped embedded cantilever retaining walls. **Ground Engineering, Vol. 14**, No. 5,London

Department of Transport (1978) Reinforced earth retaining walls and bridge abutments. Technical Memorandum (Bridges) BE3/78.London

Harr,M (1966). Foundations of theoretical soil mechanics. McGraw-Hill Book Company Ltd., London

McGown,A ,Murray,R.T. and Andrawes,K.Z.(1987) Influence of wall yielding on lateral stresses in unreinforced and reinforced fills. Dept. of Transport, TRRL Research Report 113 ,Crowthorne(Transport and Road Research Laboratory).

Murray,R.T. (1980)Fabric reinforced earth walls : development of design equations. **Ground Engineering**, October,pp 29 - 38 ,London

Murray,R.T. and McGown,A (In Press).Assessment of time dependent behaviour of geotextiles for reinforced soil applications. Proc. Seminar on Long-Term Behaviour of Geotextiles, St-Remy-les-Chevreuse, 4 - 6 November 1986, ITBTP, Paris.

Symons,I.F. (1983). Assessing the stability of a propped in-situ wall in overconsolidated clay. Proc. Instn. Civil Engnrs., Part 2.

LIST OF SYMBOLS.

A - Surface area of reinforcement
a - Cross-sectional area of reinforcement
B - Width of reinforcement
C_1 - Constant of proportionality
C_2 - Constant of integration
D - Disturbing force applied to a single element based on K_a conditions
D_M - Disturbing force applied to a single element base on KM conditions
F - Specified factor of safety
F_A - Factor of safety against local adherence
F_{AS} - Average safety factor of the soil over the depth of wall
F_R - Factor of safety against local rupture
F_S - Factor of safety of the soil
F_{MOM} - Factor of safety against overturning or bearing failure
F_{SLID} - Factor of safety against sliding failure
F_{Si}, F_{Si+1} - Factor of safety of soil after i,i+1 iterations
H - Height of reinforced soil wall
i - Number of reinforcing element counting from top of structure
Counter for number of iterations
K_a - Coefficient of active earth pressure
K_{ma} - Average mobilised coefficient of earth pressure over the height of the wall
K_{az} - Coefficient of active earth pressure variable with depth
K_M - Coefficient of earth pressure when strength of soil is not fully mobilised
K_{MZ} - Coefficient of earth pressure at depth z when strength of soil is not fully mobilised
K_O - Coefficient of earth pressure "at-rest"
L - Length of reinforcement
L_D - Design length of reinforcement
L_L - Limiting length of reinforcement
m - Coefficient defining the slope of the load versus strain of the reinforcement
n - Coefficient defining the slope of the mobilised friction angle versus lateral soil strain relation up to a strain of ϵ_{CV}
N - Number of reinforcements in a vertical column
P - Total force based on Coulomb analysis
p_L - Limiting stress in a reinforcement
p_M - Permissable stress in a reinforcement
q_L - Limiting bearing pressure in foundations
q_a - Allowable bearing pressure in foundations
R - Resisting force developed by a single element at limiting conditions
R_M - Resisting force developed by a single element at mobilised strength conditions
S_H - Horizontal spacing of reinforcing elements
S_V - Vertical spacing of reinforcing elements
T - Tension in reinforcement
T_i - Tension to be resisted by the ith element
T_M - Maximum tension in a reinforcement

T_T - Total tension to be resisted by reinforcements
T_X - Tension in reinforcement at distance x from facing
T_Z - Tension to be resisted at depth z
W - Weight of reinforced soil region
x - Distance from facing to point on reinforcement
z - Depth from top of structure to element under consideration
α - Coefficient expressing interface friction angle as a proportion of the soil friction angle
β - Orientation of failure plane from the vertical
γ - Unit weight of soil
γ_{M1} - Partial factor to take account of variations in material properties
γ_{M2} - Partial factor to take account of site damage and environmental effects
γ_{fL} - Partial factor to take account of deviations in applied loading
γ_{f3} - Partial factor to take account of construction tolerances, connection misalignment and associated effects
δL - Projected length of reinforcement beyond failure surface
ϵ - Strain in the reinforcement
ϵ_{ϵ} - Strain in reinforcement based on load versus extension test
ϵ_{CV} - Minimum lateral strain to mobilise ϕ_{CV}
ϵ_i - Strain determined after i^{th} iteration
ϵ_M - Maximum strain in reinforcement
ϵ_S - Lateral strain in soil
η - Slope of the mobilised friction angle versus depth relation for facing rotating about the top
ϕ_{CV} - Ultimate friction angle of the soil
ϕ_{KO} - Friction angle corresponding to "at-rest" conditions
ϕ_M - Mobilised friction angle of the soil
ϕ_{MZ} - Mobilised friction angle of the soil at depth z
ϕ_P - Peak friction angle of the soil
ϕ_{PZ} - Peak friction angle that can be mobilised at depth z
ϕ_{μ} - Interface friction angle
μ_L - Limiting interface friction coefficient
μ_M - Mobilised interface friction coefficient
σ_f - Vertical pressure beneath front of wall
σ_h - Horizontal earth pressure
σ_v - Vertical earth pressure
θ_O - Initial friction angle at depth z (less than or equal to ϕ_{KO}) when wall is rotated about top
θ_P - Maximum friction angle at depth z

APPLICATION OF FINITE ELEMENT TECHNIQUES TO THE ANALYSIS OF REINFORCED SOIL WALLS

R.K. ROWE and S.K. HO

GEOTECHNICAL RESEARCH CENTRE, UNIVERSITY OF WESTERN ONTARIO, LONDON, CANADA.

1. INTRODUCTION

The beneficial effect of incorporating tensile inclusions within a soil mass is well recognized and has been demonstrated by the successful construction of numerous reinforced soil walls using facings ranging from relatively rigid full face concrete panels to a flexible geotextile "skin", and reinforcement ranging from relatively stiff steel strips or meshes to geotextile sheets of low stiffness. Design methods for reinforced-soil walls (which are discussed in detail in other papers at this workshop) are typically based on limit equilibrium calculations which do not explicitly consider deformations or interaction between the inclusion and the soil.

In the case of Reinforced Earth (R) walls, there is now twenty years of empirical evidence to suggest that the approximate method of analysis, in conjunction with the normally adopted soil properties and safety factors, provides a safe design which gives acceptably small deformations under working conditions. This may also be the case for some geotextile and geogrid reinforced walls designed using current practice, however the wide range of facings, backfill materials and properties of the reinforcement leads one to question the generality of the simplified methods of analysis that are being proposed for the design of reinforced soil walls with geosynthetic inclusions. Questions that can be raised include:

- what is the effect of reinforcement extensibility;
- what is the effect of using different facings and construction techniques;
- what is the effect of soil-facing-reinforcement interaction both during construction and subsequently (over the life of the structure);
- under what circumstances might one expect strain-softening within the soil mass and what influence do the properties of the reinforcement have on strain softening;
- what <u>is</u> the "Factor of Safety" of a reinforced soil wall?

In principle, these questions could all be answered by the construction and monitoring of a large number of full scale field test walls. Unfortunately, the cost of performing and adequately monitoring a sufficiently large number of full scale walls is so large that it is not practical to perform a detailed experimental study. "Numerical experiments" or simulations provide an alternative and more cost effective means of performing such a study.

Finite Element techniques have the potential to allow us to:

P. M. Jarrett and A. McGown (eds.), The Application of Polymeric Reinforcement in Soil Retaining Structures, 541–553.

- improve our understanding of observed behaviour in field trials;
- model the complete response of a reinforced soil wall up to collapse;
- examine the effects of changes in the elements of the system (i.e. the properties of the reinforcement, the soil or the facing);
- investigate changes in construction procedures and the nature of the system; and
- study the interaction between the foundation characteristics and the performance of the reinforced soil wall (i.e. to evaluate the effect of local yield and settlement within the foundation for reinforced walls on less than ideal foundations).

The Finite Element technique is well recognized as being a very powerful tool and examples of its application to modelling reinforced soil walls (especially Reinforced Earth (R) Walls) can be found in the literature (eg. see ASCE Symposium on Earth Reinforcement, 1978). However, the use of the technique for performing studies which could answer the questions raised above is subject to some important constraints.

Firstly, if the technique is to provide answers to some fundamental questions (as opposed, say, to simply calculating the response of a structure under working conditions), then the formulation must provide a good description of all components of the reinforced soil wall system.

Secondly, a numerical study such as this should not be performed in isolation. It is essential that the formulation and technique be validated against (a) limiting analytical benchmark solutions; and (b) available data from both model and field tests. Indeed, the paucity of well documented field cases involving polymer reinforced walls which could be used to validate finite element models, has been a key factor constraining the use of Finite Element technique for the detailed study of reinforced soil wall behaviour.

Finally, although a numerical study may represent a cost-effective alternative to a full scale field study, it is not without cost. A good non-linear F.E. analysis requires an experienced "driver", considerable data and data preparation time, and considerable computer time.

Thus the objective of this paper is to review the past application of Finite Element techniques for the analysis of reinforced soil walls and to comment on some of the factors requiring consideration if the technique is to be used to answer some of the questions raised in this section.

2. REPRESENTATION OF A REINFORCED SOIL WALL SYSTEM

There are two techniques for modelling reinforced soil walls using finite elements, viz. composite and discrete representation of the constituents. Each approach has its own advantages and limitations as discussed below.

2.1 Composite representation

Composite formulations are based on an extension of continuum concepts to a macro level of observation whereby the entire reinforced soil mass is treated as an anisotropic, homogeneous material (eg. Romstad et al., 1976; Chang and Forsyth, 1977). In most formulations of this approach, the composite element stiffness is formed by superimposing the stiffness of the reinforcement and soil (assuming no slip) and so the distribution of elements need not be directly related to the distribution of reinforcement within the soil mass (eg. one or more layers of reinforcement may be included in a single composite element). This can result in a substantial decrease in the number of equations that must be solved for a given physical problem (as compared to discrete element formulations) with consequent substantial savings in computer time. Clearly, a disadvantage of the approach is that it does not provide direct information concerning the stress and strains at the interface of the reinforcement, nor does it provide information concerning localized deformation near the edge of the reinforced soil mass.

As noted above, most composite formulations assume a no-slip condition between the reinforcement and soil and hence are only applicable provided that the applied loads and distribution of reinforcement are such that little or no slip would be expected.

Attempts have been made to broaden the applicability of the approach by modelling slip in the composite representation of reinforced soil. This can be achieved by introducing extra nodal displacement variables relating the relative displacements between the soil mass and reinforcement (eg. see Herrmann and Al-Yassin, 1978; Naylor and Richards, 1978). However, the introduction of the extra variables required to model slip together with the consequent restrictions on the number of layers of reinforcement which can be included in a composite element largely eliminates the computational savings of a composite approach (i.e. which arise if it is not necessary to represent every reinforcing element) while still being subject to some of the disadvantages of the approach previously discussed.

2.2 Discrete representation

In a discrete representation of a reinforced soil system (eg. see Al-Hussani and Johnson, 1978; Andrawes et al., 1982; Andrawes et al., 1980; Banerjee, 1975; Rowe, 1984), the soil mass, the reinforcement, the facing and the interface between "structural" element and the soil are all independently represented by discrete elements. This approach provides direct information concerning the stresses and deformations at the interface of the reinforcement, the variation in stresses and strains (within the soil) between layers of reinforcement, and the localized deformations near the edge of the reinforced soil mass. However, to obtain accurate results, it is necessary to ensure that the choice of finite element and the distribution of finite elements (particularly between layers of reinforcement) provide sufficient "freedom" for realistic stress and strain distributions to be developed. To capture these details, a number of bays of elements should be provided between layers of reinforcement. The actual number of elements required will depend on the problem and the type of finite elements being used. An indication of the suitability of a given finite element mesh can be obtained by detailed comparison of stresses and strains calculated at various stages in an analysis using this mesh, with corresponding values calculated using a substantially refined mesh.

In summary, the only disadvantage arising from the use of a discrete representation of a reinforced soil system is that it requires more computing resources and effort in preparing input data than a composite approach. Weighing this disadvantage against the advantages of potentially more accurately analyzing the reinforced-soil system, the discrete approach is considered to be most appropriate for detailed investigation of reinforced soil wall behaviour. Furthermore, Herrmann and Al-Yassin's (1978) claims that discrete and composite approaches can be applied with equal accuracy (which are based on the analysis of a simple highly idealized problem) should be viewed with considerable caution since it has not been demonstrated that this is indeed true for realistic problems which involve construction simulation, non-linear elastic-plastic behaviour of the soil, either steel or strip reinforcement and different interface properties in pullout or direct shear modes of failure.

3. MODELLING OF DISCRETE COMPONENTS

Modelling of the discrete components of a reinforced soil system involves consideration of the type of finite element and the constitutive relationship that will be adopted. Table 1 summarizes the type of element and constitutive model which have been used in the past by a number of investigators analyzing reinforced soil walls.

3.1 The soil

The literature abounds with continuum elements which could be used to model the soil. Based on past experience with related problems, it would appear that many of the available elements can be used provided that sufficient attention is paid to checking the adequacy of associated finite element mesh, however it should also be noted that:

(1) particular care is required in the choice of mesh when using 3 noded (constant strain) triangles;
(2) lower order (eg. 4 noded) quadrilateral elements may give poor results and/or be computationally very inefficient;
(3) higher order quadrilateral elements (eg. 8 noded isoparametric) used in conjunction with reduced integration may give rise to physically unacceptable results under some circumstances.

The choice of element and details of the mesh design are likely to be far more critical when attempting to predict collapse than when simply calculating behaviour under working conditions.

To date, the majority of analyses which have been performed (see Table 1) have assumed an elastic or non-linear elastic (hyperbolic) constitutive relationship. Elastic models which do not consider the variation in soil stiffness with increasing stress level during construction are of doubtful validity since the pressure sensitive nature of the stress-strain characteristics of the backfill may have a significant influence on the stresses and displacements developed within the reinforced structure. The simplest way of avoiding this limitation is to adopt a non-linearity based on Janbu's equation (viz)

TABLE 1. Summary of some finite element analyses of reinforced soil walls

Reference	Type of Soil Element	Soil Model	Reinforcement Model	Soil/Reinforcement Interface Model	Construction Simulation*
Al-Hussaini & Johnson, 1978	5-node incompatible quadrilateral	hyperbolic	linear elastic-plastic bar element	joint element with hyperbolic shear stress-displacement relationship	(4) & surface loading
Banerjee, 1975	constant strain triangle	elastic isotropic	elastic bar element	no-slip	(6)
Chang & Forsyth, 1977	5-node incompatible quadrilateral	hyperbolic	composite with soil	no-slip	(3)
Herrmann & Al-Yassin	4-node isoparametric quadrilateral	hyperbolic	a) composite with soil b) elastic-plastic beam element	extra nodal displacement variables between soil & reinforcement	(3)
Naylor, 1978 Naylor & Richards, 1978	6-node quadrilateral	linear elastic	composite with soil	" "	(1)
Romstad et al., 1976	5-node incompatible quadrilateral	hyperbolic	a) composite with soil b) beam element	a) no-slip b) no-slip	(2)
Seed et al., 1986	4-node isoparametric element	a) hyperbolic b) hysteretic	linear elastic bar element	normal and shear spring	(5)
Shen et al., 1976	5-node incompatible quadrilateral	hyperbolic	composite with soil	no-slip	(3)

* (1) Assume initial K_o condition and then remove traction at wall surface. (2) Turn on gravity for unpropped system. (3) Construct and turn on gravity layer by layer. (4) As in (3) but no lateral displacement is allowed at wall face. (5) As in (3) but including consideration of compaction effect. (6) Not reported.

$$(E/P_a) = K(\sigma/P_a)^n \qquad (1)$$

where E is the Young's modulus of the soil, σ is the minor principal stress or the mean stress depending on the details of the formulation, P_a is atmospheric pressure and K and n are the material parameters. This non-linearity is included in the "hyperbolic" model and can also be readily included in non-linear elastic-plastic models (eg. Rowe, 1986). It should be recognized that the modelling of "yield" implicit in Eq. 1 is only approximate and is not appropriate for situations where there may be cyclic loading (eg. see Zylynski et al., 1978). However, there is considerable evidence to suggest that this approach can provide reasonable results for problems involving monotonic loading (as is generally the case in modelling wall construction).

Non-linear elastic (hyperbolic) models can be expected to provide acceptable results at low stress levels (eg. when there is a large "factor of safety"), however since they are based on elastic theory they can not correctly model plastic failure and plastic strains within the soil mass (it is noted that the use of a cohesion intercept c and friction angle ϕ in a hyperbolic model does not imply that the model is a plasticity model--eg. see Duncan, 1980). Numerous plasticity formulations have been proposed in the literature. The simplest of these involves Mohr-Coulomb failure surface and a non-associated flow rule. This model has been successfully applied in the analysis of geotextile reinforced embankments (eg. Rowe, 1982; Rowe, 1984; Rowe et al., 1984; Rowe and Soderman, 1984). This form of analysis can be readily modified to include the consideration of a non-linear failure envelope commonly encountered with granular materials (see Rowe et al., 1982). These models can be expected to model the soil behaviour up to and including failure. By examining the results of studies performed using this class of model it is possible to assess the magnitude of the strains to be expected prior to collapse of the structure and hence to make some initial assessment of potential significance of strain softening. However, this class of model is not suitable for modelling strain-softening behaviour and indeed the modelling of localization and strain softening even for unreinforced granular materials requires considerable additional research.

3.2 The reinforcement

The reinforcement can be modelled using a one dimensional bar element. Non-linearity of the stress strain behaviour and yield can also be readily modelled by making the element stiffness a function of stress (or strain) level. Breakage (snap) of the reinforcement can also be modelled however this involves the redistribution of stresses developed in the reinforcement prior to breaking and erroneous stress distributions can be obtained unless particular care is taken with the numerical algorithm used to redistribute these stresses.

3.3 The facing

Depending on the type of facing being considered, it may be appropriate to use continuum elements, beam elements or bar elements. The use of continuum elements or beam elements to model relatively rigid facing is quite straight forward. The more difficult problem is to correctly model wrap around facings (i.e. cases where a facing consists simply of the geotextile or geogrid reinforcement (which has been modelled with bar elements) being "wrapped around" the soil and "locked" into place by the overlying backfill). Correctly modelling the stresses and deformations resulting from the form of construction is not a trivial exercise.

3.4 The soil-reinforcement interface

The interaction between the soil mass and the reinforcement can be modelled by introducing soil-reinforcement interface elements. This can be achieved in a number of ways including the use of joint elements, nodal-compatibility slip elements or by substructuring. Common approaches to modelling the soil reinforcement interface involve three nodes at each point along the reinforcement; one attached to the soil above the reinforcement, one on the reinforcement, and one to the soil below the reinforcement. The nodal-compatibility slip element (which may be formulated initially in terms of normal and tangential springs with very high stiffnesses) (i) ensures compatible displacement between a pair of dual nodes (one attached to the soil and one attached to the reinforcement) until a Mohr-Coulomb failure criterion is reached, and (ii) replaces the compatibility conditions by a failure condition and dilatancy equation once the interface strength is exceeded. Joint elements allow relative deformation of the soil and reinforcement, prior to failure of the interface, based on some assumed constitutive relationship of what is in effect an interface layer between the reinforcement itself and the general soil continuum (eg. Andrawes et al., 1980, 1982 used a hyperbolic model to represent the interface behaviour).

In its simplest form, the joint element may be comprised of a pair of normal and tangential springs. Clearly, as the stiffness of a joint element increases, it tends to a nodal-compatibility slip element and the distinction between the two is related to the question of whether a distinct interface layer exists or whether the deformations at the interface (prior to failure) are simply due to the interaction between the reinforcement and the soil on either side of the interface. If there is good experimental data indicating that a distinct interface layer exists with experimentally defined stress-strain characteristics then this can be readily modelled as a joint element or as a thin layer of continuum element (with slip still being modelled using a nodal-compatible slip element). In the absence of this data, a nodal-compatibility slip element would seem appropriate.

Any modelling of interface behaviour must consider three possible mechanisms of failure as noted below.

(a) If there is insufficient anchorage capacity, failure will occur at the soil reinforcement interface above and below the reinforcement as the reinforcement is pulled out of the soil. This "pullout" mode involves displacement of the reinforcement relative to the soil on both sides of the reinforcement.

(b) If the shear strength of the soil reinforcement is less than the shear strength of the soil alone, then failure may occur by

sliding of the soil along the upper surface of the reinforcement and the upper soil mass moves relative to both the reinforcement and the underlying soil.
(c) The soil below the reinforcement (usually the foundation soil if one has a soft foundation) may be squeezed out from beneath the lowest reinforcement layer (and the entire reinforced soil wall). In this case, the lower soil may move relative to the reinforcement and the overlying soil.

If the reinforcement is in the form of a sheet, completely separating the soil above and below the reinforcement, then the interface resistance can be readily determined by direct shear tests (see Rowe et al., 1985). In this case, provision for slip at the interface is the same irrespective of the mechanism of failure (that is, direct shear or pullout). However, if the reinforcement takes the form of a geogrid, with openings which are large compared to the grain size of the soil, or if the reinforcement consists of separate reinforcing strips (eg. steel strips), then special care is required to correctly model the failure mechanism. For these materials, the interface shear resistance in direct shear (eg. if there is sliding of the soil along the upper surface of the reinforcement) may be substantially higher than the interface resistance in pullout (for example see Rowe et al., 1985). In modelling these materials, it is necessary for the formulation of the interface element to be such that it can automatically detect whether it is in a direct shear or pullout mode and to then select the appropriate interface parameters to model this mode of shearing. Thus the behaviour of the interface element on one side of the reinforcement is related to the behaviour of the interface element on the other side (since the mode of shearing can only be assessed by consideration of the direction of shear on either side of the reinforcement).

For planar reinforcement independent movement of the soil may occur above and below the reinforcement following either a direct shear or pullout failure. For strip reinforcement, independent movement of the soil above and below the plane of reinforcement can only occur during a direct shear mode of failure. Pullout of strips is really a three dimensional phenomenon in which the strips move relative to the soil around them but the soil between strips remains continuous. As noted by Naylor and Richards (1978), the common approach of using a conventional joint element (or nodal compatibility element) implicitly treats the strips as an equivalent two dimensional sheet and will cause serious error since it interrupts the transfer of shear stress through the soil.

Since pullout of strips does represent a truly three dimensional situation, it can only be approximately modelled in a two dimensional analysis. A number of different approaches can be adopted. For example, Naylor and Richards (1978) proposed a composite formulation which ensured continuity of shear stress in the soil after pullout by introducing a "conceptual shear zone". An alternative approach implemented by the authors in their discrete formulation involves an interface element which involves a node above the reinforcement, a node on the reinforcement and a node below the reinforcement. Prior to slip, normal and tangential compatibility between the soil and reinforcement is enforced by means of very stiff springs. The normal and shear stresses "above" and "below" the reinforcement are automatically monitored. If a pullout mode of failure occurs (as inferred by the direction of shear above and below the reinforcement together with a Mohr-Coulomb failure criterion), then the computer program automatically enforces compatibility between the soil nodes "above" and "below" the reinforcement (thereby maintaining continuous transfer of shear stress in the soil) while allowing slip between the reinforcement node and the two soil nodes. The normal force between these nodes is used to assess the normal forces acting on the strip; the corresponding shear resistance (based on a Mohr-Coulomb failure criterion) between the strip and soil is applied to both the upper and lower soil node, and as an equilibrating force to the node on the soil strip. (Since the strip only covers a small area of the soil, the Mohr-Coulomb parameters must be adjusted to take account of the actual surface area, per unit width of wall, which is in contact with the soil.)

4. SIMULATION OF LATERAL SOIL PRESSURE AND CONSTRUCTION DETAILS

To date, most finite element analyses of reinforced soil wall systems have employed classical earth pressure theory in simulating the pressure exerted on a reinforced soil wall system. One approach is to assume that the wall is constructed under "at rest, K_o" conditions, i.e., as if temporary support was provided to prevent lateral yielding during construction; loading is then provided by "removing the support" which involves applying a horizontal traction to the face of the wall equal to $K_o\gamma H$ (eg. see Naylor and Richards, 1978). Another approach is to assume the reinforced soil wall is constructed in an "active, K_a" state of failure, with equilibrium being maintained by applying a traction equal to $K_a\gamma H$ on the back of the wall (i.e. assuming the wall is free to translate, or rotate about the top or bottom; eg. see Banerjee, 1975). These approaches are simple but they also neglect the influence of construction method and, in general, cannot be expected to provide a good representation of the behaviour of the reinforced soil structure.

One factor affecting the magnitude and distribution of lateral pressures, soil movements and reinforcement strains within a reinforced soil wall system is compaction. Test data and theoretical calculations both indicate that the lateral earth pressure due to compaction of the fill behind flexible (and rigid) retaining walls can be far in excess of values predicted by classical earth pressure theory (eg. see Seed and Duncan, 1986). In a reinforced soil wall this phenomenon is likely to have a significant impact on the magnitude of the stresses and strains in the reinforcement, particularly if compaction equipment is brought close to the face of the wall.

It has also been demonstrated that compaction of the fill can have an important effect on the changes in horizontal stress which arise from applied surficial loads behind the wall (see Duncan and Seed, 1986).

In an attempt to study the effect of compaction on a reinforced soil system, Seed et al. (1986) performed finite element analyses on three case histories. Their analyses employed a hysteretic loading/unloading model in simulating compaction induced lateral pressures. They found that conventional finite element analyses which do not model compaction induced pressures gave rise to lower estimates of the tensile stress in the reinforcement compared with analyses which did model compaction; the effect of compaction being most pronounced at shallow depths. The conclusion from Seed et al.'s study was that field measurement (in terms of reinforcement force) could be better predicted when compaction effects are modelled in the analysis.

Another major factor affecting the magnitude and distribution of lateral pressure exerted by a reinforced soil wall system is the lateral restraint of the facing. McGown et al. (1987) have shown that the magnitude and distribution of earth pressures behind a model reinforced soil wall will depend on both the stiffness of the wall supports and the number of layers of reinforcement. These pressures may range from values corresponding to K_o conditions to values much lower than active earth pressures. There is also evidence to suggest that the behaviour of walls with full faced panels will be different to walls with segmented facings (where each segment can undergo some independent rotation and translation).

It may be concluded that careful finite element modelling of reinforced soil walls should include simulation of the actual or expected construction procedure. In particular, modelling of the facing support (and its removal) may be expected to provide better results than more conventional approaches which involve applying an assumed pressure distribution (be it K_o or K_a) to the back of the wall. Perhaps less important, but nevertheless deserving of consideration, is the simulation of the effects of compaction. (The importance of compaction will of course depend on the height of the wall, the type of compaction equipment used and the proximity of compaction equipment to the face).

5. FOUNDATION-WALL INTERACTION

One of the advantages of a reinforced soil wall over conventional wall systems is that it should be more tolerant of deformations and stresses induced by some yielding in the foundation, thereby allowing construction of walls on less than ideal sites. Unfortunately, conventional methods of analysis (eg. see Gourc et al., 1987) can not provide insight regarding the effect of foundation movements on the stresses and deformations of the wall. The finite element method is ideally suited for modelling the foundation-reinforced soil wall interaction which would occur when there is yielding in the foundation soils.

Modelling of this interaction will, however, require the use of a constitutive model that models plastic strains using a consistent plasticity formulation which can take account of the influence of rotations in principal stress directions which will occur near the toe of the wall. As a prerequisite, the finite element procedure adopted in these calculations should be capable of accurately predicting bearing capacity collapse loads and should be calibrated against relevant published bearing capacity solutions (eg. Davis and Booker, 1973).

6. TIME EFFECTS

An important concern in the analysis of walls reinforced with geosynthetics is the effect of the time dependent characteristics of the reinforcing material. In principle, finite element methods are well suited to modelling creep/relaxation in both the reinforcement and the soil (the latter being of particular importance if cohesive backfills are used). Numerous visco-elastic-plastic finite element formulations have been published in the literature however, as yet, these techniques have not been applied to a comprehensive study of reinforced soil wall systems. Modelling of this time dependent behaviour is an important challenge but one that can not be fully met until there is good quality field (or model scale) time dependent test data which can be used for validating the finite element calculations.

7. CONCLUSION

In this paper we have attempted to review the application of the Finite Element technique to the analysis of reinforced soil walls. There are still many unanswered questions regarding the behaviour of reinforced soil structures and the finite element method provides a very useful tool which can be used to help answer these questions. However, it has also been emphasized that there are many different types of finite element analyses and that considerable care must be exercised in both the selection of the particular finite element formulation to be adopted (eg. in the choice of constitutive model for the soil; modelling of the interface, etc.) and in the detailed application of the technique (eg. choice of elements, distribution of elements, construction simulation, etc.). There is also considerable scope for additional research in developing or adapting techniques for modelling time dependent interaction between the various components of the reinforced soil system as well as for developing techniques which model strain softening and localization (a major problem in itself) within reinforced soil systems.

ACKNOWLEDGEMENTS

This review forms part of a general programme of research into reinforced soil and geosynthetics being conducted by the Geotechnical Research Centre with funding from the Natural Sciences and Engineering Research Council of Canada under grant A1007.

REFERENCES

1. Al-Hussaini MM, Johnson LD: Numerical Analysis of a Reinforced Earth Wall. Proc. ASCE, Symposium on Earth Reinforcement, Pittsburg, pp. 98-126, 1978.
2. Andrawes KZ, McGown A, Mashhour MM, Wilson-Fahmy RF: Tension Resistant Inclusion in Soils. ASCE, J. Geotech. Eng. Div., Vol. 106, No. GT12, pp. 1313-1326, 1980.
3. Andrawes KZ, McGown A, Wilson-Fahmy RF, Mashhour MM: The Finite Element Method of Analysis Applied to Soil-Geotextile Systems. Proc., 2nd Int. Conf. on Geotextiles, Las Vegas, 2, pp. 695-700, 1982.
4. Banerjee PK: Principles of Analysis and Design of Reinforced Earth Retaining Walls. J. Inst. Highway Eng., 22, pp. 13-18, 1975.
5. Chang JC, Forsyth RF: Finite Element Analysis of Reinforced Earth Wall. ASCE, J. Geotech. Eng. Div., Vol. 103, No. GT7, pp. 711-724, 1977.
6. Davis EH, Booker JR: The Effect of Increasing Strength With Depth On the Bearing Capacity of Clays. Geotechnique, Vol. 23, No. 4, pp. 551-563, 1973.
7. Duncan JM: Hyperbolic Stress-Strain Relationships. Proc. ASCE, Workshops on Limit Equilibrium, Plasticity and Generalized Stress-Strain in Geotechnical Engineering, pp. 443-460, 1980.
8. Duncan JM, Seed RB: Compaction-Induced Earth Pressures Under K_o-Condition. ASCE, J. Geotech. Eng. Div., Vol. 112, No. 1, pp. 1-22, 1986.
9. Gourc JP, Ratel A, Gotteland Ph.: Analysis and Comparison of Existing Design Methods and Proposal For a New Approach. Nato Advanced Research Workshop, Application of Polymeric Reinforcement in Soil Retaining Structures, Royal Military College of Canada, 1987.
10. Herrmann LR, Al-Yassin Z: Numerical Analysis of Reinforced Soil Systems. Proc. ASCE Symposium on Earth Reinforcement, Pittsburg, pp. 428-457, 1978.
11. McGown A, Andrawes KZ, Murray RT: The Influence of Lateral Boundary Yielding On the Stresses Exerted by Backfills.
12. Naylor DJ: A Study of r.e. Wall Allowing Strip Slip. Proc. ASCE Symposium on Earth Reinforcement, Pittsburg, pp. 618-643, 1978.
13. Naylor DJ, Richards H: Slipping Strip Analysis of Reinforced Earth. Int. J. for Numerical and Analytical Methods in Geomechanics, Vol. 2, pp. 343-366, 1978.
14. Romstad KM, Herrmann LR, Shen CK: Integrated Study of Reinforced Earth - I. Theoretical Formulation. ASCE, J. Geotech Eng. Div., Vol. 102, No. GT5, pp. 457-471, 1976.
15. Rowe RK, Lo KY, Tham L: The Analysis of Tunnels and Shafts in Dense (Oil) Sands. Proceedings of the Fourth International Conference on Numerical Methods in Geomechanics, Edmonton, pp. 587-596, 1982.
16. Rowe RK: The Analysis of an Embankment Constructed On a Geotextile. Proc. 2nd Int. Conf. on Geotextiles, Las Vegas, 2, pp. 677-682, 1982.

17. Rowe, RK: Reinforced Embankments: Analysis and Design. ASCE, J. Geotech. Eng. Div., 110, pp. 231-246, 1984.
18. Rowe RK, MacLean MD, Soderman KL: Analysis of a Geotextile Reinforced Embankment Constructed on Peat. Can. Geotech. J., 21, pp. 563-576, 1984.
19. Rowe RK, Soderman KL: Comparison of Predicted and Observed Behaviour of Two Test Embankments. Int. J. of Geotextiles and Geomembranes, 1, pp. 157-174, 1984.
20. Rowe RK: Numerical Modelling of Reinforced Embankments Constructed on Weak Foundations. 2nd Int. J. on Numerical Models in Geomechanics, Ghent, pp. 543-551, 1986.
21. Rowe RK, Ho SK, Fisher DG: Determination of Soil-Geotextile Interface Strength Properties. Proc. 2nd Canadian Symp. on Geotextiles and Geomembranes, Edmonton, September, 1985.
22. Seed RB, Duncan JM: FE Analysis: Compaction-Induced Stresses and Deformations. ASCE, J. Geotech. Eng. Div., Vol. 112, No. 1, pp. 23-43, 1986.
23. Seed RB, Collin JG, Mitchell JK: FEM Analysis of Compacted Reinforced Soil Walls. 2nd International Symposium on Numerical Models in Geomechanics, Ghent, pp. 553-562, 1986.
24. Shen CK, Romstad KM, Herrmann LR: Integrated Study of Reinforced Earth - II: Behaviour and Design. ASCE, J. Geotech. Eng. Div., Vol. 102, No. GT6, pp. 577-590, 1976.
25. Zylynski M, Randolf MF, Nova R, Worth CP: On Modelling the Unloading-Reloading Behaviour of Soils. Int. J. Num. and Analytical Methods in Geomechanics, 2, pp. 87-93, 1978.

REVIEW OF SESSION

ANALYTICAL TECHNIQUES AND DESIGN METHODS

The chairman, Rowe, began the session by posing a number of questions that require answers when considering analysis and design and which the various speakers would address during the session. They included the following:

1. Should we be trying to estimate deformations using limit equilibrium analyses?

2. What is the effect of reinforcement extensibility?

3. Is strain-softening of the soils important?

4. How valid is it to find the level of reinforcement required without considering the interaction between it and the soil?

5. What is the Factor of Safety in our designs?

6. How does one account for construction technique in design?

7. How useful is the finite element method in analysis of reinforced soil?

8. Are any of the existing analytical methods providing the needed answers?

Jewell presented details of his ideas concerning the interpretation of the soil parameters that should be used in limit equilibrium analysis. They were based on simplifications of stress-dilatency concepts. He also addressed the problem of compatibility of strains between the soil and the reinforcement. Andrawes questioned how the dilatancy term used could always be additive to ϕ_{cv}. McGown noted the problems created in the construction process by using "ideal" spacing concepts which produced variable spacings between reinforcement layers. Leflaive questioned whether the important point of anisotropic strain was being considered and also cautioned that one must be careful to differentiate between the various polymers available when considering the effects of time and temperature on the stress-strain properties. McGown stated that although difficult it would be useful to attempt to estimate horizontal strains from laboratory tests.

Bonaparte then considered the effect of variation in reinforcement extensibility on the analytical and design processes. He considered the problem from the perspective of a design engineer who had reasonable analytical tools for stability analysis but no simple methods for predicting movements. His approach used stress path analysis for the proposed structure leading to estimates of soil strains, especially the horizontal

P. M. Jarrett and A. McGown (eds.), The Application of Polymeric Reinforcement in Soil Retaining Structures, 555–556.

strains. These could by assuming compatibility be related to strains in the reinforcement. At one point he indicated that the stiffnesses of most of the common geosynthetics used in North America may limit their use to walls of about 9 m in height if deformations in the structure need to be limited to reasonable values of around 1 inch. Jones stated that there are stiffer geosynthetics in use in Europe that allow considerably higher walls to be built without excessive deformations. Bonaparte concluded his presentation by saying that he believed that soils were usually in an active state immediately behind reinforced soil walls and that limit state methods were acceptable for simple design. He also suggested that research was needed on the influence of construction methods and facings on wall behaviour, on the development of horizontal strains with changing lateral earth pressures, on the influence of vertical spacing between reinforcement layers relative to the particle size of the soil and on full scale trials.

Murray's contribution dealt with the use of Factors of Safety in design and their implications in an analytical and practical sense concerning the mobilization of resistance in both the soil and the reinforcements. McGown pointed out that in certain of Murray's analytical developments the facing was assumed to be rigid whereas in practice there may commonly be a compressible zone of less compacted soils next to the wall that would radically alter the earth pressure distribution and the mobilization of soil strength. A general discussion ensued involving Raymond, Bonaparte, McGown, Jewell and Murray based yet again on the topics of strain compatibility between soil and both extensible and inextensible reinforcements and the need for knowledge of horizontal strains.

Gourc then compared a number of different methods of analysis indicating the effects of variations between the methods and discussing the topic of partial factors of safety. Andrawes and McGown pointed out the difficulty of applying different partial safety factors to the geosynthetic and to the soil and still expecting to obtain strain compatibility between the components. Great care would need to be exercised in selecting the different safety factors.

Rowe presented his contribution which dealt with the general methodology for developing a successful finite element analysis for reinforced soil walls. He suggested that finite element analysis was sufficiently complex that it is unlikely to become a day to day design tool but is likely to be used in the development of parametric studies and design charts. One point of discussion of importance to most analyses was the need to simulate the construction process and thus avoid the "switch on gravity" approach in which gravitational forces are applied only to an idealized complete wall model.

In summary during this and most other sessions there was general agreement that ϕ_{cv} was the best and safest strength parameter for general design usage. Limit equilibrium analyses still provide an acceptable simple method for general design and that there is hope to improve them through better comprehension of <u>all</u> the aspects involved. This includes the soil, the polymer and their interaction. Finite element analysis can be a most valuable aid in developing parametric studies and making the best use of the limited amount of case study information available. Formal contributions to this session were provided by Scott, Floss and Milligan and they follow this review.

FINITE ELEMENT ANALYSIS OF REINFORCED SOIL

R. Chalaturnyk, D.H.K. Chan, J.D. Scott

Department of Civil Engineering
University of Alberta
Edmonton, Alberta, T6G 2G7

This summary outlines the development of a finite element program capable of analyzing the performance of reinforced soil structures. SAFE (Soil Analysis by Finite Elements), a computer program developed at the University of Alberta (Chan, 1985), was selected for the development of the program. The modifications to SAFE include the implementation of a two dimensional, isoparametric bar element specifically suited to modelling the behaviour of geosynthetic reinforcing materials and an interface element for modelling the soil-reinforcement interaction. SAFE is based on a displacement formulation assuming small strains and small deformations. The program is capable of performing two dimensional, plane strain analyses using total or effective stress formulations for either fully undrained or drained soil conditions. Table 1 lists some of the soil, reinforcing and interface models available with SAFE. Load increment subdivision, program restart capability at any stage of an analysis, material property variation at any stage of an analysis and choice of stress calculation procedures are some of the standard features incorporated within the program. An element birth and death option allows incremental construction analyses to be conducted.

Several post-processing programs have been developed in order to aid in the examination of the finite element analyses results. Finite element mesh and deformed mesh plotting, stress and strain contour plotting, displacement arrow plotting, reinforcement load distribution plotting and interface normal and shear stress distribution plotting are all available. Additional development of SAFE includes no tension analyses using a crack model, an anisotropic plasticity soil model, special shear band element with a discontinuous shape function and a time dependent, strain softening soil model.

SAFE is currently being utilized in examining the stability of a steep (1:1) reinforced cohesive soil embankment constructed on a rigid foundation. A total stress, undrained finite element analysis of this slope is being conducted in order to investigate the end of construction behaviour of the reinforcement and the soil. The intent of the research is to compare the finite element analysis results with limit

P. M. Jarrett and A. McGown (eds.), The Application of Polymeric Reinforcement in Soil Retaining Structures, 557–560.

equilibrium results in order to determine whether present limit equilibrium design methods adequately assess the factor of safety of a reinforced slope.

Figure 1 illustrates the geometry of the reinforced embankment selected for the analysis. The finite element discretization of the problem, as illustrated in Figure 2, utilizes 8 node rectangular and 6 node triangular isoparametric in soil regions, 3 node isoparametric reinforcement elements for the geosynthetic material and 6 node isoparametric interface elements for the soil-reinforcement interface. The inset in Figure 2 illustrates a typical arrangement of these elements.

Linear elastic and nonlinear elastic analyses are being conducted for three cases:

1. Soil elements only (unreinforced embankments)
   - 529 elements, 1,652 nodes
2. Soil and reinforcement elements only (no interface elements)
   - 797 elements, 1,652 nodes
3. Soil, reinforcement and interface elements
   - 1,333 elements, 2,762 nodes

As a results of the large number of elements and degrees of freedom required for the analysis, solving the finite element equations necessitated the use of CDC Cyber 205 Supercomputer which is located at the University of Calgary.

Figure 3 illustrates the contours of soil strength mobilized in the embankment at a constructed height of 10 m. The strength mobilized is defined as the ratio of the deviatoric stress at failure to the mobilized deviatoric stress. The shaded zone indicates the region where the available soil strength has been exceeded. Contours of different stress and strain parameters can be plotted using SAFE's post-processing programs. Figures 4 and 5 illustrate the normal and shear stresses along a reinforcement layer, which is 3 m above the foundation, at embankment heights of 4, 6, 8 and 10 m. Both the top and bottom soil-reinforcement interface stress distributions are plotted on the graphs. It is clear from Figures 4 and 5 that the top and bottom interfaces are behaving identically throughout the embankment construction. Figure 6 illustrates the development of the tensile load along the length of the reinforcement during the embankment construction. The progression of the point of maximum reinforcement tensile load as the embankment is constructed is clearly illustrated in this figure.

REFERENCES

1. Chan, D.H.K. 1985. Finite Element Analysis of Strain Softening Material. Ph.D. Thesis, University of Alberta, Edmonton, Alberta, 345p.

Table 1 Element Material Models

Soil (Stress - Strain)	Reinforcement (Load - Strain)	Interface (Shear stress- Displacement)
1) Linear Elastic	1) Linear Elastic	1) Linear Elastic
2) Hyperbolic Elastic	2) Nonlinear Quadratic	2) Hyperbolic Elastic with linear failure envelope
3) Elastic Perfectly Plastic or Brittle Plastic; - von Mises - Tresca - Drucker - Prager - Mohr - Coulomb (associated and non-associated flow rule)	3) Elastic Polynomial	3) Hyperbolic Elastic with curved failure envelope
4) Elastic Plastic Strain hardening and softening models		
5) Elastic Hyperbolic softening model		

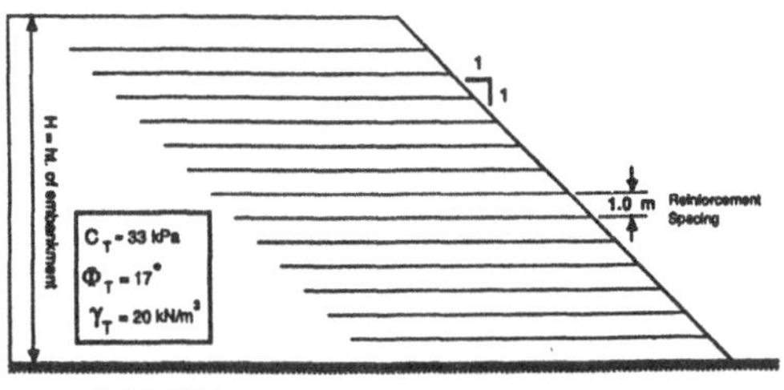

Figure 1 Reinforced Embankment Geometry

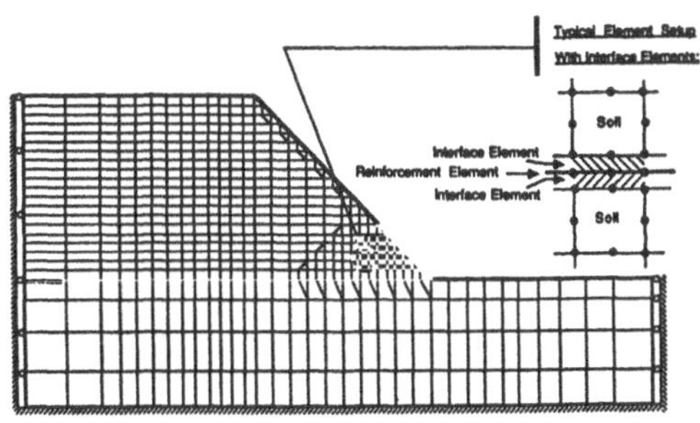

Figure 2 Finite Element Mesh of Reinforced Embankment

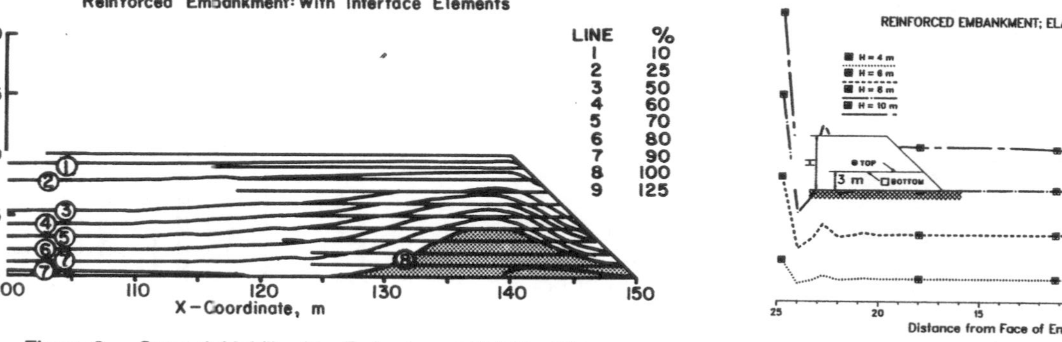

Figure 3 Strength Mobilized for Embankment Height = 10 m

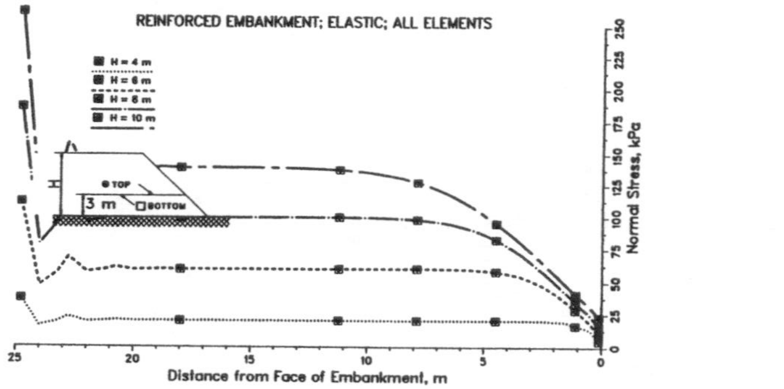

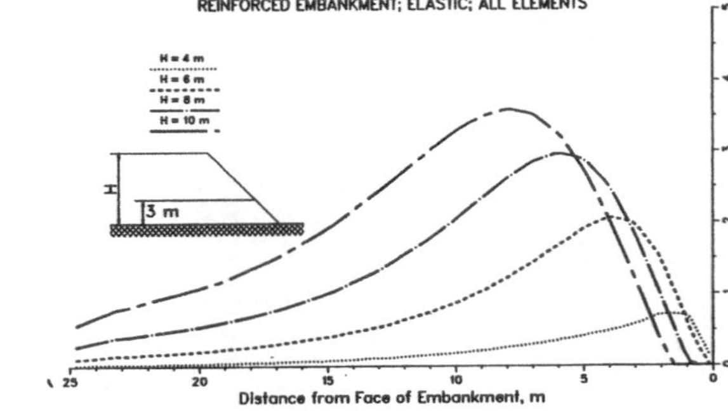

Figure 4 Interface Normal Stress Distribution

Figure 5 Interface Shear Stress Distribution

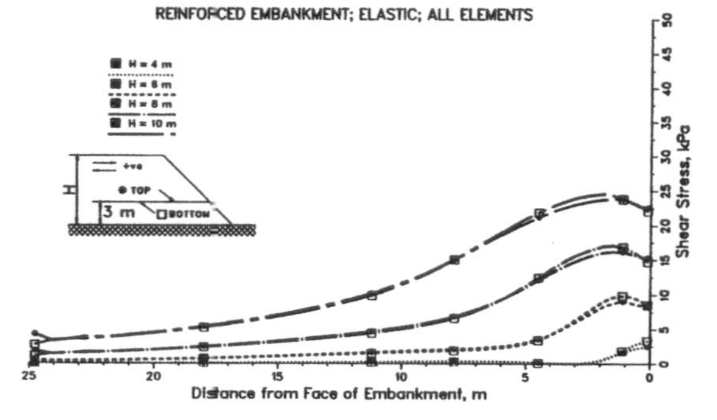

Figure 6 Load Development in Reinforcement

REINFORCING ELEMENTS IN STEEP SLOPES AND VERTICAL-FACED EARTH STRUCTURES - German State of the Art

R. FLOSS

Technical University Munich

1. GENERAL DEFORMATION AND FAILURE MECHANISM

The following basic variants are considered:

(1) Earth retaining structures with wall elements as face lining and structurally connected reinforcement similar to the "Terre Armée" construction method

(2) Same as (1), with gabion facing

(3) Fill with reinforcing mats (fabric, vleece with tensile strength), which are folded back at the face so that the soil cannot slip out; constructed (a) as so-called bolster dam (Fig. 1) or (b) as safety measure for extremely steep slopes.

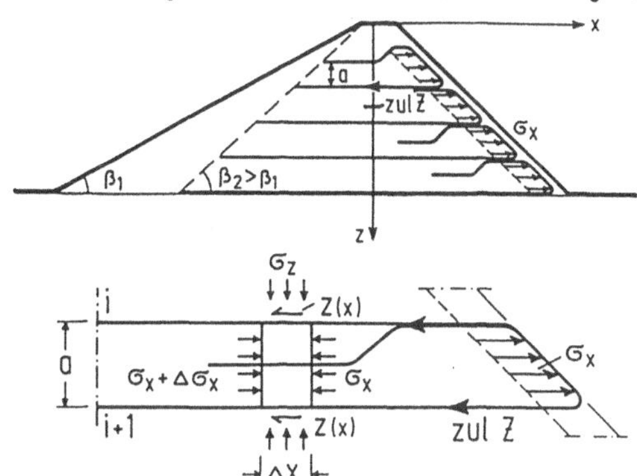

system of equations for limit state $(\Delta\tau = 0)$

(1) soil: $\sigma_1 = \sigma_z = \gamma \cdot z$

$$\sigma_3 = \sigma_x + \Delta\sigma_x = \frac{1}{\lambda_\varphi} \cdot \sigma_1 \qquad \frac{1}{\lambda_\varphi} = \frac{1 - \sin\varphi}{1 + \sin\varphi}$$

(2) reinforcement: $\Delta\sigma_x = 2 \cdot Z(x) \cdot \Delta x / a$

$$a = 2 \cdot \text{zul}\, Z \cdot \Delta x / \Delta\sigma_x$$

(3) soil-reinforcement-interaction:

$$Z(x) = \text{zul}\, Z = \sigma_z \cdot \tan\varphi_R$$

FIGURE 1 . Interactive forces for the limit state of a steep fill slope constructed according to the principle of the so-called bolster dam

P. M. Jarrett and A. McGown (eds.), The Application of Polymeric Reinforcement in Soil Retaining Structures, 561–567.

As a rule, non-cohesive soils which are not prone to deformation by creep and can easily be drained are used for the construction of earth retaining structures. For fills, however, one can also use cohesive material.

For want of a reliable empirical and measuring data pertaining to the failure and deformation behaviour of self-supporting earth bodies, whose high bearing capacity is achieved by soil reinforcement interaction, a number of questions remain unanswered. Besides, the applied construction methods may differ substantially and other conditions may have some influence, too. Previous observations and measurements of embankments constructed as bolster dam indicate no disproportionate displacements, creep or other damages to fabric or vleece in the face-area, providing there is a high degree of compaction of the non-cohesive soil. The bearing capacity is partially much higher than the conventionally determined ultimate loads. In view of these findings, it can be taken for granted that the plane-line reinforcing effect and the full plane activated shear stresses in the non-cohesive soil can produce a somewhat more effective reduction of the earth pressure in the case of the bolster dam than in cases where the system is reinforced with bands.

According to general soil-mechanical knowledge, such composite soil systems act as a slack body with internal shear, with the reinforcement layers creating the effect of an anisotropic cohesion. There is a certain analogy to the principle of the cofferdam which also acts as a slack body, where the external forces are transmitted to the base by means of shear forces between face and fill.

Compared to the monolithic, rigid body the conventionally reinforced composite body is noted for its totally different settlement, earth pressure and failure behaviour (Fig. 2):

(a) Because of its relatively low natural stiffness the reinforced composite body reacts less sensitive to differential settlement and horizontal deformation. From the cross-section it can be noted that the settlements occurring under the composite body are trough-shaped, i.e. the maximum settlements occur under the body rather than under the face-line if uniform subsoil conditions prevail.
(b) The effective, oblique eccentric force acting in the base and resulting from the dead weight of the composite body and the earth pressure, causes a bearing pressure σ_0 which - compared to the monolithic rigid body - lessens towards the outside.
(c) Contrary to the monolithic body, the earth pressure does not act at the backside. Instead, the forces resulting from earth pressure are reduced within the composite body, through shear stresses and arching effects in between the reinforcing elements so that only a residual pressure reaches the face area. This was confirmed by measuring results from

tests on supporting systems of the "Terre Armée" construction method. The borderline of the zone of active earth pressure follows the geometry of the maximum tensile force of the reinforcement; this is a curved line which embraces a much smaller zone than that of the plane sliding body according to the earth pressure theory by Coulomb.

(d) The most important factors of the stability analysis are the safety against base failure, sliding of the composite body, the body cannot tilt but would come apart in the case of an effective moment of tilt. Mathematical analyses show that the composite body by itself (without vertical load) has sufficient inner safety against failure if the stress redistribution caused by tensile forces of the reinforcement is considered and specific geometric design criteria are observed. Under these conditions, and presuming that the safety against base failure is observed, it would suffice to investigate only potential shear zones outside the reinforced body to verify the safety against slope failure. However, if b is much larger than h the slack behaviour of such a broad body would also lead to internal states of failure which start locally. Such cases require special investigation.

As mentioned earlier, the known investigations show that the bearing capacity of the composite body can be well above the conventionally calculated failure load. The following phenomena may add to this behaviour:

1. The reinforcing layers cause global and local prestressing effects in the non-cohesive soil.
2. The construction of any new reinforcing element changes the resultant main stress direction and thus the geometry of the potential failure figure.
3. Because of the restraint between reinforcing layers the soil is compacted to a higher degree.
4. The horizontal load transmission by reinforcing layers and the compaction effects lead to vault-type force transmission links in the non-cohesive soil and zones of high shear strength along the reinforcing layers; thus additional horizontal stresses σ_x induce retaining shear forces.

The influence of the cited effects is likely to be particularly high in cases of geotextile reinforcing elements because of their planelike friction and bonding effects. Besides, there is no way of relief for the soil, because of the all-round restraint. The vault-type load transmission structure within the soil strengthens layer by layer as the construction processes. Within certain areas they even relieve the reinforcement. In areas where the limit state is reached locally, these load transmission links are changed as a result of the newly induced σ_x-stresses as parts of the load are transmitted to the reinforcement, so that the plasticizing process is delayed. However, this process implies that the σ_z-stresses are not excessive.

The above-mentioned phenomena need to be verified by future in-situ investigations and models. They are still mere hypotheses, but observations indicate that they are close to reality. Considering the relevant technical and safety demands, there is little doubt that it will be possible to construct a composite body of uniform and high density.

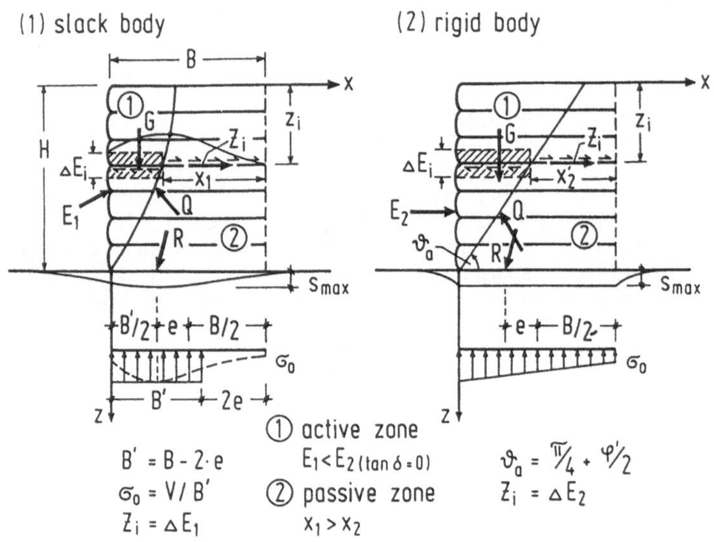

FIGURE 2 Comparison of models: (1) slack body and (2) quasi-monolithic earth body with regard to their settlements, base pressure σ_0, earth pressure E, and tensile force of reinforcement Z_i.

2. CONSERVATIVE ANALYTICAL MODEL

Safety aspects and the lack of reliable empirical data are the reasons why the effects described in Section 1 have not yet been reflected in the stability analysis. As a substitute one resorts to analytical models which are based on a quasi-monolithic composite body (Fig. 2). For the typical case, these models abide by the following design principles, which are on the safe side: (a) application of the earth pressure, according to Coulomb, to the backside of the composite body, but (in contradiction to the monolithic precondition) with an evenly limited active wedge of failure under $\vartheta_a = \pi/4 + \phi/2$ within the composite body; (b) induction of the retaining tensile forces of the reinforcement behind the shear zone; (c) trapezoidally distributed base pressure analogously to the rigid body which is subjected to effective, oblique eccentric load; (d) sustenance of the horizontal earth pressure components in the base of the composite body; (e) in the case of face elements, design of the elements with full earth pressure application.

The conventional stability analysis includes proof of the external and internal stability of the composite body.

(1) Depending on whether a steep slope or supporting earth structure is being considered, the safety of the external stability has to be proven by:

(a) Safety against sliding in the base of the composite body according to DIN 1054 and for supporting structures with stiff reinforcing elements (e.g. geo-grids), also in each reinforcement joint. Safety coefficient η_g > 1.2 ... 1.5 depending on the loading case.

(b) Limited eccentricity of the load inb the foundation joint.

(c) Safety against base failure according to DIN 4017. Safety coefficient η_p > 1.3 ... 2.0.

(d) Safety against slope failure according to DIN 4084; safety coefficient η > 1.1 ... 1.4. Depending on loading case and analytical method, also for parts of the reinforced earth body (especially for systems with vertical and traffic loads).

The investigations must also include the construction conditions. Although check-ups with a lower safety coefficient still yield a statically sufficient safety, the possibility of conspicuous deformation occurring in the system cannot be ruled out. Shear and slope failure verified according to DIN 4084 only apply to plane deformation states; applied to spatial cases they yield results which are on the safe side. During construction of the composite body higher earth pressures may occur as a result of the layer-by-layer compaction of the soil; they have to be considered in the calculation.

(2) The safety of the inner stability has to be proven by:

(a) Safety against tensile failure of the reinforcement (η > permissible Z/Z_i).

(b) Safety against pull-out of each individual reinforcement layer, with the retaining forces over the embedded length of the reinforcement behind the sliding wedge and with a reduced coefficient of friction for the interactive forces between reinforcement and soil.

(c) Safety against sliding in horizontal reinforcement levels within the composite block with reduced coefficients of friction according to soil and reinforcement type.

(d) Sustenance of tensile forces at transition points between reinforcement and outer lining (Z_i > 0.85 max Z_i).

In the working state it is essential to define the permissible tensile force (permissible Z) for the respective reinforcement. This working-tensile-force depends upon the service life and elasticity of the reinforcement and its long-term behaviour (ageing, temperature-dependent reductions etc.). Although the experience with geotextile reinforcement is still rather limited, the following guide values may be used in practice: They depend on the planned service life of the structure:

(a) Permanent reinforcement: sustaining tensile forces is 20 % to 25 % of the tensile force at 5 % permissible strain;

(b) Temporary reinforcement: sustaining tensile force is 30 % of tensile force at 10 % permissible strain.

3. MODIFIED STABILITY MODELS

Although there is a number of modified models, it is only possible to summarize their principles in this paper. Contrary to the conventional approach, however, they are all aimed at describing a bearing condition which compares more favourably to the natural state.

(1) Modification of the geometry of the active failure zone: (a) assuming a broken limiting line for the active sliding body for better adaption to the true course. The most unfavourable shear zone is found by variation of both angles of sliding surface. (b) Assuming a sliding body geometry, which approximately follows the geometry of the maximum tensile forces (e.g. Berg 1986, John 1986). (c) Using a block element model with application of tensile force at the edges of the elements (Fig. 3). Not only does this method consider the tangential component of the tensile force T_1 as retaining force but - with component T_2 - it also considers the retaining shear forces resulting from the normal stress portion which is increased by reinforcement in the failure zones.

(2) Assuming reduced earth pressure in face area, giving consideration to the reduction of earth forces within the reinforced slack earth body.

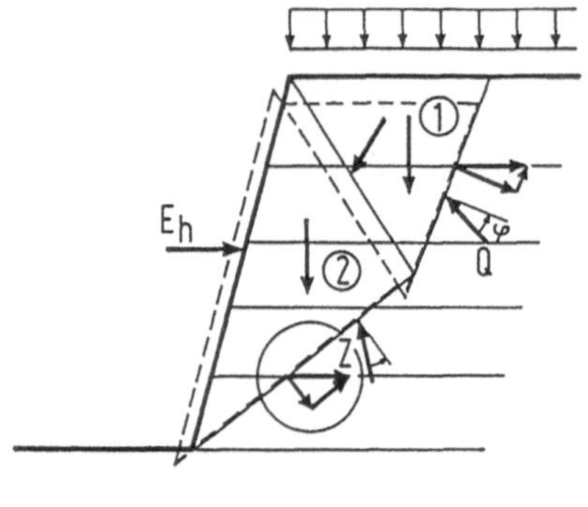

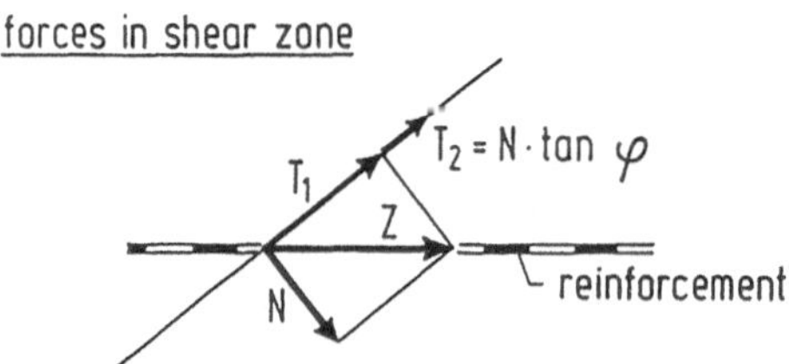

FIGURE 3 Block model for steep slope with two elements, and showing approach of tensile force distribution in the failure zone

(3) Application of higher shear parameters ϕ' and c' for the determination of the bearing capacity of the composite structure (e.g. Ingold/Miller 1982).

(4) Modification of failure mechanisms based on the described vault-theory. Instead of assuming a uniformly curved sliding zone for the entire composite body it is implied that the system fails layer by layer (Werner/Resl 1986).

REFERENCES

1. BERG, R.R., BONAPARTE, R. et al (1986): Design, Construction and Performance of Two Geogrid Reinforced Soil Retaining Walls. 3rd Int. Conf. on Geotextiles, Vienna, Vol. II, 401-406

2. INGOLD, T.S., MILLER, K.S. (1982): Analytical and Laboratory Investigations of Reinforced Clay. 2nd Int. Conf. on Geotextiles, Las Vegas, Vol. III, 587-592

3. JOHN, N.W.M. (1986): Geotextile Reinforced Soil Walls in a Tidal Environment. 3rd Int. Conf. on Geotextiles, Vienna, Vol. II, 331-336

4. WERNER, G., RESL, S. (1986): Stabilitätsmechanismen in geotextilverstärkten Erdstützkonstruktionen. 3rd Int. Conf. on Geotextiles, Vienna, Vol. II, 465-469

ANALYTICAL TECHNIQUES AND DESIGN METHODS - DISCUSSION

DR. G.W.E. MILLIGAN,

University of Oxford, England.

In considering the possible development of horizontal strain in a reinforced soil wall, it should be noted that in certain cases the soil strains may be simply related to the movement of the wall facing. Bransby and Milligan (1975) and Milligan (1983) have shown that the strains behind a retaining wall are related to the movements of the wall as shown in Figures 1 and 2. Comparisons with the results of model and full scale tests have confirmed that the analysis works well for outward translation and rotation of the wall about its base. For outward bulging, or rotation of the wall about its top, the simple analysis breaks down and the strain field becomes more complex.

A reinforced soil wall with a full height facing normally deforms by rotation of the facing about its base. The maximum strains likely to be developed in the soil may then be given approximately by the simple analysis. The presence of reinforcement within the soil will modify the strain pattern somewhat and reduce the maximum strains.

Two examples are considered here: -

(i) In the full height panel wall at RMC, the outward rotation of the facing panel at the end of the test was about 4.3×10^{-3} radians. The peak angle of dilation for the fill was about 20° or a little less (the calculation is rather insensitive to the value of ψ). The resulting tensile horizontal strain at all points in the active zone is then calculated to be about 0.6%. This compares with measured tensile strains in the reinforcement in the active zone of about 0.4% to 0.6%.

(ii) For a more typical full scale wall, the angle of dilation for the fill might be 10°. If the allowable outward rotation of the wall facing is 1%, which would normally be considered a fairly large movement, the horizontal strain in the active zone is 1.2%. The reinforcement strain cannot normally exceed this value, and this would then provide a strain limit criterion for the wall. If it is desired to allow the reinforcement to develop larger strains, either

P. M. Jarrett and A. McGown (eds.), The Application of Polymeric Reinforcement in Soil Retaining Structures, 569–570.

larger facing movements must be tolerated or additional soil strains must be induced by the presence of a compressive layer behind the facing or other similar construction expedient.

References

Bransby, P.L. and Milligan, G.W.E. (1975). Soil deformation near cantilever sheetpile walls. Geotechnique 25. N°2.

Milligan, G.W.E. (1983). Soil deformations near anchored sheet pile walls. Geotechnique 33. N°1.

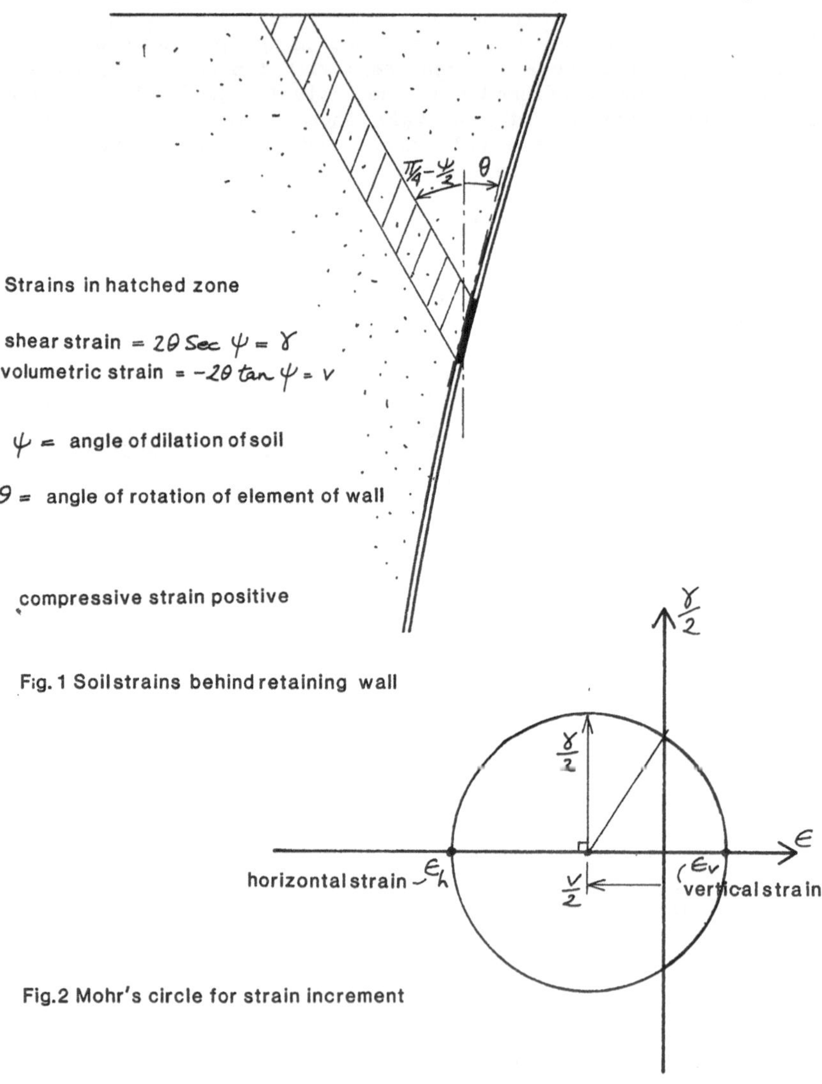

Fig. 1 Soil strains behind retaining wall

Fig.2 Mohr's circle for strain increment

Construction Methods and Economics

CONSTRUCTION METHODS, ECONOMICS AND SPECIFICATIONS

C.J.F.P. Jones, Professor, University of Newcastle upon Tyne
A. McGown, Professor, University of Strathclyde
D.J. Varney, Engineer, Netlon Ltd.

1. INTRODUCTION

The modern forms of reinforced soil have evolved from an understanding and introduction of effective construction techniques, the first of which was developed by Vidal (1966). The first structures used steel strip reinforcements and it was not until the early 1970's that polymeric reinforcements were used successfully with vertical faced structures. Prior to this, high density polythene grids had been used in the construction of railway embankments in Japan, Yamamoto (1966).

This paper considers the practical aspects of the construction of vertical or near vertical reinforced soil structures formed using polymeric reinforcements. Included are details relating to construction systems and techniques, economics and specifications. Although there are numerous other possible construction arrangements, those detailed here have all been used in practice. Where possible the potential advantages displayed by individual systems or materials are described.

2. CONSTRUCTION METHODS

Reinforced soil structures must be of a form in keeping with the assumed idealization and analysis, however, the theoretical form of the structure may be quite different from the economical prototype, and attention should be paid to the method of construction throughout the design process.

Speed of construction is usually essential to achieve economy and in part this may be achieved by the simplicity of the construction technique. Construction techniques compatible with the use of soil as a constructional material are required. The use of soil, deposited in layers to form the structure, results in deformations within the soil mass caused by gravitational and compaction forces. These deformations result in the reinforcing elements positioned on discrete planes moving together and being tensioned as the layers of soil separating the planes of reinforcement are compressed vertically and expand laterally. Construction techniques capable of accommodating this internal consolidation and straining of the fill are required.

Failure to accommodate the compression, particularly at the face, may result in loss of serviceability or even rupture of reinforcements or connections, whilst restriction of lateral expansion of the fill may prevent the tension in the reinforcements developing fully.

P. M. Jarrett and A. McGown (eds.), The Application of Polymeric Reinforcement in Soil Retaining Structures, 573–611.

2.1 BASIC TYPES

Three constructional techniques which can accommodate vertical settlements and limited lateral movements within the soil mass are shown in Fig 1. Except for some special cases, reinforced soil structures constructed above ground use one or other of these forms of construction or a combination of them.

2.1.1. Concertina Method. The constructional arrangement of the concertina method developed by Vidal (1966), is shown in Fig 1(a). Differential settlement and lateral movement within the soil mass (d1-d4) is achieved by the front or face of the structure concertinaring in a manner similar to the action of a set of bellows. Some of the largest modern reinforced soil structures have been built using this approach, and it is the form of construction frequently used with geotextiles and related reinforcing materials in both embankments and cuttings. A flexible hoop shaped steel facing unit has been used when the structure is reinforced with strip reinforcement, Fig 2(a), as have facing units formed from aluminium and glass reinforced plastic, although the latter have had limited success due to their lack of robustness.

Geotextiles and related materials often provide their own facing by employing wrap-around techniques, Fig 2(b),(c),(d) and (e). This is particularly useful for structures where distortion of the facing is acceptable.

2.1.2 Telescope Method. In the telescope method of construction developed by Vidal (1978), Fig 1(b), the deformations within the soil mass are accommodated by the facing panels closing up and moving forward an amount equivalent to the internal deformations. This is made possible by supporting the facing panels by the reinforcing elements and leaving a discrete horizontal gap between each facing panel, i.e. the facing panels hang from the reinforcing elements. The horizontal gap between each facing panel may be effected by the use of compressible gaskets, Fig 3. Failure to provide a large enough gap between facing elements can result in crushing and spalling of the units as the fill deforms under the action of gravity and compaction forces.

The closure between panels will vary from application to application depending upon the geometry of the structure, quality of fill material, size of the facing panels and the degree of compaction achieved during construction as well as the subsoil conditions. Typical movements on a steel reinforced structure, reported by Finley (1978), show vertical closures of 5-15 mm for facing panels 1.5 m high. Also the shape and form of the facing panel must be compatible with the procedure adopted, and reinforced concrete cruciform, tee-shaped, hexagonal or "Z" shaped panels covering 1 4 m2 and 150 250 mm thick are typical. The construction sequence for the telescope method is shown in Fig 4.

2.1.3 Sliding Method. In the sliding method of construction developed by Jones (1978), differential settlement and compaction within the fill forming a reinforced soil structure can be accommodated by permitting the reinforcing members to slide vertically relative to the facing. Slideable attachments can be provided by the use of grooves, slots, vertical poles, lugs or bolts. If vertical poles are used, these may form the structural elements of the facing and the facing may become non-structural providing only a

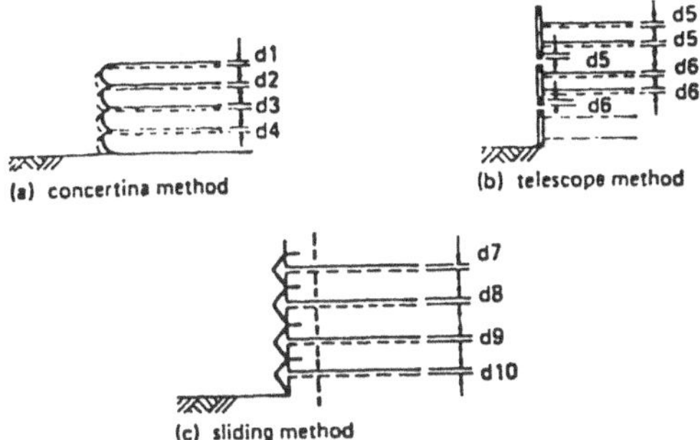

Figure 1 Methods used for constructing reinforced soil structures

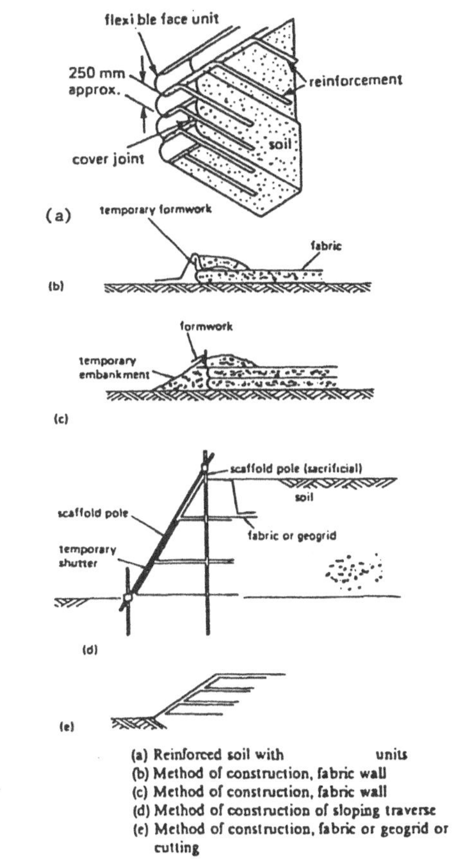

Figure 2 Concertina Method of Construction

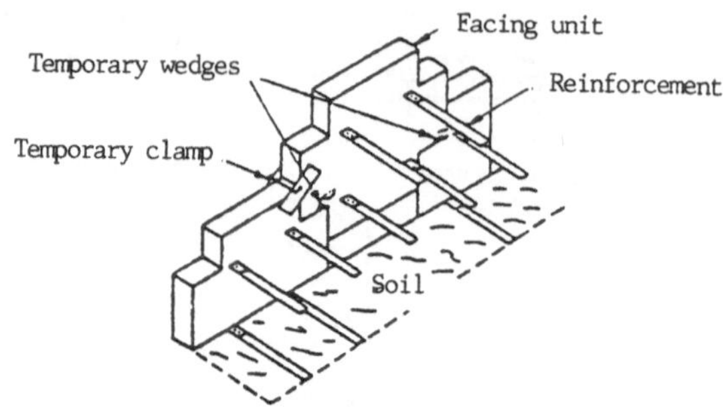

Figure 3 Telescope Method of Construction

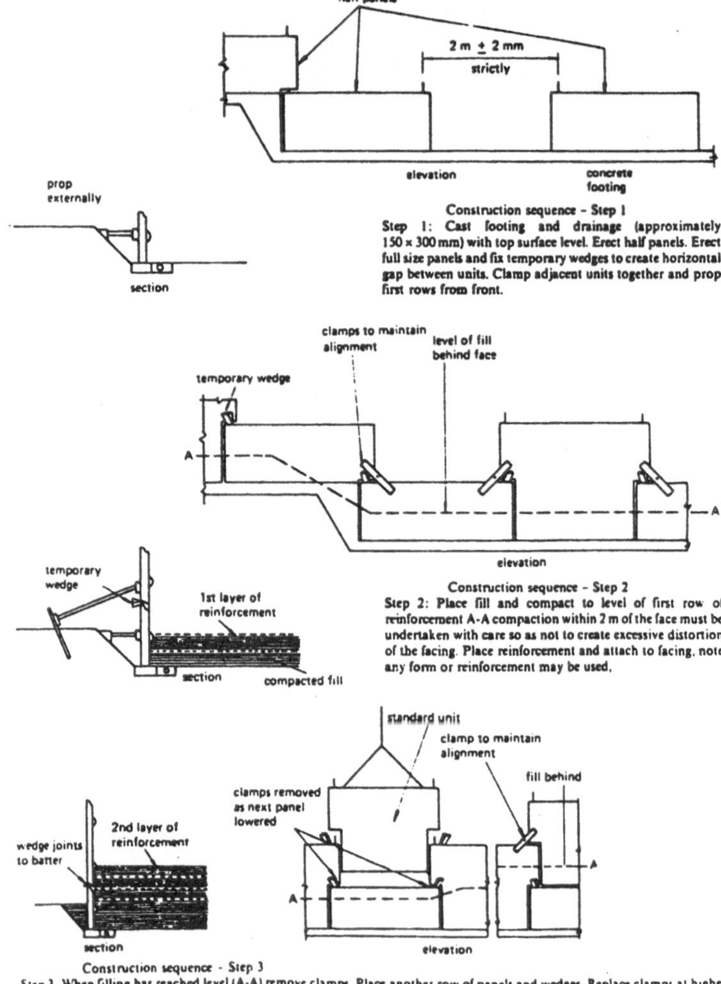

half panels
2 m ± 2 mm
strictly
elevation
concrete
footing
prop
externally
section
Construction sequence - Step 1
Step 1: Cast footing and drainage (approximately 150 × 300 mm) with top surface level. Erect half panels. Erect full size panels and fix temporary wedges to create horizontal gap between units. Clamp adjacent units together and prop first rows from front.
clamps to maintain
alignment
level of fill
behind face
temporary wedge
A
A
elevation
temporary
wedge
1st layer of
reinforcement
section
compacted fill
Construction sequence - Step 2
Step 2: Place fill and compact to level of first row of reinforcement A-A compaction within 2 m of the face must be undertaken with care so as not to create excessive distortion of the facing. Place reinforcement and attach to facing, note any form or reinforcement may be used,
standard unit
clamp to maintain
alignment
fill behind
clamps removed
as next panel
lowered
wedge joints
to batter
2nd layer of
reinforcement
A
A
section
elevation
Construction sequence - Step 3
Step 3 When filling has reached level (A-A) remove clamps Place another row of panels and wedges Replace clamps at higher level and continue cycle. As erection proceeds, remove temporary wedges to permit vertical settlement of soil mass and facing
Note When extensible reinforcement is used, a degree of prestressing is required prior to placing fill

Figure 4 Construction sequence for the Telescope Method

covering, whose role is to protect the completed structure from the elements, prevent erosion or act as an aesthetic feature. With this arrangement the type and form of the non-structural part of the facing can be chosen to suit the particular application or environment, Fig 5. The erection sequence for a non-structural facing is shown in Fig 6.

If a structural facing is used, the connecting element, the vertical pole, may be reduced in size to an appropriate form, such as a bolt; alternatively, attachment of the reinforcement to the facing can be through slotted holes. The facing may be rigid or semi rigid. Up to a height of approximately 10 m a rigid facing may be used; for heights above 10 m an elemental form of facing is appropriate. Where a full height rigid facing is used, this is erected and held in place before filling starts. The erection sequence is shown in Fig 7.

2.1.4. Hybrid Methods. Recently hybrid structures, using a combination of the Telescope and Sliding methods of construction have been used successfully with geogrid reinforcement, Fig 8. In this system the reinforcement, which passes through the facing, is attached by means of a slideable connection to king posts, whilst the facing operates in a telescope arrangement. The primary advantage is of economy and speed of construction. The system has been used to provide masonry faced structures of high aesthetic appeal. When architectural features are omitted this form of structure is particularly suited to industrial situations. In the case of unfaced structures the reinforcement is connected to the rear flanges of the king posts so as to avoid the risk of vandalism and degredation of the reinforcement.

2.1.5 Specials. In some special conditions there is no need to make allowances for consolidation of the fill, as the degree of movement involved is limited by the design. This can occur when the foundation beneath the structure is very competent (eg. rock) and when the backfill material used is of very high quality and well compacted. In these special circumstances the differential movements (d1-d4) in Fig 1, are very small and may be safely accommodated within the slackness of the connections and the extensibility inherent in polymeric reinforcement. The use of stiff reinforcement and rigid connections is not recommended with this form of construction.

2.2 SYSTEMS IN USE

2.2.1 Reinforced Fill Behind Conventional Walls. The benefit of reinforcing the backfill of conventional structures has been demonstrated by Saran et al (1979). In this technique the active pressure on the structure is reduced. However, the structure is reliant upon the facing to prevent erosion and for stability of the fill immediately adjacent to the wall.

2.2.2 Independently Reinforced Soil Mass and Facing Panels. An advance on the use of a reinforced backfill with conventional structures can be achieved by substituting an incremental panel to act as the facing. In a recent structure built at Lithonia, Georgia, small precast concrete cruxiform units into which were embedded two 1.2m long geogrid "tabs" were used as the facing. The "tabs" were used to secure the facings to the independently reinforced soil mass, and the main geogrid elements were simply laid horizontally in the fill without any connections to the facings.

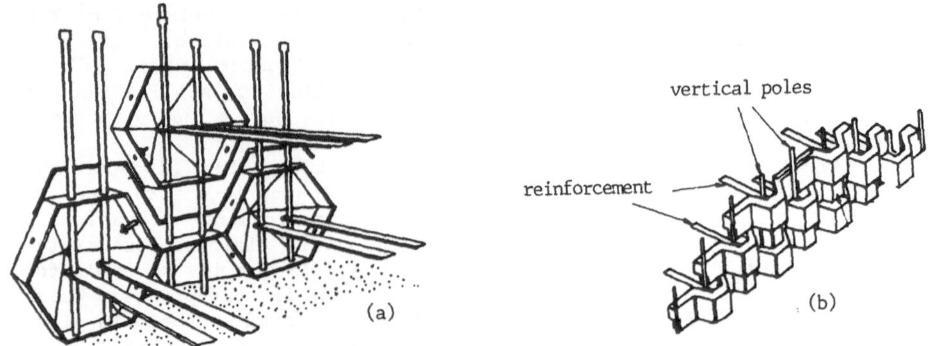

Figure 5 Sliding Method of Construction

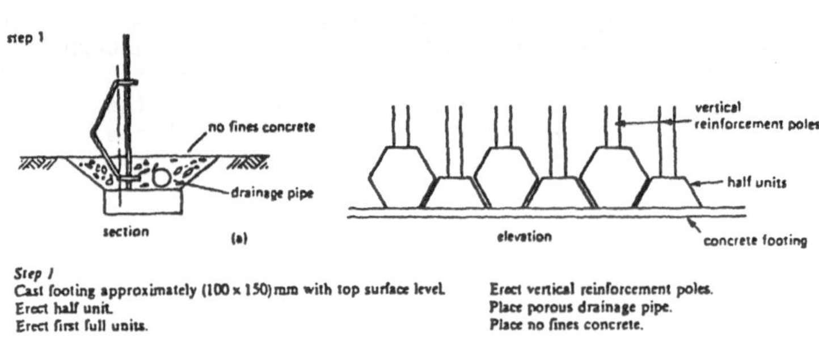

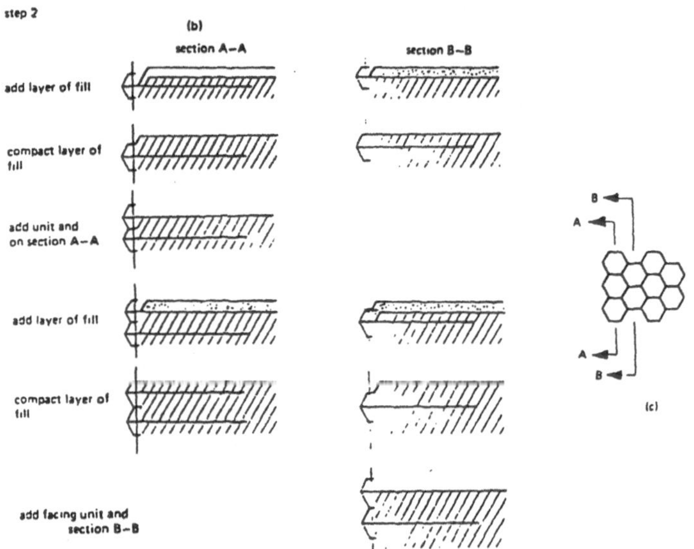

Step 2

Note: Reinforcement position is at mid-height of facing unit
Any form of reinforcement may be used.
Compaction of the fill close to the face is restricted to small plant so as not to distort the facing
Speed of construction is dependent upon the speed of placing the fill
A bold facing is usually used to disguise any inconsistencies or distortions caused during construction
Construction can be stopped at any level or position without any fear for the safety of the workers
No proping is used

Figure 6 Construction Sequence of Sliding Method (non-structural facing)

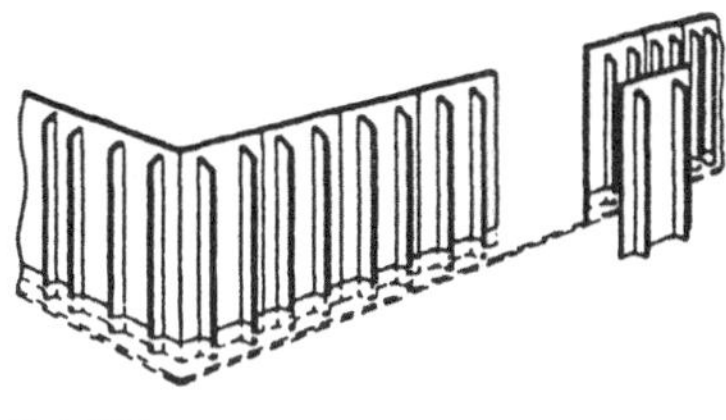

(a) exploded view

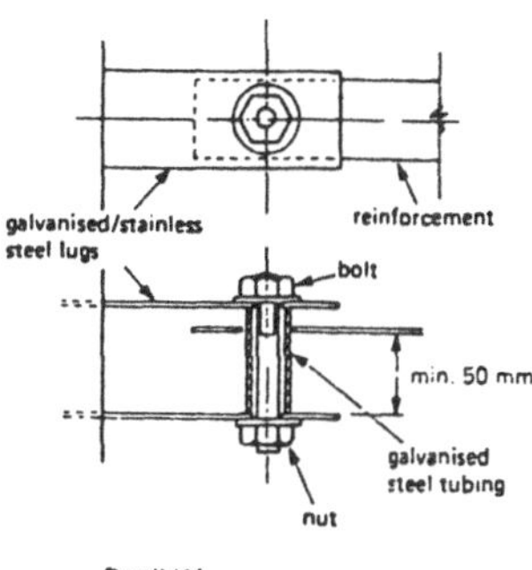

Detail 'A'

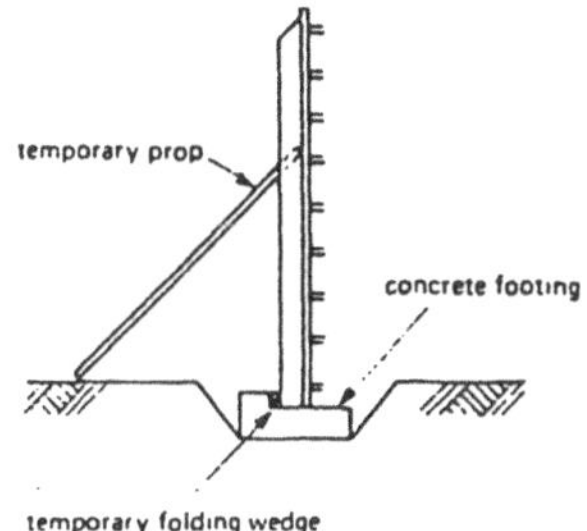

(b) stage 1

Stage 1:
Cast footing (approx 150 × 300 mm) plus upstand. Erect and prop facing panels

Stage 2:
Construct drainage.
Place fill layers and compact.
When fill level with top of first pair of connecting lugs attach first level of reinforcement.
Continue filling.
When filling is complete, or when sufficient fill has been placed to stabilise facing remove props and folding wedges.

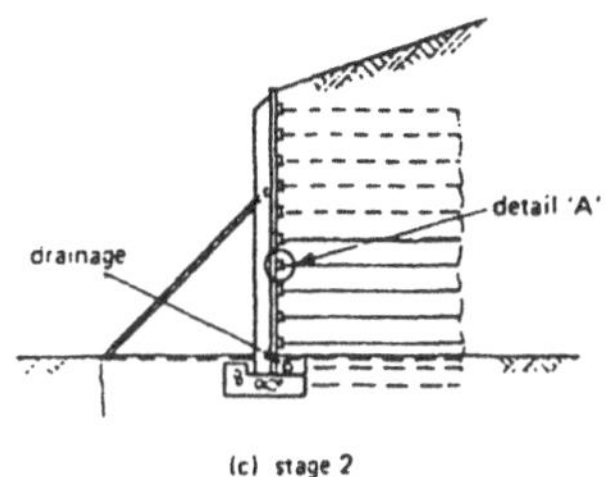

(c) stage 2

Sliding method of construction - rigid facing

Figure 7 Sliding Method of Construction - rigid facing

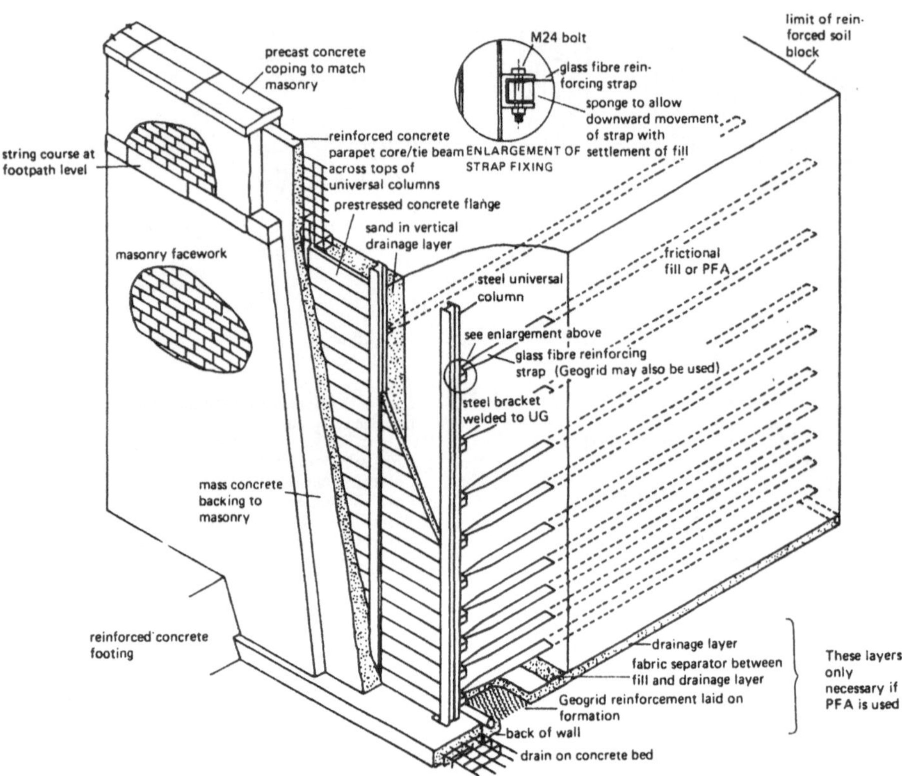

precast concrete coping to match masonry
M24 bolt
glass fibre reinforcing strap
sponge to allow downward movement of strap with settlement of fill
ENLARGEMENT OF STRAP FIXING
limit of reinforced soil block
string course at footpath level
reinforced concrete parapet core/tie beam across tops of universal columns
prestressed concrete flange
sand in vertical drainage layer
masonry facework
frictional fill or PFA
steel universal column
see enlargement above
glass fibre reinforcing strap (Geogrid may also be used)
steel bracket welded to UG
mass concrete backing to masonry
reinforced concrete footing
drainage layer
fabric separator between fill and drainage layer
Geogrid reinforcement laid on formation
back of wall
drain on concrete bed
These layers only necessary if PFA is used

Figure 8 Hybrid Method of Construction - Telescope and Sliding

This structure has a maximum height of 6.1 metres and forms part of a material handling platform supporting 450kN dumptrucks. A cross section through the wall is shown in Fig.9. The design and construction of the wall has been described by Berg et al (1986).

2.2.3 Incremental Construction with Discrete Panels.
The most common form of reinforced soil structure is the telescope method using incremental panels, Fig. 3. A number of proprietary systems are based upon this technique, including the familiar Reinforced Earth Company's system which uses a reinforced concrete cruciform facing unit and the Georgia Department of Transportation GASE method, which uses steel grid reinforcement. The method has been used with polymeric reinforcement in the United Kingdom, the Middle East, and recently in Japan, using Tee shaped and 'Z' shaped concrete facing panels.

The Soil Structures International Limited 'Websol' Fig.4 A number of the most sucessful proprietary methods which uses polymeric reinforcement. This is based on a precast concrete facing and tape reinforcement, Paraweb, formed from ten lanes of high tenacity polyester fibres encased in a sheet of low density polythene. The Paraweb reinforcement, which is connected to the facing panels by toggles, is laid using an entire roll at a time, typically 100m. long. This is acheived by lacing the reinforcement between the facing and a restraining member laid in the fill, Figure 10. In some structures layers of 'Terram' geotextile are laid on top of the reinforcement, Kempton et al (1985). Recently, an anchored earth system using paraweb has been introduced in Austria. This system is now being used in the United Kingdom, Lazlo et al (1995).

Geogrid reinforcement has also been used with the incremental construction system. Pigg and McCafferty (1985) have described an application in the UK, whilst Berg et al (1986) have detailed the use of geogrid reinforcement with the Georgia GASE system. Recently geogrid has been used with the Japanese 'Z' panel, Yamanouchi (1986).

2.2.4 Full Height Panels. The use of full height panels is often an attractive form of construction for walls of limited height (<10 metres). However, since the possibility of articulation by either the concertina or telescopic methods is lost, consideration must be given to providing sliding connections, Figure 7.

In the United Kingdom a number of agencies have successfully used full height panels for highway and industrial projects. Consolidation of the reinforced fill has been accommodated by use of sliding connections between the reinforcement and facing panel. Three polymeric reinforcing materials have been used namely, "Fibretain" glass reinforced plastic stips, "Tensar" geogrids and "Paraweb" tape. The system of full height panels was also used for the construction of the first Anchored Earth structure constructed on the Otley by Pass in Northern England (Jones et al, 1985).

In the United States some full height panel wallsconstructed with rigidly connected steel strip reinforcement have exhibited severe and unacceptable distortion of the facings on account of there being no provision for relative movement between the reinforcement and panels as the fill settles and deforms laterally. Crushing of steel facing panels has also been observed in Japan on Anchored Earth structures when no provision has been made for movement of the reinforcing anchors relative to the facing. Observations on full height

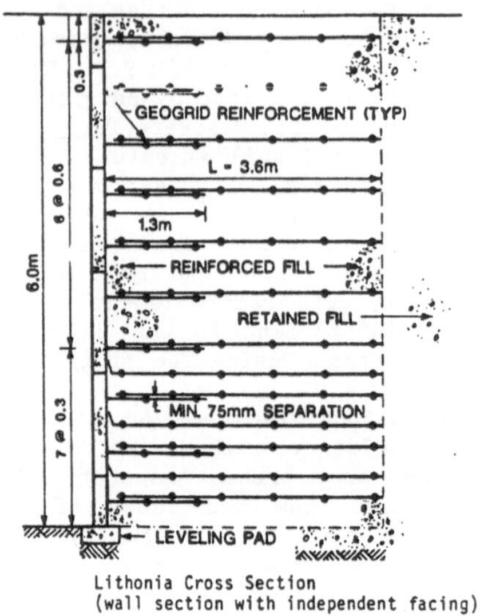

Lithonia Cross Section
(wall section with independent facing)

Figure 9

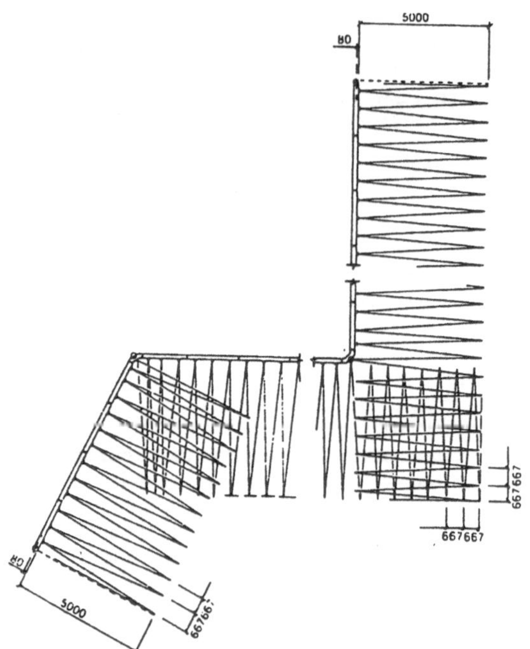

Figure 10

panel structures in the United Kingdom have shown that the reinforcement regularly settles 20-50mm with a sliding type construction (Jones et al 1987). Analysis of this condition shows that as resistance to this movement develops, so significant stresses will be imposed on the facing panels.

One of the common construction techniques using geogrids consists of full height facing panels into which short "tails" of reinforcement are cast. The "tails" are later joined to the main reinforcement as shown in Fig. 11. Whilst the relatively low bending stiffness of geogrids offers more flexibility to accommodate settlement than does steel strip reinforcement, the system requires the use of good quality fill which can be well compacted to reduce settlement and ensure good performance in service. The system has been used successfully on projects in the United Kingdom and in the United States at Tuscon, Berg et al (1986).

The 'hybrid' system currently used in the United Kingdom is a form of the full height system which can be used to provide structures of high aesthetic quality, Figure 8. As described in paragraph 2.1 (iv) the system is a combination of the telescope and sliding methods of construction. Structures using glass fibre, polymeric tape and polymeric grid have been constructed, Jones et al (1987).

2.2.5 Tronderblock System. Tronderblock precast concrete facing blocks have been used in Norway for the construction of low height gravity retaining walls for some years, Figure 12. The blocks are 500mm long x 250mm high x 600mm deep and require a small crane to lift them into place. Once placed they are stable and self supporting and provide a suitable facing formwork against which fill can be placed and compacted. Recently the versatility of the system has been improved by introducing horizontal layers of "Tensar" geogrid reinforcement between the rows of blocks which are connected to the facings. The result is a hybrid structure, part gravity wall, part reinforced soil, details of which are shown in Figure 13. The use of reinforcement extends the height to which the system can be used and structures of 5.5m in height have been completed at Karmay in Norway. A current development is the introduction of smaller facing blocks which can be handled without the need for cranage.

A separate development in the United Kingdom is the Porcupine retaining retaining wall block in which walls exceeding 3m in height are reinforced with the geotextile "Stabilenka". In poor ground conditions the use of a 'Nicobag' formation is advised.

2.2.6 Wrap-around Facings. Many permanent structures have been built which utilise geotextile sheets, grids, nets and mesh reinforcements as the facings. These may be termed wrap-around walls which conform to the concertina method of construction. Wrap-around walls are constructed simply by folding an extended reinforcement element through 180 degrees to form the face and anchoring it back into the fill or to another element at a higher elevation. Fill is usually placed and compacted against external, temporary formwork, Figs. 14 and 15. The reinforcing elements of wrap-around structures are usually pretensioned to limit post construction distortions.

Wrap-around facings are a very versatile method of construction and accordingly a wide range of temporary facing support methods have emerged, particularly on the initiative of Contractors, who stand to gain from reduced costs derived from the use of innovative site

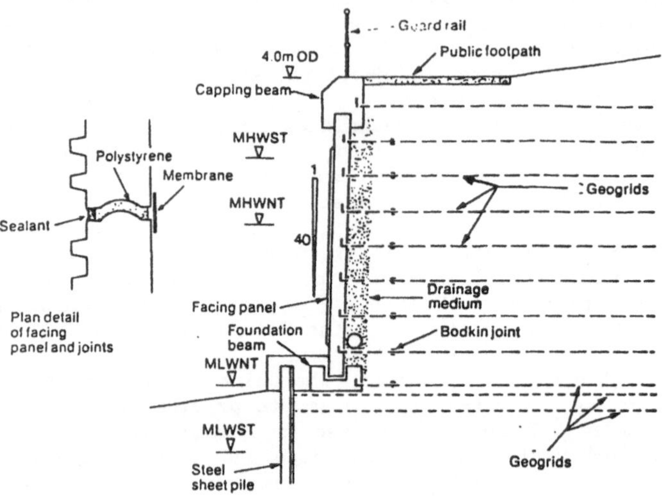

Figure 11 Full Height Panel Construction

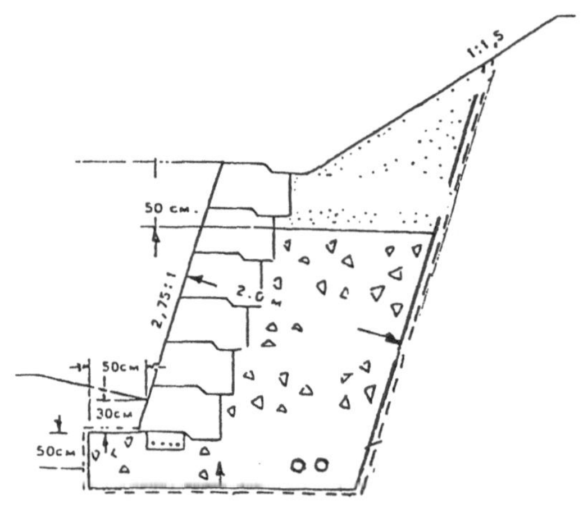

Figure 12

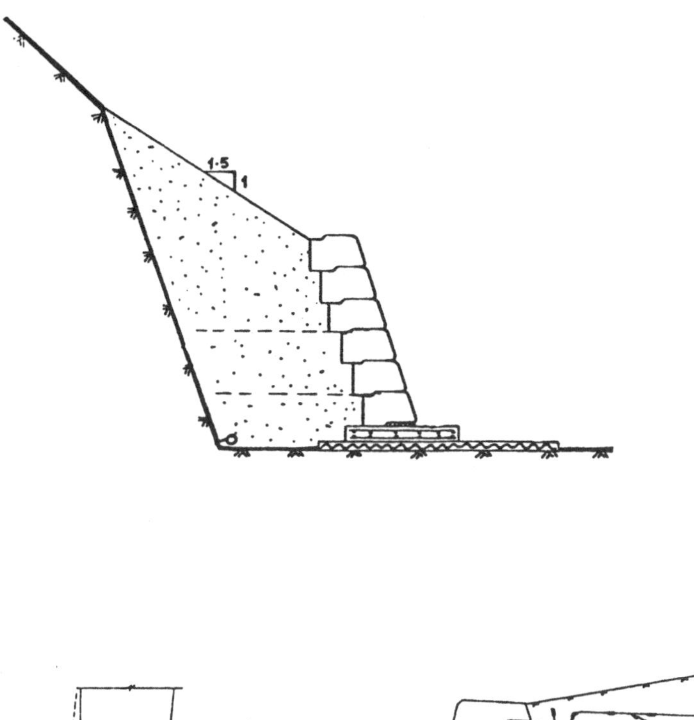

Figure 13

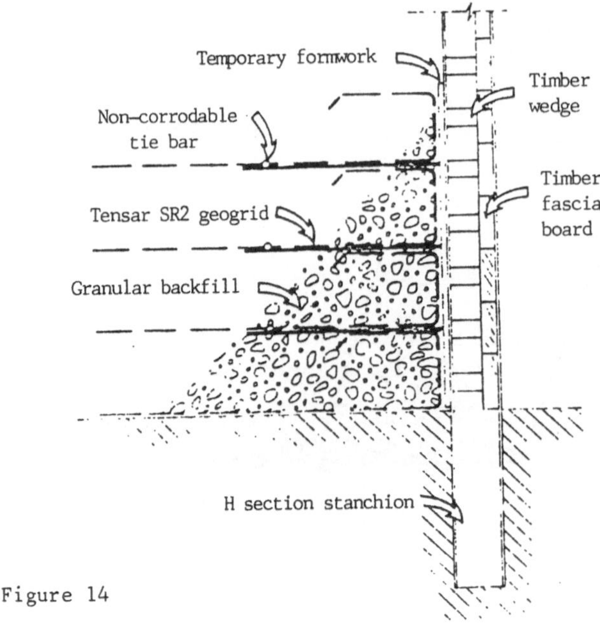

Figure 14

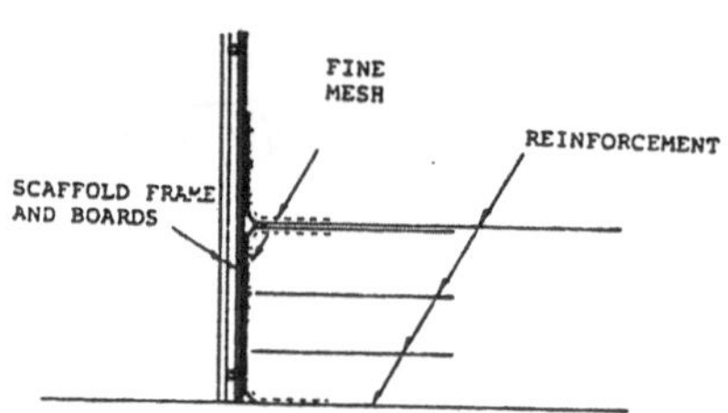

Figure 15

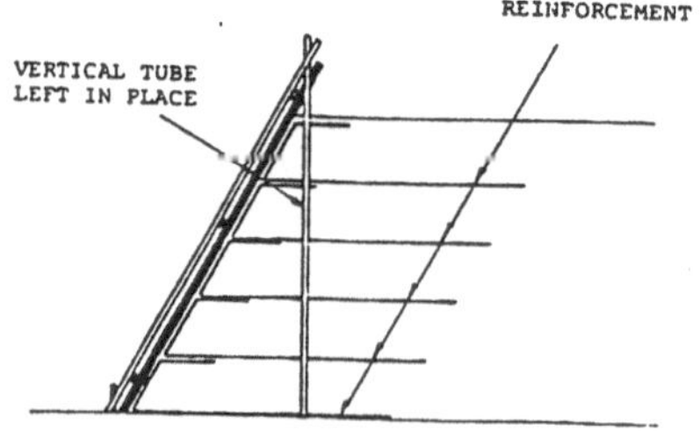

Figure 16

specific techniques. Lightweight formwork, comprising a grid of scaffold tubes and boards has been used for the construction of steep and vertical walls. Fig. 16 shows the arrangement used for the temporary works of a missile barrier in the United Kingdom. Sacrificial tubes have also been used to support formwork but these should be avoided in favour of external bracing, as this reduces cost and eliminates the difficulties of filling and compaction around obstructions.

Climbing or sliding formwork has been successfully used both on steep and on stepped faces, Figs. 17 and 18. These methods assume the structure will be self supporting as each stage is completed.

In order to avoid surface erosion the facing must be capable of fill retention once the support is removed. Where open grids, nets or meshes are used, a turf or seeded mat system with a topsoil lining is recommended, Fig. 19. This provides a root system which binds the surface of the fill together; after a relatively short period of time, the polymeric facing becomes hidden by a mat of grass which offers protection against UV degredation and damage by fire. Where aesthetics do not command priority (eg. temporary structures) retention of fill may be achieved using a geotextile, natural fibre liner, or coarser fill.

Where geotextile wrap-around structures are specified, account has to be taken of the facial distortion which may develop. Most permanent structures are constructed with an outer protective facing, and the most common solution recommended by geotextile manufacturers is to provide a 40-75mm thick sprayed concrete facing.

This solution also overcomes the problems of vandalism and durability, but is not particularly aesthetic and may be expensive. An example of this method of construction is shown in Fig. 20 which is a cross section of a 5.2m high geotextile reinforced wall constructed by the New York Department of Transport. The structure was finished by applying a 40mm thick sprayed concrete external facing, Douglas (1981).

Whilst wrap around walls are suitable for permanent works, their use as a construction expedient is equally common, particularly on account of the high rates of construction which have been achieved. Construction times have been reduced by between 30 and 60 per cent using this technique.

The creation of a neat edge to a wrap around structure can pose problems. One solution is to use a series of sand or grout bags to facilitate shaping of the edge. The bags are built up on the inside of the facing material and act as permanent formwork against which the fill can be placed and compacted. The necessity for temporary shuttering is avoided and there are usually no problems of fill retention. The content of the bags can be constituted so as to provided a growing medium for grass or quick rooting plants, such as ivy.

The system has been used in the United Kingdom, and similar structures in Japan are described by Fukuda, (1986).

2.2.7 Contained Facings. Contained facings are similar in concept to the sandbag detail, except the facing and support element is formed using the inherent reinforcing material. Gabions, cubic containers and pillows fall into this this category. In each case the facing acts compositely with the reinforcing elements and is

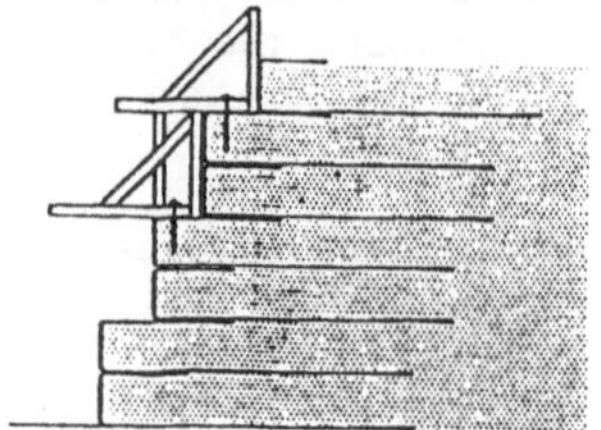

Figure 17

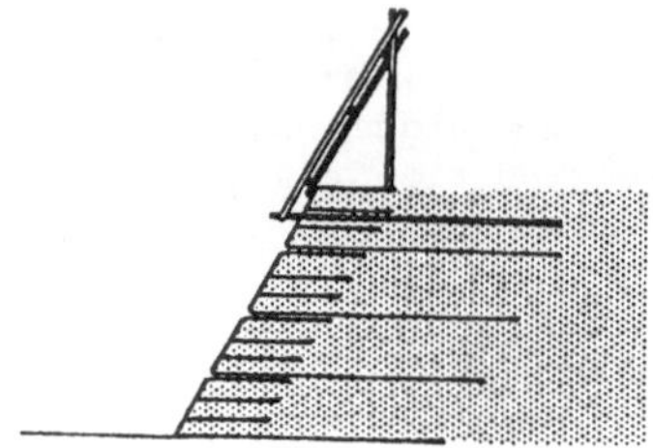

Figure 18

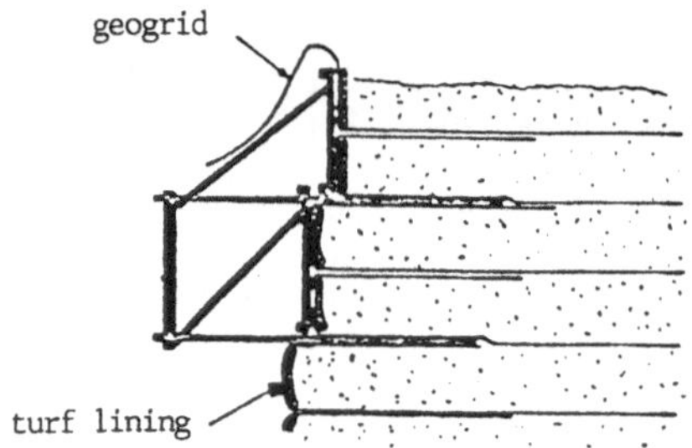

Figure 19

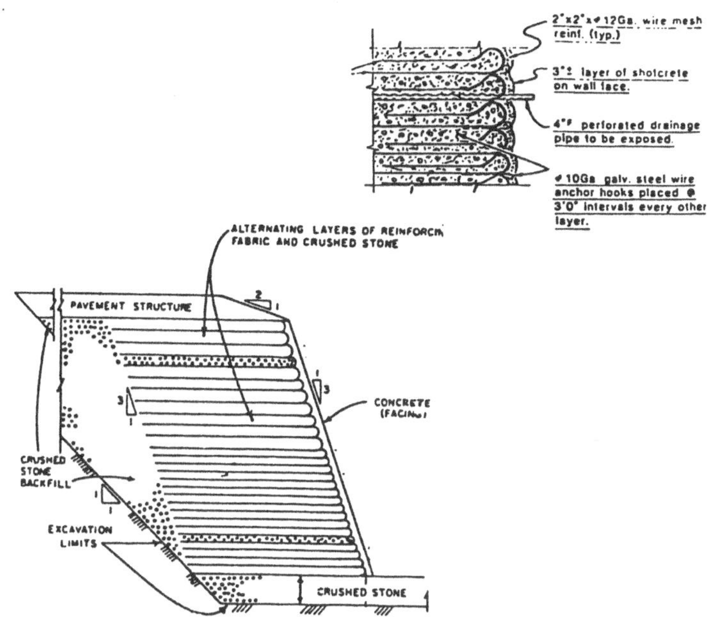
2"x2"x# 12Ga. wire mesh reinf. (typ.)
3"± layer of shotcrete on wall face.
4"Ø perforated drainage pipe to be exposed.
10Ga galv. steel wire anchor hooks placed @ 3'0" intervals every other layer.
ALTERNATING LAYERS OF REINFORCM FABRIC AND CRUSHED STONE
PAVEMENT STRUCTURE
CONCRETE (FACING)
CRUSHED STONE BACKFILL
EXCAVATION LIMITS
CRUSHED STONE

Figure 20

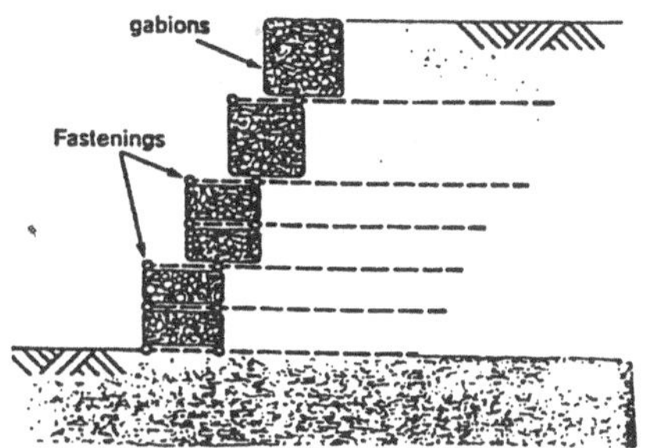
gabions
Fastenings

Figure 21

usually joined to them.

Reinforced or "tailed" gabion structures developed by Templeman and Jones (1979) are constructed using gabions connected to horizontal layers of geogrid reinforcement, Fig. 21. In this system the gabion facing is erected conventionally and the "tails" are laid in position and tied to the base of each row. Structures up to 8m in height have been built in England.

Recent work by LCPC in France has produced a system of construction using geotextile gabions or "cubic containers/pillows". This is similar to the "tailed" gabion concept except that the reinforcing material is a geotextile. This technique was first used to construct a 5.0m high steep stepped faced structure at Trouville Sur Mer, France, Perrier et al, (1986).

2.3. CHOICE OF SYSTEM

A number of interacting factors can be shown to influence the choice of one system over another and to determine the form of the construction itself. It is convenient to group the various influences under three headings:

a) Technical Requirements
b) Market Forces
c) Working Practices

2.3.1 Technical Requirements.

The technical elements which influence the selection, design construction and form of a reinforced soil structure are shown in Table 1.

It is difficult to determine the degree of importance of separate elements on the construction system adopted. Table 2 provides an indication of potential weightings which might occur, although individual conditions will produce alternative conditions

(i) Function.

The ultimate use of the proposed structure probably has the major influence on the choice of structure. As an example some materials can be used to provide the facing as well as the reinforcement although their appearance is often not acceptable other than for temporary, industrial or military structures. Similarly, fill properties and the durability of reinforcing materials may not be important with temporary contruction but are critical with permanent structures.

(ii) Subsoil Conditions.

Subsoil conditions are important in all reinforced soil constructions and often the primary technical reason for the choice of this type of construction. All of the three basic construction systems can be used on weak subsoils, although some facings and constructional details are better than others at accommodating significant differential settlements. In particular, the wrap-around and gabion style facings, used with the concertina method of construction, are able to accept major distortions, Godfrey, (1984). Subsoil conditions also influence whole body rotations and distortion, a factor discussed later.

(iii) Rate of Construction.

Construction of reinforced soil stuctures is normally rapid. Construction rates of 40-200 m^2 per day per man may be expected and usually the speed of construction is determined by the rate of

TABLE 1

Reinforcement	Soil	Construction
Composition	Particle Size	Construction System
Durability	Grading	Compaction
Form	Index Properties	
Surface Properties	Mineral Content	Facing
Dimensions	Durability	
Strength	Availability	
Stiffness		
Reinforcement Distribution	**Soil State**	**Structures**
Location	Density	Geometry
Spacing	Confinement	End Use
Orientation	State of Stress	Aesthetics
	Degree of Saturation	
	Drainage	

TABLE 2

APPLICATIONS

Construction techtural Factor	Temporary	Short Life	Permanent	Industrial	Military	Archi-
Drainage	***	***	***	***	***	***
Distortion	*	*	** (***)	*	*	** (***)
Subsoil	**	**	**	**	***	**
Rate of Construction	***	**	**	**	***	*
Fill Properties	*	**	***	*	*	**
Reinforcement properties	*	**	***	**	**	**
Facing	*	*	***	*	**	***
Aesthetics	*	*	**	*	*	***
Durability	*	*	***	**	*	***

* Secondary Importance
** Important
*** Very Important

placing and compaction of the fill. On a recent UK contract the rate of production of the industrial waste used as fill became of the limiting criteria on the construction, Jones et al, (1987). In some cases the economic rate for the production of facing units may determine the construction rate, particularly if an original or unique facing is required. It is essential to determine this aspect of any project early in the design cycle. Construction is normally unaffected by weather except in extreme situations.

(iv) Durability and Degredation.

In all situations care should be taken to minimise physical damage to facing elements and reinforcement, and wherever possible corrosion must be avoided.

Normally vehicles are restricted so as not to run over exposed reinforcement and cause damage. Polymeric reinforcement is not subject to corrosion and is usually very durable in conventional fills, however, it is susceptible to ultra voilet light and must be stored out of direct sunlight. In some Codes of Practice the properties of polymeric reinforcing materials are factored downwards to accommodate possible damage caused by fill/reinforcement interaction. The use and selection of fill which is benign to polymeric reinforcement in this context can be advantageous. An example is the use of pulverised fuel ash as a structural fill with "Tensar" geogrids, where a material properties reduction factor (γm) = <0.9 is taken to be appropriate. At present it is not considered acceptable to use pulverised fuel ash with metallic reinforcement because of the risk of severe corrosion.

(v) Aesthetics.

The appearance of most structures is closely associated with use. With permanent structures appearance is usually of considerable importance. In the case of structures in urban environments the need for a high standard of finished appearance can influence the construction system chosen. In the case of the hybrid system of construction, masonry faced structures have been found to be the most economic. Wraparound structures on the faces of which vegetaion is encouraged can be very attractive, similarly hand placed stone in gabions has aesthetic appeal.

(vi) Distortion

Some reinforced soil structures may be prone to distortion, particulary during construction. Many of the construction details adopted in practice are chosen to minimize distortion and its effects.

a) Concertina Construction.

Structures built using the concertina method from geotextiles and geogrids are particularly prone to distortion of the face. The degree of distortion cannot be accurately predicted. An accepted method of overcoming the problem is to cover the face of the structure either with soil or with some form of facing. An alternative is to provide a rolling block or sand bag facing against which the compaction plant can act, to minimise distortions and make them as uniform as possible, Fig.22.

b) Telescopic Construction.

An estimate of the internal movements and distortion of the facing can be made from observations of prototype and trial structures.

Typical vertical movements within the soil mass which are

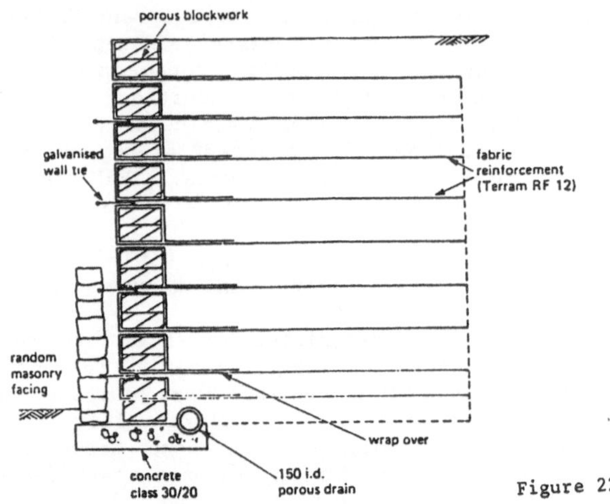

Figure 22

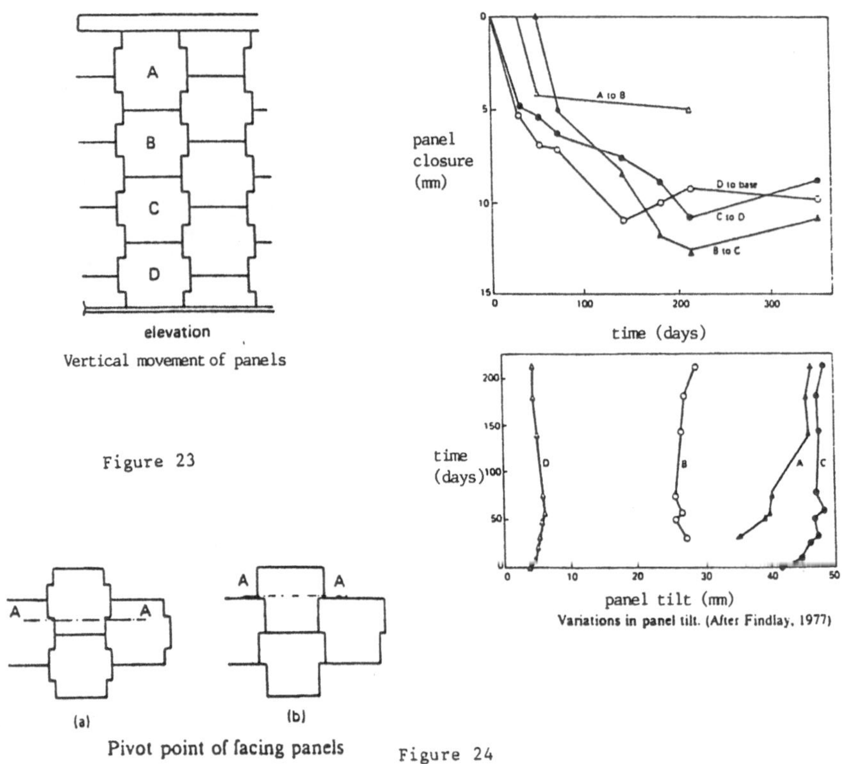

Vertical movement of panels

Variations in panel tilt. (After Findlay, 1977)

Figure 23

Pivot point of facing panels

Figure 24

transmitted to the facing are made up of three components:

1. horizontal movements at the connections
2. tilt of the facing units
3. time dependent strains in polymeric reinforcements.

Providing good practice is followed and the reinforcement is slightly pre-tensioned, movement of connections during construction is not normally significant and is likely to be 2-5mm. All facing panels in this form of construction tilt, the pivot point depending upon the geometry of the facing, Fig. 24. The movement of individual panels is normally that of an outward rotation (ie. the top of the panels moving away from the fill). In some special cases the face movement is determined by whole body rotation which can be a movement of the structure into the fill. A prediction of the whole body movement should be part of the design.

c) Sliding method of construction.

Non-Structural Facing

When a non-structural facing is used, distortion of the facing is likely to occur, the degree being dependent upon compaction. Distortion can be controlled by:

1. Using light plant in the 2m zone adjacent to the facing.
2. Using bold architectural features to mask the distortion, such as in Fig. 5(a). These can disguise forward rotations and major bulges by creating a face without a natural sight line.

Structural Facing

When a full height structural facing is used and the construction method is as in Fig.7, the horizontal movement of the facing will be limited to the movement capacity provided by the reinforcement/facing connections and the stiffness of the reinforcing material. Movements of 2-5mm for walls 5-8m high have been observed with reinforcements with a high modulus. With more extensible reinforcements, movements of 5-20mm are typical. The facing will normally rotate outwards about its toe and the movement is usually sufficient to reduce the lateral soil pressures behind the facings to the active state. Under some conditions associated with very weak subsoils whole body rotation of the structure may occur. This may result in a backward lean of the facing. The propensity of the structure to rotate and the degree of rotation is a function of subsoil condition and the geometry of the structure, it has been suggested that narrow, double sided (back to back) structures are not prone to whole body rotations, Jones and Edwards (1980).

(vii) Logistics.

The speed of construction must be catered for if the full potential of the use of reinforced soil structures is to be realized. Normally this will cause little or no problems with the reinforcing materials, but the production and delivery rate of the facing units may cause problems, particularly if multiple use of formworks is required for economy. The delevery of fill has also been shown to be a deciding factor associated with construction rate.

The choice of structural form and construction technique may depend ultimately upon the ease and economy of moving constructional materials. As an example, the light weight of the geotextiles and related materials makes them suitable for air freight, whereas concrete and steel elements can have significant transport problems.

(viii) Construction Sequence.
Reinforced soil structures encourage the use of non-conventional construction sequences and technical innovations and may permit a reduction in the number of construction steps as illustrated in Fig 25.

Alternatively it is possible to change round a construction sequence as in Fig 26, where the backfill of a structure is placed before the structure itself. This is achieved by forming the backfill into a temporary reinforced soil structure and using the face of this structure as the rear formwork for the permanent works. This technique has been used sucessfully in bridgeworks construction in the United Kingdom.

2.3.2 Market Forces. Market forces frequently determine the form of engineering structures. These forces are not necessarily synonymous with base or prime cost, but rather with what the market will bear. At a time when there is general over capacity in the geotextile industry, cost may fall at times, to levels which are not necessarily economic.

The development of proprietary materials and systems can be a major influence on the choice of structure. Development costs are always large, but through use in a large number of applications the costs may appear modest, whereas structures specifically tailored with special one off details and facings can prove very costly. In addition, the latter may not be well designed. As a result there is an obvious incentive to duplicate methods and techniques and rationalise details.
Monopolies in the form of patents have also played a significant role not only in championing reinforced soil, but also in determining the form used. Patented systems, usually represent innovation, however, they can themselves be a hindrance to further innovation in that, superior elements or systems may be positively discriminated against. The development of market choice and the acceptance of alternatives by Commissioning Agencies is the means to eradicate this negative element.

2.3.3 Working Practices. The civil engineering profession is inherently very conservative, and innovations are scrutinised with rigour. In addition the life expectancy of most civil engineering works makes for a cautious approach. As a consequence the introduction of new techniques and materials can be difficult, particularly if the use of the new system requires an element of learning from the user. The development of polymeric products for use in civil engineering has come about through market forces, operating within both the textile and civil engineering industries. The acceptance of these materials has been due to skilled marketing within the civil engineering profession rather than as a demand from the profession.

National Agencies are frequently the key or catalyst to innovation. In the United Kingdom, developments within the highway field have occurred largley as a result of the encouragement of the Department of Transport and the Transport and Road Research Laboratory; without Technical Memorandum BE3/78 relating to reinforced soil structures, this form of construction would still be experimental.

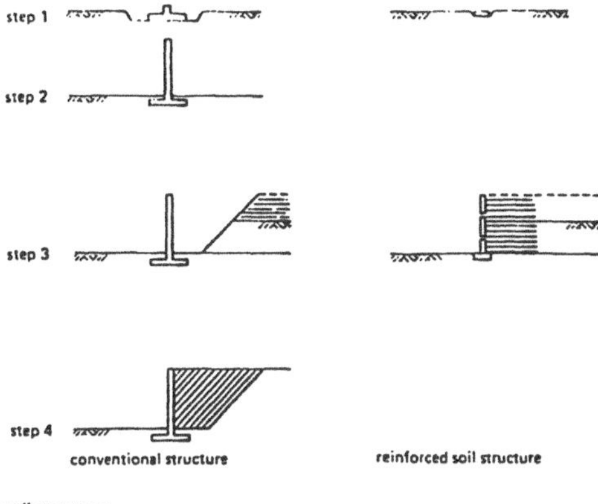

Figure 25 Construction sequence for conventional structures and reinforced soil structures

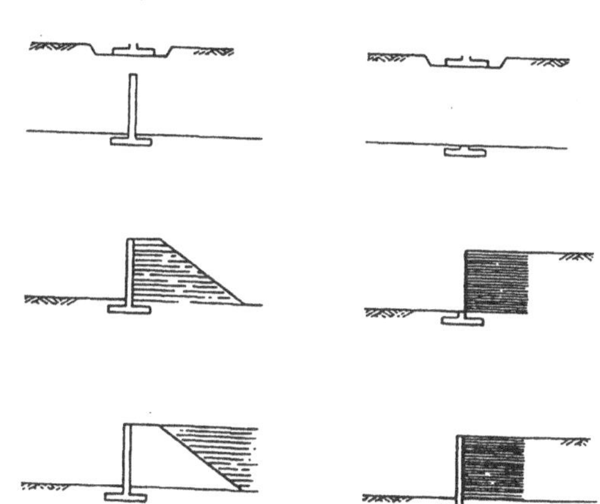

Figure 26

The greatest encouragement to the use of reinforced soil is the development of recognised National Design Standards. In the conservative civil engineering profession Codes of Practice have become an essential element to many designers. In the absence of a Code of Practice, acceptance criteria such as the British Board of Agrement Certificate became essential to progress.

A second influence on the choice of reinforced soil systems follows from the structure of the professions in different countries. These differences are important and have been indirectly responsible for legal difficulties in the reinforced soil field.

The Designer or Consultant is often reluctant to consider alterations or alternatives to his original design often looking upon these as a form of criticism. Changing a "conventional" design to a reinforced soil structure is seen by some designers as very radical. Even suggesting a change from one form of reinforced soil to another or the use of a different polymeric product can be resisted. This resistance may be legitimate as the suggestion of an alternative can often only be made during the tender or bid period. As acceptance of the alternative places responsibility on the Designer, resistance to change can be understandable, with the argument being that all the technical consequences are not known. In Scotland a method of overcoming the problem of alternative designs has had some success, Varley (1984).

In the United States, the deliberate action of the Federal Highway Authority and some State Highway Authorities to encourage competition has had the desired effect, not only of reducing costs but also of making possible a wide range of reinforced soil alternatives. The different structure of the civil engineering industry in Canada and the United States to that in Europe encourages competition from specialists each offering their own proprietory system. To ensure the best choice is made, many Highways Agencies in North America prefer to provide complete plans for all recognised alternatives. This permits the Agency to maintain control of the engineering and at the same time it provides the best suited structures for each particular site. The extra engineering effort in this approach is often mitigated by the willingness of many suppliers to provide complete plans to the Agency, Leary and Klinedinst, (1984).

Designers themselves can compete successfully in developing effective and economical reinforced soil structures. In the United States of America, Georgia State Highway Authority has developed its own system (GASE, Georgia Stabilised Embankment) based partly upon the work pioneered by the California Highway Authority, Forsyth (1978). In the United Kingdom, the West Riding/West Yorkshire Metropolitan County Council developed sufficient expertise and experience to tailor reinforced soil structures to specific situations, Jones (1985).

The lesson from California, Georgia and West Yorkshire is that expertise develops with application. Reinforced soil is a relatively recent reintroduction of an ancient technique and experience in the field is on a steep growth curve. Accordingly the choice of reinforced soil system or style will be influenced by the level of expertise available and the past working practices adopted by the designer.

3. ECONOMICS

The primary advantage gained from the use of reinforced soil structures may be the improved idealisation which the concept permits; thus, structural forms which would have been impossible to contemplate become feasible and economic. It is generally accepted that the use of reinforced soil walls or bridge abutments will produce significant savings; costs of (30-50 per cent) relative to conventional construction.

The savings produced through the use of reinforced soil are influenced by a range of factors, not least of which is the level of competition available. Without competition the cost of reinforced soil is heavily influenced by the cost of the conventional structure alternative, and savings can be reduced in line with what the market will bear. Without competition from alternative soil reinforcement systems costs may be 70 to 80 per cent of conventional structures.

Before embarking upon a reinforced soil structure design an estimate of the possible cost is required, the most reliable of which are based upon previous work. The base cost of any reinforced soil structure is very difficult to determine, in some cases it may only be determined by the use of a form of direct labour contract. However, the major factors which influence costs for any practice can be identified, Table 3

3.1 TOTAL COST

The total cost of a reinforced soil structure is made up of the following cost elements:

- Soil Fill, (Cs)
- Reinforcement and connections (CR)
- Facing Elements (if required) (CF)
- Labour for transport and construction (Ct)
- Transport of materials (CT)
- Construction (including all ancillary items such as drainage, copings and facings),(Cc)
- Material testing (CMT)
- Profit, (P)

Thus total cost,

$$TC = (CS + CR + CF + CL + CT + CC + CMT + P) \text{ ----- } 1$$

For contractual purposes equation (1) may be reduced to the first three elements, but with the labour, transportation and the ancillary costs included:

$$TC = (C'S + C'R + C'F) \text{ ------------------------- } 2$$

Where C's represents the cost of the soil fill, including transport, placing compaction and material testing

C'R represents the cost of the reinforcement, including transport and fixing.

C'F represents the cost of the facing, including transport and erection. Profit is included.

The elements included in equation 1 are inter-related and the minimum total cost of a structure may be produced by a combination of the most compatible elements in any particular situation. For example, if the particular materials testing systems are unavailable, then the use of reinforcing elements exempt from these testing requirements may provide the economic solution even if

Table 3

Technical Factors	Use, design life, quality of detailing, quality and suitability of fill, testing requirements, subsoil conditions, parapet details, analytical models.
Market Forces	Competition, size of structure, use of proven components, familarity of contractor with reinforced soil techniques logistics.
Working Practices	Conforming to Codes of Practice, available specifications, Experience of designers.

these reinforcing elements are highly priced. Similarly a combination of construction elements permitting the use of an indigenous fill or waste fill such as colliery shale or pulverised fuel ash may be economical. In other cases technical factors may predominate; in a recent case the connection detail proved to be the deciding factor between the use of polymeric grid or strip reinforcement, Jones et al (1987).

3.2. DISTRIBUTION OF COSTS. Using the elemental breakdown given in equation 2 it is possible to illustrate that the distribution of the cost elements vary not only with the relative costs of the constituent materials but also with the dimensions of the structure. Assuming that the relative cost of the three elements of equation 2 are in the ratios of:

Soil fill, per unit volume (m3)	1.0
Reinforcing elements per unit area of face (m2)	1.5
Facing elements per unit area of facing	10.0

and where the width of the structure is B, the height is H and (B=H), then the distribution of costs with respect to the height of a vertically faced retaining wall are shown in Fig.27. It can be seen that the relative costs of the three basic elements for a 10 m structure are approximately:

Soil fill	30 percent
Reinforcing Elements	40 percent
Facing elements	30 percent

If the relative costs of the three basic elements of the same structure are changed to:

Soil fill, per unit volume (m^3)	0.5
Reinforcing elements per unit area of face (m^2)	2.0
Facing elements per unit of facing	10.0

then, although the total cost of the structure remains the same, the distribution of the costs is very different, Fig.27.

These data also illustrate the influence that scale or size of construction may have on costs and show that at lower heights the influence of the cost of the facing on the overall costs becomes dominant. With small structures, the material requirements for the facing may be of the same order as the material from a conventional structure, a point reflected in Fig. 28, which also shows the imprecise nature of cost comparisons. At low heights, particular attention may be required to reduce the costs of the facing in order to retain the economic benefit of a reinforced soil structure; one method known to be successful is to use masonry or brick facing normally associated with small scale construction or building techniques. Experience has shown that this form of structure is compatible with the use of polymeric reinforcements, Jones (1982).

A second influence on overall cost, associated with the scale of the project, is the contractural arrangements under which the structure is built. For some individual structures and structures under 3 m in height the labour requirements are low and the use of specialist sub-contractors may prove uneconomic, in which case economic construction of the soil structure may be attained only by the main contractor, through local contractors, or by a direct labour organisation.

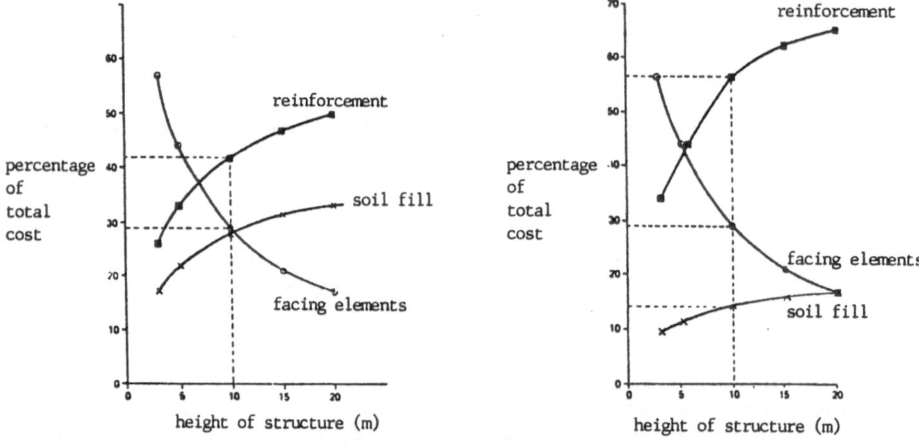

Figure 27

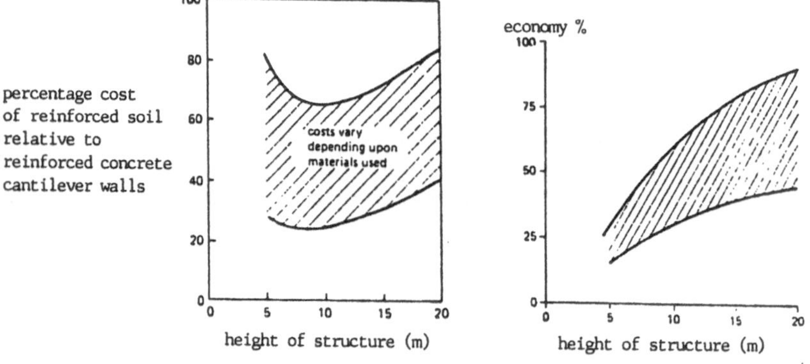

Figure 28

<u>3.3. CONSTRUCTION</u>. In conventional structures the placing and compaction of fill may not be associated with the construction. If reinforced soil structures are treated similarly, the fill element costs may be removed and the distribution of construction costs change dramatically. The costs established in one study are as shown in Fig.29 The volume of structural fill required for use with a conventional structure may exceed that used in a reinforced soil structure, Fig. 30.

<u>3.4. COST DIFFERENTIALS</u>. The cost of fill materials is dependent upon local availability and haulage rates. Similarly, the cost of facing materials is a function of locality and custom. Reinforcement materials have different properties and costs vary. Thus, even though the theoretical cost for different reinforcements may indicate financial preference for one material, market conditions may give a different trend. Overriding all considerations is the requirement to obtain the minimum cost, equation 1.

<u>3.5. ECOLOGY AUDIT</u>. An alternative method of assessing the benefits of earth reinforced soil systems, is to use an ecology audit. An advantage of this approach is that it is immune from the commercial distortions which are associated normally with new constructional systems, and therefore it may produce a more realistic assessment of the true costs.

The increase in energy costs has led to an interest in energy calculations including the energy content of building materials. However, energy is only one of the ecological parameters needed to determine the complete effects, (short-term, long-term and associated) of engineering works. Of growing importance and interest are the problems created by scarcity of raw materials, the environmental problems created by pollution, both of the atmosphere as well as the land from mining activities, the increase in manpower costs and transportation costs and the cost of maintenance. The choice of structural form used for any scheme influences all of these parameters. Determination of the complete costs of a structure to society may be attempted by studying the ecological parameters represented in the whole cycle necessary for its production, including:

* mining
* raw materials
* process industry
* basic materials
* product/construction industry
* users maintenance
* waste recycling.

In practical terms the ecological parameters associated with a reinforced soil structures are:

* energy content of the materials forming the structure
* quantity of process water required to manufacture the materials.
* despoiling of land necessary to produce the materials
* pollution caused during manufacture and construction.

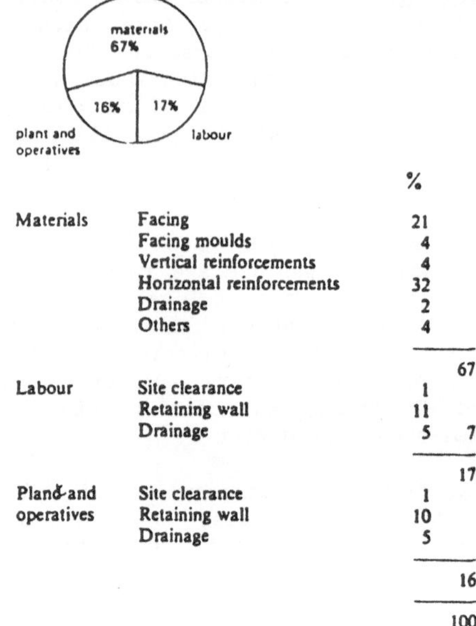

		%	
Materials	Facing	21	
	Facing moulds	4	
	Vertical reinforcements	4	
	Horizontal reinforcements	32	
	Drainage	2	
	Others	4	
			67
Labour	Site clearance	1	
	Retaining wall	11	
	Drainage	5	7
			17
Plant and operatives	Site clearance	1	
	Retaining wall	10	
	Drainage	5	
			16
			100

Figure 29 Breakdown of construction costs

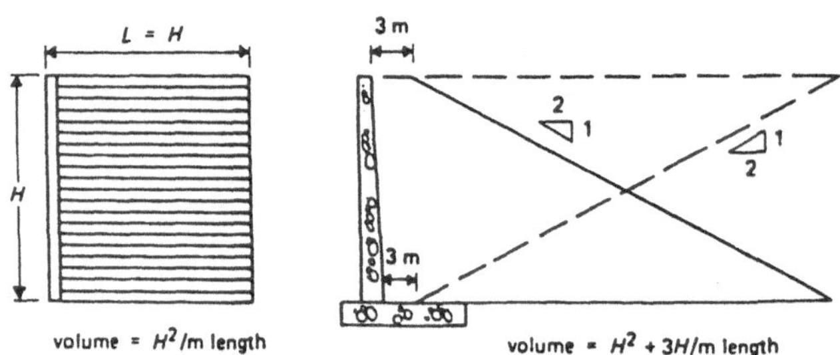

Figure 30 Volume of fill required in reinforced soil and reinforced concrete retaining walls

* labour costs for material manufacture, transport, construction and maintenance.
* demolition requirements.

The ecological parameters associated with the construction of reinforced soil structures formed using reinforced concrete, steel reinforcement and cohesionless soils are illustrated in Table 4.

Figure 31 shows ecological parameter values for a 64 m prototype reinforced soil structure compared with an equivalent reinforced concrete cantilever retaining wall. The reinforced soil structure is significantly more efficient in ecological terms. Arguably economic factors have as their ultimate base the ecological parameters; accordingly Fig.31 is a potent argument that reinforced soil structures are efficient and economic. In addition, polymeric reinforcement appears to be competitive when compared with steel, Table 4.

4. SPECIFICATIONS

Many civil engineering projects worldwide are undertaken by contract. The system requires that the client and his appointed representatives have the means with which to adequately design and properly specify their needs and that the Contractor is fully aware of these requirements and their implications at the time of tender.

The Specification describes how aspects of the work shall be executed, referring when appropriate to approved Codes of Practice and technical guidelines.

The use of polymers as reinforcing materials for soil structures raises a number of problems with the development of specifications. Unlike established materials such as concrete and steel, the properties and characteristics of polymer materials are less well understood. In addition, because of their relatively recent introduction and the range of available polymeric materials, standard tests to evaluate their properties are only now emerging.

For these reasons, the development of Codes of Practice, design guidelines and contract documentation, for use with polymeric materials in reinforced soil structures, have tended to divide into two categories - those for methods and those for materials, indeed in some countries only the former is presently underway.

4.1. METHOD RELATED DOCUMENTATION. There are two main sources of design and construction information for reinforced soil systems,

(a) Technical Memorandum (Bridges) BE 3/78
"Reinforced Earth Retaining Walls and Bridges Abutments for Embankments".
Department of Transport (UK) 1978

(b) "Les Ouvrages en terre armee - Recommendations et regles l'art". Laboratoire Central des Ponts et Chausees. (LCPC), France 1979.

These documents were originally formulated for use with metallic reinforcement and both have been adopted as the basis for standards of design and construction by other countries worldwide. The United Kingdom, Memorandum BE3/78 caters for the use of non-metallic reinforcement. However, the increasing availability and use of polymeric materials has left both documents deficient for the general case. Accordingly, the tendency has been to develop

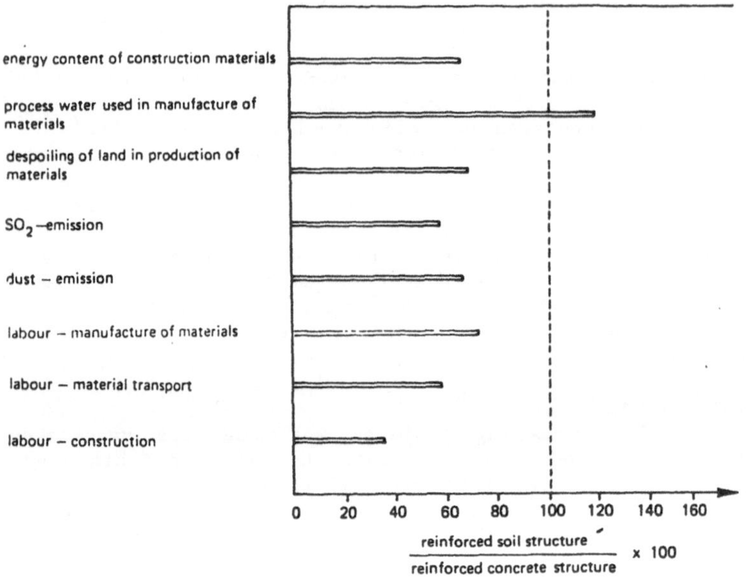
energy content of construction materials
process water used in manufacture of materials
despoiling of land in production of materials
SO_2–emission
dust – emission
labour – manufacture of materials
labour – material transport
labour – construction
0
20
40
60
80
100
120
140
160
reinforced soil structure / reinforced concrete structure x 100

Figure 31

Table 4 Energy consumption of construction materials

MATERIAL	GJ/ton
Gravel	0.104
Sand	0.128
Ordinary Portland cement (OPC)	8.2
Blastfurnace cement (HC)	3.0
Water	0.004
Mixing concrete	0.058
Mild steel reinforcement (bar, grid)	22.8
Prestressing stell (bar, strand)	28.3
Plastic (high density polythene sheet)	84.0
Concrete (340 kg HC/m^3)	2.18 GJ/m^3
Concrete (360 kg OPC/m^3)	3.28 GJ/m^3
Concrete tiles and bricks	3.18 GJ/m^3
CONSUMPTION OF PROCESS WATER IN MANUFACTURE	
Concrete	6.30 l/m^3
Steel (reinforcement, prestressing,	55 m^3/ton
DESPOILING FROM PRODUCTION OF MATERIALS	
Concrete	0.69 m^2/m^3
Steel	5 m^2/ton
POLLUTION - SO2 EMISSION	
Concrete	0.37 kg/m^3
Steel	2.00 kg/ton
DUST EMISSION	
Concrete	1.29 kg/m^3
Aggregate/fill (sand, gravel)	1.1 kg/m^3
Steel	2.7 kg/ton
LABOUR - MATERIAL MANUFACTURE AND TRANSPORT	
Concrete	1 man-h/m^3 manufacture 1.45 man-h/m^3 transport
Steel	10 man-h/ton manufacture; 0.6 man-h/ton transport

guidelines derived from manufacturers recommendations and local experience.

A number of National documents which cover the use of polymeric reinforcement are known to exist or be in the course of preparation including:

(i) "Model Specification for Reinforced Fill Structures"
Geotechnical Control Office, Hong Kong.
ii) "Design Guidelines for the use of Extensible Reinforcements for Mechanically Stabilized Earth Walls in Permanent Applications".
AASHTO-AGC-ARTBA, Joint Committee Task Force 27. (Draft 7/86).
(iii)"Guide to Ground Improvement"
Construction Industry Research and Information Association (CIRIA). UK.
(iv)"Model Specification for the use of Geotextile and Related Products".
Institution of Civil Engineers, London.
(v)"Code of Practice for Soil Reinforced Retaining Structures".
Indian Standards Institution Sub-Committee for Geotextiles, BDC 23.8 (expected July 1987)
(vi)"Strengthened/Reinforced Soils and other Fills"
UK Code of Practice, British Standards Institution
Committee CSB/56.

4.2. MATERIAL RELATED DOCUMENTATION. There are several different agencies involved in the development of testing methods and many countries now have national committees dealing with the development of test methods for geotextiles and related materials e.g. B.S.I. and A.S.T.N. and these national test method committees are co-ordinated internationally through I.S.O.. Several other professional bodies also attempt to co-ordinate their work, either directly or indirectly, e.g. R.I.L.E.M., P.I.A.R.C. and I.C.O.L.D.

Progress to date has not been as rapid as first anticipated with very few test methods agreed which provided data that can be used directly in design.

For this reason no national or international material specifications to deal with all aspects of geotextiles and related materials used as reinforcements is available.

The specification being produced in the United Kingdom. by the Institution of Civil Engineers is probably the most advanced.

5. DISCUSSION

The number of reinforced soil structures using polymeric reinforcement is growing. The rate of growth is likely to increase as the technical and economic advantages of the technique become more widely appreciated.

From current usage it is possible to identify areas where improvements can be made, and which, if introduced would assist both designers and contractors, and also lead to further economies.

5.1. CONSTRUCTION METHODS. The existing construction methods can be improved, and the following developments have been identified as likely to be cost effective:

a) The development of standard reinforcement layouts
b) The introduction of more standard facings and details
c) The identification of optimum plant to comply with construction specifications.

In addition, consideration of the following points may lead to developments;

d) The need for and effectiveness of pretensioning of polymeric reinforcement.
e) The logical arrangement of joints.

5.2 ECONOMICS

A better understanding of the economics of reinforced soil structures is required. Advances in this important area would be greatly assisted by:

a) The publication of case studies which detail actual costs, including background data.
b) Studies of the long term and maintenance costs, including further consideration of the durability of polymeric materials.
c) Identification of the factors which influence the economics of reinforced earth soil structures in different countries.

5.3. SPECIFICATIONS. Many potential designs are being hampered by the lack of adequate specifications, although the widespread development of Codes of Practice will resolve many questions. Even so, it is possible to conclude that:

a) Specifications should be broadened
b) There is a need for standardised tendering procedures
c) There is a need for accepted methods of assessing different products and for comparing structures developed using different polymeric reinforcements.

5.4 FUTURE. Likely future developments will come from two areas. The first is through the introduction of new advanced polymeric materials, such as the Directionally Structured Fabrics (D.S.F.) which have recently been announced.

The second area of future development will result from a general growth of the reinforced soil technique. This will produce more diverse applications. It is already possible to discern the development of different "styles" in reinforced soil technology based upon need and frequently influenced by local topology and local practice. An example of the latter is the development of the Norwegian Trondblock walling system to incorporate polymeric grid.

REFERENCES

1. Berg, R. R., Boneparte R., Anderson R.P, and Chouery V.E, (1968). "Design Construction and Performance of Two Geogrid Reinforced Soil Retaining Walls". Proc. 3rd Int. Conf. on Geotextiles, Vienna.
2. Cantelli R., and Munfakh G., (1986). "Geotextile Walls in Mountainous Terrain". Proc. 3rd. Int Conf. Geotextiles, Viennna.
3. Department of Transport (1978). Reinforced Earth Retaining Walls and Bridges Abutments for Embankments. Tech. Memo. (Bridges) BE 3/78, London.
4. Douglas,G.E., (1981). "Design and Construction of Fabric-Reinforced Retaining Walls by New York State". Trans Research Record 872.
5. Findlay, T.W. (1978). "Performance of a Reinforced Earth Strucutre at Granton". Ground Engineering 2, No. 7, 42-44.
6. Forsyth R.A., (1987). "Alternative Earth Reinforcements". ASCE Spring Convention, Pittsburg.
7. Fukuoka M., (1986). "Fabric Retaining Wall with Multiple Anchors". Proc. 3rd Int. Conf. Geotextiles, Vienna.
8. Fukuda N, Yamanouchi T, and Miura N., (1986). "Comparative Studies on Design and Construction of Steep Reinforced Embankment". Proc. 3rd Int. Conf. Geotextiles, Vienna.
9. Godfrey K.A. Jnr. (1984). "Retaining Walls: Competition or Anarchy". ASCE Civil Eng. Dec.
10. Jones, C.J.F.P. (1978). "The York Method of Reinforced Earth Construction". ASCE, Spring Convention, Pittsburg.
11. Jones C.J.F.P. (1985). "Earth Reinforcement and Soil Structures". Butterworths, England.
12. Jones C.J.F.P. (1987). "Practical Construction techniques for Retaining Structures using Fabrics and Geogrids". 2nd Int. Conf. Geotextiles, Las Vegas.
13. Jones C.J.F.P., Murray R.T., Temporal J. and Mair R.J. (1985). "First Application of Anchored Earth". XI ISSMFE San Francisco, August.
14. Jones C.J.F.P., Jamison W, and Garner D. (1987). "Design Cônstruction and Economics of Dewsbury Retaining Walls". In print.
15. Jones C.J.F.P., and Edwards L.W. (1980). "Reinforced Earth Structures Situated on Soft Foundations". Geotechnique, June.
16. Kempton G.T, Entwistle, R.W. and Barclay M.J. (1985). "An anchored fill harbour wall using synthetic fabrics". Proc. Inst. Civil Eng. Part 1, vol.78 April pp 327-347
17. Laboratoire Central des Ponts et Chausees (1979). "Les Onulages en terre armee - Recommendation et regles d'art". Paris.
18. Leary R.M., and Klinedinst G.L., (1984) "Reinforced Wall Alternatives". FHWA, Washington.
19. Perrier, H., Blivet, J-C, and Khay, M, (1986). "Experimental and Actual use of Geotextile Reinforcement of a Slope". 3rd Int. Conf. Geotextiles, Vienna.
20. Pigg D.Z., and McCafferty W.R., (1985). "The Design and Construction of Reinforced Soil Retaining Wall at Low Southwick, Sunderland". Polymer Grid Reinforcement, Thomas Telford Ltd.

21. Saran, S., Talwar D.V., and Prakash, S. (1979). "Earth pressure distribution on retaining walls with reinforced earth backfill". C.R. Col. Int. Renforcement des Sols, Paris.
22. Templeman J., and Jones C.J.F.P. (1979). "Soil Structures using high tensile plastic grids". U.K. patent No. 7941627.
23. Varley W.R., (1979). "Aspects of Alternative Tendering". BCSA Nat. Struct. Steel Conf. Part 2, London.
24. Vidal, H (1978). "The development and future of Reinforced Earth". Keynote address. ASCE, Spring Convention, Pittsburg.
25. Vidal, H. (1978). "La terre armee". Annales de L'Institut Technique du Batiment et des Travaux Publics, 19, Nos. 223-224, July-August, France.
26. Yamanouchi, T (1986). Private Communication.

REVIEW OF SESSION

CONSTRUCTION METHODS AND ECONOMICS

Jones opened this session with a general appreciation of the many factors affecting the economics of geosynthetic construction. Apart from regional market force aspects, which in many cases control cost and methodology, he pointed out that working practices may be the dominant cost factor. The actual cost of the reinforcement is in most cases a small proportion of the total cost of the structure and as such may have little influence on overall economics. He then discussed the concept of using an "Ecology Audit" to establish comparable base costs for different forms of reinforced soil. An ecology audit sums the overall cost to society of a particular form of construction especially with respect to the energy used throughout the process. It therefore avoids to a great extent market factors and other artificial variables. Such an audit indicates that geosynthetic reinforcement is a cost effective means of construction. Despite this economic advantage designers are still hesitant to use this form of construction due primarily to lack of experience. Therefore the keys to market growth are Specifications and Codes of Practice for geosynthetic reinforced soil as such standards relieve the "tension" from the design situation by providing set terms of reference. Work is progressing in a number of countries on these standards.

The focus of the session then changed to discuss variations in construction methods. Jones presented a series of slides showing aspects of construction for a number of reinforced soil structures. Many interesting questions, comments and points of amplification were contributed during this stage of the session. The following sections attempt to summarize the topics discussed:

The advantages and methodology of providing horizontal and vertical compliance in the connection between the reinforcement and the facing element were addressed by Bonaparte, Christopher and Jewell. The method of connection is vital as it is a construction detail often not considered in the basic earth pressure design calculations. Problems may arise if a rigid face and rigid connection exist and vertical settlement occurs in the soil behind the wall. Reinforcement failure at the connection and even crushing of the facing have been observed in such circumstances as the vertical forces caused by the settling backfill far exceed the lateral earth pressure derived forces for which the reinforcement had been designed.

The problems of bidding for jobs, where the fill to be used is tightly specified to allow steel reinforcing to be used, when in fact much cheaper fills could be safely used with geosynthetics. This problem mentioned by Delmas, Bonaparte, McGown and Christopher pointed once again to the need for Specifications and Codes specifically for geosynthetics. Jones pointed out that the majority of the large

P. M. Jarrett and A. McGown (eds.), The Application of Polymeric Reinforcement in Soil Retaining Structures, 613–615.

number of case studies that he had presented could in fact not have been constructed using steel reinforcement due to the corrosive nature of many of the fills.

One aspect of marketing that found general consensus was the impression that where good technical design support was available from particular manufacturers it does seem to bear fruit in terms market share.

The provision of architecturally acceptable facings was identified as a vital need for geotextile walls by Leflaive.

It was pointed out by Jones that strong, inextensible materials such as "Paraweb" were available and had been successfully used on structures up to 28 m high. In addition similar yet stronger materials could be developed using Kevlar. One slide of the ziggurats in ancient Babylon left Christopher musing about the possibilities of performing wide width tensile tests on the reed mats from which they were constructed and Jones hoping that one day he will build something as high that lasts as long.

At the end of this session, formal presentations were made by four speakers. Bush discussed a case history in which a void had formed around the connection between a facing unit and the reinforcement. He used this case to make the cautionary point that the "in-isolation" strength of the reinforcement is the weakest state and that care must be taken if one were to use higher strengths from "in soil" tests to ensure that the construction procedures do in fact give "in soil" conditions. Leflaive discussed a case history of a motorway fill that was constructed with a temporary vertical face up to 8 m high using a woven polyester geotextile and no formwork. The face itself was stabilized using Texsol. McGown discussed results obtained in a 1 m high testing facility that allows varying degrees of horizontal stiffness at the wall facing elements. He concluded by indicating that greater soil resistance is mobilized as larger horizontal movements occur leaving less force to be resisted by the wall or the reinforcement. In practical terms the movements necessary for development of the Active earth pressure state may be encouraged by limiting the backfill compaction immediately behind the wall or by slackness in the facing to reinforcement connection. However he suggested that one may more effectively guarantee and control such compliance at the face by providing a thin compressible polymeric drainage layer directly next to the facing. Finally Milligan discussed the soil movements and strains that occur behind retaining walls as the walls move or rotate. He indicated that for reasonable practical levels of wall movement that horizontal soil strains of approximately 1% would occur and that such levels of strain represent the probable working strains for the reinforcing materials.

In summation this session stressed the urgent need for Standard Codes and Specifications directed at geosynthetic reinforced soils. A major benefit from such standards will be the allowable use of cheaper, lower quality fills with geosynthetic reinforcement than are presently allowed with metallic reinforcement. At present many authorities in the absence of such Codes insist on having geosynthetic reinforcement bid directly against metallic reinforcement based on the use of a common high grade backfill. In terms of design detail both the stiffness of the facing and the compliance of the connection between the facing and reinforcement must be considered during design. The connection can be a point of stress concentration if the backfill settles. The stresses developed in a rigid connection may easily exceed those estimated from lateral earth pressure considerations.

Research Needs

RESEARCH NEEDS, REVIEW OF SESSION

DR. K. ANDRAWES, University of Strathclyde, Scotland

DR. P.M. JARRETT, Royal Military College of Canada

1. INTRODUCTION

This chapter is intended to identify the areas which require further work and research in order to improve the understanding of geosynthetic reinforced soil behaviour. From such enhanced comprehension should come the establishment of safe yet more economic design methods and further advances of the construction technology for such structures. The material presented is based primarily on the extensive discussion which took place in the final session of the workshop and which involved all those participating. In addition various speakers and authors during the workshop identified topics requiring further study. These too have been included. The statement therefore represents to some extent the product of the research workshop as it records the questions still unanswered and felt by the participants to be of importance.

To provide a framework for the presentation, the needs will be presented under the following five main headings even though it is realized that the areas are inter-related and some repetition may arise:

Testing soils and reinforcements
Testing systems
Design
Construction
Developments of materials and techniques

One final aspect of introduction is to point out that there are two distinct levels of needed research. Initially and of greatest urgency is the need to provide information that will allow the development of accepted, simple, day to day design methods that will in appropriate circumstances encourage the use of geosynthetic reinforcement in soils. Secondly there is the need to provide a fully developed rational explanation of the behaviour of geosynthetic reinforced soils. In suggesting the research needed for this second goal it is most important that it be realized that the initial goal can be reached prior to completion and even without many aspects of that work.

2. TESTING SOILS AND REINFORCEMENTS

2.1 Testing Soils

(1) There is an urgent need to define the environmental working conditions for geosynthetic materials in the field. The following soil descriptions are needed for a complete range of soils and conditions likely to arise in backfills:

P. M. Jarrett and A. McGown (eds.), The Application of Polymeric Reinforcement in Soil Retaining Structures, 619–625.

(i) Chemical environment
(ii) Biological environment
(iii) Temperature regime.

The first two conditions are needed in order to define the parameters for which the durability of the polymeric materials should be assessed. The third aspect will set representative temperature ranges to be considered in design and polymer testing. Of special interest is the possibility of high temperatures directly behind facing elements.

(2) Stress-Strain relationships: For most simple design, the direct shear box may be used to define the value of ϕ and the interface friction between the geosynthetic and the soil. However for more accurate prediction of reinforced soil behaviour, knowledge of the manner in which strains develop, especially horizontally, is important. Unfortunately conventional apparatus, such as the shear box and triaxial test do not give acceptable relationships involving strains in the soil, especially post peak. Hence it may be necessary to use the simple shear or other apparatus to gain better information on strain behaviour. This research need for a better comprehension of strain development in soils has long been a fundamental and rather intractable problem for soils analysts.

2.2 Testing Polymeric Materials

(1) The most urgent need is the development of comprehensive durability testing methods. These should assess the potential change in properties, especially strength properties, caused by the following effects:

(i) Chemical
(ii) Biological
(iii) Ultra violet
(iv) Ageing under stress
(v) Construction damage
(vi) Temperature and humidity

(2) Stress-Strain relationships: The simplest method for determining these relationships is to conduct "in-isolation" tests under constant temperatures that are relevant to field conditions. More data regarding creep, fatigue and dynamic behaviour are required. Ideally however these properties should be measured using "in-soil" tests. This represents a far greater undertaking and so for simple work it would be useful if a relationship between in-isolation and in-soil behaviour could be established for the major groupings of geosynthetic reinforcements.

(3) Communications with Polymer and Textile specialists: The range and complexity of polymeric materials available as reinforcement is continually increasing. Civil engineers researching the use of these materials must learn more about their most basic properties and methods of manufacture. There is therefore a serious need to develop communications and cooperative research with polymer and textile specialists especially with regard to the durability problem and the development and design of new products. Please note that this is not the usual pious platitude but a very real and serious need! Civil engineers must have dialogue with specialists who can assist them in dealing with such questions as the effect of chemicals on the behaviour of different polymers, what environmental conditions should be avoided, is biological attack most likely

under anaerobic or aerobic conditions, etc. In fact the dialogue is vital in terms of even defining what questions should be asked concerning the long term behaviour of polymers.

2.3 Testing Composite Materials

(1) A better understanding of the composite behaviour and the effects of interaction between the constituent parts of multi-layer soil-geosynthetic systems is required.

(2) Very little information exists concerning the reinforcement of fine-grained soils. Data is required concerning frictional development between the geosynthetic reinforcement and such soils. Frictional development is dependent on the pore water pressure regime in the compacted backfill. When thick geotextiles that allow inplane drainage are used in clays then rapid consolidation may occur near the geotextile leading to an improved strength and frictional resistance. This interactive aspect requires investigation.

(3) Pull out testing can be a useful method of estimating soil-geosynthetic interaction especially with regard to anchorage. Standardization of test size and methodology is needed however to enable meaningful exchange and comparison of information.

(4) Testing composite materials can mean testing samples of soils containing polymers with each sample considered as a unit. The results obtained are not divided between the constituent materials but refer to the unit as a whole. Such an approach was used in the early years of the development of soil reinforcement, and although many researchers have abandoned this technique in favour of testing the constituent materials separately, the composite testing approach may still offer advantages in certain situations. Recently there has been a revival of this type of testing due to the introduction of new methods of soil reinforcement using randomly oriented polymers as in Texsol and the mesh element technique. It seems that a large scale triaxial apparatus is most suitable for the determination of the properties of the composites. However, other types of testing should be explored and the parameters relevant to design should be identified.

3. TESTING SYSTEMS

Systems is meant to infer complete structures in which the reinforcements are embedded in the soil and the behaviour of the structure and its components are investigated under both self weight and external loading. Such tests play a crucial role in enhancing the understanding of the interaction between the soil and the reinforcements, examining the validity of design assumptions and exploring the effects of various parameters on the overall behaviour. Systems may be scale models or full sized structures.

(1) There is a need for all scales of systems to be tested preferably with a progressive development of the scale up to full size. The problems of scale and time effects should not be overlooked.

(2) Centrifuge testing is also proving to be a useful means of modelling and should be further explored.

(3) For all scales of systems tested, measurements of stresses and strains should be made in <u>both</u> the soil and the reinforcement and the effects on behaviour of boundary conditions and construction techniques should be assessed. For scale models, the testing of an unreinforced system may often help in analysing and isolating boundary effects.

(4) Parametric studies in physical or numerical model tests are needed to explore the effects of the stiffness of the foundation soil, the stiffness of the facing, the stiffness of the reinforced soil mass and spacing of the reinforcement relative to soil size.

(5) Earthquake and dynamic loading of reinforced soil structures should be examined and should include consideration of likely pore water pressure effects.

(6) Stiffer and stronger geosynthetic reinforcements are being developed and should allow the construction of higher walls. These should be evaluated.

(7) Full scale tests or case histories are, as usual, of inestimable value. Very few fully documented case histories have been published and so there is an urgent need to fully instrument and report the behaviour of full scale structures. Factors to be studied include:

- (i) Stresses and strains in the soil both horizontally and vertically
- (ii) Stresses and strains in the reinforcement
- (iii) Forces at the reinforcement to facing connection
- (iv) Vertical earth pressures to confirm any variation from uniform overburden pressure
- (v) Movements of the wall facing
- (vi) Temperatures throughout the backfill

(8) Of specific interest from field structures besides their general performance is information on:

- (i) Damage caused to the reinforcement during construction
- (ii) Effect of damage to facing elements on the overall stability. Such damage may arise from fire, vandalism or traffic impact.
- (iii) Long term creep and durability of the polymeric materials.

(9) It is most desirable to assemble a well documented data bank of case histories of both successful structures and, if they occur, failures. The case histories should be prepared to some standardized format. The information tabulated by Yako and Christopher in this volume is a useful starting point for the scope of information required.

4. DESIGN

4.1 General Problems

The design of soil reinforcement systems offers a great challenge to geotechnical engineers due to the numerous inter-related factors involved. The obvious problems of a general nature that require attention are:

(1) Lack of a suitable definition for failure of the system. For extensible polymeric reinforcements failure should take into account both servicability criterion for excessive movement and collapse.

(2) What in fact do the factors of safety of reinforced soil structures and the partial factors of safety of the constituent materials mean? How realistic are the Factors of Safety presently being used? Factors of Safety must be defined starting from the basic premise that F = 1 represents a failure condition. To enable this to be achieved much more needs to be learned about failure in these structures.

(3) Lack of clear design assumptions, concepts, specifications and guidelines which are applicable to all situations and take into account all the parameters involved.

(4) Choice of soil and reinforcement parameters to be used in design.

(5) Distinction between the design of a safe structure and the prediction of its behaviour in terms of deformations and stresses.

Although real progress in design technique will only be achieved by solving these problems, it seems advantageous at this stage of development to consider two levels of design, simplified design and rigorous design. The former will allow the construction of safe simple structures using an easily understood design methodology and at the same time the latter will permit researchers to explore complicated concepts in order to achieve coherent and valid rigorous techniques.

4.2 Simplified Design

(1) For small uncomplicated structures this should be based on existing limit equilibrium methods which are, in general, safe and very conservative. The soil and reinforcement parameters should be obtained from simple tests and used in an easy to understand design procedure. In most cases the use of ϕ_{cv}, obtained from shear box tests, and the assumption of active earth pressures seems appropriate for the soil except when relatively inextensible polymeric materials are employed. The design should allow for the calculation of the global factors of safety and the approximate determination of the working strains.

(2) Design methods are needed to consider Earthquake and Dynamic Loadings.

4.3 Rigorous Analysis and Design

This can be performed using various techniques such as a discrete constituent materials approach, a system simulation approach, an equivalent composite material approach and finite element models. Preference for rigorous analysis is for "transparent" mathematical models that logically relate test results for the basic soil and polymeric materials to the performance of the structure.

In addition to the general problems mentioned earlier, rigorous analysis should consider and encompass the following:

(1) Calculations and relevance of earth pressure coefficients.

(2) Strain development in the soil including strain anisotropy.

(3) Compatability of strains between the soil and the reinforcement.

(4) Effect of reinforcement extensibility and spacing.

(5) Stiffness of the reinforced soil mass, of the facing elements and of the foundation soil.

(6) Prediction of wall movements.

(7) Effects of creep and relaxation.

(8) Analysis of complex structures including three dimensional problems.

(9) Simulation of compaction induced stresses.

(10) Simulation and allowance for construction detail and procedures, ensuring that the assumed design boundary conditions are reasonable.

5. CONSTRUCTION

A wide variety of construction techniques are presently being employed, and it is felt that these would benefit from a greater measure of standardization and general specification. Work should be directed towards:

(1) Preparing a comprehensive summary of details and variates of different existing construction methods.

(2) Identification of desirable and undesirable practice and establishing the limits of applicability of polymeric products.

(3) Establishing a number of standard forms of construction. These would include specifications for standard reinforcement layouts, standard facings and connections and model construction practice and equipment.

(4) Examination of the effects of construction methods and construction details on boundary conditions and the assumptions used in the design process.

Standards, Codes of Practice and Specifications created directly for polymerically reinforced soil structures are vital for the expeditious development of this promising form of construction.

6. DEVELOPMENT OF MATERIALS AND TECHNIQUES

The success of polymer reinforced soil systems may be attributed in part to the imagination and development work carried out by the workers in this field. Although complete understanding of the system behaviour has not yet been achieved, development work must continue in order to ensure safe and economical design. Such work should include:

(1) Development of new composite materials, such as Texsol and mesh element reinforced soil.

(2) Development of methodology and techniques necessary for the use of low cost, fine grained back-fills.

(3) Development of aesthetic facing units and methods of attaching them to the reinforcements.

(4) Development of construction techniques to both minimize and allow for post construction deformations.

(5) Interaction of civil engineers with polymer experts and chemical engineers for the understanding and development of new materials especially stiffer and stronger products.

(6) A forum to act as a focus and coordinating body for international research and development efforts. This could be organized through IGS or ISSMFE. It should encourage further workshops of the nature of this NATO meeting, cooperative research projects and the establishment of a data bank of case histories and other information necessary for the progressive development of polymerically reinforced soil structures.

Participants

PARTICIPANTS

NATO-ARW, "Application of Polymeric Reinforcement in Soil Retaining Structures"

June 1987, Royal Military College of Canada, Kingston

LIST OF PARTICIPANTS

(1) H. Miki
(2) J. Smith-Meyer
(3) P. Delage
(4) M. deGroot
(5) J. Scott
(6) R. Berg
(7) K. Andrawes
(8) B. Christopher
(9) G. Richardson
(10) R. Bonaparte
(11) G. Raymond
(12) J. Lafleur
(13) S. Hermann
(14) R. Floss
(15) P. Jarrett
(16) J. Studer
(17) D. Bush
(18) R. Jewell
(19) G. Werner
(20) J. Perfetti
(21) E. Leflaive
(22) R. Bathurst
(23) C. Jones
(24) F. Schlosser
(25) Ph. Delmas
(26) M. Sotton
(27) S. Christoulas
(28) R. Koerner
(29) R. Gopal
(30) K. Rowe
(31) R. Murray
(32) A. Rollin
(33) A. McGown
(34) P. Rimoldi
(35) J-P. Gourc
(36) D. Cazzuffi

Missing from photograph:

G.W.E. Milligan

J.M. Rigo

ADDRESS LIST OF PARTICIPANTS, ARW 588/86

Dr. K. Andrawes (Kamal)	Dept. of Civil Engineering, Strathclyde University, 107 Rottenrow, Glasgow, G4 ONG Scotland
Dr. R.J. Bathurst (Richard)	Dept. of Civil Engineering, Royal Military College of Canada, Kingston, Ontario, Canada, K7K 5LO
Mr. R.R. Berg (Ryan)	The Tensar Corporation, 1210 Citizens Parkway, Morrow, GA, 30260, U.S.A.
Dr. R. Bonaparte (Rudy)	GeoServices Inc., Suite 204, 1200 S. Federal Highway, Boynton Beach, FL, 33435, U.S.A.
Dr. D.I. Bush (David)	Netlon Limited, Kelly Street, Mill Hill, Blackburn, Lancashire BB2 4PJ England
Mr. D.A. Cazzuffi (Daniele)	ENEL - Centro Ricerca Idraulica, e Strutturale, Via Ornata 90/14, I-20162 Milano, Italy
Mr. B. Christopher (Barry)	STS Consultants, 111 Pfingsten Road, Northbrook, IL, 60062, U.S.A.
Dr. S. Christoulas (Stavros)	Director Soils Division, Research Center of Public Works, Secretariat of Public Works, Ymittou 81, 155 62 Holargos, Greece
Ir. M.T. de Groot, M.Sc. (Max)	Delft Geotechnics, Stieltjesweg 2, P.O. Box 69, 2600 AB Delft, The Netherlands
Dr. P. Delage (Pierre)	Ecole National des Ponts et Chaussées, CERMES, Central 2 La Courtine, Boite 105, 93194 Noisy le Grand Cedex, France
Dr. Ph. Delmas (Philippe)	Laboratoire Central des Ponts, et Chaussées, 58 boulevard Lefebvre, 75732 Paris Cedex 15, France

Professor Dr.-Ing. R. Floss (Rudolph)	Ordinarius für Grundbau, Bodenmechanik, und Felsmechanik, Technischen Universität München, Baumbachstrasse 7, 8000 München 60, West Germany
Dr. J-P. Gourc (Jean-Pierre)	IRIGM, Université de Grenoble, BP 68, 38402 St. Martin d'Heres Cedex, France
Mr. S. Hermann (Steinar)	Norwegian Geotechnical Institute, P.O. Box 40, Taasen, 0801 Oslo 8, Norway
Dr. P.M. Jarrett (Peter)	Dept. of Civil Engineering, Royal Military College of Canada, Kingston, Ontario, Canada, K7K 5L0
Dr. R. Jewell (Richard)	Dept. of Engineering Science, University of Oxford, Parks Road, Oxford OX1 3PJ England
Professor C.J.F.P. Jones (Colin)	Dept. of Geotechnical Engineering, Drummond Building, University of Newcastle Upon Tyne, Newcastle Upon Tyne NE1 7RU England
Dr. Robert M. Koerner (Bob)	Dept. of Civil Engineering, Drexel University, Philadelphia, PA, 19104, U.S.A.
Dr. J. Lafleur (Jean)	Ecole Polytechnique, CP6079, Station A, Montreal, Quebec, Canada, H3C 3A7
Dr. E. Leflaive (Etienne)	Laboratoire Central des Ponts et Chaussées, Direction des Programmes et Applications, Orly Sud No. 155, 94396 Orly Aerogare Cedex, France
Professor A. McGown (Alan)	Dept. of Civil Engineering, Strathclyde University, 107 Rottenrow, Glasgow, G4 0NG Scotland
Mr. H. Miki (Hiroshi)	Dept. of Civil Engineering, Drexel University, Philadelphia, PA, 19104, U.S.A.
Dr. G.W.E. Milligan (George)	Dept. of Engineering Science, University of Oxford, Parks Road, Oxford, OX1 3PJ England
Dr. Richard T. Murray (Dick)	Transport and Road Research Laboratory, Old Wokingham Road, Crowthorne, Berkshire, RG11 6AU England

Mr. J. Perfetti (Jacques)	Rhône Poulenc Fibres, Département Nontissé BIDIM, 44 rue Salvador Allende, B.P. 80, 95871 Bezons Cedex France
Dr. G.P. Raymond (Gerry)	Dept. of Civil Engineering, Queen's University, Kingston, Ontario, K7L 3N6, Canada
Dr. G. Richardson (Greg)	Soil and Material Engineers, Inc., 1903 Harrison Avenue, Box 609, Cary, North Carolina, 27511, U.S.A.
Dr. J-M. Rigo (Jean-Marie)	Civil Engineering Material Department, Civil Engineering Institute, University of Liège, Quai Banning 6-4000 Liège, Belgium
Ing. P. Rimoldi (Pietro)	Technical Director, Civil Engineering Applications, RDB Plastotecnica S.p.A., 22060 Vigano Brianzi (CO), Italy
Dr. A.L. Rollin (André)	Ecole Polytechnique, CP 6079, Station A, Montreal, Quebec, Canada, H3C 3A7
Dr. R.K. Rowe (Kerry)	Geotechnical Research Center, The University of Western Ontario, London, Ontario, Canada, N6A 5B9
Professor F. Schlosser (Francois)	Terrasol, Bureau d'Ingénieurs-Conseils en Géotechnique, Tour Horizon, 52, quai de Dion Bouton, 92806 Puteaux Cédex, France
Dr. J.D. Scott (Don)	Dept. of Civil Engineering, The University of Alberta, Edmonton, Alberta, Canada, T6G 2G7
Mr. J. Smith-Meyer (Johan)	Konglungveien 143, N-1392 Vettre, Norway
Dr. M. Sotton (Michel)	Directeur de l'Institut Textile, de France Section LYON, Avenue Guy de Collongue, BP 60, 69132 Ecully Cédex, France
Dr. J. Studer (Jost)	GSS Consulting Engineers Ltd., Witikonerstrasse 15, CH-8032 Zurich, Switzerland
Dipl.Ing G. Werner (Gerhard)	Chemie-Linz AG, St. Peterstrasse 25, 4021 Linz, Austria

Index

INDEX